Lecture Notes in Computer Science   14804

Founding Editors

Gerhard Goos
Juris Hartmanis

AF166228

The series Lecture Notes in Computer Science (LNCS), including its subseries Lecture Notes in Artificial Intelligence (LNAI) and Lecture Notes in Bioinformatics (LNBI), has established itself as a medium for the publication of new developments in computer science and information technology research, teaching, and education.

LNCS enjoys close cooperation with the computer science R & D community, the series counts many renowned academics among its volume editors and paper authors, and collaborates with prestigious societies. Its mission is to serve this international community by providing an invaluable service, mainly focused on the publication of conference and workshop proceedings and postproceedings. LNCS commenced publication in 1973.

Elisa H. Barney Smith · Marcus Liwicki ·
Liangrui Peng
Editors

Document Analysis and Recognition - ICDAR 2024

18th International Conference
Athens, Greece, August 30 – September 4, 2024
Proceedings, Part I

 Springer

Editors
Elisa H. Barney Smith
Luleå Tekniska Universitet
Luleå, Sweden

Marcus Liwicki
Luleå Tekniska Universitet
Luleå, Sweden

Liangrui Peng
Tsinghua University
Beijing, China

ISSN 0302-9743   ISSN 1611-3349 (electronic)
Lecture Notes in Computer Science
ISBN 978-3-031-70532-8   ISBN 978-3-031-70533-5 (eBook)
https://doi.org/10.1007/978-3-031-70533-5

This Springer imprint is published by the registered company Springer Nature Switzerland AG
The registered company address is: Gewerbestrasse 11, 6330 Cham, Switzerland

If disposing of this product, please recycle the paper.

Foreword

We are honoured to welcome you to the proceedings of ICDAR 2024, the 18th IAPR International Conference on Document Analysis and Recognition, which took place in Athens, the beautiful and historic capital of Greece. ICDAR 2024 marked the start of the annual basis for the ICDAR series.

ICDAR 2024 was the 18th edition of a longstanding conference series that has come of age, sponsored by the International Association for Pattern Recognition (IAPR). It is the premier international event for scientists and practitioners in document analysis and recognition. This field continues to play an important role in document understanding and recognition.

The IAPR TC10/11 technical committees endorse the conference. The very first ICDAR was held in St. Malo, France in 1991, followed by Tsukuba, Japan (1993), Montreal, Canada (1995), Ulm, Germany (1997), Bangalore, India (1999), Seattle, USA (2001), Edinburgh, UK (2003), Seoul, South Korea (2005), Curitiba, Brazil (2007), Barcelona, Spain (2009), Beijing, China (2011), Washington, DC, USA (2013), Nancy, France (2015), Kyoto, Japan (2017), Sydney, Australia (2019), Lausanne, Switzerland (2021) and San Jose, USA (2023).

Keeping with its tradition from past years, ICDAR 2024 featured a three-day main conference, including several competitions to challenge the field and a pre-conference slate of workshops, tutorials, and a doctoral consortium. The conference was held in Athens, Greece on September 2–4, 2024, and the pre-conference tracks on August 30 till September 1, 2024.

The highlights of the conference included keynote talks by the recipient of the IAPR/ICDAR Outstanding Achievements Award, and distinguished speakers: Jürgen Schmidhuber, Director of the AI Initiative at KAUST, Swiss AI Lab IDSIA, and the University of Lugano, Switzerland; Maria Kamilaki, Acting Director-General of e-Administration, Library & Publications of the Hellenic Parliament, Greece; and Cheng-Lin Liu, State Key Laboratory of Multimodal Artificial Intelligence Systems, Institute of Automation of the Chinese Academy of Sciences, China.

A total of 263 papers were submitted to the main conference (plus 35 papers to the ICDAR-IJDAR journal track), with 52 papers accepted for oral presentation (plus 17 IJDAR track papers) and 92 for poster presentation. We would like to express our deepest gratitude to our Program Committee Chairs, featuring three distinguished researchers from academia, Elisa Barney Smith, Liangrui Peng, and Marcus Liwicki, who did a phenomenal job in overseeing a comprehensive reviewing process and who worked tirelessly to put together a very thoughtful and interesting technical program for the main conference. We are also very grateful to the members of the Program Committee for their high-quality peer reviews. We extend our gratitude to our competition chairs, George Retsinas and Xiang Bai, for overseeing the competitions.

The pre-conference featured 6 excellent workshops, 4 value-filled tutorials, and the doctoral consortium. We would like to thank Harold Mouchère and Anna Zhu, the workshop chairs, Vincent Christlein and Alicia Fornés, the tutorial chairs, and KC Santosh and Andreas Fischer, the doctoral consortium chairs, for their efforts in putting together a wonderful pre-conference program. We would like to thank and acknowledge the hard work put in by our Publication Chairs, Giorgos Sfikas and Christophoros Nikou, who worked diligently to compile the camera-ready versions of all the papers and organize the conference proceedings with Springer. Many thanks are also due to our sponsorship, awards, industry, and publicity chairs for their support of the conference.

Finally, we would like to thank our many financial sponsors for their support and the conference attendees and authors, for helping make this conference a success. We sincerely hope that all attendees had an enjoyable conference, a wonderful stay in Athens, and fruitful academic exchanges with their colleagues.

August 2024

Basilis Gatos
Vassilis Katsouros
Foteini Simistira Liwicki

Preface

Welcome to the proceedings of the 18th International Conference on Document Analysis and Recognition (ICDAR 2024). ICDAR is the premier international event for scientists and practitioners involved in document analysis and recognition. This iteration is the first iteration in an even year, marking the beginning of the series becoming an annual event.

This year, we received 263 conference paper submissions. In order to create a high-quality scientific program for the conference, we recruited 159 regular and 32 senior program committee (PC) members. Each paper received at least 2 single-blind reviews, with most papers receiving 3 or more reviews. In addition, senior PC members who oversaw the review phase for typically 8–10 submissions took care of consolidating reviews and suggested paper decisions in their meta-reviews. Based on the information provided in both the reviews and the prepared meta-reviews we PC Chairs then selected 144 submissions (54.8%) for inclusion in the scientific program of ICDAR 2024. From the accepted papers, 52 were selected for oral presentation, and 92 for poster presentation.

In addition to the papers submitted directly to ICDAR 2024, we continued the tradition of teaming up with the International Journal of Document Analysis and Recognition (IJDAR) and organized a special journal issue. The journal-track submissions underwent the same rigorous review process as regular IJDAR submissions. The ICDAR PC Chairs served as Guest Editors and oversaw the review process. From the 35 manuscripts submitted to the journal track, 17 were accepted and were published in a Special Issue of IJDAR entitled "Advanced Topics of Document Analysis and Recognition." In addition, all papers accepted in the journal track were included as oral presentations in the conference program with a slightly extended presentation time (25 instead of 20 minutes including Q&A).

With the transition of ICDAR to an annual conference, the alternate year workshops DAS and ICFHR also made some changes. DAS decided to now be colocated with ICDAR. Handwritten text recognition is now a prominent topic in the DAR community, so the organizers and attendees of ICFHR 2023 decided to become part of the main ICDAR conference. Two sessions in ICDAR 2024 were on Frontiers of Handwriting Recognition and a third was devoted to Chinese text recognition. The popular Workshop on Historical Document Imaging and Processing (HIP) decided to remain an alternate year workshop, but in odd years. The main ICDAR conference had 2 sessions devoted to Historical Document Analysis. Scene text recognition, music, tables and charts continue to be topics of interest. Visual Question Answering, Document Understanding and NLP are rising topics gaining enough papers to devote sessions to them. As over the years many new machine learning techniques have been developed by researchers in the DAR community, a track also was devoted to papers that focused more on the methods than on any particular application. This year, nine scientific competitions were held in conjunction with ICDAR. A session was devoted to presenting the results.

As ICDAR 2024 was held with in-person attendance, all papers were presented by their authors during the conference. Exceptions were only made for authors who could not attend the conference for unavoidable reasons. Such oral presentations were then provided by synchronous video presentations. Posters of authors that could not attend were presented by recorded teaser videos, in addition to the physical posters.

Three keynote talks were given by Jürgen Schmidhuber, Director of the AI Initiative at KAUST, Swiss AI Lab IDSIA, and University of Lugano, Switzerland; Maria Kamilaki, Acting Director-General of e-Administration, Library & Publications of the Hellenic Parliament, Greece; and Cheng-Lin Liu, State Key Laboratory of Multimodal Artificial Intelligence Systems, Institute of Automation of the Chinese Academy of Sciences, China and the recipient of the IAPR/ICDAR Outstanding Achievements Award. We thank them for the valuable insights and inspiration that their talks provided for participants.

Finally, we would like to thank everyone who contributed to the preparation of the scientific program of ICDAR 2024, namely the authors of the scientific papers submitted to the journal track and directly to the conference, reviewers for journal-track papers, and both our regular and senior PC members. We also thank the Springer staff and the ICDAR 2024 Publication Chairs, who oversaw the creation of these proceedings.

August 2024

Elisa H. Barney Smith
Marcus Liwicki
Liangrui Peng

Organization

Organizing Committee

General Chairs

Basilis Gatos	NCSR "Demokritos", Greece
Vassilis Katsouros	Athena Research Center, Greece
Foteini Simistira Liwicki	Luleå University of Technology, Sweden

Program Committee Chairs

Elisa Barney Smith	Luleå University of Technology, Sweden
Marcus Liwicki	Luleå University of Technology, Sweden
Liangrui Peng	Tsinghua University, China

Workshop Chairs

Harold Mouchère	Nantes Université, France
Anna Zhu	Wuhan University of Technology, China

Competition Chairs

George Retsinas	National Technical University of Athens, Greece
Xiang Bai	Huazhong Univ. of Sci. & Technology, China

Tutorial Chairs

Vincent Christlein	University of Erlangen-Nuremberg, Germany
Alicia Fornés	Universitat Autònoma de Barcelona, Spain

Publication Chairs

Giorgos Sfikas	University of West Attica, Greece
Christophoros Nikou	University of Ioannina, Greece

Doctoral Consortium Chairs

Andreas Fischer	Univ. of App. Sci. & Arts Western Switzerland, Switzerland
K. C. Santosh	University of South Dakota, USA

Awards Chairs

Michael Blumenstein	University of Technology Sydney, Australia
Ioannis Pratikakis	Democritus University of Thrace, Greece

Posters/Demo Chairs

Umapada Pal	Indian Statistical Institute, India
Momina Moetesum	National University of Sciences & Technology, Pakistan
Kenny Davila	DePaul University, USA

Sponsorship Chairs

Markus Weber	Wacom, USA
Xu-Cheng Yin	University of Science and Technology Beijing, China

Industry Chairs

Dimosthenis Karatzas	Universitat Autònoma de Barcelona, Spain
Srirangaraj Setlur	University at Buffalo, USA
Errui Ding	Baidu Inc., China

Local Organization Chairs

Anastasios Kesidis	University of West Attica, Greece
Kosmas Kritsis	Athena Research Center, Greece
Elena Galifianaki	NCSR "Demokritos", Greece
Pelagia Drosaki	NCSR "Demokritos", Greece

Publicity Chairs

Panagiotis Kaddas	NCSR "Demokritos", Greece
Vassilis Papavassiliou	Athena Research Center, Greece
Elena Galifianaki	NCSR "Demokritos", Greece

Program Committee

Senior Program Committee Members

Apostolos Antonacopoulos	University of Salford, UK
Anurag Bhardwaj	eBay Research Labs, USA
Michael Blumenstein	University of Technology, Sydney, Australia
Jean-Christophe Burie	L3I - Université de La Rochelle, France
Bertrand Coüasnon	Irisa/Insa, France
Mickaël Coustaty	Laboratoire L3i - La Rochelle Université, France
David Doermann	University at Buffalo, USA
Véronique Eglin	LIRIS-INSA de Lyon, France
Gernot Fink	TU Dortmund, Germany
Andreas Fischer	University of Fribourg, Switzerland
Alicia Fornés	Computer Vision Center, UAB, Spain
Liangcai Gao	Peking University, China
Nicholas Howe	Smith College, USA
C. V. Jawahar	CVIT, IIIT, Hyderabad, India
Lianwen Jin	South China University of Technology, China
Dimosthenis Karatzas	CVC, Universitat Autónoma de Barcelona, Spain
Koichi Kise	Osaka Metropolitan University, Japan
Bart Lamiroy	Université de Reims Champagne-Ardenne, France
Cheng-Lin Liu	CASIA, China
Lu Liu	Lazada, Singapore
Josep Llados	CVC, Universitat Autónoma de Barcelona, Spain
Daniel Lopresti	Lehigh University, USA
R. Manmatha	University of Massachusetts, Amherst, USA
Angelo Marcelli	Università di Salerno, Italy
Simone Marinai	University of Florence, Italy
Jean-Marc Ogier	University of La Rochelle, France
Wataru Ohyama	Tokyo Denki University, Japan
Marçal Rusiñol	AllRead Machine Learning Technologies, Spain
Robert Sablatnig	TU Wien, Austria
Faisal Shafait	National University of Sciences and Technology, Pakistan
Seiichi Uchida	Kyushu University, Japan
Jerod Weinman	Grinnell College, USA
Richard Zanibbi	Rochester Institute of Technology, USA
Yu Zhou	Nankai University, Japan

Program Committee Members

Irfan Ahmad
Alireza Alaei
Musab Al-Ghadi
Eric Anquetil
Vlad Atanasiu
Muhammad Naseer Bajwa
Byron Bezerra
Ujjwal Bhattacharya
Jean-Luc Bloechle
Alceu Britto
Rina Buoy
Jorge Calvo-Zaragoza
Cristina Carmona-Duarte
Sukalpa Chanda
Clément Chatelain
Bidyut B. Chaudhuri
Joseph Chazalon
Shanxiong Chen
Jin Chen
Youssouf Chherawala
Vincent Christlein
Christian Clausner
Mark Clement
Florence Cloppet
Kenny Davila
Claudio De Stefano
Abhisek Dey
Sounak Dey
Antoine Doucet
Fadoua Drira
Mounîm A. El Yacoubi
Jonathan Fabrizio
Francesco Fontanella
Yasuhisa Fujii
Akio Fujiyoshi
Rajib Ghosh
Romain Giot
Lluis Gomez
Petra Gomez-Krämer
Daichi Haraguchi
Sheng He
Nina S. T. Hirata
Qiang Huo

Donato Impedovo
Brian Kenji Iwana
Maham Jahangir
Aashi Jain
Mohammed Javed
Jobin K. V.
Ehsanollah Kabir
Karim Kalti
Lei Kang
Slim Kanoun
Christopher Kermorvant
Yousri Kessentini
Florian Kleber
Pramod Kompalli
Aurélie Lemaitre
Hongjun Li
Zhouhui Lian
Lingyu Liang
Minghui Liao
Laurence Likforman
Rafael Lins
Chang Liu
Yuliang Liu
Muhammad Muzzamil Luqman
Nam Tuan Ly
Sriganesh Madhvanath
Nishatul Majid
Carlos David Martinez Hinarejos
Maroua Mehri
Carlos Mello
Ronaldo Messina
Evangelos Milios
Zuheng Ming
Tomo Miyazaki
Momina Moetesum
Hussein Mohammed
Ajoy Mondal
Harold Mouchère
Shobharani N.
Nibal Nayef
Clemens Neudecker
Hung Tuan Nguyen
Shinichiro Omachi

Umapada Pal
Shivakumara Palaiahnakote
Thierry Paquet
Mohammad Tanvir Parvez
Antonio Parziale
Marco Peer
Dezhi Peng
Vincent Poulain D'Andecy
Ioannis Pratikakis
Irina Rabaev
Jean-Yves Ramel
Oriol Ramos-Terrades
Frédéric Rayar
Kaspar Riesen
Christophe Rigaud
Verónica Romero
Henry A. Rowley
Joan Andreu Sanchez
Ravi Kiran Sarvadevabhatla
Martin Schall
Amina Serir
Anuj Sharma
Ying Sheng
Nicolas Sidère
Steven Simske
Sukhdeep Singh
Daniel Stoekl Ben Ezra
Tonghua Su
Xiangdong Su
Suresh Sundaram
Salvatore Tabbone
Sandeep Tata
Christopher Tensmeyer
Kengo Terasawa
Iuliia Tkachenko
Ruben Tolosana
Alejandro Toselli
Xiao Tu
Oliver Tüselmann
Huy Quang Ung
Szilard Vajda
Ernest Valveny
Ekta Vats
Ruben Vera-Rodriguez
Enrique Vidal

Lars Vögtlin
Yanwei Wang
Qiu-Feng Wang
Da-Han Wang
Yang Xue
Chun Yang
Mingkun Yang
Berrin Yanikoglu
Fei Yin
Qi Zeng
Heng Zhang
Yanming Zhang
Guangwei Zhang
Yuchen Zheng
Anna Zhu
Majid Ziaratban
Chandranath Adak
Oluwatosin Adewumi
Akshay Agarwal
Peeta Basa Pati
Khadiravana Belagavi
Asma Bensalah
Mohammad Idrees Bhat
Mélodie Boillet
Victoria Bourgeais
Iheb Brini
Francisco J. Castellanos
Francesco Castro
Apurba Chakraborty
Xu Chen
Denis Coquenet
Simon Corbillé
Aravinda Cv
Tiziana D'Alessandro
Avijit Dasgupta
Julien Delaunay
Vincenzo Dentamaro
Shubhang Desai
Alessandra Scotto di Freca
Moises Diaz
Ray Ding
Kalvin Dobler
Biyi Fang
Yuhang Fu
Gilad Fuchs

Cristiano Garcia
Vincenzo Gattulli
Loann Giovannangeli
Nathalie Girard
Tongkun Guan
Ahmed Hamdi
Raphaela Heil
Andre Hochuli
Kai Hu
Ludvig Hult
Syed Mohammad Baqir Husain
Nushrat Hussain
Aman Jaiswal
Mahdi Jampour
Nanfeng Jiang
Wang Jiawei
Michael Jungo
Wafa Khlif
Florian Kordon
Omar Krichen
Ahana Kundu
Songze Li
Zhixin Liu
Dongliang Luo
Puneet Mathur
Lin Meng
Elmokhtar Mohamed Moussa
Omar Moured
Emanuele Nardone
Emanuel Orler

Glen Pouliquen
Zhidong Qiao
Xingming Qu
Sachin Raja
Bulla Rajesh
Yann Ricquebourg
Antonio Ríos-Vila
Hugo Romat
Anna Scius-Bertrand
Gianfranco Semeraro
Mathias Seuret
Yilin Shi
Yongxin Shi
Mohamed Ali Souibgui
Yann Soullard
Maksym Taranukhin
Solène Tarride
Stacey Taylor
Vishvesh Trivedi
David Villanova-Aparisi
Manuel Villarreal Ruiz
Jiawei Wang
Xuewen Wang
Minghui Xia
Yejing Xie
Fuxiang Yang
Zhenhua Yang
Yan Zheng
Peijun Zou

Contents – Part I

Document Understanding and NLP

Transformers

Charts and Tables

Business Documents

A Multiclass Imbalanced Dataset Classification of Symbols from Piping and Instrumentation Diagrams

Laura Jamieson, Carlos Francisco Moreno-García(✉)(iD), and Eyad Elyan(iD)

Robert Gordon University, Aberdeen AB10 7QB, Scotland, UK
c.moreno-garcia@rgu.ac.uk

Abstract. Engineering diagrams provide rich source of information and are widely used across different industries. Recent years have seen growing research interest in developing solutions for processing and analysing these diagrams using wide range of image-processing and computer vision techniques. In this paper, we first, present a new multiclass imbalanced dataset of symbols extracted from Piping and Instrumentation Diagrams (P&IDs). The dataset contains 7,728 instances representing 48 different types of engineering symbols and it is considered the first of its kind in the research community. Second, we present a new method for handling multiclass imbalance classification based on class decomposition by means of unsupervised machine learning methods. Experiments using Convolutional Neural Networks showed that using class decomposition significantly improves the classification performance that can be achieved, without causing information loss, as it is the case with other class imbalance data sampling approaches.

Keywords: Piping and Instrumentation Diagrams · Class Imbalance · Convolutional Neural Networks

1 Introduction

Engineering diagrams are used in a wide range of industries including electronics [5], architecture [25] and the oil industry [9]. It is common for engineering diagrams to be stored in an undigitised format [1, 23] and consequently, extracting data from these diagrams is a significant task. Piping and Instrumentation Diagrams (P&IDs) are one type of complex engineering diagram. P&IDs are used to show equipment, connectors and associated technical details. A section of a P&ID is shown in Fig. 1. Extracting information from undigitised P&IDs can be a very time-consuming task, even for engineering specialists [27]. For example, the task of identifying the locations of one equipment type in a set of undigitised P&IDs requires exhaustive manual search through each P&ID in the set [31].

Supported by DNV.

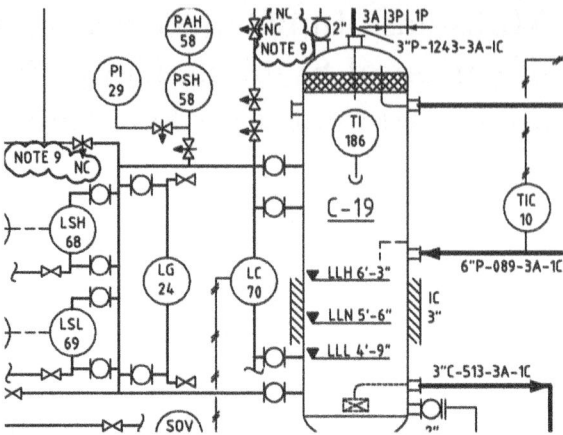

Fig. 1. Section of a P&ID showing different symbol types.

Class imbalance is another research challenge in engineering diagram digitisation [19]. Class imbalance occurs when one or more classes is over represented or underrepresented in a dataset. Typically, supervised learning models trained on imbalanced datasets are biased towards the majority class [32], leading to missclassifications of minority samples as majority samples [33]. Research into class imbalance has predominantly focused on binary classification rather than the multiclass scenario [34]. Additionally, although extensive research has been carried out on class imbalance in traditional machine learning, less research has considered the problem in deep learning [3,14].

In this paper, the two research challenges of lack of engineering diagram datasets and multiclass imbalance are addressed. A new multiclass symbol dataset is presented to further research in this important area. The dataset comprises 7,728 symbols distributed across 48 classes from two P&ID drawing standards. The dataset contains symbols extracted from a set of P&IDs using an object detection method [7] and symbols from the Symbols in Engineering Drawings (SiED) dataset [9]. In addition, this paper also presents a technique to handle multiclass imbalanced data classification. The technique is extended from CDSMOTE [8], which handles class imbalance in binary datasets. CDSMOTE uses class decomposition to reduce dominance of the majority class and synthetic oversampling to increase representation of the minority class. The multiclass imbalance data method uses class decomposition and involves synthetic oversampling of multiple minority classes to rebalance the dataset. CNN classification experiments are also presented.

The rest of this paper is organised as follows: In Sect. 2, related work in digitisation of engineering drawings is discussed. Section 3 presents the symbol dataset and the multiclass imbalance data method. Section 4 describes the classification experiments and discusses the results. Then, the conclusions and suggestions for future work are presented in Sect. 5.

2   Related Work

2.1   Symbol Digitisation in P&IDs

The demand to digitise engineering diagrams, and in particular P&IDs, is evident in the literature [12, 22, 30]. Engineering diagrams represent dense and complex arrangement of equipment using drawing elements of symbols, text and pipelines. Published research in engineering diagram digitisation often focuses on digitisation methods for one of these drawing elements. For example, a number of research works focused on recognising text within engineering drawings [13, 18, 20, 28]. Detection and classification methods for engineering symbols is another active research area [1, 2, 6, 9, 30].

The earliest published research in digitisation of engineering diagrams used a traditional computer vision approach [1, 5, 20, 30]. Digitising engineering diagrams using traditional methods had partial success, however limitations of these methods were reported. One drawback is that traditional rule based methods require extensive fine tuning for each unique use case. This is problematic for engineering diagrams as drawing element appearance can vary for several reasons. For example, symbol appearance in engineering diagrams may be altered due to the presence of overlapping and/or closely located drawing elements. In addition, undigitised images may be of low quality or contain image noise, which can alter original drawing element representations [19].

Aiming to improve on traditional digitisation methods, several recent works used deep learning for engineering diagram digitisation [7, 11, 30]. Studies that focus on symbol detection within P&IDs include [6, 7, 30]. In [7], authors developed a symbol detection method based on object detection model You Only Look Once (YOLO) [26]. The method did not require diagram preprocessing. An average class accuracy of 94% across 29 symbol classes was recorded.

In [18], authors approached digitisation of P&IDs using a CNN for symbol detection and graph search to determine component connectivity. A dataset of two constantly sized symbol types, (tags and locally mounted instruments), was annotated from 18 P&IDs and used to train the CNN. A sliding window approach, that filtered out windows identified as more than 90% blank, was used with the CNN to classify image windows. The resulting symbol detections were post-processed using non-maximum suppression. The method was tested on 11 P&IDs and reported a precision of 85% and above and a recall 90% and above for the two symbol types.

Also using deep learning to extract data from P&IDs, in [24] a Fully Convolutional Network (FCN) [17] was used to detect ten different symbol classes. Symbols were annotated from a set of P&IDs to create a training dataset. To increase the size of the training dataset, data augmentation methods of translation and rotation were used on the annotated data. The method was evaluated on four sheets of P&IDs and achieved an F1-score of 0.86 for symbol detection with reported missclassification between visually similar classes. The authors highlighted varying image qualities and noisy textual information contained within the diagrams as a challenge for digitisation methods for real world P&IDs [24].

Focusing on symbols classification, in [6] classification of an imbalanced engineering symbols dataset was examined. Random Forest (RF), Support Vector Machine (SVM) and CNN were used for classification. Each classification method was evaluated with and without class decomposition. The k-means clustering algorithm was used for class decomposition. Classification experiments reported comparable results with RF, SVM and CNN methods. Class decomposition was found to benefit RF and SVM classification. Results reported a slight decrease in CNN classification performance using class decomposition, which was attributed to the limited number of instances per class in the decomposed dataset.

The literature also highlights remaining challenges in the development of deep learning models for engineering diagram digitisation. A recent review [21] of trends in this domain identified one barrier to deep learning model development for engineering diagrams as a lack of publicly available annotated engineering symbol datasets. A well-defined labelled symbol dataset was identified as a requirement to fully benefit from deep learning models for engineering diagram digitisation. Acquiring a representative symbols dataset through manual symbol annotation is a significant task, considering a P&ID can contain in excess of 100 symbols of multiple types. An additional consideration in obtaining a symbol dataset relevant to multiple sets of P&IDs, is that multiple drawing standards are used for equipment symbols representation [21].

2.2   Class Imbalance

Existing literature highlighted that uneven data distribution is inherent in P&IDs [19], meaning that class imbalance is a problem in this domain. Class imbalance occurs when classes are not approximately equally represented in a dataset [4]. In an imbalanced binary dataset, the minority (positive) class is underrepresented compared to the majority (negative) class. Multiclass imbalanced datasets consist of more than two classes, in which the majority class(es) is over represented compared to the minority class(es). Imbalanced datasets were reported as a challenge in obtaining accurate deep learning models [6,32] and are known to lead to deep learning models biased towards majority classes [8].

Class imbalance in engineering diagrams has been discussed in published literature [6,7,9]. The Symbols in Engineering Drawings (SiED) dataset was presented in [9]. To obtain the dataset, symbols were extracted from P&IDs using a combination of interactive and traditional image processing methods. The method extracted 2432 symbols from 39 symbol classes, however the method required fine tuning to extract symbols with any change in representation. Classification experiments using a CNN model showed that underrepresented classes in the dataset recorded comparatively lower classification performance compared to over represented classes. The class imbalance problem was also observed in engineering symbol detection [7]. In [7], whilst good performance was observed overall, a significantly lower class accuracy was obtained for the symbols that were underrepresented in the training dataset.

A range of methods address class imbalance by introducing new data samples. These methods oversample minority classes by adding synthetic samples, instead of resampling instances from the original dataset. One popular synthetic over-sampling method is the Synthetic Minority Oversampling Technique (SMOTE) [4]. SMOTE is designed to create artificial data based on similarities between existing minority samples. The method generates an artificial sample by inter-polating between a minority class sample and one of its k-nearest neighbours [4]. SMOTE creates a broader decision region for the minority class, compared to oversampling original minority class instances [4].

CDSMOTE [8] is a technique to handle class imbalance in binary datasets. The technique reduces the impact from the original majority class while retaining all available information for training, unlike other undersampling approaches. In the technique, the k-means clustering algorithm is used to decompose the majority class into smaller sub-classes. To increase minority class representation, the minority class is then oversampled using SMOTE.

3   Methodology

3.1   Dataset

The dataset consists of P&ID symbols obtained from two different drawing standards. An object detection method extracted 5,296 symbols representing 23 classes from one drawing standard. To represent a different drawing stan-dard, 2,432 symbols from 39 symbol classes were acquired from the Symbols in Engineering Drawings (SiED) dataset [9]. In total the dataset contains 7,728 instances representing 48 symbol classes. To extract the symbols from the first drawing standard, a set of 137 P&IDs was obtained from an industry part-ner. An object detection method based on YOLO v3 [26] was developed. The method localises most symbols in the diagrams, for additional details the reader is referred to [7]. Following object detection, a post-processing step verified extracted regions as true or false positives. Each extracted region was consid-ered a true positive if it contained the whole engineering symbol. Partial symbol detections were considered as false positives. The missclassified symbols were reassigned with the correct class. The method resulted in the extraction of 5,296 symbols representing 23 classes.

To increase the applicability of the symbol dataset across a range of P&ID drawing standards, the symbols extracted using the object detection method [7] were combined with the Symbols in Engineering Drawings (SiED) dataset [9]. Symbol class names are noted as Type 1 or Type 2 where the representation of the symbol varied in the two different symbol standards. Each symbol in the dataset is a row vector of 100×100 features which are pixel values. One instance of each class is presented in Fig. 2.

The class distribution within the dataset is highly imbalanced, as observed in Fig. 3. The majority class, *Sensor*, contains 2,845 instances and represents 36.8% of the dataset. In comparison, there are 3 classes, (*Barred Tee, Ultrasonic Flow Meter* and *Valve Butterfly Type 2*), that are each represented by one instance, and each represents 0.01% of the total dataset.

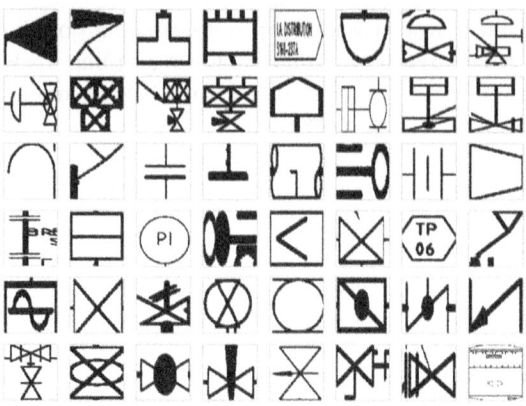

Fig. 2. One instance of each class in the dataset.

In total 60% of the entire dataset consists of only 3 of the 48 classes present in the dataset. To quantify the level of imbalance in the dataset, the imbalance ratio ir and fraction of minority classes fm are used as defined in [3] and [10].

The imbalance ratio, ir, is defined in Eq. 1:

$$ir = \frac{max_i M}{min_i m} \tag{1}$$

where $max_i M$ is the maximum number of instances in a majority class and $min_i m$ is the minimum number of instances in a minority class. The fraction of minority classes, fm is defined in Eq. 2:

$$fm = \frac{mt}{t} \tag{2}$$

where mt is the total number of minority classes and t is the total number of classes in a dataset. The imbalance ratio ir of the symbol dataset is 1422.5. The fraction of minority classes, fm, is 0.978, which reflects the presence of multiple minority classes in the dataset. Considering the low absolute sizes of multiple minority classes, undersampling to obtain a fully balanced dataset would significantly reduce the amount of data available for learning. Therefore, an approach that incorporated oversampling as opposed to undersampling was used.

3.2 Multiclass Imbalance Handling Method

To address class imbalance, data resampling methods which undersample majority classes and/or oversample minority classes can be used. In the case of undersampling, information is lost from the dataset. To avoid this, the proposed multiclass method is extended from CDSMOTE [8], which is a method for handling class imbalance in binary datasets, that adjusts the data distribution whilst avoiding information loss. CDSMOTE uses class decomposition to decompose

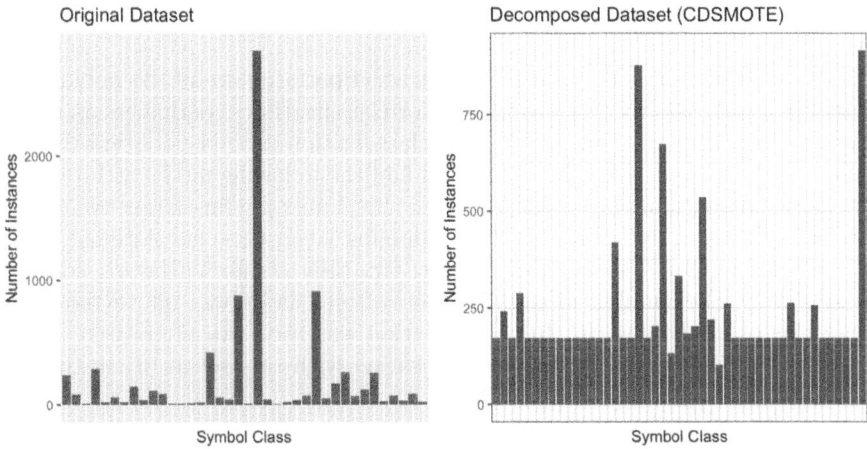

Fig. 3. Class distribution in the original dataset (left) and the decomposed (new) dataset (right). Note that y-axes have different scales on the two subplots.

the majority class into sub-classes using the k-means clustering algorithm. Afterwards, oversampling by means of the Synthetic Minority Over-Sampling Technique (SMOTE) [4] is applied to the remaining classes to increase representation of the minority classes. In the multiclass variant of this method, the majority class (i.e. *Sensor*) is decomposed into multiple sub-classes given the vast imbalance ratio with respect to the other classes. Then, synthetic oversampling using SMOTE [4] is applied to multiple minority classes with the aim of rebalancing the data distribution.

Due to the use of SMOTE as the oversampling algorithm, both the binary and multiclass versions of CDSMOTE require at least 5 instances per class for the synthetic oversampling. Therefore, classes that contained fewer than 5 instances were excluded from the dataset. To apply CDSMOTE in a multiclass scenario, we first located the majority class in the dataset (i.e. *Sensors* with 2845 samples). Afterwards, we decomposed this class into 10 different subclasses, given that this way we could generate a better spread of samples amongst classes. Although other classes e.g. *Valve* still presented considerable majority over the rest (i.e. 915 samples), we did not decompose those as we simply wanted to test whether by decomposing the majority class would suffice to improve results. Afterwards, we calculated the new average number of samples per class (i.e. 173) and applied SMOTE to all classes with fewer samples than this value to generate synthetic samples. Notice that although the use of SMOTE is typically reserved for tabular data rather than image data, an inspection of the obtained samples showed that the quality of the synthetic samples is acceptable given that symbols have a definite shape which is easy to replicate by synthetic generation methods.

3.3 Classification Method

A CNN was used for evaluation of classification performance of the imbalanced symbol dataset. A CNN [16] was chosen for classification experiments as significant advancements in the field of image classification have been obtained using CNNs in recent years [15]. The CNN used had the following model architecture: 100×100 convolutional layer with 32 filters; 2×2 max pooling layer; two convolutional layers with 64 filters; 2×2 max pooling layer; two convolutional layers with 128 filters; 2×2 max pooling layer; two fully connected layers; softmax output layer. The number of units in the softmax layer was set to the number of classes in the dataset. ReLU activation was used. All filters were 3×3. Dropout [29] was used for model regularisation with rate set at 0.5 in the two fully connected layers.

3.4 Performance Metrics

Multiclass CNN classifiers are most often evaluated using overall accuracy [3], however this does not necessarily reflect true usefulness of a classifier on imbalanced data. Take for example, an imbalanced dataset where the majority class represents 95% of the dataset. In this scenario, an accuracy of 95% would be obtained using a classifier that predicts the majority class for all instances. Therefore, in addition to overall accuracy, class wise values for precision, recall and F1-score are presented.

4 Experiments and Results

Three experiments were performed to analyse data level methods for imbalanced data classification. The first experiment uses the CNN to classify the imbalanced symbol dataset. The second experiment trains the CNN with a fully balanced symbol dataset. The third experiment uses the multiclass imbalanced data method, prior to CNN classification. A demo of the generation and comparison of performance between the original and the CDSMOTE version of the symbol dataset can be found here[1]

4.1 Setup

The dataset was split, using stratification, 70:30 into training and test sets respectively. Significantly under-represented classes (i.e. that contained one instance) were excluded from the experiment. The batch size was set to 64. The CNN was trained for 10 epochs. This training setting was selected as the experiment is designed to provide a comparative analysis of the imbalance method against the baseline, as opposed to optimise classifier performance.

---

[1] https://github.com/carlosfmorenog/CDSMOTE-NONBIN-Symbols.

4.2   Baseline Experiment

The CNN trained on the original dataset obtained a classification accuracy of 90.3% on the test data. Of the 2305 symbols in the test set, 222 were miss-classified. A weighted average precision of 0.871, recall of 0.904 and F1-score of 0.882 was reported on the symbols in the test set. The precision, recall and F1-score per class in the test set are presented in Table 1. It was observed that classification performance decreases as the number of instances per class in the training set decreases, as shown in Table 1. This finding is consistent with observations from the literature, as it was reported that classification models trained on imbalanced datasets can be biased towards the majority classes [8]. For example, the majority class, *Sensor*, contained 1991 instances in the training set. On the majority class, the CNN reported precision, recall and F1-score of 0.98, 1.00 and 0.99 respectively. In contrast, a precision, recall and F1-score of 0.00, 0.00 and 0.00, respectively was recorded for all 15 classes with fewer than 42 instances in the training set.

Table 1. CNN classification performance using Baseline, SMOTE (fully balanced) and Multiclass imbalance. **Bold** shows the best precision, recall and F1-score for each symbol. <u>Underline</u> indicates when the best result was obtained for a single dataset.

Class	No. in test	Baseline				SMOTE Fully Balanced				CDSMOTE Multiclass Imbalanced			
		No. in train	Precision	Recall	F1-score	No. in train	Precision	Recall	F1-score	No. in train	Precision	Recall	F1-score
Sensor	854	1991	0.98	**1.00**	0.99	1991	**1.00**	**1.00**	**1.00**	1991	**1.00**	**1.00**	**1.00**
Valve	275	640	0.93	0.98	0.96	1991	**0.99**	**0.99**	**0.99**	640	0.98	0.98	0.98
Reducer	263	614	0.96	1.00	0.98	1991	0.97	1.00	0.98	614	**0.98**	1.00	**0.99**
Flange Joint	126	293	0.89	**0.99**	0.94	1991	**0.98**	0.98	**0.98**	293	0.93	0.95	0.94
Continuity Label	87	201	0.95	1.00	0.97	1991	0.99	1.00	**0.99**	201	**1.00**	0.95	0.98
Valve Ball Type 2	79	183	**0.98**	1.00	**0.99**	1991	0.93	1.00	0.96	183	0.96	0.97	0.97
Valve Globe Type 1	77	179	0.91	**0.97**	0.94	1991	0.99	0.96	0.97	179	**1.00**	0.96	**0.98**
Arrowhead	72	169	0.99	0.97	0.98	1991	**1.00**	0.99	**0.99**	169	0.93	0.99	0.96
Valve Ball Type 1	52	121	0.70	0.98	0.82	1991	0.91	0.98	0.94	121	**0.98**	0.98	**0.98**
DB&BBV	43	101	0.53	0.93	0.68	1991	**1.00**	**1.00**	**1.00**	126	0.98	0.91	0.94
Valve Check	38	88	0.92	0.63	0.75	1991	0.97	**1.00**	**0.99**	126	**1.00**	0.89	0.94
DB&BPV	34	79	0.54	0.56	0.55	1991	**1.00**	**1.00**	**1.00**	126	0.82	0.97	0.89
Valve Plug	26	62	0.61	0.65	0.63	1991	**0.96**	0.92	**0.94**	126	0.79	0.85	0.81
ESDV Valve Ball	26	62	0.37	0.88	0.52	1991	0.89	**0.92**	0.91	126	**0.95**	0.81	0.88
Arrowhead + Triangle	25	58	0.75	0.72	0.73	1991	1.00	0.84	0.91	126	1.00	0.84	0.91
Valve Needle Type 1	23	54	0.64	0.78	0.71	1991	**0.96**	1.00	**0.98**	126	0.85	1.00	0.92
Triangle	22	52	0.88	0.64	0.74	1991	**0.92**	1.00	**0.96**	126	0.86	0.82	0.84
Valve Butterfly Type 1	21	50	**1.00**	0.86	0.92	1991	1.00	**0.90**	**0.95**	126	1.00	0.86	0.92
Flange Single T-Shape	19	45	**1.00**	0.89	0.94	1991	0.90	0.95	0.92	126	1.00	0.95	**0.97**
Control Valve	18	42	**1.00**	0.22	0.36	1991	0.94	**0.83**	0.88	126	0.92	0.67	0.77
Valve Angle	15	36	0.00	0.00	0.00	1991	0.82	0.60	0.69	126	**0.83**	**0.67**	**0.74**
Injector Point	13	30	0.00	0.00	0.00	1991	0.92	0.92	0.92	126	0.92	0.92	0.92
Tie In Point	12	29	0.00	0.00	0.00	1991	**0.92**	**0.92**	**0.92**	126	0.89	0.67	0.76
Spectacle Blind	13	29	0.00	0.00	0.00	1991	**1.00**	1.00	**1.00**	126	0.93	1.00	0.96
DB&BBV + Valve Check	12	27	0.00	0.00	0.00	1991	0.92	1.00	0.96	126	**1.00**	1.00	**1.00**
Valve Needle Type 2	10	23	0.00	0.00	0.00	1991	0.73	**0.80**	0.76	126	**1.00**	0.70	**0.82**
Valve Globe Type 2	8	20	0.00	0.00	0.00	1991	**0.88**	**0.88**	**0.88**	126	0.86	0.75	0.80
Valve Slab Gate	7	17	0.00	0.00	0.00	1991	**0.71**	0.71	**0.71**	126	0.35	**0.86**	0.50
Three Way Valve	7	17	0.00	0.00	0.00	1991	0.83	**0.71**	0.77	126	**1.00**	0.57	0.73
Control Valve Globe	7	16	0.00	0.00	0.00	1991	**1.00**	0.71	**0.83**	126	0.50	**0.86**	0.63
Control	6	14	0.00	0.00	0.00	1991	1.00	0.67	0.80	126	1.00	0.67	0.80
Flange + Triangle	5	12	0.00	0.00	0.00	1991	**1.00**	0.80	**0.89**	126	0.67	0.80	0.73
Exit to Atmosphere	4	10	0.00	0.00	0.00	1991	1.00	0.50	0.67	126	1.00	0.50	0.67
Rupture Disc	3	7	0.00	0.00	0.00	1991	1.00	0.67	0.80	126	1.00	0.67	0.80
ESDV Valve Slab Gate	3	6	0.00	0.00	0.00	1991	0.67	0.67	0.67	126	0.67	0.67	0.67

4.3   Fully Balanced Dataset

Classification performance of the CNN trained on a dataset with equal class distribution was evaluated. To equally balance the dataset, the minority class sizes were increased to the majority class size (see Fig. 3). To obtain equal class distribution required 64308 additional samples. The SMOTE algorithm [4] was used to generate additional samples. The resulting dataset is fully balanced across all classes, compared to the original dataset with ir of 332. The fm is 0, compared to 0.972 for the original dataset.

The SMOTE generated samples resembled realistic symbols in many cases, as can be seen in Fig. 4. Visually acceptable synthetic symbols were generated in two scenarios, Fig. 4a. In the first scenario, the SMOTE algorithm selected two visually similar minority class samples to interpolate between. In the second scenario, the selected sample and the chosen sample from the original sample's 5 nearest neighbours are not visually close in appearance. However the randomly chosen point between the two samples is selected such that a realistic instance is generated. Synthetic instances that showed more interpolation between two visually dissimilar samples were also generated, as shown in Fig. 4b.

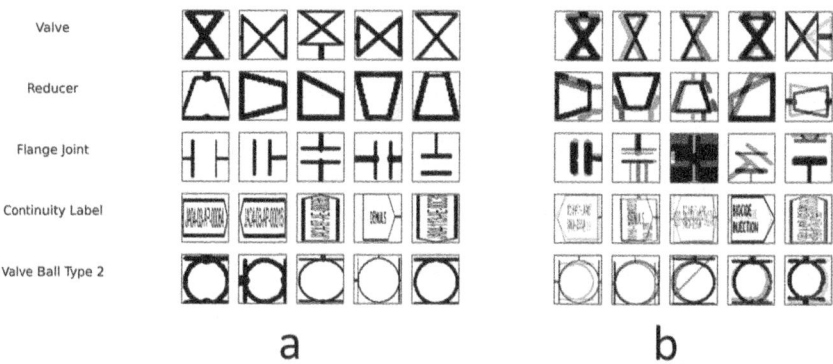

Fig. 4. Synthetically generated minority class samples for some classes. a) realistic samples b) less realistic samples.

The CNN trained on the fully balanced dataset obtained an overall classification accuracy of 97.7% on the test data. Of the 2,305 symbols in the test set, 54 were missclassified. A weighted average precision of 0.977, recall of 0.977 and F1-score of 0.976 was reported on the symbols in the test set. A precision, recall and F1-score of 1.00 was observed for 4 classes *Sensor, DB&BBV, DB&BPV* and *Spectacle Blind*. A precision, recall and F1-score of 0.50 or above was obtained across all classes in the dataset. Results showed slightly lower performance for the classes with lower original class sizes. This finding may result from the CNN overfitting to the training data. The lowest values of F1-score were obtained for classes where 1,981 or higher synthetic samples were generated using information from 10 or fewer original samples.

4.4 Multiclass Imbalance Experiment

To evaluate the multiclass imbalance method, it was applied to the symbol dataset. The original majority class was decomposed into 10 subclasses. Synthetic oversampling of 26 minority classes occurred using SMOTE. The resulting dataset consisted of 7,667 instances compared to 5,377 in the original dataset. A more evenly balanced class distribution was obtained as a result of applying the multiclass imbalance data technique, as seen in Fig. 3. The method has altered the ir of the dataset considerably from 332 to 16. The fm remains the same as the original dataset at 0.972. The CNN obtained a classification accuracy of 96.1% on the decomposed dataset, with 88 out of 2,305 symbols missclassified. A prediction of the majority class was considered correct if the prediction was of any one of the decomposed majority subclasses. A weighted average precision, recall and F1-score of 0.938, 0.905 and 0.913 respectively was reported on the decomposed dataset.

Classification performance across the minority classes has been improved as a result of applying the multiclass imbalance data technique, as can be observed in Table 1. Examining the results obtained for the decomposed dataset, with one exception, the precision, recall and F1-scores of 0.50 or above were observed for all classes in the dataset. Compared to the fully balanced dataset, the multiclass imbalance method obtains higher F1-score for eight classes, whereas the fully balanced dataset obtains a higher F1-score for 21 classes (Table 1). The reported classification performance is equal using the fully balanced training dataset compared to the multiclass imbalance method for five classes in the dataset. These classes are *Exit to Atmosphere*, *Rupture Disc*, *ESDV Valve Slab Gate*, *Control* and *Arrowhead + Triangle*.

To compare the classification results obtained with the multiclass imbalance method against baseline results, the t-test was completed. A *p-value* of 0.0000013 was obtained which shows statistically significant improvement was achieved using the proposed multiclass imbalance technique. The training dataset size impacted the CNN model training time. The model training time increased by a factor of 1.4 using the multiclass imbalance method (7,667 instances), compared with the baseline (5,377 instances). Required model training time increased by a factor of 14.5 for the fully balanced dataset (69,685 instances), compared to the baseline model.

5   Conclusions

This paper addresses the lack of engineering symbol datasets by presenting a labelled dataset of symbols from P&IDs. The dataset consists of 7,728 symbols, distributed across 48 classes, from two distinct P&ID drawing standards. Engineering equipment is unevenly represented within P&IDs and as a result, the symbol dataset is significantly imbalanced. A new method for handling multiclass imbalanced classification is described. The method redistributes the data distribution within the dataset, whilst avoiding information loss and is an extension of CDSMOTE. Classification experiments on the symbols dataset demonstrated that the multiclass imbalanced data method improves CNN classification performance across multiple minority classes.

Future research will utilise the engineering symbols dataset for development and evaluation of deep learning models to digitise engineering diagrams. In particular, a symbol detection model would represent a key component of diagram digitisation. Additionally, further methods could be investigated to avoid biased learning models for multiclass imbalanced data, which is inherent within engineering diagrams. Generation of artificial samples to rebalance the dataset using GANs is suggested as one possible direction.

References

1. Arroyo, E., Hoernicke, M., Rodríguez, P., Fay, A.: Automatic derivation of qualitative plant simulation models from legacy piping and instrumentation diagrams. Comput. Chem. Eng. **92**, 112–132 (2016)
2. Banerjee, P., Choudhary, S., Das, S., Majumdar, H., Roy, R., Chaudhuri, B.B.: Automatic hyperlinking of engineering drawing documents. In: 2016 12th IAPR Workshop on Document Analysis Systems (DAS), pp. 102–107 (2016). https://doi.org/10.1109/DAS.2016.76
3. Buda, M., Maki, A., Mazurowski, M.A.: A systematic study of the class imbalance problem in convolutional neural networks. Neural Netw. **106**, 249–259 (2018). https://doi.org/10.1016/j.neunet.2018.07.011
4. Chawla, N.V., Bowyer, K.W., Hall, L.O., Kegelmeyer, W.P.: Smote: synthetic minority over-sampling technique. J. Artif. Int. Res. **16**(1), 321–357 (2002)
5. Datta, R., Mandal, P.D.S., Chanda, B.: Detection and identification of logic gates from document images using mathematical morphology. In: 2015 Fifth National Conference on Computer Vision, Pattern Recognition, Image Processing and Graphics (NCVPRIPG), pp. 1–4 (2015). https://doi.org/10.1109/NCVPRIPG.2015.7490040
6. Elyan, E., Moreno-García, C.F., Jayne, C.: Symbols classification in engineering drawings. In: 2018 International Joint Conference on Neural Networks (IJCNN), pp. 1–8 (2018). https://doi.org/10.1109/IJCNN.2018.8489087
7. Elyan, E., Jamieson, L., Ali-Gombe, A.: Deep learning for symbols detection and classification in engineering drawings. Neural Netw. **129**, 91–102 (2020). https://doi.org/10.1016/j.neunet.2020.05.025
8. Elyan, E., Moreno-García, C.F., Jayne, C.: CDSMOTE: class decomposition and synthetic minority class oversampling technique for imbalanced-data classification. Neural Comput. Appl. (2020). https://doi.org/10.1007/s00521-020-05130-z
9. Elyan, E., Moreno-García, C.F., Johnston, P.: Symbols in engineering drawings (SiED): an imbalanced dataset benchmarked by convolutional neural networks. In: Iliadis, L., Angelov, P.P., Jayne, C., Pimenidis, E. (eds.) EANN 2020. PINNS, vol. 2, pp. 215–224. Springer, Cham (2020). https://doi.org/10.1007/978-3-030-48791-1_16
10. Fajardo, V.A., et al.: On oversampling imbalanced data with deep conditional generative models. Expert Syst. Appl. **169**, 114463 (2021). https://doi.org/10.1016/j.eswa.2020.114463

11. Gellaboina, M.K., Venkoparao, V.G.: Graphic symbol recognition using auto associative neural network model. In: 2009 Seventh International Conference on Advances in Pattern Recognition, pp. 297–301 (2009). https://doi.org/10.1109/ICAPR.2009.45

12. Gupta, G., Swati, Sharma, M., Vig, L.: Information extraction from hand-marked industrial inspection sheets. In: 2017 14th IAPR International Conference on Document Analysis and Recognition (ICDAR), vol. 6, pp. 33–38 (2017). https://doi.org/10.1109/ICDAR.2017.346

13. Jamieson, L., Moreno-García, C.F., Elyan, E.: Deep learning for text detection and recognition in complex engineering diagrams. In: 2020 International Joint Conference on Neural Networks (IJCNN), pp. 1–7 (2020). https://doi.org/10.1109/IJCNN48605.2020.9207127

14. Johnson, J.M., Khoshgoftaar, T.M.: Survey on deep learning with class imbalance. J. Big Data **6**(1), 27 (2019)

15. Krizhevsky, A., Sutskever, I., Hinton, G.E.: Imagenet classification with deep convolutional neural networks. Commun. ACM **60**(6), 84–90 (2017). https://doi.org/10.1145/3065386

16. Lecun, Y., Bottou, L., Bengio, Y., Haffner, P.: Gradient-based learning applied to document recognition. Proc. IEEE **86**(11), 2278–2324 (1998). https://doi.org/10.1109/5.726791

17. Long, J., Shelhamer, E., Darrell, T.: Fully convolutional networks for semantic segmentation. In: 2015 IEEE Conference on Computer Vision and Pattern Recognition (CVPR), pp. 3431–3440 (2015). https://doi.org/10.1109/CVPR.2015.7298965

18. Mani, S., Haddad, M.A., Constantini, D., Douhard, W., Li, Q., Poirier, L.: Automatic digitization of engineering diagrams using deep learning and graph search. In: Proceedings of the IEEE/CVF Conference on Computer Vision and Pattern Recognition (CVPR) Workshops, pp. 1–7 (2020)

19. Moreno-García, C.F., Elyan, E.: Digitisation of assets from the oil gas industry: challenges and opportunities. In: 2019 International Conference on Document Analysis and Recognition Workshops (ICDARW), vol. 7, pp. 2–5 (2019). https://doi.org/10.1109/ICDARW.2019.60122

20. Moreno-García, C.F., Elyan, E., Jayne, C.: Heuristics-based detection to improve text/graphics segmentation in complex engineering drawings. In: Engineering Applications of Neural Networks, vol. CCIS 744, pp. 87–98 (2017)

21. Moreno-García, C.F., Elyan, E., Jayne, C.: New trends on digitisation of complex engineering drawings. Neural Comput. Appl. (2018). https://doi.org/10.1007/s00521-018-3583-1

22. Moreno-García, C.F., Johnston, P., Garkuwa, B.: Pixel-based layer segmentation of complex engineering drawings using convolutional neural networks. In: 2020 International Joint Conference on Neural Networks (IJCNN), pp. 1–7 (2020). https://doi.org/10.1109/IJCNN48605.2020.9207479

23. Nurminen, J.K., Rainio, K., Numminen, J.-P., Syrjänen, T., Paganus, N., Honkoila, K.: Object detection in design diagrams with machine learning. In: Burduk, R., Kurzynski, M., Wozniak, M. (eds.) CORES 2019. AISC, vol. 977, pp. 27–36. Springer, Cham (2020). https://doi.org/10.1007/978-3-030-19738-4_4

24. Rahul, R., Paliwal, S., Sharma, M., Vig, L.: Automatic information extraction from piping and instrumentation diagrams. CoRR (2019). http://arxiv.org/abs/1901.11383

25. Ravagli, J., Ziran, Z., Marinai, S.: Text recognition and classification in floor plan images. In: 2019 International Conference on Document Analysis and Recognition

Workshops (ICDARW), vol. 1, pp. 1–6 (2019). https://doi.org/10.1109/ICDARW.2019.00006

26. Redmon, J., Farhadi, A.: Yolov3: an incremental improvement. CoRR (2018). http://arxiv.org/abs/1804.02767

27. Rica, E., Alvarez, S., Moreno-García, C.F., Serratosa, F.: Zero-error digitisation and contextualisation of piping and instrumentation diagrams using node classification and sub-graph search. In: Krzyzak, A., Suen, C.Y., Torsello, A., Nobile, N. (eds.) S+SSPR 2022. LNCS, vol. 13813, pp. 274–282. Springer, Cham (2022). https://doi.org/10.1007/978-3-031-23028-8_28

28. Sinha, A., Bayer, J., Bukhari, S.S.: Table localization and field value extraction in piping and instrumentation diagram images. In: 2019 International Conference on Document Analysis and Recognition Workshops (ICDARW), vol. 1, pp. 26–31 (2019). https://doi.org/10.1109/ICDARW.2019.00010

29. Srivastava, N., Hinton, G., Krizhevsky, A., Sutskever, I., Salakhutdinov, R.: Dropout: a simple way to prevent neural networks from overfitting. J. Mach. Learn. Res. **15**, 1929–1958 (2014). http://jmlr.org/papers/v15/srivastava14a.html

30. Tan, W.C., Chen, I.M., Tan, H.K.: Automated identification of components in raster piping and instrumentation diagram with minimal pre-processing. In: 2016 IEEE International Conference on Automation Science and Engineering (CASE), pp. 1301–1306 (2016). https://doi.org/10.1109/COASE.2016.7743558

31. Toral, L., Moreno-García, C.F., Elyan, E., Memon, S.: A deep learning digitisation framework to mark up corrosion circuits in piping and instrumentation diagrams. In: Barney Smith, E.H., Pal, U. (eds.) ICDAR 2021. LNCS, vol. 12917, pp. 268–276. Springer, Cham (2021). https://doi.org/10.1007/978-3-030-86159-9_18

32. Vuttipittayamongkol, P., Elyan, E.: Neighbourhood-based undersampling approach for handling imbalanced and overlapped data. Inf. Sci. **509**, 47–70 (2020). https://doi.org/10.1016/j.ins.2019.08.062. http://www.sciencedirect.com/science/article/pii/S0020025519308114

33. Wang, S., Liu, W., Wu, J., Cao, L., Meng, Q., Kennedy, P.J.: Training deep neural networks on imbalanced data sets. In: 2016 International Joint Conference on Neural Networks (IJCNN), pp. 4368–4374 (2016). https://doi.org/10.1109/IJCNN.2016.7727770

34. Wang, S., Yao, X.: Multiclass imbalance problems: analysis and potential solutions. IEEE Trans. Syst. Man Cybern. Part B (Cybern.) **42**(4), 1119–1130 (2012). https://doi.org/10.1109/TSMCB.2012.2187280

Weakly Supervised Training for Hologram Verification in Identity Documents

Glen Pouliquen[1,2](✉) ⓘ, Guillaume Chiron[1] ⓘ, Joseph Chazalon[2] ⓘ,
Thierry Géraud[2] ⓘ, and Ahmad Montaser Awal[1] ⓘ

[1] IDnow AI & ML Center of Excellence, Cesson-Sévigné, France
{glen.pouliquen,guillaume.chiron,ahmad.awal}@idnow.io
[2] EPITA Research Laboratory (LRE), Le Kremlin-Bicêtre, France
{glen.pouliquen,joseph.chazalon,thierry.geraud}@epita.fr

Abstract. We propose a method to remotely verify the authenticity of Optically Variable Devices (OVDs), often referred to as "holograms", in identity documents. Our method processes video clips captured with smartphones under common lighting conditions, and is evaluated on two public datasets: MIDV-HOLO and MIDV-2020. Thanks to a weakly-supervised training, we optimize a feature extraction and decision pipeline which achieves a new leading performance on MIDV-HOLO, while maintaining a high recall on documents from MIDV-2020 used as attack samples. It is also the first method, to date, to effectively address the photo replacement attack task, and can be trained on either genuine samples, attack samples, or both for increased performance. By enabling to verify OVD shapes and dynamics with very little supervision, this work opens the way towards the use of massive amounts of unlabeled data to build robust remote identity document verification systems on commodity smartphones. Code is available at https://github.com/EPITAResearchLab/pouliquen.24.icdar.

Keywords: Know Your Consumer (KYC) · Identity Documents · Hologram Verification · Weakly Supervised Learning · Contrastive Loss

1 Introduction

Often called KYC (Know Your Customer), remotely verifying the authenticity of identity documents is a critical point for building online trust. This is an increasingly regulated process which relies on identity documents, among other proofs, to establish the link between an online identity and a real state-backed one. This linking usually requires checking that the document is original and was not altered. The photography of the bearer is of paramount importance here to ensure that the user of a remote system is the intended one. Optically variable devices (OVDs), commonly referred to as "holograms" and illustrated in Fig. 2, are powerful tools to secure physical documents in line with the recommendation of the EU council [14], among others. Built using elaborated and undisclosed optical techniques,

E. H. Barney Smith et al. (Eds.): ICDAR 2024, LNCS 14804, pp. 17–33, 2024.
https://doi.org/10.1007/978-3-031-70533-5_2

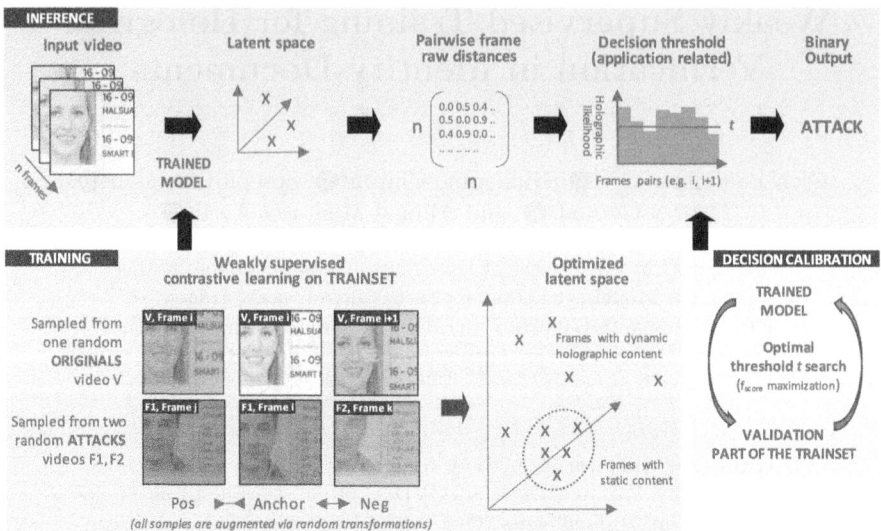

Fig. 1. Proposed approach overview, involving 1) the weakly supervised TRAINING with a specific data selection strategy over the trainset; 2) the INFERENCE pipeline extracting optimized features used afterward to compute the final "Original/Attack" decision based on a thresholding of pairwise distances. The DECISION illustrates how the threshold is calibrated over the validation part of the train set.

these devices exhibit rich visual behaviors when viewing and/or illuminating conditions (angle, light color, etc.) change. They are embedded in a wide spectrum of sensitive elements, i.e. not only identity documents but also banknotes or tamper-proof labels (e.g. for medical drugs), and contribute to ensure: 1. **Integrity:** they cannot be removed without altering their properties, making tampering very challenging; 2. **Authenticity:** they are very difficult to forge, making the creation of fraudulent documents equally hard.

Despite the widespread use of holograms, automating their remote verification in the context of an automatic enrollment, whether it is to open an account in an online bank or to contract a loan, poses many challenges. Indeed, such visual objects were primarily designed for manual inspection, sometimes using special tools like magnifiers or dedicated light sources. As a result, automated remote validation is limited in many aspects: acquisition is often performed using commodity smartphones under uncontrolled ambient light to capture macroscopic and visible patterns, while following simple interactive scenarios. Nevertheless, verifying holographic devices from a video is possible to some extent, and many recent works and datasets contributed to this effort.

While, in the general forgery detection literature, several approaches try to detect falsification clues [16], others follow the opposite (yet complementary) direction of checking whether clues of authenticity and integrity are present [7]. This work contributes to the latter: we propose a method to control the presence

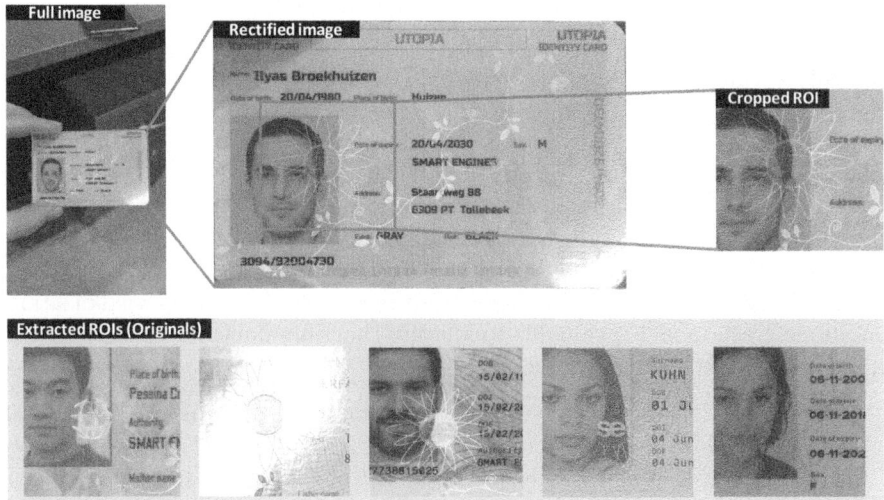

Fig. 2. In MIDV Holo dataset, documents are captured in different places involving various backgrounds and lightning conditions (left). Document quads are annotated on all images allowing rectifications (center). Additionally, we propose to define a region of interest containing part of the face and the holograms in charge of securing it (right). Extracted Regions of Interest (ROIs) from sampled labeled as "Originals" (below) contain more or less visible holographic content. *Identities* (names and faces) are synthetic.

of some holographic content at specific positions of a document (e.g. photography area), and address the problem of photography replacement, which was introduced in MIDV-Holo [18] but not yet addressed (to our knowledge). After a detailed review of related approaches and datasets (Sect. 2), we introduce our key contribution: a new method to detect and validate holographic content, whose feature extraction is trained in a weakly-supervised fashion (i.e., not requiring a precise labeling of each video frame with the particular visual appearance of a hologram), and which outperforms the original approach on public datasets (Sect. 3). For practical reasons, we also propose an updated experimental protocol which specifies, among others, training, validation and test sets for the MIDV-Holo dataset, as well as a public, open-source reimplementation of the approach proposed in the original MIDV-Holo publication [18], with systematic optimization of the calibration of the decision function (Sect. 4). Our approach (illustrated in Fig. 1) is carefully evaluated on several public datasets, over several runs, and an ablation study is conducted to challenge the benefits of every aspect of our method (Sect. 5). The code to reproduce our results is publicly available at https://github.com/EPITAResearchLab/pouliquen.24.icdar.

2   Related Works

Optically Variable Devices (OVDs) are often built using polarized inks or diffraction grating—a network of microscopic reflective structures engraved in the thickness of some transparent layer. We refer the reader to the MIDV-Holo publication [18] for a concise introduction on the optical design of these objects. These OVDs exhibit continuous, sometimes rapid, transitions among a virtually infinite set of visual states when changing the relative positions of the light source(s), the camera and the document. We can consider such space of visual appearances as a sort of manifold which we navigate by changing visualization conditions. Such model is valuable to identify the 3 fundamental visual features an automated method can check:

1. **Appearance Conformity:** *Can a particular visual appearance (shape and color) be generated by a genuine OVD?*— This can be viewed as assessing how far a particular sample (usually an image) is from the real manifold of a given hologram, and use a distance threshold as verification criterion. Implementing this control enables to detect attacks with no hologram or with a different hologram shape but, if used alone, would be tricked by simple static copies of the expected hologram.
2. **Appearance Coverage:** *How well do a particular set of visual appearances (usually captured as a video) matches the set of possible ones?*— This can be viewed as measuring how well the samples obtained cover the real manifold of a given hologram. Implementing this control enables to detect attacks with static holograms but, if used without control of state conformity, would be tricked by any random holographic layer. Approaches checking only color distributions are vulnerable to this attack.
3. **Transitions Validity:** *Are the transitions between observed visual appearances consistent with expected ones?*— This can be viewed as checking whether samples obtained describe valid paths on the real manifold of a given hologram. Implementing this control enables to detect attacks with imperfect hologram imitations or rapid swapping of static holograms. The low frame rates utilized in current real remote authentication applications present a significant challenge that has not been adequately explored in the research literature.

Early approaches like the one of Hartl et al. [2] identified a discrete set of visual appearances to check for during a manual inspection of identity documents. Expected visual states were acquired and validated using a robotic arm with controlled light. While lacking automated verification, this approach proposed a practical protocol to assist a human operator during its work to validate 1. visual state conformity thanks to visual comparison, and 2. visual behavior completeness by checking every expected visual state is seen.

The work of Chapel et al. [17] proposes a way to automate the verification of visual states conformity thanks to a learned classifier based on local binary patterns (LBP) features. However, training this system requires to label each video frame with target class (visual state), which is both too expensive for real application in our experience, and also challenging because of the frequent "mixing" of visual shapes in real OVDs. Furthermore, robustness of the static feature extractor may be a concern when background are not constant like in the case of face photos: a learned feature extractor seems necessary here.

To overcome the need for labeled frames when training a visual state classifier, some approaches relaxed the control on visual state conformity to focus instead on the validation of visual behavior completeness. The work of Kada et al. [15] opened an interesting direction by restating the problem as a semantic segmentation problem where images of documents captured as a video are first carefully registered, then a pixel-level classification is performed to predict whether a particular region belongs to a hologram. Such prediction is mainly based on statistics on the distribution of pixel color values. While the final segmentation map may be used as some visual appearance clue, this method does not check whether inter-pixel behavior is consistent, nor visual appearance conformity for a particular frame, and lacks a global decision stage.

A major step was made thanks to the work of Koliaskina et al. [18]. The authors not only reuse the same idea of semantic segmentation (also based on a static, handcrafted feature extraction) to produce a map of pixels which exhibit some "holographic behavior", but also provide a global decision stage based on a variance threshold and a first public dataset, MIDV-Holo, containing document with holographic contents along with several presentation attacks, as illustrated in Fig. 2. While validated on full-size documents, the approach was not tested on the particular case of the face photo region, and was not evaluated against the photo replacement attack. Furthermore, as detailed in Sect. 4, this milestone publication required us to reimplement the proposed approach and specify training, validation and test splits to conduct a fair comparison with our proposed approach.

Another family of approaches tackled the problem of learning a useful embedding space, thanks to which it may be possible to both check visual state conformity and visual behavior completeness. A first example is the work of Soukup et al. [4] targeting hologram verification on banknotes. Their approach extract representations from video frames using a Convolutional Neural Network (CNN) trained with a supervised classification task. Target classes represent different visual appearances, and were captured using a LED ring that illuminated the hologram in various directions to automate the annotation process. Because of the changing nature of the background for some OVDs in identity documents (such as in the area of the face picture), such approach may need an important amount of training data to be applied in our context, which makes it impractical since it requires physical access to real documents.

Finally, a last related work is the one of Ay et al. [12] which proposed to train a Generative Adversarial Network (GAN) to capture the visual properties of some hologram. While the generative properties of such approach are very attractive, successfully training such architecture to model thin holograms on non-constant backgrounds like face pictures is a great challenge, as the network may more easily capture and generate facial feature which cover a larger proportion of the area of interest. Furthermore, as final decision is performed using the discriminator network, a proper calibration would require attack samples. Another limitation of using the discriminator is that, despite the rich representation learned, this approach is limited to controlling the visual state conformity of isolated frames, and cannot check visual behavior completeness as differences between visual states cannot be measured.

Looking at these existing works, we can sketch out desirable features an automated verification method should provide, which our proposed approach tends to incorporate:

1. It should be based on a learned representation which can be tied to a particular OVD, in order to be able to capture both visual appearance and behavior information, as opposed to handcrafted, static, pixel-based feature extraction techniques.
2. Such representation should be learned in a weakly supervised way to avoid requiring manual labelling of existing frames, or physical access to a large amount of original documents and presentation attacks.
3. The learning objective should be able to guide the training even in the presence of non-constant backgrounds and thin holographic objects, like in the case of the face picture area.

3   Contrastive Learning of Hologram Representation

This section introduces our key contribution, summarized in Fig. 1: a new method to detect and validate holographic content, whose representation (feature extraction) is trained in a weakly-supervised fashion; i.e. it only requires a single label ("original" or "attack", as per MIDV-Holo [18] terminology) for each video clip. For this purpose, we use a particular kind of contrastive loss which enables the training to be driven by intrinsic data properties. This relies on certain assumptions about the video clips, such as their ability to capture varied perspectives of the document. It also involves various transformations to enhance the data. The resulting representation can be shown to effectively focus on hologram regions, and can be used to assess both appearance consistency and coverage in a final decision stage considering as many video frames as necessary.

3.1   Learning Objective

To avoid the need for assigning labels to every video frame of the training set, we employ a contrastive learning objective which enables us to drive model training using intrinsic data properties (described in the next subsection). More specifically, we use a triplet loss [1] defined on a minibatch of N elements as

$$\mathcal{L}(a, p, n) = \frac{1}{N} \sum_{i}^{N} l_i(a_i, p_i, n_i), \quad l_i(a_i, p_i, n_i) = \max\left(d(a_i, p_i) - d(a_i, n_i) + m, 0\right) \quad (1)$$

where a_i is the projected representation of an *anchor* sample whose distance from the representation of a *positive* (similar) sample p_i is minimized, while the distance to the representation of a *negative* (dissimilar) sample n_i is maximized. Each of these representations are computed from augmented inputs to improve training. An extra margin term $m = 1$ is used to enforce a minimal distance to negative samples. We use $d(x_i, y_i) = ||xi - y_i||_2$ as distance function, and train using an AdamW optimizer [8] with default PyTorch parameters.

3.2   Triplet Sample Selection Strategy

The selection of the samples which constitute the triplets is the cornerstone of our approach. It is guided by weak labels provided at the video clip level, i.e. "original" or "attack", which exhibit different properties in the MIDV-Holo dataset. In the case of video clips labeled as "originals", we assume the visual appearance of the hologram throughout a major part of the recording. Conversely, for video clips labeled as "attacks", we assume that there will be no change in its visual appearance. This requires to remove cases of "photo replacement" attacks from our training set, as they exhibit the behavior of the real hologram except at the position of the replaced face picture. These assumptions led to the following selection process, illustrated in Fig. 3:

- **Original**: Given that the document is always moving in the videos of the dataset (at 5 frames per second), we assume that the hologram is changing. Thus, frame t and frame $t+1$ are expected to contain two different visual states of the hologram. The anchor and the positive samples are generated from the same frame t with different augmentations. Frame $t + 1$, with augmentation, is used as the negative sample.
- **Attack**: In the case of an attack, we know that all the frames from a same video contain the exact same visual state of the hologram. Therefore, we take uniformly selected frames from this video as anchor and positive sample. For the negative sample, we select a uniformly selected frame from another video with the same identity.

In both cases, the anchor, positive, and negative frames all represent the same identity (face picture). Consequently, the network aims to minimize the distance between the embeddings of two frames depicting the same visual state of a hologram while maximizing the difference between a frame showing the same face but with different hologram states. The assumption that the viewpoint continuously changes is generally valid in the MIDV-Holo dataset and can be easily enforced in a real industrial scenario. This is because a document detection stage is typically required during capture to localize, classify, and rectify documents, providing indications about the camera pose relative to the document.

3.3   Augmentations

To diversify anchor, positive, and negative samples (initially resized to 256 px), we apply transformations with specified probabilities: – Rotation, horizontal or vertical mirroring ($p = 0.5$) applied equally to anchor, positive and negative

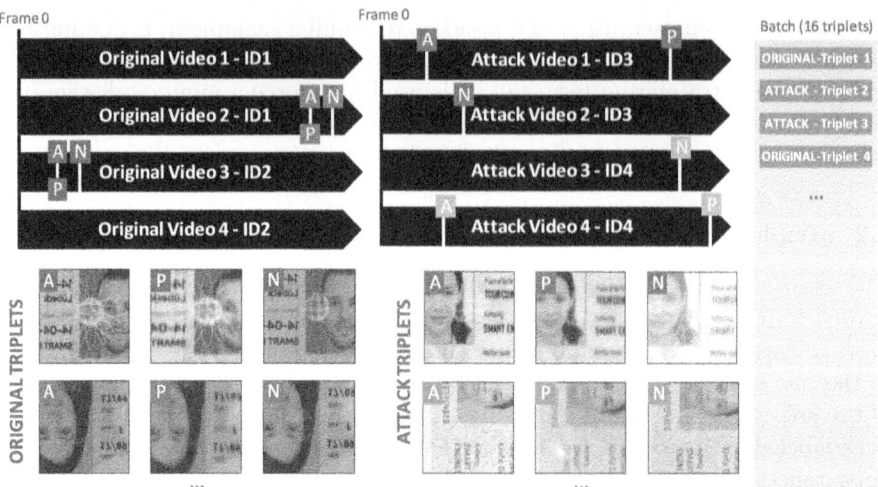

Fig. 3. Frame sampling strategy for building triplets [A]nchor/[P]ositive/[N]egative from *original* and *attack* videos. Each *original* triplet is sampled from a unique *original* video (A and P at t, N at $t+1$). Each *attack* triplet is sampled from 2 different videos of a same identity with A and P, both belonging to a common video, and N belonging to a different one. All samples are transformed with uniformly selected augmentations.

samples. – Crop by a random 80% ROI, resized to 224 px ($p = 1$). – Gaussian blur ($p = 0.4$, kernel: $3 <$ kernel < 11, $2 < \sigma < 10$). – Color jittering ($p = 0.4$, $0.7 <$ brightness < 1.3, $0.9 <$ contrast < 1.1, $0.95 <$ saturation < 1.05). Images are then normalized to ImageNet's mean and standard deviation.

3.4   Qualitative Validation: Feature Attribution Maps

We employed the Integrated Gradients method by Sundararajan et al. [5], implemented in Captum [10], to identify the focal elements of our approach. The results, illustrated in Fig. 4, showcase the efficacy of our weakly supervised training method. Specifically, the model ($mobilevit_{xxs}$) trained using this approach assigns significant importance to the hologram area. This stands in stark contrast to the same network architecture trained exclusively on ImageNet, which lacks a similar level of focus on the hologram. This observation underscores the value of our training strategy in guiding the model's attention towards pertinent features.

3.5   Final Decision Stage

Finally, thanks to the representation produced by the feature extraction network, learned in a weakly-supervised manner as previously presented, we can extract and compare vector representations for each frame of a video clip to control.

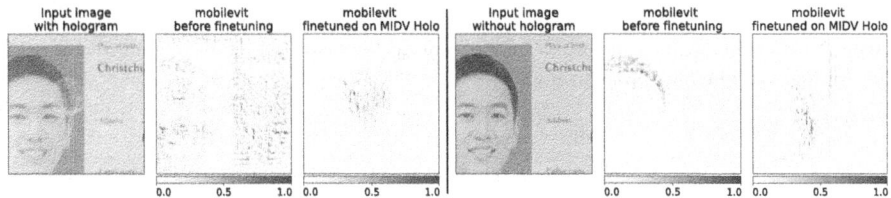

Fig. 4. Integrated Gradients [5] visualizes a training sample, emphasizing our method's effectiveness in directing the network's attention towards the hologram. In contrast, the ImageNet-trained model lacks this focused attribution to the hologram, highlighting the significance of our training approach.

Subsequently, we compute pairwise *cosine* similarity between the representations of the frames. We consider two scenarios: analyzing the full video (more resource-intensive but theoretically more reliable), or adopting the incremental cumulative mode introduced in the original MIDV-Holo publication. This last mode deems a video clip as original as soon as it meets the acceptable criterion. By computing the mean of these differences, it becomes possible to obtain an indicator of the expected visual behaviors' coverage. We use a single threshold calibrated on the validation set to make the final decision on accepting the video clip as original or triggering an alert for a potential attack. It is important to note that this approach does not directly inspect each individual visual appearance of the hologram. However, as the representation is trained to project non-hologram content to the same representation, such content tends to be constant while frames containing hologram content, on the other hand, typically exhibit variability. This results in a simultaneous control of *Appearance Conformity* and *Appearance Coverage*.

4   Extensions to MIDV-Holo

This section introduces extensions to the original MIDV-Holo dataset and evaluation protocol that we needed to benchmark our contribution. To compare the performance of our proposed approach with the MIDV-Holo baseline, we re-implemented and open-sourced the latter for the "no tracking" mode, i.e., without the frame alignment preliminary stage. Then, we propose an improved protocol enabling cross-validation to cope with the variance we observe in experimental results. This requires to revise the metrics used and to ways to separate training, validation, and test sets, over several runs.

4.1   Reproduction of the MIDV-Holo Baseline Approach

The authors of the original MIDV-Holo publication [18] introduced a baseline approach for semantic segmentation of video frames, identifying pixels within a holographic area and computing their ratio as a proxy to verify the hologram's shape, resulting in a binary decision: – *negative*: the video clip is deemed to

Table 1. Reproduction of Table 1 from the original MIDV-Holo [18] publication, comparing ROC AUC values in "no tracking" mode between our re-implementation and the original, validating its accuracy.

S_{thresh}	30			40			50		
h_{thresh}	0.01	0.02	0.03	0.01	0.02	0.03	0.01	0.02	0.03
Original MIDV-Holo [18]	0.795	0.825	0.832	0.828	0.841	0.832	**0.847**	0.838	0.807
Our re-implementation	0.838	0.846	0.844	0.855	0.844	0.831	**0.857**	0.826	0.790

contain the expected hologram, considered "original"; – *positive*: no hologram is found, raising an alert for a potential "attack". No public implementation of this baseline approach existed, so we created a public, open-source re-implementation following the authors' guidance. We evaluated our approach using the same conditions and metrics outlined in Table 1 before integrating it into our experiments.

The baseline approach, originally calibrated and tested on the entire MIDV-Holo dataset without clear separation between calibration and test sets, relied on three parameters: S_{thresh}, h_{thresh} and T. Our implementation differs from the original in two notable aspects: 1. resizing images to 1123×709 pixels and 2. imposing a minimum buffer of five frames before returning a result, compared to the original's theoretical requirement of two frames. We recalculated Table 1 of the original paper for the "no tracking" mode and observed nearly identical performance in terms of ROC AUC as reported in Table 1. Specifically, we identified the same optimal threshold configuration ($S_{\text{thresh}} = 50$ and $h_{\text{thresh}} = 0.01$) across values of the T parameter.

4.2 Enabling Cross-validation on MIDV-Holo

While being an important contribution with a first public dataset with "holograms" in identity documents, MIDV-Holo still contain little data: 700 video clips which can be broken down as follows: – 2 types of documents, equally shared: identity card-like and passport-like, – 10 model variants for each type, also equally shared, – 5 "identities", i.e., fake holder for each model variant, and – 3 originals and 4 presentation attacks (actual video clips) for each "identity". These presentation attacks can either contain static content, in the case of the "copy without holo" (no hologram at all), "pseudo holo copy" (static imitation using an image editor), and "photo holo copy" (photocopy of the document) attacks; or dynamic content as in the case of the "photo replacement" attack where an original document is physically altered to change the face picture. In this latter case, no hologram is visible over the face picture, but it is still present on the rest of the document. All original document variants exhibit the *same hologram*, with a small translation between identity card and passport documents.

Because we need to be able to compare methods trained and calibrated on this dataset and reduce variance in the experiments, we propose to specify training, validation and test sets for 5 different splits. The process for generating such

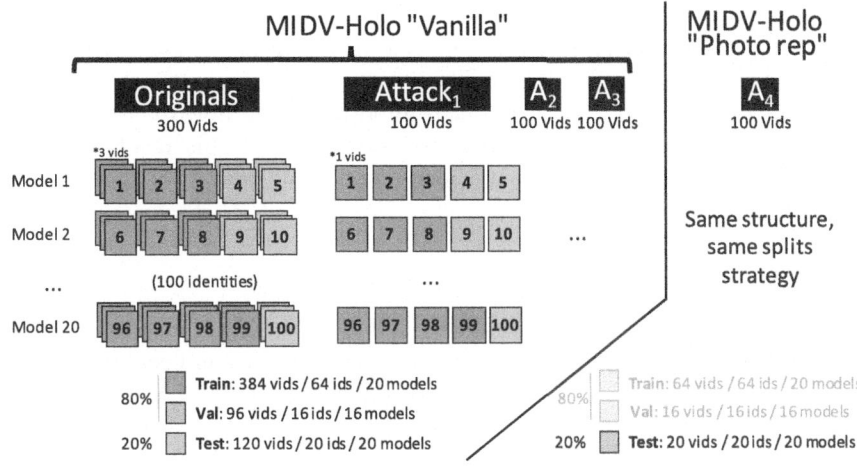

Fig. 5. Proposed split over the MIDV-Holo dataset (64% train, 16% validation and 20% test). MIDV-Holo "Vanilla" refers to the part tackled in the original paper. "Photo replacement" attacks are exclusively used for testing in our experiments.

splits is illustrated in Fig. 5, and aims at challenging the generalization to new identities rather than the generalization to new documents. Therefore, all identities in the train, validation, and test sets are distinct, while the document models (i.e. identity cards, passports) are common across the different sets. We proceed as follows: we stratify the dataset by document model (20 cases), then for each document model, we select 1 out of 5 identity for testing, the 4 remaining ones being used for training, except for 1 document every 5 items where an identity is kept for validation instead of training. All video clips for the selected identities go into the same target subset for a given split. This results in the following data partition: 64% training, 16% validation and 20% test. No identity can be present in two subsets for a given split. The *"photo replacement"* attack case is handled specially as we never use the corresponding samples for training or validation, and only use them for testing.

A last modification to the original protocol is that we use the F_{score} (harmonic mean of Precision and Recall) as the final metric, while MIDV-Holo authors preferred to report Recall values for a false positive rate close to 10%. Reporting ROC curves computed on the test set would be possible, but would only give a hint about the expected performance in production while hiding calibration uncertainty. In order to avoid an extra level of complexity during the training of a learned feature extractor (to favor Recall over Precision), we encourage the use of a simpler metric which provides a total ordering.

5 Experiments and Results

This section proposes an experimental evaluation of our proposed approach (described in Sect. 3) compared to the MIDV-Holo baseline [18], along with an ablation study. Contrary to the MIDV-Holo original experiment applied to the whole rectified documents (see reproduced results in Table 2), our experiments exclusively focus on the critical region of the document containing the face picture of the bearer. We believe it is important to be able to leverage prior knowledge about the documents controlled, and challenged this idea by cropping registered document image to a particular region of interest for each document model, as illustrated in Fig. 2.

5.1 Experimental Protocol

Our experiments utilized three publicly available datasets. Initially, both our proposed method and the MIDV-Holo baseline were trained and calibrated using the MIDV-Holo "Vanilla" training and validation sets, as defined in the preceding section. Subsequently, they were assessed on three distinct test sets from the MIDV series to gauge their generalization capabilities:

– **MIDV-Holo "Vanilla"** (120 test videos) originally introduced in [18], selecting only test set elements as defined in Sect. 4.2.
– **MIDV-Holo "Photo Replacement"** (20 test videos) is a specific subset of MIDV-Holo, and represents a distinct task from the "Vanilla" set, as it was not addressed in the original paper. While the complete set comprises 100 videos, our experiments solely involve the 20 test videos at each run, as our approach does not entail training on this dataset.
– **MIDV 2020 "Clips"** (1000 videos, test only): To assess the method's generalization to various document types, we utilized images from the "Clips" category of the MIDV 2020 [13] dataset. Following document rectification, similar Regions of Interest (ROIs) were extracted as for MIDV-Holo. As clips were sampled at 10 frames per second (fps), we dropped one frame out of two to match the frame rate of MIDV-Holo (5 fps).

It is important to note that the MIDV-Holo dataset features a single form of holographic layer, consistent across all 20 document models, with minor translations between identity-card-like and passport-like models. Consequently, our system effectively trains to detect and validate this specific holographic device.

We compared several variants of the approach. For the feature extraction stage, we tested the following models, all implemented using the timm library [9]: $resnet_{18}$ [3], $mobilevit_{xxs}$ [11] and $mobilenet_{small0.5}$ [6]. By default, all experiments utilized models initialized with weights pretrained on ImageNet. Regarding the global binary decision stage, we considered the following strategies as mentioned in Sect. 3.5 :

– **Whole video**: Decision made using all video frames. It is a greedy strategy which prevents any eventual bias related to frame selection or video duration.
– **Cumulative**: This strategy, utilized by MIDV-Holo [18], involves iteratively updating a cumulative metric over the sequence. If the metric surpasses a predefined threshold, the video is deemed original, potentially leading to an early stop. Otherwise, if the threshold is not met by the sequence's end, the video is classified as attack.

Table 2. Comparison of results between the MIDV-Holo baseline (reimplemented by us) and our proposed method. Metrics (F_{score} or *Recall* for attacks-only datasets) are presented across three distinct test datasets, for 2 decision strategies: Whole video and Cumulative. Both methods utilize exclusively our proposed MIDV-Holo "Vanilla" train/validation sets for training and calibration. * denotes the original MIDV-Holo configuration (applied on full rectified documents), albeit using train-validation and test splits. "MIDV-Holo ROI"and our method both focus on the same ROI.

Decision	Metric → Method ↓	MIDV-HOLO "Vanilla" (120 mixed vids) F_{score} (%)	MIDV-HOLO "Photo repl." (20 attack vids) Recall (%)	MIDV 2020 "Clips" (1k attack vids) Recall (%)
Whole video	MIDV-Holo ROI	80 ± 3	63 ± 10	92 ± 2
	OUR - $mobilevit_{xxs}$	90 ± 2	87 ± 14	93 ± 6
Cumulative	MIDV-Holo FULL DOC *	77 ± 1	27 ± 12	84 ± 5
	MIDV-Holo ROI	82 ± 4	66 ± 10	93 ± 0
	OUR - $mobilevit_{xxs}$	86 ± 5	84 ± 11	94 ± 4
Dummy	Perfectly random	50	50	50
	Always positive (attack)	67	100	100
	Always negative (original)	0	0	0

For our weakly supervised approach, network features are trained on the train set, the best epoch is selected based on the validation set, and the decision threshold is calibrated on the validation set. Calibration involves selecting the best F_{score} for both the *whole video* and *cumulative* decision strategies. For the MIDV-Holo baseline reproduction, parameters are calibrated on the union of training and validation sets. Each operation is repeated 5 times with different train/validation/test splits, utilizing various random seeds for generation to mitigate potential biases. Results presented in Tables 2 and 3 represent averages and standard deviations across these 5 runs.

5.2   Results and Ablation Study

Table 2 presents the outcomes achieved with the various configurations. For brevity, only the result for the best feature extraction architecture is reported here. The final rows of the table serve as a baseline, indicating the performance metrics for completely random and constant decision processes. Notably, the constant prediction of attacks reaches an F_{score} of 67% on MIDV-Holo "Vanilla", setting a lower bound for acceptable results.

These first results show the superiority of our proposed approach on MIDV-Holo test sets for both decision strategies. While achieving similar performance to the baseline on the MIDV 2020 test set, the consistently high scores may suggest a bias towards predicting attacks, contrasting with our method's robustness shown in the mixed dataset.

Finally, we conducted an ablation study to challenge the benefits of various aspects of our method. Table 3 summarizes the results, with the second row being the reference for our non-ablated approach (Augmentations, Contrastive, Full train set). The ablated components are described here below.

Data Augmentation: *What is the impact of data augmentations?* The triplet loss, along with our sampling strategy, essentially relies on augmentations. The difference in performance between the first two rows of Table 3 confirms that this key component helps generalize across different datasets. Let's note that for "Originals" triplets, disabling augmentations nullifies the term $d(a_i, p_i)$ in Equation (1) as a_i and p_i are equal.

Training Strategy: *Is a contrastive loss competitive against direct decision optimization?* We trained a simple classifier under the same conditions (pre-trained on ImageNet, same augmentations) to distinguish between original and attack frames. Then, the evaluation was done at the video level based on the average prediction for each frame, and the final outcome was calculated using a threshold calibrated on the validation set, similar to other methods. Results were surprisingly good on the MIDV-Holo "Vanilla" test set but showed a significant drop on other datasets. This underscores that while MIDV-Holo is useful as it is the first academic one of its kind, it must be handled with care. It also emphasizes the necessity of using multiple datasets to demonstrate the relevance of each method. Furthermore, it must be noted that a binary classifier cannot individually control the coverage of expected visual appearances of a hologram.

Training Set: *Are attack samples required to train our approach?* The proposed method operates on the assumption that there are equal numbers of frauds and origins. However, in a real-world scenario, it is challenging to access attack samples. Thus, it makes sense to study the impact of training only on original samples. For this specific experiment, during training, the best model was selected based on a minimum validation loss criterion (over Originals only). However, the final decision threshold calibration was computed on the extended validation set (Originals and Attacks). Training only with MIDV-Holo "Vanilla" Originals results in slightly lower performance on the test set, but still remains better than the MIDV-Holo baseline.

Model Architecture and Tuning: *How important is the model architecture and are pretrained weights sufficient?* As our approach is not tied to a particular architecture, we trained and tested several lightweight ones that can match industrial processing speed requirements. We also checked whether fine-tuning actually improved performance, as pretrained weights can already exhibit sensitivity to saturated colors present in holograms. Results show similar performance for the architectures tested, with $mobilevit_{xxs}$ and $mobilenet_{small0.5}$

Table 3. Ablation study showing the contribution of 3 essential components of the proposed method: 1) data augmentation, 2) contrastive learning strategy, 3) training data. All configurations are tested on 3 different model architectures. * best configuration with all the features enabled (reported in Table 2).

					MIDV-Holo "Vanilla" (120 mixed vids)	MIDV-Holo "Photo replace" (20 fake vids)	MIDV 2020 "Clips" (1k fake vids)
ABLATED ELEMENTS OF THE PIPELINE				Test dataset →			
Data aug.	Train strategy	Training Dataset	Decision	Metric → Archi ↓	F_{score} (%)	Recall (%)	Recall (%)
On	Contrast (triplet loss)	MIDV-Holo "Vanilla" full train set (Originals & Attacks)		mobilenetv3$_{s50}$	88 ± 3	93 ± 8	92 ± 5
				mobilevit$_{xxs}$ *	90 ± 2	87 ± 14	93 ± 6
				resnet18	88 ± 2	91 ± 7	93 ± 5
Off				mobilenetv3$_{s50}$	83 ± 6	75 ± 17	86 ± 7
				mobilevit$_{xxs}$	87 ± 12	65 ± 20	87 ± 7
				resnet18	88 ± 6	81 ± 13	83 ± 5
	Classifier (softmax)		Whole video	mobilenetv3$_{s50}$	89 ± 3	77 ± 12	44 ± 7
				mobilevit$_{xxs}$	94 ± 3	85 ± 11	59 ± 4
				resnet18	92 ± 1	76 ± 10	76 ± 14
On	Contrast (triplet loss)	Originals only		mobilenetv3$_{s50}$	82 ± 7	89 ± 11	94 ± 4
				mobilevit$_{xxs}$	84 ± 4	87 ± 18	89 ± 9
				resnet18	83 ± 2	84 ± 13	87 ± 8
None (pretrained weights)				mobilenetv3$_{s50}$	73 ± 6	81 ± 15	61 ± 19
				mobilevit$_{xxs}$	67 ± 1	92 ± 10	82 ± 7
				resnet18	77 ± 7	76 ± 19	59 ± 16

being slightly superior when trained on mixed samples and originals only, respectively. Furthermore, the poor performance in the last row of Table 3 proves that generic features do not provide an adequate representation for our problem.

6 Conclusion

We have presented a novel approach for verifying both Appearance Conformity and Appearance Coverage of Optically Variable Devices (OVDs, or "holograms") in identity documents using video clips recorded from commodity smartphones. This approach leverages a feature extraction network trained with a contrastive loss, which can be specialized to a given hologram while requiring only video-level annotations, rather than individual frame labels. Furthermore, we have demonstrated that this approach can achieve attractive results using original video samples alone, which are abundantly obtained in industrial pipelines. Thanks to the separate calibration of its decision stage, our approach can be easily tuned to specific security requirements.

The evaluation of this approach necessitated the introduction of several extensions to the original MIDV-Holo dataset and the reimplementation of the proposed baseline. Our experiments have revealed the superiority of our approach over the previous baseline and its robust generalization capabilities across both the MIDV-Holo "Photo Replacement" and MIDV 2020 "Clips".

Lastly, our ablation study has uncovered a significant insight: while the MIDV-Holo "Vanilla" dataset yields intriguingly good results when tested with a simple binary classifier trained at the frame level, its generalization to other datasets is poor, as expected. This raises the question: *"What does the binary classifier actually learn?"* for future investigation.

Acknowledgements. The SOTERIA project, partially supporting this work, was funded by the European Union's Horizon 2020 research and innovation program under grant agreement No 101018342.

References

1. Balntas, V., Riba, E., Ponsa, D., Mikolajczyk, K.: Learning local feature descriptors with triplets and shallow convolutional neural networks. In: Proceedings of the British Machine Vision Conference 2016, pp. 119.1–119.11. British Machine Vision Association, York (2016). ISBN: 978-1-901725-59-9. https://doi.org/10.5244/C.30.119
2. Hartl, A.D., Arth, C., Grubert, J., Schmalstieg, D.: Efficient verification of holograms using mobile augmented reality. IEEE Trans. Visualiz. Comput. Graph. **22**(7), 1843–1851 (2016). https://doi.org/10.1109/TVCG.2015.2498612. ISSN:1941-0506
3. He, K., Zhang, X., Ren, S., Sun, J.: Deep residual learning for image recognition. In: 2016 IEEE Conference on Computer Vision and Pattern Recognition (CVPR), pp. 770–778 (2016). https://doi.org/10.1109/CVPR.2016.90
4. Soukup, D., Huber-Mörk, R.: Mobile hologram verification with deep learning. In: IPSJ Transactions on Computer Vision and Applications 9 (2017). https://doi.org/10.1186/s41074-017-0022-7
5. Sundararajan, M., Taly, A., Yan, Q.: Axiomatic attribution for deep networks. In: Proceedings of the 34th International Conference on Machine Learning (ICML2017), Sydney, vol. 70, pp. 3319–3328. JMLR.org (2017). arXiv: 1703.01365
6. Howard, A., et al.: Searching for MobileNetV3. In: 2019 IEEE/CVF International Conference on Computer Vision (ICCV), pp. 1314–1324 (2019). https://doi.org/10.1109/ICCV.2019.00140
7. Lin, B., Li, X., Yu, Z., Zhao, G.: Face liveness detection by rPPG features and contextual patch-based CNN. In: Proceedings of the 2019 3rd International Conference on Biometric Engineering and Applications (ICBEA 2019), pp. 61–68. Association for Computing Machinery, Stockholm (2019). https://doi.org/10.1145/3345336.3345345. ISBN:9781450363051
8. Loshchilov, I., Hutter, F.: Decoupled weight decay regularization. arXiv: 1711.05101 (2019)
9. Wightman, R.: PyTorch Image Models (2019). https://github.com/rwightman/pytorch-image-models. https://doi.org/10.5281/zenodo.4414861
10. Kokhlikyan, N., et al.: Captum: a unified and generic model interpretability library for PyTorch (2020). arXiv:2009.07896 [cs.LG]
11. Mehta, S., Rastegari, M.: MobileViT: light-weight, generalpurpose, and mobile-friendly vision transformer. In: International Conference on Learning Representations (2021). arXiv: 2110.02178
12. Ay, B.: Open-set learning-based hologram verification system using generative adversarial networks. IEEE Access **10**, 25114–25124 (2022). https://doi.org/10.1109/ACCESS.2022.3155870. ISSN:2169-3536

13. Bulatov, K., et al.: MIDV-2020: a comprehensive benchmark dataset for identity document analysis. In: Computer Optics **46**(2) (2022). https://doi.org/10.18287/2412-6179-CO-1006. arXiv:2107.00396. ISSN:0134-2452, 2412-6179
14. Council of the European Union. PRADO - Public Register of Authentic Travel and Identity Documents Online, v. 12344/22 (2022). https://www.consilium.europa.eu/prado/en/prado-glossary.html. Accessed 25 Jan 2024
15. Kada, O., Kurtz, C., van Kieu, C., Vincent, N.: Hologram detection for identity document authentication. In: El Yacoubi, M., Granger, E., Yuen, P.C., Pal, U., Vincent, N. (eds.) Pattern Recognition and Artificial Intelligence. LNCS, pp. 346–357. Springer, Cham (2022). https://doi.org/10.1007/978-3-031-09037-0_29. ISBN:978-3-031-09037-0
16. Nirkin, Y., Wolf, L., Keller, Y., Hassner, T.: DeepFake detection based on discrepancies between faces and their context. IEEE Trans. Pattern Anal. Mach. Intell. **44**(10), 6111–6121 (2022). https://doi.org/10.1109/TPAMI.2021.3093446
17. Chapel, M.-N., Al-Ghadi, M., Burie, J.-C.: Authentication of holograms with mixed patterns by direct LBP comparison. In: IEEE 25th International Workshop on Multimedia Signal Processing (MMSP) 2023, pp. 1–6. (2023). https://doi.org/10.1109/MMSP59012.2023.10337669. ISSN: 2473-3628
18. Koliaskina, L.I., et al.: MIDV-Holo: a dataset for ID document hologram detection in a video stream. In: Fink, G.A., Jain, R., Kise, K., Zanibbi, R. (eds.) Document Analysis and Recognition - ICDAR 2023, pp. 486–503. Springer, Cham (2023). https://doi.org/10.1007/978-3-031-41682-8_30. ISBN:978-3-031-41682-8

Multi-task Learning for License Plate Recognition in Unconstrained Scenarios

Zhen-Lun Mo[1]($\boxtimes$), Song-Lu Chen[1], Qi Liu[1], Feng Chen[2], and Xu-Cheng Yin[1]

[1] University of Science and Technology Beijing, Beijing, China
zhenlunmo@xs.ustb.edu.cn, {songluchen,xuchengyin}@ustb.edu.cn
[2] EEasy Technology Company Ltd., Zhuhai, China

Abstract. The recognition of license plates in natural scenes often face challenges such as multi-directional and multi-line variations. Additionally, previous studies have treated license plate detection and recognition as separate tasks, resulting in inefficiencies and error accumulation. To address these challenges, we propose an end-to-end method for license plate detection and recognition using multi-task learning. Firstly, we introduce two parallel branches to detect the horizontal bounding box and the four corners of the license plate, enabling multi-directional license plate detection in a multi-task manner. The outputs from these branches are combined to enhance recognition accuracy. Secondly, we propose to extract global features to perceive character layout and utilize reading order to spatially attend to characters for recognizing multi-line license plates. Finally, we combine detection and recognition using the same backbone, with the detection branch based on multiple deep layers and the recognition branch based on multiple shallow layers, thereby constructing an end-to-end detection and recognition network. Comparative experiments on CCPD and RodoSol datasets validate that our method significantly outperforms state-of-the-art methods, particularly in scenarios involving multi-directional and multi-line license plates.

Keywords: License plate recognition · Multi-task · Multi-directional · Multi-line · End-to-end

1 Introduction

Automatic License Plate Recognition (ALPR) is a crucial technology in intelligent transportation systems, essential for traffic monitoring, electronic toll collection (ETC), and smart parking lots [8,9,27]. While effective in specific operational contexts, ALPR faces challenges in unconstrained scenarios, as shown in Fig. 1. Recognizing multi-directional and multi-line license plates poses significant obstacles for conventional ALPR systems.

As depicted in Fig. 2(a), previous license plate detectors [4,10,18,30] have shown limitations in effectively handling multi-directional license plates due to their inherent restriction to detecting only the horizontal bounding box. Consequently, this limitation can introduce significant background noises into

Horizontal single-line plate

Multi-directional plate

Multi-line plate

Fig. 1. Recognizing license plates in unconstrained scenarios presents challenges, notably with multi-directional and multi-line configurations.

subsequent recognition processes, thereby impeding accuracy. To address this constraint, certain studies [5,17,26] have proposed a method to predict the four corners of multi-directional license plates, as illustrated in Fig. 2(b). However, when faced with horizontally oriented license plates, detecting the four corners may yield inferior results compared to detecting the bounding box. This discrepancy arises from the inherently more complex nature of corner detection, constituting an 8-DoF (Degree of Freedom) task involving both horizontal and vertical coordinates of the four corners, whereas detecting a horizontal bounding box involves a more straightforward 4-DoF task, considering only the coordinates of the center along with the height and width. To effectively overcome these challenges, our proposed method strategically integrates the strengths of both horizontal and multi-directional detectors. This integration is achieved through a multi-task framework that concurrently detects both the horizontal bounding box and the four corners. By adopting this approach, our method aims to optimize detection accuracy across diverse plate orientations, effectively navigating the intricacies associated with multi-directional license plate recognition.

In contrast to single-line license plates, the recognition of multi-line license plates poses greater challenges due to their inherently complex character layout. As depicted in Fig. 2(c)–(f), previous methodologies for multi-line license plate recognition encompass character segmentation [33], line segmentation [3], and segmentation-free methods [16,20]. Character segmentation methods involve the identification of individual characters followed by the aggregation into character strings. However, these approaches often rely heavily on manual annotation, which can be labor-intensive and prone to errors. Line segmentation techniques horizontally divide the license plate into multiple segments and subsequently concatenate them horizontally. However, horizontal lines may intersect with characters on the license plate, leading to incomplete character recognition. Segmentation-free methods utilize convolutional neural networks (CNNs) or recurrent neural networks (RNNs) to spatially attend to license plate characters without explicit segmentation. While CNN-based methods excel at extracting local features, they may lack a holistic perception of character layout. On the other hand, RNN-based methods leverage their intrinsic characteristics of

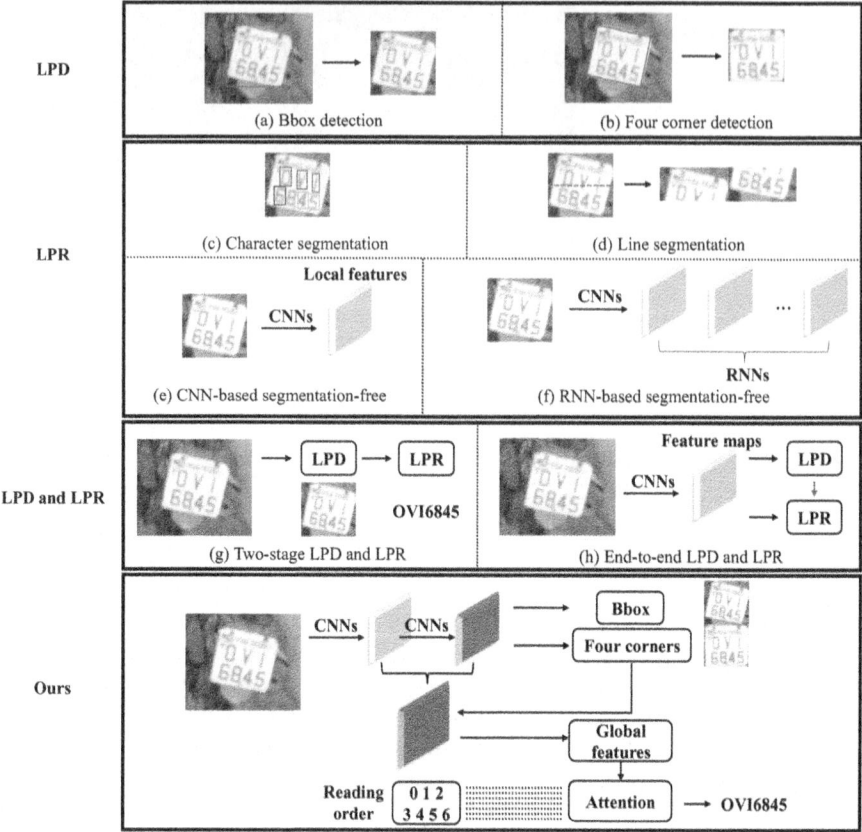

Fig. 2. Comparison of different schemes for license plate detection (LPD) and license plate recognition (LPR).

state computation for character recognition. However, the sequential nature of RNNs makes them challenging to parallelize, resulting in computationally intensive operations and prolonged processing times. Our proposed method leverages local features extracted by CNNs to derive comprehensive global features, enabling a nuanced perception of the character layout. Subsequently, our approach utilizes the reading order to efficiently attend to characters in parallel, thereby optimizing inference time and computational efficiency.

Furthermore, an ALPR system typically comprises two primary subtasks: license plate detection (LPD) and license plate recognition (LPR). Previous research [12,23,25,31,36] has predominantly focused on LPD and LPR in isolation, overlooking the intrinsic correlation between these tasks and potentially leading to error accumulation. Errors originating from the LPD stage can propagate to the LPR stage, thereby impacting the overall accuracy. Recent advancements in ALPR research have introduced single-stage end-to-end methodologies

[13,15,42] designed to jointly optimize both LPD and LPR processes. However, it's imperative to acknowledge that while recognition features are typically extracted from the deep layers of the backbone network, repeated downsampling operations may lead to the loss of critical feature information, such as character shape and texture. [4] proposes utilizing the shallowest layer of the backbone network to preserve crucial features for recognition; however, a single layer may lack diverse features for accurate recognition. To address this limitation, we propose to combine multiple shallow layers of the backbone network for recognition. By incorporating diverse features with larger resolutions, our method aims to preserve crucial feature details essential for accurate license plate recognition.

In summary, our contributions can be delineated as follows:

- We devise a multi-task framework that intelligently combines horizontal and multi-directional detectors. This integration enhances detection accuracy across diverse plate orientations, effectively addressing the challenges inherent in multi-directional license plate recognition.
- By leveraging global features, we accurately perceive character layout and employ reading order to efficiently process characters in parallel, thereby overcoming hurdles associated with multi-line license plate recognition.
- We adopt a multi-task approach integrating detection and recognition to minimize error accumulation and enhance efficiency. Additionally, by incorporating multiple shallow layers of the backbone network, we retain diverse features with larger resolutions, thereby improving recognition performance.
- Extensive experiments on CCPD [37] and RodoSol [14] datasets showcases the superior performance of our method. Particularly in scenarios involving multi-directional and multi-line license plates, our approach outperforms state-of-the-art methods. For instance, on CCPDv2, our method achieves an average recognition accuracy of 88.75%, surpassing the previous state-of-the-art method by more than 15%.

2 Related Work

2.1 License Plate Detection

The primary goal of License Plate Detection (LPD) is to accurately localize the bounding box encompassing license plates within input images. Traditional object detection methodologies, exemplified by FPN [18] and YOLO [30], typically approach the detection region as a horizontal bounding box. However, when confronted with the task of detecting and recognizing multi-directional license plates, this reliance on horizontal bounding boxes often leads to the inadvertent inclusion of substantial background noise, consequently hampering recognition accuracy. To tackle this issue, various methods [17,26] have been developed, focusing on the detection of license plate corners and subsequent correction through perspective transformation. Yet, corner detection introduces inherent complexities compared to bounding box detection, potentially resulting in diminished performance. Even with endeavors like those documented in [5], which incorporated constraints between corners to augment performance, the efficiency of direct bounding box detection may not be matched, particularly concerning horizontally oriented license plates. In response to these challenges, our proposed solution advocates for the simultaneous detection of both bounding boxes and corners. This multi-task detection approach substantially amplifies recognition accuracy while incurring only a marginal increase in parameters and runtime.

2.2   License Plate Recognition

License Plate Recognition (LPR) is aimed at accurately identifying characters from a license plate image. Previous methodologies [41] predominantly rely on extracting one-dimensional features, which, however, encounter challenges in effectively recognizing multi-line license plates with intricate character layouts. To address this issue, researchers have proposed various techniques for multi-line license plate recognition, including character segmentation [33], line segmentation [3], and segmentation-free methods [16,20]. Character segmentation [33] involves the detection of individual characters followed by their aggregation into character strings. However, these methods often necessitate extensive manual labeling. Line segmentation [3] divides the license plate horizontally into segments, which are then connected horizontally. However, this approach may result in incomplete recognition due to potential overlap with license plate characters by horizontal lines. Segmentation-free methods leverage CNNs or RNNs to spatially process license plate characters without explicit segmentation. While CNN-based methods [20] excel at extracting local features, they may lack a comprehensive understanding of character layout. RNN-based methods [16] utilize their inherent state-dependent computational properties for character recognition. However, the sequential nature of RNNs poses challenges in parallelization, leading to computationally intensive operations and longer processing times. Our proposed methodology capitalizes on the local features extracted from CNNs to derive holistic global features, enabling detailed perception of character layout. Subsequently, our approach optimizes inference time and computational efficiency by efficiently processing characters in parallel, leveraging the reading order.

2.3   End-to-End License Plate Detection and Recognition

Numerous prior studies [12,23,25,31,36] approach Automatic License Plate Recognition (ALPR) as a two-stage process. These methods typically partition LPD and LPR into separate, independent steps, where the output of LPD serves as the input for LPR. However, by treating them as disjoint tasks, their inherent relationship is often overlooked, potentially resulting in error accumulation. In contrast, end-to-end methods [13,15,42] integrate detection and recognition tasks within a unified framework, facilitating training of a single model. While such approaches leverage deep-layer features from the backbone network for recognition, these features may lack detailed license plate character information,

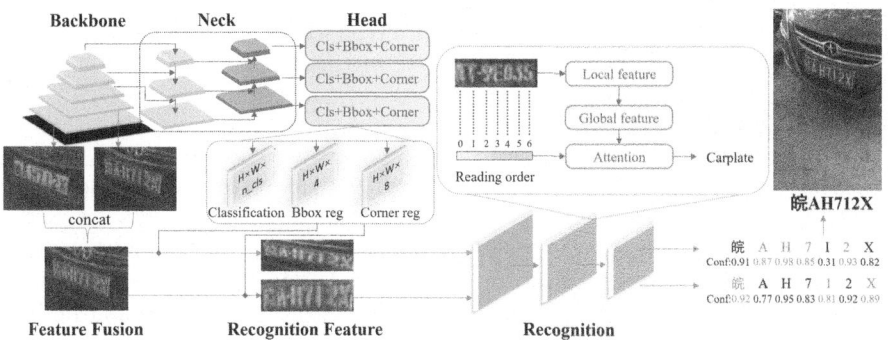

Fig. 3. Overall structure of our network.

hindering accurate recognition. To address this limitation, [4] proposes extracting license plate features from the shallowest layer of the network to construct an end-to-end detection and recognition network. However, a single layer of shallow features may offer limited information at a single scale. In our approach, we advocate leveraging multiple shallow features for feature fusion, thereby extracting more comprehensive details and features from license plates and enhancing recognition accuracy within the end-to-end network.

3   Method

Figure 3 illustrates the proposed network, including the backbone, neck, detection head, feature fusion, recognition feature, and recognition. It integrates detection and recognition models into a single network trained in an end-to-end approach. Compared to the two-stage approach, the recognition module in our proposed network reuses the features extracted by the detection module as inputs to the LPR. As a result, we do not need to re-extract LP features in the recognition module, which saves significant computational cost. Second, sharing features between the detection and recognition modules strengthens the intrinsic connection between the two tasks. Intuitively, the license plate features contain character features, which are also very important for determining whether it is a license plate or not. Therefore, by integrating both license plate detection and recognition modules into a unified framework, the performance of each module can be further improved.

First, we use CSPDarknet [2] as the backbone to extract shared features, which are used by the neck and the recognizer for feature fusion, respectively, to be used as inputs for the two subsequent tasks. Then, the detection head accurately detects the bounding box and the four corners of the license plate using the features from PAN [21]. For the license plate with bounding box, RoIAlign [11] is used to crop and resize the shared feature map, and for the license plate with four corners, the perspective transform is used to correct and resize the shared feature map. Finally, the license plate characters will be recognized by recognizer.

3.1 Feature Extraction

We use CSPDarknet [2] as the backbone to extract shared features for detection and recognition. Specifically, five feature maps with different scale sizes (the input size of the backbone is 640×640) are generated by five convolutions each containing conv2d (kernel $= 3$, stride $= 2$), bn and silu. Then we use PAN [21] to connect them into a path by up-sampling the feature maps starting from the low-resolution ones, while down-sampling them starting from the high-resolution ones. In this process, the information of each layer of feature maps will be fused with the feature maps of the neighboring layers above and below, but unlike FPN [18], PAN [21] will cascade the results of the fused feature maps of different layers instead of summing them up. This avoids the loss of information in the summation process, thus improving the detection.

3.2 Detection Head

In order to solve the problem of multi-directional license plate recognition, we design an anchor-free detection head that predicts the four corners and the bounding box of the license plate at the same time, as shown in Fig. 3. There are three sub-branches, including bounding box regression, four corners regression, the last is classification and conference of the license plate. The bounding box information of the license plate will be combined with the shared features of the license plate to crop and resize the shared feature graph through RoIAlign [11]. The corners information of the license plate will be corrected and the size of the shared feature map will be adjusted by perspective transformation.

We use BCELoss as the loss function for classification, and CIoU [40] as the loss function for bbox regression because the CIoU [40] loss considers the overlap region, central point distance, and aspect ratio for better and faster regression. Use cocoeval's keypoint IoU as loss function for four corners regression. The training loss comprises the classification loss L_{cls}, bounding box regression loss $L_{CIoU}(B, B_{gt})$, and keypoint loss $L_{kpIoU}(KP, KP_{gt})$.

$$L_{kpIoU}(KP, KP_{gt}) = \frac{(KP - KP_{gt})^2}{2 * sigmas * (area + 1e - 9) * 2} \tag{1}$$

$$L_{det} = L_{cls} + L_{CIoU}(B, B_{gt}) + L_{kpIoU}(KP, KP_{gt}) \tag{2}$$

where L_{cls} is implemented by BCEloss. B and B_{gt} denote the predicted box and ground-truth box, KP and KP_{gt} denote the predicted keypoint and ground-truth keypoint, area denote the ground-truth area.

3.3   Shared Feature Fusion

For LPD, deep features have stronger semantics and larger receptive fields [19], which can better help detect the target location. However, deep features are not favorable for LPR because the character features needed for LPR belong to small targets, such as character shapes and textures, which lose a lot of information after multiple down-sampling operations. Whereas shallow features are semantically weak and retain a lot of information about the characters. In our backbone, the first convolutional layer and the second convolutional layer still retain a large amount of character information from two different scales. Therefore, we fuse two feature layers as recognition features to further improve the recognition performance.

3.4   Recognition

We resize the recognition features to 96×32 as input for the recognizer. Our recognizer consist of local feature extraction, global feature generation, and parallel character attention. Specifically, we first obtain 24×8 local features through multiple convolutional layers. Then, we stack two transformers units to extract global features, as shown in Fig. 4. Because local features cannot perceive the type of license plate and character layout, while global features can better enhance character attention. In sequence recognition [7,32], attention is used to generate N features, each corresponding to a character in the text. Existing methods typically use the hidden state H_{t-1} as a query to generate the t-th feature. To enable parallel computation, we encode the character reading order by parallel embedding as the query, while the key-value set is obtained from the global features. Finally, we use a fully-connected layer to map to class probabilities $P_n \times cls$, where n represents the number of characters in the license plate, cls represents the number of character categories. We use cross-entropy loss as the recognition loss to maximize the prediction confidence of each character.

4   Experiment

We evaluate our proposed end-to-end ALPR model through experiments on several datasets. Section 4.1 introduced the experimental setup and ALPR dataset. Sections 4.2 and 4.3 report the performance comparison of the proposed ALPR model with existing methods. In addition, the ablation study is explained in Sect. 4.4 to analyze the effect of our proposed model.

4.1   Dataset and Experimental Setup

Dataset. CCPD [37] is the Chinese City Parking Dataset, which includes two versions, CCPDv1 and CCPDv2. Both versions comprise over 250,000 car images with a resolution of 720×1160 pixels. Each license plate in the dataset is a single-line license plate. The detailed information of the CCPD dataset is presented in the table. CCPDv1 is divided into subsets: Base, DB, FN, Rotate, Tilt,

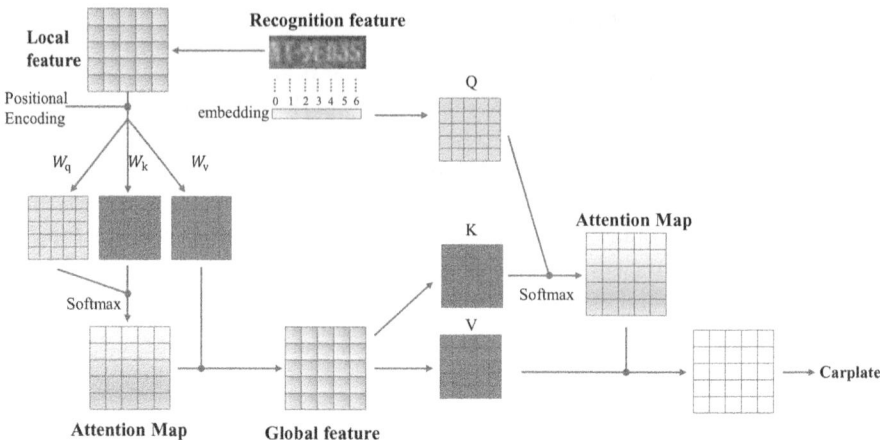

Fig. 4. The architecture of recognition network.

Challenge, and Weather. CCPDv2 excludes the Weather subset but includes an additional Blur subset. The Base subset is used for training and validation, while the other subsets are used for testing. The test set of CCPDv2 is more diverse and challenging compared to CCPDv1. Table 1 provides an overview of the characteristics of the CCPD dataset.

RodoSol [14] is collected at toll stations on highways in Brazil and consists of 20,000 car images with a resolution of 1280×720 pixels. The dataset is divided into training, testing, and validation sets according to a 4:4:2 ratio, with an equal number of single-line and double-line license plates in each set. Notably, in all the mentioned datasets, each image contains only one license plate.

Table 1. Description of the CCPD dataset, including CCPDv1 and CCPDv2

	CCPDv1	CCPDv2	Description
Base	200k	200k	Normal license plate
DB	20k	10k	License plate under low or high lighting conditions
Blur		20k	Blur license plate
Weather	20k		License plate in rainy, snowy, or foggy weather
FN	10k	20k	License plate at a distant or close proximity to the camera
Rotate	10k	10k	Horizontally tilted at 20–50°, vertically tilted at −10 to 10°
Tilt	10k	30k	Horizontally tilted at 15–45°, vertically tilted at −15 to 45°
Challenge	10k	50k	Challenging license plate

Experimental Setup. When training the end-to-end network, considering that some data augmentations commonly used in LPD might impact the performance of LPR, we opted to utilize only random scaling and the image augmentation function straug [1] designed for Scene Text Recognition (STR). The network is using

SGD optimizer, momentum 0.9, and batch size 64, with learning rates set to 1e-2 for the detection branch and 1e-3 for the recognition branch. The experiments were conducted on an Nvidia GeForce RTX 4060 GPU with 8 GB GDDR6 RAM.

Evaluation Criterion. Firstly, the accuracy of bounding box detection is determined by the Intersection-over-Union(IoU) metric. Given a predicted bounding box B' and a ground truth bounding box B, their IoU is calculated as:

$$IoU = \frac{area(B' \cap B)}{area(B' \cup B)} \tag{3}$$

where area represents the area of a region. Based on existing research [15,37], we consider a detection to be accurate only if $IoU > 0.7$. Subsequently, a license plate is considered correctly recognized only when all predicted characters match the ground truth values. We use accuracy to measure the effectiveness of different algorithms. TP represents samples that are correctly predicted. FP represents samples that are incorrectly predicted. The accuracy of LPD/LPR algorithms are defined as follows:

$$accuracy = \frac{TP}{TP + FP} \tag{4}$$

4.2 Accuracy Evaluation for Multi-direction License Plate

We test the dataset CCPD to verify the generality of our method. As shown in Table 2, we compare the two-stage method and end-to-end method, where the detectors of the two-stage method include SSD [22], SSD(CIOU) [40], and Chen et al. [6], and the recognizers include the line segmentation method [3], the 1D-sequence recognition methods LPRNet [41], RNN-based segmentation-free method RNN-Attention [39], and CNN-based segmentation-free method SALPR [20]. our method can achieve the best recognition performance with a fast inference speed on all subsets of CCPDv2, proved that our method can effectively recognize multi-directional license plates and horizontal license plates.

Moreover, as shown in Table 3, our method can achieve state-of-the-art average recognition accuracy, significantly outperforming all end-to-end methods. However, our method does not achieve the best inference speed because our method spends more time in recognizing the two branches of the detected head. In future work, we will further optimize the recognition time of our method.

4.3 Accuracy Evaluation for Double-Line License Plate

We compared the effectiveness of double-line license plate recognition on the RodoSol dataset, as shown in Table 4. In the comparative method, the detectors for the two-stage method including vanilla SSD, SSD(CIOU) and Chen et al. [6], SSD(CIoU) denotes using CIoU loss for bounding box regression. The recognizer for the two-stage method including CNN-based segmentation-free method SALPR [20] and RNN-based segmentation-free method RNN-Attention [39]. Our method obtains a significant improvement in double-line license plate recognition performance.

Table 2. Comparative experiments on CCPDv2. E2E means the end-to-end network. Our method can achieve the best recognition accuracy and fastest inference speed.

Method	E2E	Avg	DB	Blur	FN	Rotate	Tilt	Cha	FPS
SSD [22]+LPRNet [41]		27.74	21.59	20.59	33.57	20.19	14.34	34.57	40
SSD(CIOU) [40]+LPRNet [41]		28.47	22.45	21.49	34.50	20.65	14.48	35.48	40
SSD [22]+Cao et al. [3]		28.95	24.05	19.63	32.60	24.64	19.38	33.74	32
SSD(CIoU) [40]+Cao et al. [3]		29.82	25.32	20.63	34.06	25.30	20.09	34.46	32
Chen et al. [6]+Cao et al. [3]		41.43	26.55	17.51	40.03	69.45	53.50	34.69	27
Chen et al. [6]+LPRNet [41]		41.53	24.05	18.00	40.06	67.87	52.36	35.54	35
SSD [22]+SALPR [20]		43.34	35.50	26.67	45.78	46.15	41.18	45.29	36
SSD [22]+HC [34]		43.42	34.47	25.83	45.24	52.82	52.04	44.62	11
SSD(CIoU) [40]+SALPR [20]		44.45	36.85	27.87	47.51	48.83	42.29	46.10	36
SSD [22]+RNN-Attention [39]		47.23	39.66	27.84	48.98	53.70	47.40	47.35	33
SSD(CIoU) [40]+RNN-Attention [39]		47.89	40.72	28.82	49.66	54.59	47.71	48.10	33
Chen et al. [6]+SALPR [20]		51.91	34.78	23.25	53.13	82.43	68.14	43.54	31
Chen et al. [6]+RNN-Attention [39]		57.23	38.81	24.11	58.89	88.02	78.05	46.66	28
Chen et al. [4]	✓	73.48	63.85	63.00	75.18	75.70	73.40	74.63	37
Ours	✓	**88.75**	**84.90**	**80.27**	**92.20**	**97.17**	**94.04**	**86.68**	**41**

Table 3. Comparative experiments on CCPDv1. E2E means the end-to-end network. Our method can achieve the best recognition accuracy.

Method	E2E	Avg	DB	Blur	FN	Rotate	Tilt	Cha	FPS
LPRNet [41]		93.0	92.2	91.9	79.4	85.8	92.0	69.8	56
TE2E [15]	✓	94.4	94.8	94.5	87.9	92.1	86.8	81.2	3
Silva et al. [33]		94.6	86.5	85.2	94.5	95.4	94.8	91.2	31
RPNet [37]	✓	95.5	96.9	94.3	90.8	92.5	87.9	85.1	**61**
DAN [35]		96.6	96.1	96.4	91.9	93.7	95.4	83.1	19
MANGO [28]	✓	96.9	97.1	95.5	95.0	96.5	95.9	83.1	8
HomoNet [38]	✓	97.5	96.9	95.9	97.1	98.0	97.5	85.9	19
ILPRNet [43]		97.5	98.1	98.5	90.3	95.2	97.8	86.2	-
Qin et al. [29]	✓	97.5	93.3	93.7	98.2	95.9	98.9	92.9	26
MORAN [24]		98.3	98.1	98.6	98.1	98.6	97.6	86.5	55
AttentionNet [39]		98.5	98.8	98.8	96.4	97.6	98.5	88.9	40
Chen et al. [4]	✓	98.8	99.1	98.8	98.4	98.5	98.8	90.0	37
Ours	✓	**99.1**	**99.7**	**99.5**	**99.7**	**99.7**	**99.8**	**95.1**	41

Table 4. Comparative experiments on RodoSol. E2E means the end-to-end network. Our method has a significant improvement in doube-line license plate recognition.

Method	E2E	RodoSol			FPS
		Avg	Single	Double	
SSD [22]+SALPR [20]		91.65	94.15	89.15	36
SSD(CIoU) [40]+SALPR [20]		91.98	93.95	90.00	36
Chen et al. [6]+SALPR [20]		92.04	93.95	90.13	31
SSD [22]+RNN-Attention [39]		92.79	95.10	90.48	33
SSD(CIoU) [40]+RNN-Attention [39]		93.48	95.78	91.18	33
Chen et al. [6]+RNN-Attention [39]		93.84	95.98	91.70	28
Ours	✓	**97.06**	**97.78**	**96.35**	41

4.4 Ablation Studies

Ablation Studies on Recognizer Components. As shown in Table 5, we conduct ablation experiments of different recognizer components on CCPD [37] and RodoSol [14], including global feature generation (GFG) and parallel character attention (PCA). After removing GFG, our proposed method uses local features instead of global features as input to PCA, which significantly reduces the recognition accuracy, especially for CCPDv2 and double-line license plate of Rodosol, however, inference speed almost unchanged. This suggests that global features can better enhance character attention and perceive character layout, thus improve recognition accuracy. Without PCA, our proposed method will degrade to the RNN based segmentation-free model. RNN introduces temporal information of sequences, however, there is no or minor sequential relationship between license plate characters, which as noise affects the recognition accuracy. Moreover, RNN can only encode sequentially, which is time-consuming.

Table 5. Ablation study of different recognizer components.

Input	CCPDv1	CCPDv2	RodoSol			FPS
			Avg	Single	Double	
Ours	**99.17**	**88.75**	**97.06**	**97.78**	**96.35**	41
w/o GFG	98.91	84.17	95.71	97.27	94.15	41
w/o PCA	98.65	81.31	94.58	96.57	92.60	37

The Impact of Recognition Features on LPR. The choice of different recognition features significantly impacts the accuracy of the LPR module, especially as deep features lose a considerable amount of character information, which is detrimental to LPR. We conducted tests on three datasets to assess the impact of selecting different recognition features as inputs for the recognizer.

Here, "Image" represents a non-end-to-end two-stage detection and recognition, using the same detector and recognizer as the end-to-end network. This means that the input to the recognizer is the original image. "Layer x" denotes selecting the feature map output of the xth convolutional layer in the backbone as the input for the recognizer. Performance was notably enhanced on all three datasets when choosing the first feature map as the recognition feature, compared to the two-stage detection and recognition. This improvement is attributed to the joint optimization network, which reduces error accumulation and improves recognition rates, while shallow features retain abundant character information. However, selecting deeper features as recognition features led to a significant decrease in recognition accuracy. This is because deep features lose character information after multiple down-sampling operations, negatively impacting LPR. As shown in Table 6, selecting the first feature map from backbone achieved the best recognition performance across all datasets (Fig. 5).

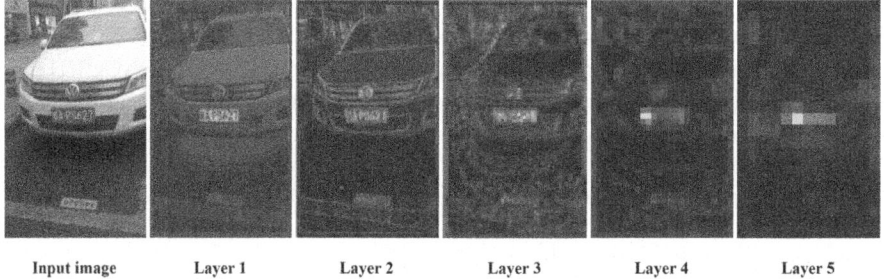

| Input image | Layer 1 | Layer 2 | Layer 3 | Layer 4 | Layer 5 |

Fig. 5. Feature visualization of the input image and different convolutional layers.

Table 6. Ablation study of different convolutional layers as recognition inputs. Image means use the original image as recognition input. LayerX means constructing end-to-end network using different convolutional layers as recognition features.

Input	size	CCPDv1	CCPDv2	RodoSol		
				Avg	Single	Double
Image		98.76	84.90	96.73	97.65	95.80
layer1	320	**99.17**	**88.30**	**96.93**	**97.73**	**96.13**
layer2	160	97.69	83.38	83.21	84.65	81.76
layer3	80	87.28	54.62	3.25	3.75	2.75
layer4	40	60.17	23.01	0	0	0
layer5	20	10.48	3.01	0	0	0

However, when choosing the second feature map as the recognition feature, recognition accuracy did not decrease significantly, indicating that the second feature map still retains some larger-scale information, such as character layout.

Therefore, we propose concat the first and second feature maps as the recognition feature, encompassing feature information at different scales and further enhancing the recognition accuracy. As shown in Table 7, experimental results demonstrate that we achieved the highest recognition accuracy when using the first and second feature maps for feature fusion.

Table 7. Ablation study of feature fusion. LayerA+B means concat layerA and layerB as recognition feature. Two shallowest feature fusion achieved the best performance.

	CCPDv1	CCPDv2	RodoSol		
			Avg	Single	Double
layer1	99.17	88.30	96.93	97.73	96.13
layer1+2	99.17	**88.75**	**97.06**	**97.78**	**96.35**
layer1+2+3	**99.18**	88.45	95.14	96.38	93.90

Selection of Bounding Box and Four-Corners Correction. As shown in Table 8, in the FN, Rotate, and Tilt subsets, corners detection achieves better performance, addressing the challenges of recognizing multi-direction license plates. However, on subsets of license plates with smaller rotation angles or horizontal, such as DB, Blur, Challenge, corners detection performs less effectively than bounding box detection. Proved that the corner detector cannot effectively detect horizontal license plates. We further improve the recognition performance by combining horizontal and multi-directional detectors.

Table 8. corner/bbox selection in CCPDv2

	Avg	DB	Blur	FN	Rotate	Tilt	Cha	FPS
Corner	85.84	81.45	75.66	**89.09**	**96.46**	**91.61**	83.95	53
Bbox	86.14	**83.13**	78.76	88.61	96.17	88.73	**85.18**	75
C/B selection	**88.75**	**84.90**	**80.27**	**92.20**	**97.17**	**94.04**	**86.68**	41

We validated the generalizability of our approach on the CCPDv1, CCPDv2 [37], and RodoSol [14]. As shown in Table 9, the results indicate that the selection method performs the best across all three datasets.

Table 9. Ablation study of corner/bbox selection

	CCPDv1		CCPDv2		RodoSol	
	layer1	layer1+2	layer1	layer1+2	layer1	layer1+2
corner	99.02	98.99	85.62	85.84	95.74	96.46
bbox	99.04	98.94	86.47	86.14	95.95	95.83
c/b selection	**99.17**	**99.17**	**88.30**	**88.75**	**96.93**	**97.06**

5 Conclusion

In this work, we propose a multi-task framework that intelligently combines horizontal and multi-directional detectors, recognizing double-line license plates by utilizing reading order and leveraging global features, then reduce error accumulation through end-to-end network. Extensive experiments on three datasets prove our method can achieve state-of-the-art recognition performance with a fast inference speed especially on CCPDv2, our method achieves an average recognition accuracy of 88.75%, surpassing the previous state-of-the-art method by more than 15%.

References

1. Atienza, R.: Data augmentation for scene text recognition. In: International Conference on Computer Vision Workshops, Montreal, BC, Canada, pp. 1561–1570. IEEE (2021)
2. Bochkovskiy, A., Wang, C., Liao, H.M.: Yolov4: optimal speed and accuracy of object detection. CoRR abs/2004.10934 (2020)
3. Cao, Y., Fu, H., Ma, H.: An end-to-end neural network for multi-line license plate recognition. In: International Conference on Pattern Recognition, Beijing, China, pp. 3698–3703. IEEE (2018)
4. Chen, S., Liu, Q., Chen, F., Yin, X.: End-to-end multi-line license plate recognition with cascaded perception. In: Fink, G.A., Jain, R., Kise, K., Zanibbi, R. (eds.) ICDAR 2023. LNCS, vol. 14191, pp. 274–289. Springer, Cham (2023). https://doi.org/10.1007/978-3-031-41734-4_17
5. Chen, S., Tian, S., Liu, Q., Chen, F., Yin, X.: Vertex adjustment loss for multidirectional license plate detection and recognition. In: International Conference on Ubiquitous Intelligence and Computing, Haikou, China, pp. 285–292. IEEE (2022)
6. Chen, S., et al.: End-to-end trainable network for degraded license plate detection via vehicle-plate relation mining. Neurocomputing **446**, 1–10 (2021)
7. Cheng, Z., Bai, F., Xu, Y., Zheng, G., Pu, S., Zhou, S.: Focusing attention: towards accurate text recognition in natural images. In: International Conference on Computer Vision, Venice, Venice, pp. 5086–5094. IEEE (2017)
8. Datondji, S.R.E., Dupuis, Y., Subirats, P., Vasseur, P.: A survey of vision-based traffic monitoring of road intersections. IEEE Trans. Intell. Transp. Syst. **17**(10), 2681–2698 (2016)
9. Fan, X., Zhao, W.: Improving robustness of license plates automatic recognition in natural scenes. IEEE Trans. Intell. Transp. Syst. **23**(10), 18845–18854 (2022)

10. Girshick, R.B., Donahue, J., Darrell, T., Malik, J.: Rich feature hierarchies for accurate object detection and semantic segmentation. In: IEEE Conference on Computer Vision and Pattern Recognition, Columbus, OH, USA, pp. 580–587. IEEE (2014)
11. He, K., Gkioxari, G., Dollár, P., Girshick, R.B.: Mask R-CNN. In: International Conference on Computer Vision, Venice, Italy, pp. 2980–2988. IEEE (2017)
12. Henry, C., Ahn, S.Y., Lee, S.: Multinational license plate recognition using generalized character sequence detection. IEEE Access **8**, 35185–35199 (2020)
13. Huang, Q., Cai, Z., Lan, T.: A single neural network for mixed style license plate detection and recognition. IEEE Access **9**, 21777–21785 (2021)
14. Laroca, R., Cardoso, E.V., Lucio, D.R., Estevam, V., Menotti, D.: On the cross-dataset generalization in license plate recognition. In: International Joint Conference on Computer Vision, Imaging and Computer Graphics Theory and Applications, pp. 166–178. SCITEPRESS, Online Streaming (2022)
15. Li, H., Wang, P., Shen, C.: Toward end-to-end car license plate detection and recognition with deep neural networks. IEEE Trans. Intell. Transp. Syst. **20**(3), 1126–1136 (2019)
16. Li, H., Wang, P., Shen, C., Zhang, G.: Show, attend and read: a simple and strong baseline for irregular text recognition. In: AAAI Conference on Artificial Intelligence, Honolulu, Hawaii, USA, pp. 8610–8617. AAAI Press (2019)
17. Li, Z., Chen, S., Liu, Q., Chen, F., Yin, X.: Anchor-free location refinement network for small license plate detection. In: Yu, S., et al. (eds.) PRCV 2022. LNCS, vol. 13537, pp. 506–519. Springer, Cham (2022). https://doi.org/10.1007/978-3-031-18916-6_41
18. Lin, T., Dollár, P., Girshick, R.B., He, K., Hariharan, B., Belongie, S.J.: Feature pyramid networks for object detection. In: IEEE Conference on Computer Vision and Pattern Recognition, Honolulu, HI, USA, pp. 936–944. IEEE (2017)
19. Lin, T., Maji, S.: Visualizing and understanding deep texture representations. In: IEEE Conference on Computer Vision and Pattern Recognition, Las Vegas, NV, USA, pp. 2791–2799. IEEE (2016)
20. Liu, Q., Chen, S.-L., Li, Z.-J., Yang, C., Chen, F., Yin, X.-C.: Fast recognition for multidirectional and multi-type license plates with 2D spatial attention. In: Lladós, J., Lopresti, D., Uchida, S. (eds.) ICDAR 2021. LNCS, vol. 12824, pp. 125–139. Springer, Cham (2021). https://doi.org/10.1007/978-3-030-86337-1_9
21. Liu, S., Qi, L., Qin, H., Shi, J., Jia, J.: Path aggregation network for instance segmentation. In: IEEE Conference on Computer Vision and Pattern Recognition, Salt Lake City, UT, USA, pp. 8759–8768. IEEE (2018)
22. Liu, W., et al.: SSD: single shot MultiBox detector. In: Leibe, B., Matas, J., Sebe, N., Welling, M. (eds.) ECCV 2016. LNCS, vol. 9905, pp. 21–37. Springer, Cham (2016). https://doi.org/10.1007/978-3-319-46448-0_2
23. Lu, Q., Liu, Y., Huang, J., Yuan, X., Hu, Q.: License plate detection and recognition using hierarchical feature layers from CNN. Multim. Tools Appl. **78**(11), 15665–15680 (2019)
24. Luo, C., Jin, L., Sun, Z.: MORAN: a multi-object rectified attention network for scene text recognition. Pattern Recognit. **90**, 109–118 (2019)
25. Masood, S.Z., Shu, G., Dehghan, A., Ortiz, E.G.: License plate detection and recognition using deeply learned convolutional neural networks. arXiv abs/1703.07330 (2017)
26. Meng, A., Yang, W., Xu, Z., Huang, H., Huang, L., Ying, C.: A robust and efficient method for license plate recognition. In: International Conference on Pattern Recognition, Beijing, China, pp. 1713–1718. IEEE (2018)

27. Paidi, V., Fleyeh, H., Håkansson, J., Nyberg, R.G.: Smart parking sensors, technologies and applications for open parking lots: a review. IET Intel. Transport Syst. **12**(8), 735–741 (2018)
28. Qiao, L., et al.: MANGO: a mask attention guided one-stage scene text spotter. In: AAAI Conference on Artificial Intelligence, pp. 2467–2476. AAAI Press, Virtual Event (2021)
29. Qin, S., Liu, S.: Towards end-to-end car license plate location and recognition in unconstrained scenarios. Neural Comput. Appl. **34**(24), 21551–21566 (2022)
30. Redmon, J., Divvala, S.K., Girshick, R.B., Farhadi, A.: You only look once: unified, real-time object detection. In: IEEE Conference on Computer Vision and Pattern Recognition, Las Vegas, NV, USA, pp. 779–788. IEEE (2016)
31. Selmi, Z., Halima, M.B., Alimi, A.M.: Deep learning system for automatic license plate detection and recognition. In: International Conference on Document Analysis and Recognition, Kyoto, Japan, pp. 1132–1138. IEEE (2017)
32. Shi, B., Wang, X., Lyu, P., Yao, C., Bai, X.: Robust scene text recognition with automatic rectification. In: IEEE Conference on Computer Vision and Pattern Recognition, Las Vegas, NV, USA, pp. 4168–4176. IEEE Computer Society (2016)
33. Silva, S.M., Jung, C.R.: License plate detection and recognition in unconstrained scenarios. In: Ferrari, V., Hebert, M., Sminchisescu, C., Weiss, Y. (eds.) ECCV 2018. LNCS, vol. 11216, pp. 593–609. Springer, Cham (2018). https://doi.org/10.1007/978-3-030-01258-8_36
34. Spanhel, J., Sochor, J., Juránek, R., Herout, A., Marsik, L., Zemcík, P.: Holistic recognition of low quality license plates by CNN using track annotated data. In: International Conference on Advanced Video and Signal Based Surveillance, Lecce, Italy, pp. 1–6. IEEE Computer Society (2017)
35. Wang, T., et al.: Decoupled attention network for text recognition. In: AAAI Conference on Artificial Intelligence, New York, NY, USA, pp. 12216–12224. AAAI Press (2020)
36. Wang, W., Yang, J., Chen, M., Wang, P.: A light CNN for end-to-end car license plates detection and recognition. IEEE Access **7**, 173875–173883 (2019)
37. Xu, Z., et al.: Towards end-to-end license plate detection and recognition: a large dataset and baseline. In: Ferrari, V., Hebert, M., Sminchisescu, C., Weiss, Y. (eds.) ECCV 2018. LNCS, vol. 11217, pp. 261–277. Springer, Cham (2018). https://doi.org/10.1007/978-3-030-01261-8_16
38. Yang, Y., Xi, W., Zhu, C., Zhao, Y.: HomoNet: unified license plate detection and recognition in complex scenes. In: Gao, H., Wang, X., Iqbal, M., Yin, Y., Yin, J., Gu, N. (eds.) CollaborateCom 2020. LNICST, vol. 350, pp. 268–282. Springer, Cham (2021). https://doi.org/10.1007/978-3-030-67540-0_16
39. Zhang, L., Wang, P., Li, H., Li, Z., Shen, C., Zhang, Y.: A robust attentional framework for license plate recognition in the wild. IEEE Trans. Intell. Transp. Syst. **22**(11), 6967–6976 (2021)
40. Zheng, Z., Wang, P., Liu, W., Li, J., Ye, R., Ren, D.: Distance-IoU loss: faster and better learning for bounding box regression. In: AAAI Conference on Artificial Intelligence, New York, NY, USA, pp. 12993–13000. AAAI Press (2020)
41. Zherzdev, S., Gruzdev, A.: LPRNet: license plate recognition via deep neural networks. arXiv abs/1806.10447 (2018)
42. Zhou, X., Cheng, Y., Jiang, L., Ning, B., Wang, Y.: Fafenet: a fast and accurate model for automatic license plate detection and recognition. IET Image Process. **17**(3), 807–818 (2023)
43. Zou, Y., et al.: License plate detection and recognition based on yolov3 and ILPR-NET. Signal Image Video Process. **16**(2), 473–480 (2022)

Recurrent Few-Shot Model for Document Verification

Maxime Talarmain[1], Carlos Boned[1], Sanket Biswas[1,2],
and Oriol Ramos Terrades[1,2(✉)]

[1] Computer Vision Center, Universitat Autònoma de Barcelona, Catalunya, Barcelona, Spain
{mtalarmain,cboned,sbiswas,oriolrt}@cvc.uab.cat
[2] Computer Science Department, Universitat Autònoma de Barcelona, Catalunya, Barcelona, Spain

Abstract. General-purpose ID, or travel, document image- and video-based verification systems have yet to achieve good enough performance to be considered a solved problem. There are several factors that negatively impact their performance, including low-resolution images and videos and a lack of sufficient data to train the models. This task is particularly challenging when dealing with unseen class of ID, or travel, documents. In this paper we address this task by proposing a recurrent-based model able to detect forged documents in a few-shot scenario. The recurrent architecture makes the model robust to document resolution variability. Moreover, the few-shot approach allow the model to perform well even for unseen class of documents. Preliminary results on the SIDTD and Findit datasets show good performance of this model for this task.

1 Introduction

The increased of remote identity authentication systems, incorporating biometrics and the verification of ID and travel documents, has surged and become widespread in the wake of the COVID-19 pandemic. These authentication systems have empowered citizens to engage in work and business activities outside traditional office settings. Public administration, banks, productive industries, and numerous services have integrated these systems seamlessly into their routine workflows. These services provide an online enrollment option, eliminating the need for users to physically attend by requesting a selfie and an image of their ID document for authentication. Nevertheless, cybercrime has exploited societal vulnerabilities, evolving towards increasingly sophisticated threats. Therefore, it is necessary to develop techniques that allow us to detect this type of fraudulent actions that expose citizens' data and their privacy to the general public.

A current trend is to detect ID documents, passports or driving licenses that have physically or digitally been modified from images acquired from mobile devices. In [4], the authors applied texture descriptors to image patches and then applied BoW followed by an SVM classifier to classify each patch as genuine or fake. In that work, the most performant CNN architectures of that time (AlexNet, VGG and Inception) were also

This content reflects only the authors' view. The European Agency is not responsible for any use that may be made of the information it contains.

used as general descriptor extractors. The results obtained in terms of F1-Score were very good, obtaining results between 0.99 and 1 in many cases. The main problem of that kind of approach was the need for more sufficient examples of forged documents and the poor generalization capacity for unseen classes of documents during the training process.

The results in [4] are consistent with recent works using more advanced deep learning techniques. In [15], Residual networks are applied to detect forged documents at different scales. In [21], the authors also apply Residual networks to study the impact of using different scanners and printers to detect images of genuine documents compared to recaptured documents. On the other hand, the authors in [6] propose the use of Siamese networks for the detection of the recapture of ID documents (student ID Cards in their experiments). Encoder-Decoder architectures with attention modules have also been used to address this problem in [14]. Recently, spatial and spectral features have been used for the localization of modified regions [7]. In all these works, although the reported results show good yields (above 0.90 in AUC and similar metrics), with a certain capacity for generalization depending on the recapture devices used. When they have to be applied to different types of documents, like for instance ID cards from nationalities not seen in training, the performance of these methods (and similar ones) methods drops dramatically [5].

Consequently, and due to the inability to obtain sufficient representative learning data, both in terms of enough samples of forged documents and the paramount variability of genuine data, the transfer of this technology to products working in real scenarios is still far from being a reality. The results reported in the above-mentioned works, and others, show that the problem is not in the discrimination power of current models but of the incapacity of models to generalize to a real environment, where the indefinability of the types of documents and the unpredictability of the type of forgeries is the common scenario.

In this paper, we go a step beyond the generalization capacity of document verification models. We combine pre-trained models with a recurring network in a very simple way. The weights of this network are trained in a few-shot learning context. The results obtained are surprisingly good, even when we apply the models trained with the ID Documents dataset to a ticket images dataset. The achieved performance of the proposed model opens new perspectives for this technology.

2   Methodology

As we mentioned in the Introduction, the proposed model is relatively simple and, to some extent, unoriginal as it combines components that are well known in the community. Figure 1 shows the architecture used to train the model. Essentially, it is a recurring many-to-one network. The input of each recurring unit, in addition to the previous state of the network, is a feature vector calculated in patches of the image. The output of the network is an m-dimensional vector that is used to classify the document image as genuine or fake.

Document images have been partitioned into W-dimensional square patches that overlap to avoid contour effects. Each patch is processed by a pre-trained backbone

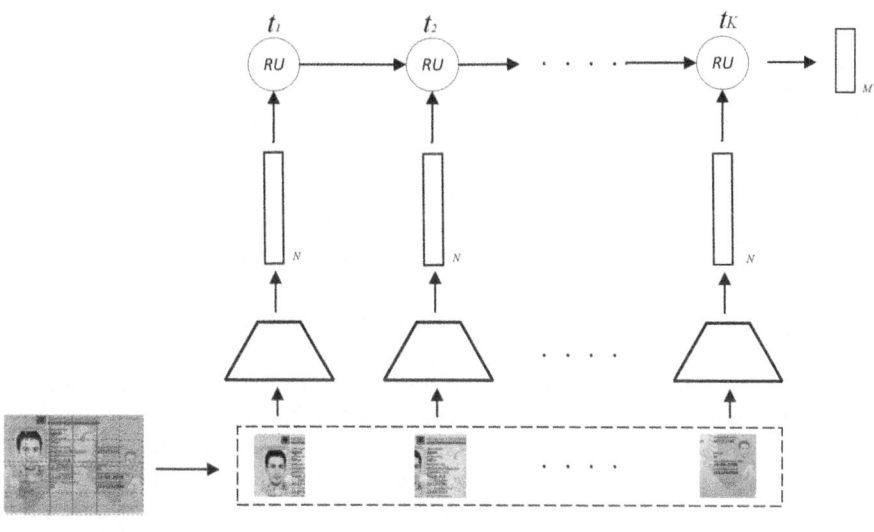

Patch Extraction   Sequence of Patches

Fig. 1. Data modelling architecture: The sequence of patches are extracted for each document. Each patch is then fed into a CNN feature extractor. The vectors from the CNN are passed to a RU at time t_k depending of the patch position. The overall vector of the document is extracted by the RU at t_K

that is used as an universal feature extractor. The output of the backbone is a vector of dimension $n > m$ that feeds the Recurrent Unit (RU).

We apply the model in Fig. 1 to both the support set and the query set to apply a few-shot learning (FSL) strategy to classify the documents based on their vector representation. FSL training is structured around the concept of episodes. At each episode, the model is learning from a support set composed of k-shots and n-ways, and then, after N_1 episodes where $N_1 < N$, we do M episodes of evaluation on the query set with q-shots and n-ways. n-ways here represents the number of class labels present in the support or query set at each episode, therefore n will always be equal to 2. The parameter k-shots (eq. q-shots) represents the number of images allocated to each class label within the support set (eq. query set) during every episode. Hence, at each episode on support set, the model is learning from $n \times k$ images, and evaluating on $n \times q$ images on query set. At each episode, the batch of images are grouped by class label, and the order of class label is shuffled at each episode, which means, the first k images (or q images for query set) could be either forged or genuine depending on the episode. It is worth to mention that support set and query set do not contain the same meta-class, as the main objective is to build a model able to generalize on unseen meta-classes. In this work, we defined meta-classes based on the country of issuance of the ID document.

From the support set vectors and the query set vectors, we learn a metric space based on Prototypical Network (PN) method, that will classify document images in the corresponding class, either genuine or faked. The loss used during training is a cross-entropy loss. We trained the whole architecture using two distinct strategies: Conditional FSL and Unconditional FSL. The Conditional FSL assumes that the system knows before-hand the document class (the meta-class in a few-shot scenario). Conversely, the Uncon-ditional FSL does not knows the document class and must detect it is genuine or not.

C-FSL: Conditional Few-Shot Learning

This approach involves training the model on a support set and a query set derived from the same document meta-class. As depicted in Fig. 2, different meta-classes are distinguished by distinct colors.

The final prediction for a query vector is given by its closest distance to the support set vectors within the same meta-class. By adopting this approach, the model is relieved from the burden of accommodating various document types in the space, allowing it to concentrate solely on detecting fake documents, see Fig. 2.

In this scenario, as shown in the figure, triangles represent fake documents, while circles represent genuine documents. Each pentagon represents the mean vector for each meta-class, and the square represents the query vector computed from the recurrent network model. The learnt distance is computed from the query vector to the support vectors belonging to the same meta-class. If the closest meta-class distance belongs to support vectors of fake documents, the query is classified as a fake document.

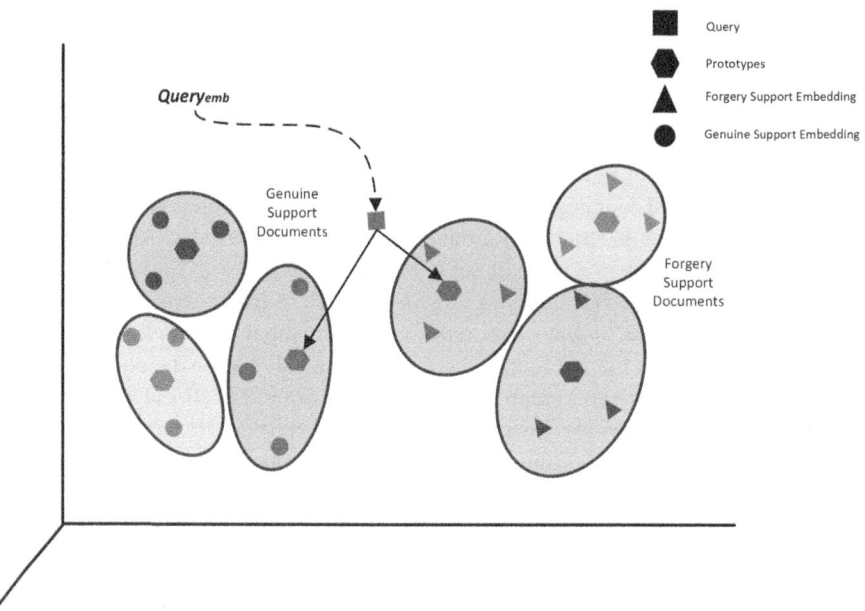

Fig. 2. Example of conditioned prototypical classification with k-shot=3

U-FSL: Unconditional Few-Shot Learning

As explained above, in the Unconditional few-shot strategy, all meta-classes are included in both the support and query sets without considering the country of origin. This ensures that each support vector exclusively represents either genuine or fake documents, regardless of the document meta class. This approach requires models to focus solely on identifying fake document and inconsistencies, while also minimizing the impact of meta-class variations in document representations. Figure 3 illustrates this concept, depicting data distribution without conditioning on meta-classes, where both genuine and fake classes remain. In this case the query will be classified as genuine as the distance to the genuine prototype is closer as the distance to the support vectors of fake documents.

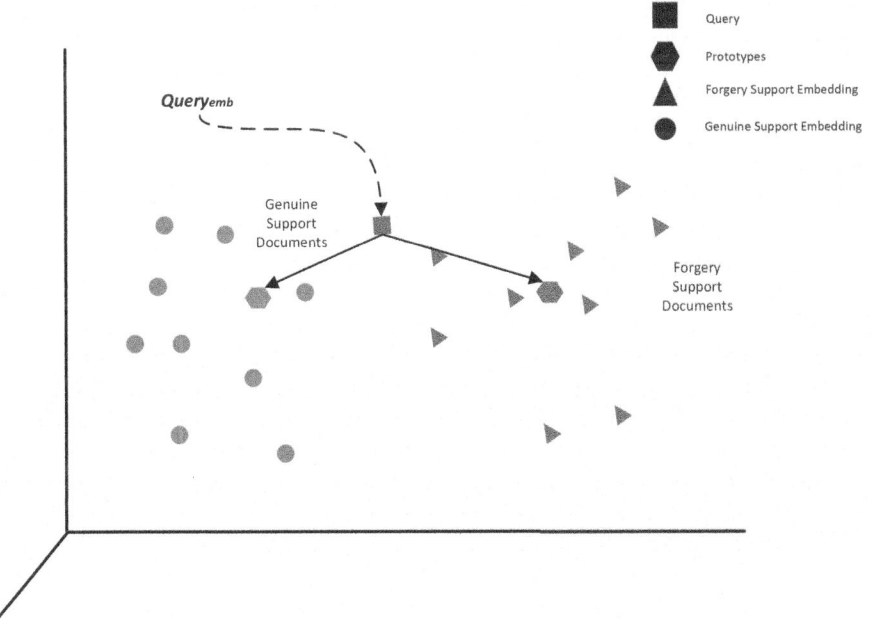

Fig. 3. Example of Unconditioned Prototypical classification with k-shot=10

3   Related Work

Document verification is essentially a binary classification task. We have seen in the Introduction that in a classical supervised context almost any classification method would perform well if it were not for the lack of sufficient quality data to provide these models enough generalization capacity. In this section, we will review the key architecture components that closely related to the proposed model. Therefore, we will briefly review the main convolution architectures that can be used as backbones in our model, the usual recurring units, and the main FSL strategies.

The proposed architecture, illustrated in Fig. 1, applies a *convolutional* module as an universal feature extractor. Although there are many models that can be used as backbone and are trained on large databases such as ImageNet, we use EfficientNet-B3 [17], ResNet50 [12], Vision Transformer Small Patch 16 (ViT-S/16) [9] and TransFG [11], since we want to evaluate the impact to these pre-trained models in the final performance of the proposed model. EfficientNet and ResNet belong to the commonly used architectures for image classification tasks while Vision Transformer and TransFG architectures are transformer-based architectures. Moreover, the TransFG model is an architecture designed for fine-grained classification task.

Although we have used attention-based models as a backbone, the main element of the proposed architecture is the Recurrent Unit (RU) that we apply to the output of them. As the patches order within the sequence is completely arbitrary we perform some preliminary experiments to evaluate the impact in the model performance, depending on the patches order. The impact on the model performance was negligible and thus, a natural choice would be to apply Attention-based models [18]. However, since recurrent models perform equally, we choose simpler models that require less data to train. The RUs used and evaluated are the usual RNN, LSTM [13] and GRU [8].

To address the problem of having to estimate models able of generalizing with a limited set of data, the authors in [16] developed the method Prototypical Networks. In this work, prototypes are computed from a reduced set of samples per class and are used to estimate a metric space that allows samples classification based on the square Euclidean distance between prototypes and the query. In [19], the authors proposed a paradigm called Matching Networks (MN), which used a differentiable nearest neighbors algorithm for FSL classification. Also, the authors in [10] introduced the Model-Agnostic Meta-Learning (MAML) technique, a strategy that updates model parameters with a small number of gradient steps and limited train data points from a new task to produce a good generalization performance. These approaches have shown promise in domains, including image recognition and natural language processing, yet their application in document ID verification remains underexplored.

4   Experiments and Discussion

As we have pointed from the beginning of this paper, the biggest weakness of document verification methods in general is their unreliability when they have to process document classes not seen in training. Consequently, the goal of the experiments carried out was to evaluate as much as possible the generalization capacity of the proposed model. To this end, we have repeated 10 times the same experiments with randomly chosen metaclasses, queries sets and support sets. We reported the performances of the evaluated models in terms of the accuracy rate and the area under the curve, respectively denoted by Accuracy and AUC in the results tables. Together with the mean of the values of these metrics, we have computed their standard deviation. Below, we briefly describe the datasets used and the experiments set up. We conducted an ablation study to assess the impact of the backbone and the recurrent units in the model performance.

Then, we analyse and discuss the generalization capacity of the proposed models under harder conditions.

(a) Genuine example   (b) Fake example

Fig. 4. Example of cropped images from the SIDTD clips. (a) corresponds to a genuine sample of a synthetic Spanish ID document and (b) to a fake sample of the same nationality.

Datasets

We trained models on the SIDTD dataset [2]. The SIDTD dataset is an extension of the MIDV2020 dataset [3] and it contains ID cards in three different formats: templates, videos, and clips. Each format includes both genuine and fake documents. For the evaluation purposes we used the clips images, which are the video frames, and cropped to remove image background and to rectify the perspective effects, see Fig. 4. In addition to the SIDTD dataset we use two more datasets for testing purposes. The first dataset, named in this paper as *UAB dataset*, includes real Spanish National ID Cards voluntarily provided by people. To generate forgeries of these documents and create fake examples, we employed the same techniques used in generating forgeries for the SIDTD dataset. The second dataset used for testing purposes was the Findit dataset, which is another fraud detection dataset based on French receipts [1]. The Findit dataset provides 1,180 images of captured receipts (240 altered and 940 genuine receipts) modified by real people from common and widespread material such as Window 10® to obtain closer real-life forgery, see Fig. 5.

Experimental Set Up

For training, and following the FSL methodology [20], we have divided the meta-train and meta-test based on the origin of the document. The training set is composed of 6 document nationalities, and in the test set there are the other 4 nationalities. We iterated ten times the same training procedures, with distinct document nationalities for both the training and testing sets, by randomly permuting the meta-classes at each iteration. Each document nationality has an average of 100–140 images.

Images are initially resized to the average shape found in the SIDTD dataset, which is 1047×1564. Then, we have set the patch size 299 width, generating 70 patches per

Fig. 5. Example of forgeries on prices, left image is a genuine receipt and the right image is altered from the FindIT dataset.

image. The batch composition for training varies based on the chosen strategy. For the C-FSL strategy, we adopted a 5-shots approach, wherein 5 fakes and 5 genuine samples serve as support, and another set of 5 fakes and 5 genuine samples act as queries. In the U-FSL strategy, we defined a 10-shots approach with the same distribution of fakes and genuine samples as described before.

Training was conducted on an NVIDIA A40 with 40GB of memory for 5,000 episodes and using the Adam optimizer. For the backbone, we trained on diverse standard CNN Backbones, including EfficientNet-B3 [17], ResNet50 [12], Vision Transformer Small Patch 16 (ViT-S/16) [9] and TransFG [11], all pretrained on ImageNet. The recurrent cell was trained on three standard recurrent cells: RNN, LSTM, and GRU.

Discussion

The results obtained are quite good overall. If we simply compare the performance metrics with the recently published ones, it can give the impression that there is not that much difference. However, remember that in the results reported in [6,7,14], among others, the experimental framework corresponds to the usual supervised learning one. On the other hand, our experimental framework corresponds to a FSL framework. In this sense, the results reported in [5] achieved an AUC close to 0.90 in the most restrictive scenario with the best of the reported methods. In this paper we obtain AUC above 0.95 in the worst-case scenario in all cases.

Experiments comparing the impact of the RU on overall performance have shown us that, surprisingly, LSTMs perform better than GRUs, when this is not the norm in other contexts. Likewise, in relation to the FSL strategies, it has been with the simplest one, the PN, that the best results have been obtained. In what follows, we will discuss the results obtained by evaluating both U-FSL and C-FSL models for each of the backbones used.

Table 1 provides a comprehensive overview of the results obtained for each backbone. There are some evidences that, for both C-FSL and U-FSL strategies, the best results are consistently achieved when using ResNet as the backbone. Again, simplest models achieves the best performance. Additionally, we observe that the results of the C-FSL strategy are slightly better to those of the U-FSL strategy. This is not surprising since the conditional model, by using the information from the metaclass, reduces the uncertainty of the model and significantly simplifies the task. What is surprising is that the decay in performance, looking at the standard deviation values of the experiments, does not seem significant for the U-FSL strategy.

Table 1. PN and LSTM model performances in terms of Accuracy and AUC scores with standard deviation. Models are trained and tested on the clip cropped SIDTD images. Training is performed with 5-shot for C-FSL models and with 10-shot with U-FSLt models.

Clip cropped SIDTD				
Models	U-FSL		C-FSL	
	Accuracy	AUC	Accuracy	AUC
EffNet	98.50 ± 0.97	99.95 ± 0.05	99.34 ± 0.47	99.98 ± 0.03
ResNet	99.65 ± 0.25	100 ± 0.0	99.98 ± 0.07	100 ± 0.0
ViT	98.48 ± 0.08	99.90 ± 0.01	99.91 ± 0.11	100 ± 0.0
TransFG	98.20 ± 0.46	99.92 ± 0.04	99.89 ± 0.19	100 ± 0.0

Previous experiments have been conducted by scaling the images to the average size of the images. However, one of the main benefits of recurrent networks is we can process sequences of different lengths. In this case, we can process document images of different sizes (or resolution). In the following experiment, with the models trained in the previous experiments, we have evaluated the robustness of our model when processing images of documents of different resolution. To that end, the images in the SIDTD clip partition have been cropped and rectified but preserving the scale. The dimensions of the processed images range from 211 pixels wide (the size of a patch) to 4,228 pixels. Table 2 presents the results obtained and are equally satisfactory for both, the C-FSL model and the U-FSL model. Again, the C-FSL model achieves consistently better performance than U-FSL models but still, the performance of the U-FSL model are quite good. What deserves further analysis is the standard deviation values, since overall it has increase with respect the first set of experiments. We must analyse whether there is a correlation between the performance and the document resolution, or not, which will make sense.

Table 2. PN and LSTM model performances in terms of Accuracy and AUC scores with standard deviation. Models are tested on clip cropped SIDTD images not rescaled. Inference is performed with 5-shot for the C-FSL model and with 10-shot for the U-FSL model.

clip cropped SIDTD No Rescaled				
Models	U-FSL		C-FSL	
	Accuracy	AUC	Accuracy	AUC
EffNet	95.21 ± 3.67	99.13 ± 0.93	97.18 ± 6.03	99.53 ± 1.14
ResNet	97.17 ± 2.79	99.79 ± 0.31	99.99 ± 0.01	100 ± 0.0
ViT	92.96 ± 4.40	98.31 ± 1.16	99.92 ± 0.12	100 ± 0.0
TransFG	95.87 ± 0.29	99.44 ± 0.50	96.75 ± 4.01	99.68 ± 0.37

The last set of experiments seeks to evaluate the generalization capacity of the proposed models in a more challenging scenario. The models trained in the first experiment have been evaluated on two different datasets. As we have previously described, the UAB dataset is made up of real ID documents that have been forged with the same techniques used to generate the fake data of the SIDTD dataset. The second dataset has nothing to do with the type of documents that have been used to train the models and is made up of images of tickets that have been altered by different means. For the latter dataset, since there is only one metaclass, the U-FSL and C-FSL models are the same. Results are reported in Table 3. Again the best performance is achieved with the ResNet as backbone. Moreover, the performance achieved in terms of Accuracy and AUC for both datasets is aligned with the performance achieved in previous experiments.

Table 3. PN and LSTM model performances in terms of Accuracy and AUC scores with standard deviation. Models are tested on UAB Dataset and FindIt. Inference is performed with 5-shot for C-FSL models and with 10-shot for U-FSL models.

Dataset	UAB Dataset				Findit	
Models	U-FSL		C-FSL		U-FSL	
	Accuracy	AUC	Accuracy	AUC	Accuracy	AUC
EffNet	96.09 ± 1.69	99.65 ± 0.27	99.98 ± 0.02	100 ± 0.0	89.93 ± 7.83	94.60 ± 7.78
ResNet	98.77 ± 1.49	99.39 ± 0.10	100 ± 0.0	100 ± 0.0	98.56 ± 1.69	99.93 ± 0.10
ViT	90.08 ± 7.17	95.85 ± 6.57	99.62 ± 0.60	99.99 ± 0.03	89.87 ± 8.99	95.26 ± 8.43
TransFG	97.65 ± 0.82	99.83 ± 0.09	99.79 ± 0.24	99.99 ± 0.01	96.50 ± 1.54	99.65 ± 0.27

As we mentioned at the beginning of this subsection, overall the results obtained are quite good. The ablation study performed by varying backbones, recurrent cells and FSL strategies shows that there are no big differences between the use of some components with respect to others. However, we observed that simpler components tend to perform better than more complex ones. In particular, the relative poor performance of pre-trained attention-based backbones (ViT and TransFG) is striking, when the general trend in other areas of application is just the opposite. Finally, the idea of making use of a recurring network along with a FSL strategy seems to have been successful. The combination of both techniques in meta-training has resulted in more robust representation that allows us to generalize well to the proposed model.

5   Conclusion

In this paper, we have presented a recurrent network model combined with FSL strategies to verify whether document images are genuine or fake. Despite of not introducing any new component in this architecture, the proposed architecture, by itself, is original and has not been applied to this task and similars. Moreover, the results obtained, as we have already discussed in the previous section, support the proposed strategy.

The proposed model seems to be able to learn good document representations. That representations are what would allow our proposed model to generalize well. However, there are still elements that deserve further study. FSL models still require few examples of fake and genuine documents and it is knows that in practice is not always feasible. Therefore, it is necessary to develop models that allow us to move towards zero-shot models.

Acknowledgements. This work has been partially supported by the SOTERIA project, which has received funding from the European Union's Horizon 2020 research and innovation program under grant agreement No 101018342. It is also partially supported by the Spanish project PID2021-126808OB-I00 (GRAIL), Ministerio de Ciencia e Innovación, CNS2022-135947 (DOLORES) and the AGAUR SGR project 2021-SGR-01559. The authors acknowledge the support of the Generalitat de Catalunya CERCA Program to CVC's general activities.

References

1. Artaud, C., Sidere, N., Doucet, A., Ogier, J.-M., Poulain D'Andecy, V.: Find it! fraud detection contest report. In: 2018 24th International Conference on Pattern Recognition (ICPR), pp. 13–18 (2018)
2. Boned, C., et al.: Synthetic dataset of id and travel document (2024)
3. Bulatov, K.B., et al.: Midv-2020: a comprehensive benchmark dataset for identity document analysis. Comput. Optics **46**(2) (2022)
4. Berenguel Centeno, A., Ramos Terrades, O., Lladós Canet, J., Cañero Morales, C.: Evaluation of texture descriptors for validation of counterfeit documents. In: 14th IAPR International Conference on Document Analysis and Recognition (ICDAR 2017), Kyoto, 9–15 November 2017, pp. 1237–1242 (2017)
5. Berenguel Centeno, A., Ramos Terrades, O., Lladós Canet, J., Cañero Morales, C.: Recurrent comparator with attention models to detect counterfeit documents. In: 2019 International Conference on Document Analysis and Recognition (ICDAR 2019), Sydney, 20–25 September 2019, pp. 1332–1337. IEEE (2019)
6. Chen, C., Zhang, S., Lan, F., Huang, J.: Domain-agnostic document authentication against practical recapturing attacks. IEEE Trans. Inf. Forensics Secur. **17**, 2890–2905 (2022)
7. Chen, C., Zhao, L., Yan, J., Li, H.: A distortion model-based pre-screening method for document image tampering localization under recapturing attack. Signal Process. **200**, 108666 (2022)
8. Cho, K., et al.: Learning phrase representations using RNN encoder-decoder for statistical machine translation. arXiv preprint arXiv:1406.1078 (2014)
9. Dosovitskiy, A., et al.: An image is worth 16x16 words: transformers for image recognition at scale. arXiv preprint arXiv:2010.11929 (2020)
10. Finn, C., Abbeel, P., Levine, S.: Model-agnostic meta-learning for fast adaptation of deep networks. In: International Conference on Machine Learning, pp. 1126–1135. PMLR (2017)
11. He, J., et al.: TransFG: a transformer architecture for fine-grained recognition. Proc. AAAI Conf. Artif. Intell. **36**, 852–860 (2022)
12. He, K., Zhang, X., Ren, S., Sun, J.: Deep residual learning for image recognition. In: Proceedings of the IEEE Conference on Computer Vision and Pattern Recognition, pp. 770–778 (2016)
13. Hochreiter, S., Schmidhuber, J.: Long short-term memory. Neural Comput. **9**(8), 1735–1780 (1997)
14. Liang, W., Dong, L., Wang, R., Yan, D., Li, Y.: Robust document image forgery localization against image blending. In: 2022 IEEE International Conference on Trust, Security and Privacy in Computing and Communications (TrustCom), pp. 810–817 (2022)

15. Saire, D., Tabbone, S.: Documents counterfeit detection through a deep learning approach. In: 2020 25th International Conference on Pattern Recognition (ICPR), pp. 3915–3922 (2021)
16. Snell, J., Swersky, K., Zemel, R.S.: Prototypical networks for few-shot learning (2017)
17. Tan, M., Le, Q.: Efficientnet: rethinking model scaling for convolutional neural networks. In: International Conference on Machine Learning, pp. 6105–6114 (2019)
18. Vaswani, A., et al.: Attention is all you need. Adv. Neural Inf. Process. Syst. **30** (2017)
19. Vinyals, O., Blundell, C., Lillicrap, T., Wierstra, D., et al.: Matching networks for one shot learning. Adv. Neural Inf. Process. Syst. **29** (2016)
20. Wang, Y., Yao, Q., Kwok, J.T., Ni, L.M.: Generalizing from a few examples: a survey on few-shot learning. ACM Comput. Surv. **53**(3) (2020)
21. Yan, J., Chen, C.: Cross-domain recaptured document detection with texture and reflectance characteristics. In: 2021 Asia-Pacific Signal and Information Processing Association Annual Summit and Conference (APSIPA ASC), pp. 1708–1715 (2021)

Document Specular Highlight Removal with Coarse-to-Fine Strategy

Xin Yang[1,2], Fei Yin[1,2(✉)], Yan-Ming Zhang[1,2], Xudong Yan[3], and Tao Xue[3]

[1] MAIS, Institute of Automation of Chinese Academy of Sciences, Beijing, China
yangxin2022@ia.ac.cn, {fyin,ymzhang}@nlpr.ia.ac.cn
[2] School of Artificial Intelligence, University of Chinese Academy of Sciences, Beijing, China
[3] T Lab, Tencent Map, Tencent Technology (Beijing) Co., Ltd., Beijing, China
{owenyan,emmaxue}@tencent.com

Abstract. Specular highlight detection and removal are fundamental challenges in computer vision and image processing, with the detection results serving as a precursor to guide the model in achieving better removal of specular highlights. This paper introduces a novel highlight removal model, which presents an efficient end-to-end deep learning framework designed to automatically remove specular highlights from a single image. Our architecture comprises three key modules: the Coarse Predictor (CP), Refinement Predictor (RP), and Global Discriminator (GD). The CP utilizes a novel Transformer-based Unet architecture to recover the primary content, while the GD incorporates a discriminator to ensure the coarse result is more feasible in a global context. Lastly, the RP is based on conditional Denoising Diffusion Probabilistic Models (DDPM) and is responsible for predicting the residual information between the ground-truth and the CP-predicted image. Experimental results on four public benchmark images demonstrate that our method surpasses state-of-the-art methods in the task of highlight removal.

Keywords: Document Specular Highlight Removal · Transformer · Conditional Diffusion Models

1 Introduction

In various life scenes, specular highlights are a common occurrence. When photographing objects, the reflective materials and the play of light on their surfaces can result in specular highlights being visible in the final image. As a result, these highlights have been shown to significantly impact the accuracy of various image processing algorithms including illumination estimation [1,2], image segmentation [3], and text recognition [4,5].

Specular Highlight Removal, a crucial aspect of image processing and computer vision, aims to enhance image quality and address common challenges in

E. H. Barney Smith et al. (Eds.): ICDAR 2024, LNCS 14804, pp. 63–78, 2024.
https://doi.org/10.1007/978-3-031-70533-5_5

image acquisition and processing. Its primary goal is to minimize the distracting effects of bright reflections and improve object visibility. While existing methods heavily rely on the detection of highlight regions to guide the network for better removal results [4,6,7], they come with a significant computational cost and only yield marginal improvements in the removal process. Furthermore, these methods often rely on simplistic assumptions related to specific scenarios or materials, such as dependence on Temporal Dark Prior [8], fruit surfaces [9], and simple backgrounds [9,10], rendering them unsuitable for complex real-world image solutions. In the backdrop of deep learning's rapid advancements, convolutional neural networks [7,11,12] have emerged as crucial means for Specular Highlight Removal. However, it is important to consider the overall color that surrounding the image with highlights, instead of just focusing on the small highlighted area, during the recovery process. Wu et al. [13] addressed this by using a self-attention mechanism based on Transformer [14] to emphasize global information in highlight images. Despite the Transformer's powerful capability to capture global information, it remains essential to also pay attention to the local details of the image.

In this study, we introduce a two-stage Coarse-to-Fine method that utilizes the global modeling capability of Transformer to recover the main content of an image with highlights, and the local modeling capabilities of CNN to restore the image's local details, without relying on a detection module. This model comprises a Coarse Predictor (CP), Refinement Predictor (RP), and Global Discriminator (GD). The CP utilizes the innovative Transformer-based Unet architecture to recover the primary global content, while the RP is based on conditional Denoising Diffusion Probabilistic Models (DDPM) to predict the residual local information between the ground-truth and CP-predicted image. The GD is a discriminator to improve the accuracy of CP's output on a global scale. Our method achieves highlight removal with satisfactory quality in most scenarios and surpasses prevailing learning-based techniques. The effectiveness of the proposed method's highlight removal capabilities is demonstrated in Fig. 1. The main contributions of this paper are threefold:

- We present a unified two-stage highlight removal framework capable of handling synthetic images and document scene images.
- We propose a Transformer-based Coarse Predictor and DDPM-based refinement module, which effectively execute highlight removal.
- Experimental results across multiple image datasets demonstrate the superiority of our approach over existing state-of-the-art algorithms.

2 Related Works

Highlight detection. The dichromatic reflection model, based on the assumption that image intensity can be represented by a linear combination of diffuse and specular reflections [15], has served as a fundamental framework for understanding and analyzing reflectance properties in computer vision. Building upon

(a) Light image   (b) OCR Result   (c) Ours   (d) OCR Result

Fig. 1. Visual results on selected single image with specular highlight of our method. "Ours" are the outputs of our two-stage networks.

this model, Lee et al. [16] conducted research aimed at developing a model specifically for detecting specular highlight regions using color images from varying viewing angles. Concurrently, Yuan et al. [17] proposed a technique for color analysis that involves estimating the separation of specular reflections by identifying and filtering out pixels in the specular area as outliers, while concurrently matching the remaining diffuse portions in other views. Furthermore, based on color distribution, Smith et al. [18] introduced an approach for endoscopic highlight detection. Additionally, Gang et al. [19] made a significant contribution by introducing a context-contrasted feature module, designed specifically for extracting contrast features. They also designed a multi-scale context-contrasted feature to aid in the precise localization of highlights. Despite the efficiency of these methods in detecting highlight reflections, their advancement has not been accompanied by similar advancements in other algorithms.

Multi-task framework of highlight removal. Recent Highlight Removal methods have prioritized Highlight Detection [4,12,13]. Hou et al. [4] explored text-aware single-image specular highlight removal, while Huang et al. [20] introduced a novel highlight feature extractor and contextual highlight attention module for detection and removal. Wu et al. [12] utilized a Unet with two output heads to carry out adversarial training for highlight detection and removal. Additionally, Wu et al. [13] used detection results to mitigate the effects of chromatic aberration and proposed an efficient Unet-Transformer Network for capturing global context. Ge et al. [21] recently proposed a parallel streams approach for decomposing and aggregating encoded deep features, featuring a module for feature decomposition and aggregation between encoders. Furthermore, Hao et al. [22] discussed a mask-guided single image specular highlight detection and removal framework, using detection results as a guide for the removal module. However, it is important to note that the neural networks have the capability to restore the original appearance of highlight images, and the incorporation of additional modules may result in increased calculation costs.

Image deblurring. In contrast to highlight removal, image deblurring aims to remove noise in order to restore the original content of clean areas. Deep learning techniques, including adversarial training, have been explored for blind deblurring. Generative adversarial network methods have achieved great performance [23–25], with Mehrdad et al. [26] solving the image deblurring problem using Cycle-GAN even without the requirement of paired noisy/clean images.

Diffusion Probabilitistic Models(DPMs). [27] have been widely used for conditional image and text generation [28–30]. Saharia et al. [31] introduced SR3, which adapts diffusion models to image super-resolution. While Palette, proposed by Saharia et al. [32], is a multi-task image-to-image diffusion model that has shown excellent performance in conditional image generation tasks, such as colorization, inpainting, and JPEG restoration. Additionally, in [28,33,34], diffusion models were implemented in a prediction-refinement framework to predict the residual, although they were not specifically designed to make good use of global and local information.

3   Proposed Method

3.1   Motivation

The removal of the highlight region in a natural image is a crucial task in recovering local detail information. However, despite the high intensity of the bright spots in the specular highlight region, it occupies a very small portion of the overall image. Therefore, the effective removal of the highlight region necessitates the full utilization of both global and local information of the image. To address this, motivated by DocDiff [34] which has powerful capability of deblurring, we also make a two stage network that combines Unet based on the Transformer architecture and Diffusion-based image refined module with full convolution structure. By leveraging both Transformer to model the overall context of global information and convolution to restore the details of local information, our model aims to produce more satisfactory highlight removal results with a coarse-to-find strategy.

3.2   Overview

Our architecture follows the overall structure illustrated in Fig. 2, comprising three modules: the coarse prediction module, refinement prediction module, and discriminator module. To leverage the global information of the image fully, we have employed a coarse prediction model based on Restormer [35], with the RP module modified from [27]. A discriminator structure has then been added to train against the coarse prediction model in order to bring the highlight removal image closer to the ground truth. In summary, it is a model structure that integrates Transformer and diffusion models and utilizes pixel loss with end-to-end training.

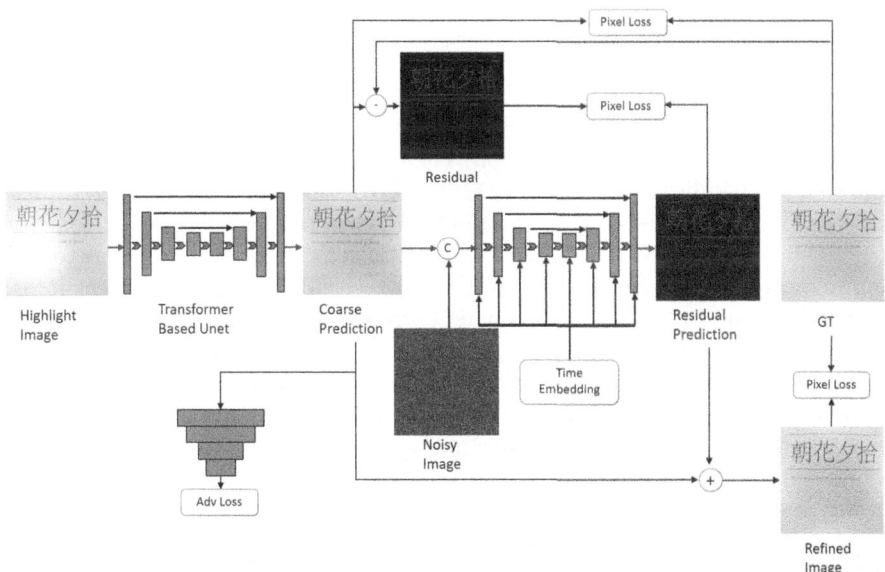

Fig. 2. The architecture of the proposed network.

3.3 Transformer Based Coarse Predictor

To better address the challenge of specular highlight removal in images, traditional methods [4,21] have primarily employed a full convolutional structure for the image-to-image transformation task. Nonetheless, it is important to acknowledge that the specular highlight region is intricately linked with its surrounding region, necessitating comprehensive global contextual reasoning. Consequently, the limitations of the full convolutional structure pose difficulties in achieving desirable results. To overcome this, our approach leverages the potent capability of the Transformer's self-attention mechanism in capturing global context, thereby enhancing the efficacy of image specular highlight removal.

We used a powerful transformer network called Restormer, as proposed by Syed et al. [35], which captures the self-attention in channels dimension among feature maps and has been applied in image deblurring and leverages transformer blocks.

The coarse predictor serves to enhance the specular highlight image by producing an initial version of x_{gt}. This enhancement is evaluated by measuring the pixel loss, which represents the disparity between x_{gt} and the enhanced image x_c. After capturing the global information, although the coarse predictor is successful in restoring the primary features of the highlight area during the removal process, it may sacrifice certain crucial local characteristics, resulting in underfitting.

3.4   Refinement Predictor

In response to the aforementioned issue, we propose the implementation of the Refinement Prediction (RP) module, aimed at generating high-quality samples from the learned posterior distribution. The RP module incorporates a Denoising Diffusion Probabilistic Model (DDPM) to predict the distribution of residuals between the Ground Truth (GT) and images generated by the Coarse Predictor (CP). Unlike training the CP and RP modules separately, the joint end-to-end training approach enables RP to dynamically learn more effective local information from the residuals. In the architecture of our RP module, we have integrated a compact U-Net structure, inspired by Ho et al. [27]. Specifically, we have replaced the self-attention layer of the middle block with a 4-layer inflated convolutional structure to restore the highlight region at the maximum of the receptive field, making it a full convoluntional Unet, thereby utilizing the local information of the image more comprehensively. Through extensive experimentation, we have demonstrated that this training strategy significantly enhances the clarity of specular highlight removal results and diminishes the discrepancy with GT.

The generative model known as the diffusion model is built upon a Markov chain structure. It takes the complex data distribution $x_0 \sim p_{\text{data}}$ and transforms it into a simpler noise distribution $x_t \sim p_{\text{dlatent}} = \mathcal{N}(0, I)$. The model then reconstructs the original data from the noise, with $\mathcal{N}$ representing a Gaussian distribution. The **Diffusion** and **Inverse** processes are the two main components of the DDPM framework.

Diffusion process, which is a crucial part of the model, consists of a Markov chain that progressively introduces noise to the original data x_0, ultimately leading to the convergence of the data to a Gaussian noise at T diffusion time steps. The corrupted data, denoted as $x_1, \ldots, x_T$, is sampled from p_{data}. More specifically, the diffusion process is defined as a Gaussian leap.

$$q(x_1, \ldots, x_T \mid x_0) = \prod_{t=1}^{T} q(x_t \mid x_{t-1}), \tag{1}$$

where t is the diffusion step, $x_0 = x_{res} = x_{gt} - x_c$, x_{gt} is the Ground Truth, and x_c is the result of CP. $q(x_t \mid x_{t-1}) = \mathcal{N}(x_t; \sqrt{\alpha_t} x_{t-1}, \sqrt{1 - \alpha_t} I)$, where α_t is a weighting parameter that decreases as t increases. An important feature of the forward diffusion process is that x_t of any step size can be obtained by sampling directly from x_0 using the following equation:

$$x_t = \sqrt{\overline{\alpha}_t}\, x_0 + \sqrt{1 - \overline{\alpha}_t}\, \epsilon, \tag{2}$$

where $\epsilon \sim \mathcal{N}(0, I)$, $\alpha_t = 1 - \beta_t$, $\overline{\alpha}_t = \prod_{i=1}^{t} \alpha_t$. Ho et al. [27] proved the existence of a closed-form expression for $q(x_t \mid x_0)$, which gives us $q(x_t \mid x_0) = \mathcal{N}(x_t; \sqrt{\overline{\alpha}_t}\, x_0, (1 - \overline{\alpha}_t)I)$, where $\overline{\alpha}_t$ tends to 0 for larger t, and $q(x_t \mid x_0)$ tends to the potential distribution p_{data}.

Inverse process is a Markov chain that iteratively samples from Gaussian noise to get a clean image. The inverse process starting from the noise $x_t \sim \mathcal{N}(0, I)$ to the clean data $x_0(x_{res})$ is defined as:

$$p_\theta(x_0, \dots, x_{T-1} \mid x_T) = \prod_{t=1}^{T} p_\theta(x_{t-1} \mid x_t), \tag{3}$$

$$p_\theta(x_{t-1} \mid x_t) = \mathcal{N}(x_{t-1}; \mu_\theta(x_t, t), \epsilon_\theta T),$$

in their study, Ho et al. [27] demonstrated that the mean $\mu_\theta(x_t, t)$ is the desired estimation target by the neural network. Concurrently, ϵ_θ serves as a function approximator for predicting ϵ from x_t.

$$\mu_\theta(x_t, t) = \frac{1}{\sqrt{\alpha_t}} \left(x_t - \frac{\beta_t}{\sqrt{1 - \overline{\alpha}_t}} \epsilon_\theta(x_t, t) \right), \tag{4}$$

in fact, $\widetilde{x}_0$ is usually predicted from x_t through swapping by Eq. 2 and then x_{t-1} is sampled using $\widetilde{x}_0$ and x_t:

$$\widetilde{x}_0 = \frac{x_t}{\sqrt{\overline{\alpha}_t}} - \frac{\sqrt{1 - \overline{\alpha}_t}\,\epsilon_\theta(x_t, t)}{\sqrt{\overline{\alpha}_t}}, \tag{5}$$

$$q(x_{t-1} \mid x_t, \widetilde{x}_0) = \mathcal{N}\left(x_{t-1}; \widetilde{\mu}_t(x_t, \widetilde{x}_0), \widetilde{\beta}_t \right), \tag{6}$$

where $\widetilde{\mu}_t(x_t, \widetilde{x}_0) = \frac{\sqrt{\alpha_t}(1-\overline{\alpha}_{t-1})}{1-\overline{\alpha}_t} x_t + \frac{\sqrt{\overline{\alpha}_{t-1}}\beta_t}{1-\overline{\alpha}_t} \widetilde{x}_0$, $\widetilde{\beta}_t = \frac{1-\overline{\alpha}_{t-1}}{1-\overline{\alpha}_t}\beta_t$.

The input of denoiser f_θ is Gauss noise, x_c and timestep. f_θ is capable of being trained to predict either x_0 or ϵ. In unconditional generation, the prediction of either x_0 or ϵ is considered equivalent due to the capability of converting them through the Eq. 2. Existing methods [27,33,34] typically predict the additional noise ϵ in order to enhance the diversity of generated natural images. But in our residual prediction task, this is an Conditional Generation. Since we want RP to restore x_{res}, which is x_0 in the Inverse Process. This is a certain target when x_c is given. Therefore we train the RP to predict x_0 instead of ϵ in the case of reducing uncertainty of sampled result. We concat Gauss noise with x_c and then send it into RP. Therefore, we get the inverse result by repeatedly use the follow equation:

$$x_{t-1} = \frac{\sqrt{\alpha_t}(1 - \overline{\alpha}_{t-1})}{1 - \overline{\alpha}_t} \mathbf{x}_t + \frac{\sqrt{\overline{\alpha}_{t-1}}\beta_t}{1 - \overline{\alpha}_t} \widetilde{x}_0. \tag{7}$$

3.5   Global Discriminator

To enhance the realism of specular de-emphasis results produced by the coarse prediction model C_θ, we employ a global discriminator D_θ. This discriminator is responsible for evaluating the overall image and motivating C_θ to generate more realistic outputs. D_θ is designed to accept input images of shape (C, H, W) and is comprised of multiple convolutional blocks. Each block includes a convolutional layer, an optional Instance Normalization layer, and a LeakyReLU activation function. These convolutional blocks progressively reduce the spatial dimensions of the input image and extract features for the feature map. Additionally, depending on the height of the input image, we incorporate a Fully Connected (FC) layer to process the final output of the discriminator. This output is then mapped to the range $[0, 1]$ through a Sigmoid activation function, which represents the probability that the input image is a real image.

During the training stage, our approach involved utilizing the global discriminator to engage in confrontation training with the coarse prediction model to enhance the performance of C_θ's outcome. However, it is important to note that in the subsequent inference stage, we opted to discontinue the use of D_θ. As illustrated in Algorithm 1, the methodology we put forth for the inference stage employs a two-stage scheme integrating a coarse-to-fine strategy.

3.6   Loss Function

We integrate the three modules into a unified network system for joint training. The network training is supervised by an effective loss function with three components: Coarse Prediction loss, Refirement Prediction loss, and Adversarial loss.

The choice of loss function for measuring pixel differences between two images involves using pixel loss, which encompasses L1 loss and L2 loss. Given that the pixel value of the specular highlight region tends to be close to white in the image, with values near 255, the selection of L1 loss function is apt. L1 loss's reliance on absolute values makes it less susceptible to larger outliers, aligning with the characteristics of the specular highlight region's pixel values.

Given the input image with specular highlight, the CP loss is defined as follow:

$$\mathcal{L}_{\mathrm{CP}} = E\|x_{\mathrm{gt}} - C_\theta(x)\|_1. \tag{8}$$

To enhance the quality of the highlight removal image, the discriminator is implemented to counteract the loss by distinguishing the positivity or falseness of generated image produced by the coarse prediction model. Consequently, the loss function incorporates the Binary Cross Entropy (BCE) Loss.

$$\mathcal{L}_{\mathrm{Adv}} = -\frac{1}{n}\sum_i [\log(D_\theta(x_c)) + \log(1 - D_\theta(x_c))]. \tag{9}$$

The diffusion model is then utilized to predict the residual, which represents the difference between the coarse prediction and the ground truth (GT). We training the RP by minimizing this objective function:

$$\mathcal{L}_{\text{RP}} = E\| \left(x_{res} - f_\theta \left(\sqrt{\overline{\alpha}_t}\, x_c + \sqrt{1 - \overline{\alpha}_t}\, \epsilon, t, \epsilon \right) \right\|_1 . \tag{10}$$

Finally, in order to further reduce the gap between the final result and GT, we set GT as the learning goal of final image:

$$\mathcal{L}_{\text{Ref}} = E\| \left(x_c + f_\theta \left(\sqrt{\overline{\alpha}_t}\, x_c + \sqrt{1 - \overline{\alpha}_t}\, \epsilon, t, \epsilon \right) \right) - x_{\text{gt}}\|_1 . \tag{11}$$

Hence, the joint loss function for C_θ, f_θ, and D_θ is formulated as follows:

$$\mathcal{L}_{\text{total}} = \mathcal{L}_{\text{CP}} + \mathcal{L}_{\text{RP}} + \mathcal{L}_{\text{Adv}} + \mathcal{L}_{\text{Ref}}. \tag{12}$$

Algorithm 1 outlines the complete inference process for our model.

Algorithm 1

Require: f_θ: Denoiser network, C_θ: Coarse predictor, D_θ: Discriminator,
$\quad\quad$ y: Highlight input image, $\alpha_{0:T}$: Noise schedule

1: $x_c \leftarrow C_\theta(y)$ $\quad\quad\quad\quad\quad\quad\quad\quad\quad\quad\quad\quad\quad\quad\quad\quad$ ▷ Coarse prediction
2: $x_t \leftarrow \sqrt{\overline{\alpha}_t}\, x_c + \sqrt{1 - \overline{\alpha}_t}\, \epsilon$ $\quad\quad\quad\quad\quad\quad\quad\quad$ ▷ Forward diffusion
3: **for** t = T, ..., 1 **do**
4: $\quad\quad \widetilde{x}_{res} \leftarrow f_\theta(x_t, t, x_c)$
5: $\quad\quad x_{t-1} \leftarrow \frac{\sqrt{\alpha_t}\,(1-\bar{\alpha}_{t-1})}{1-\bar{\alpha}_t} \mathbf{x}_t + \frac{\sqrt{\bar{\alpha}_{t-1}}\,\beta_t}{1-\bar{\alpha}_t} \widetilde{x}_{res}$ $\quad\quad$ ▷ Residual prediction
6: **end for**
7: **return** $x_c + x_0$ $\quad\quad\quad\quad\quad\quad\quad\quad\quad\quad\quad$ ▷ Return refinement result

---

4 Experiment

4.1 Datasets and Settings

We utilized the RD [4], SD1 [4], SD2 [4] and Specular Highlight Image Quadru-ples (SHIQ) dataset [36]. The RD dataset includes synthesized ID cards and driver's licenses, containing rich textual information. Initially, a transparent plastic film is placed on the image, light is turned on, and the camera captures it to obtain the highlight image. Then, the light is turned off to obtain the image without the reflection of highlight. The position of the plastic film is adjusted to obtain highlight images of different shapes and intensities. The RD dataset includes a training set of 1,800 images and a test set of 225 images. The images in the SD1 dataset were captured from supermarkets and streets with textual information. Highlight mask images were used for highlight image synthesis, where the shapes of the highlights included circles, triangles, ellipses, and rings to simulate the illumination conditions of the real scene. Images without high-light reflections in RD, including ID cards and driver's licenses, are synthesized with the above highlight mask for highlight image synthesis to obtain SD2. The image contents of RD and SD2 are the same. The SD1 dataset contains a 12k training set and a 2k test set, whereas SD2 contains a 12k training set and a 1.7k test set. SHIQ [36] is a collection of real-world, high-quality specular highlight removal datasets that include a 10k training set and a 1k test set.

Our model is implemented using PyTorch on NVIDIA A6000. The input dimensions are in the native resolution of these datasets, so input size of SHIQ is 200×200, and the others is 512×512. We set AdamW optimizer to train the network with batch size of 16 and 2. Training for each dataset is carried out for 1 million iterations and timestep is set 1000 to ensure that networks' parameters are fully optimized. In our experiments, we set learning rate 1e-4 and cosine decrease.

4.2 Qualitative Evaluation

We compared the performance of our method with JSHDR [36], M2-Net [20] and MEF-SHDR [21] on the aforementioned four datasets. The comparison results are illustrated in the following Fig. 3, revealing that our method outperforms the aforementioned state-of-the-art methods in removing highlights and accurately restoring the details of the Ground Truth. Take Fig. 3 for example, although M2-Net [20], MEF-SHDR [21] and ours can improve the OCR result, our method successfully better removes the majority of highlights and restores the word's detail in the image. This comparison highlights our method's superior performance compared to current state-of-the-art conventional methods [20,21,36]. Surprisingly, our method is not only efficient in removing specular highlights from document images, but also performs well on natural scene datasets as shown in Fig. 4.

Table 1. Quantitative comparison of highlight removal methods on SD1, SD2 and RD datasets. Evalution metrics are PSNR and SSIM, and the best result are shown in bold.

Datasets	SD1		SD2		RD	
Method/Metric	PSNR	SSIM	PSNR	SSIM	PSNR	SSIM
Multi-class [37]	26.29	89	28.99	91.81	17.17	64.23
SPEC [6]	15.61	69	9.66	53.95	14.82	52.49
TASHR [4]	22.65	88.23	28.21	90.67	21.62	77.19
JSHDR [36]	24.59	85	-	-	-	-
M2-Net [20]	33.44	92	-	-	-	-
MEF-SHDR [21]	29.53	94.78	31.63	95.62	25.26	86.85
DocDiff [34]	34.21	96.50	35.19	96.96	30.59	91.31
Ours	**34.81**	**96.96**	**38.58**	**97.42**	**31.09**	**91.73**

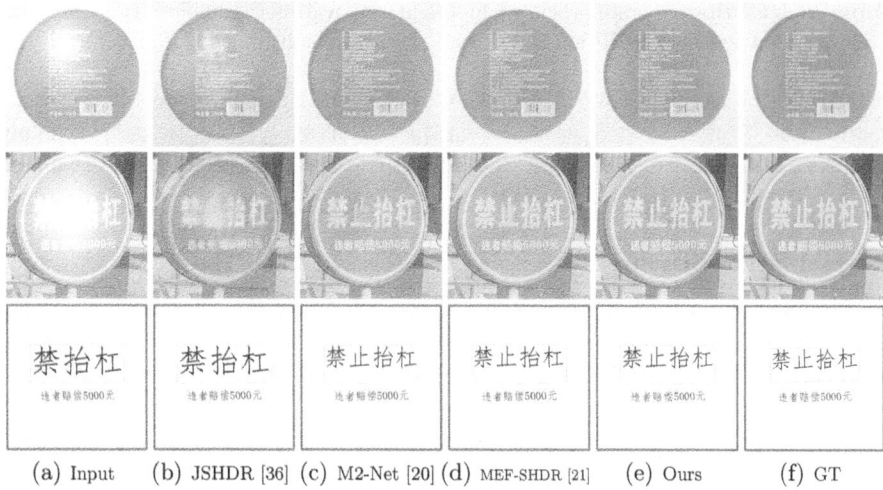

(a) Input   (b) JSHDR [36] (c) M2-Net [20] (d) MEF-SHDR [21]   (e) Ours   (f) GT

Fig. 3. Qualitative comparisons with state-of-the-art algorithms on text image data. The last row shows the results of recognition of the highlight removal image using PaddleOCR(https://github.com/PaddlePaddle/PaddleOCR).

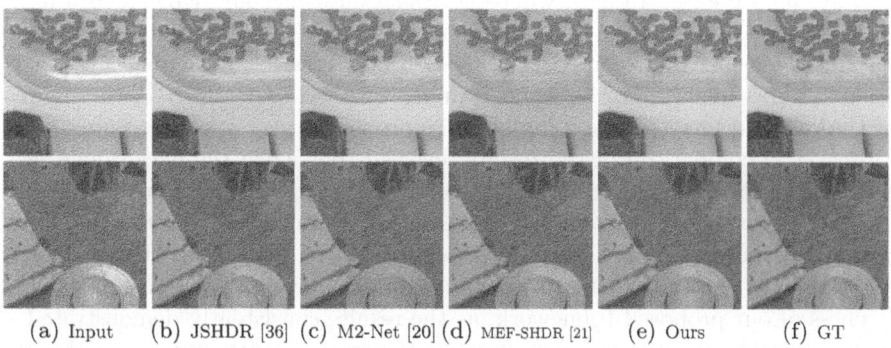

(a) Input   (b) JSHDR [36] (c) M2-Net [20] (d) MEF-SHDR [21]   (e) Ours   (f) GT

Fig. 4. Qualitative comparisons with state-of-the-art algorithms on SHIQ dataset.

4.3   Quantitative Evaluation

We quantitatively compared the specific performance of the methods mentioned above in highlight removal task by utilizing PSNR and SSIM metrics for quantitative visual quality assessment. Table 1 shows the quantitative evaluation results for SD1, SD2, and RD datasets, while Table 2 presents the results for the SHIQ dataset. Our comparison is based on PSNR and SSIM assessment metrics with existing methods. Notably, our proposed coarse-to-fine strategy demonstrates the best performance in the task of specular highlight removal both on document and natural images. Particularly, our method outperforms M2-Net [20] by more than 1 dB in PSNR on the SD1 dataset and exceeds MEF-SHDR [21] by over 3 dB in PSNR on the SHIQ dataset.

Table 2. Quantitative comparison of highlight removal methods on SHIQ datasets. Evalution metrics are PSNR and SSIM, and the best result are shown in bold.

Method	SPEC [6]	DeepFillv2 [38]	JSHDR [36]	M2-Net [20]	MEF-SHDR [21]	DocDiff [34]	Ours
PSMR	19.56	32.19	34.30	35.72	35.89	37.38	**38.90**
SSIM	69	84	86	91	96	97.57	**98.02**

4.4   Ablation Studies

We conducted a set of ablation experiments to assess the effectiveness of the proposed network components. The ablation experiments involved modifying specific components of the network to evaluate their impact. These modifications consisted of: first, replacing the attention block of the middle block with a 4-layer inflated convolutional layer, as described in the "Baseline" (see Sect. 3.4). Second, training the DPM [27] solely to remove highlights ("DPM only"). Third, training the CP without the RP ("w/o RP"). And finally, training the CP and the RP without utilizing the discriminator ("w/o Discriminator").

The results of the ablation experiments, including the PSNR and SSIM values, are presented in Table 3, while the visual outcomes are depicted in Fig. 5. Notably, it is important to highlight that the network was trained on the same dataset to ensure a fair comparison. Our observations from the ablation experiments are as follows: firstly, when training with only CP and RP, the details of the restored images were found to be more accurate ("w/o Discriminator"). Moreover, a comparison of the results revealed that when the background was filled with a single color, the DPM performed poorly in removing highlights in contrast to the other components. Additionally, although the discriminator marginally improved evaluation indicators, it led to a final result that was visually closer to the ground truth. These findings collectively support the effectiveness of our proposed framework, as the results consistently demonstrated a closer resemblance to the visual effects of the ground truth and yielded higher evaluation indicators.

Table 3. Quantitative comparison of highlight removal methods on SHIQ, SD1, SD2 and RD datasets. The best results are shown in bold.

Datasets	SD1		SD2		RD		SHIQ	
Method/Metric	PSNR	SSIM	PSNR	SSIM	PSNR	SSIM	PSNR	SSIM
Baseline	32.92	96.01	31.42	96.56	27.60	90.08	35.46	97.32
DPM only	30.23	95.72	29.35	95.80	27.14	89.70	36.38	97.43
w/o RP	32.07	96.35	36.45	97.26	28.03	90.39	38.07	97.93
w/o Discriminator	34.43	96.55	38.37	97.42	30.39	91.37	38.90	98.00
Ours	**34.81**	**96.96**	**38.58**	**97.42**	**31.09**	**91.73**	**38.90**	**98.02**

4.5   Limitations

We successfully applied our method to remove specular highlights from a variety of images. However, the use of Transformer and DDPM in our network, along with the 1000 timesteps of the diffusion model, results in a prolonged training time for our model. As a result of the network's complexity, the inference time of the model may become a critical limitation in application scenarios that require rapid processing.

Our models exhibit varying levels of parameter sizes: the CP consists of 6.23M parameters, the RP comprises 4.17M parameters, and the GD includes 2.76M parameters. Notably, based on the Traisformer architeture, the CP with $O(N^2)$ time complexity necessitates significant computational resources during training. Specifically, when scaling up the input image size from 200 to 512, the total floating-point calculations surge from 2.83K GFLOPs to 121K GFLOPs. This substantial increase in computational demand renders training on lower-performance computing cards challenging.

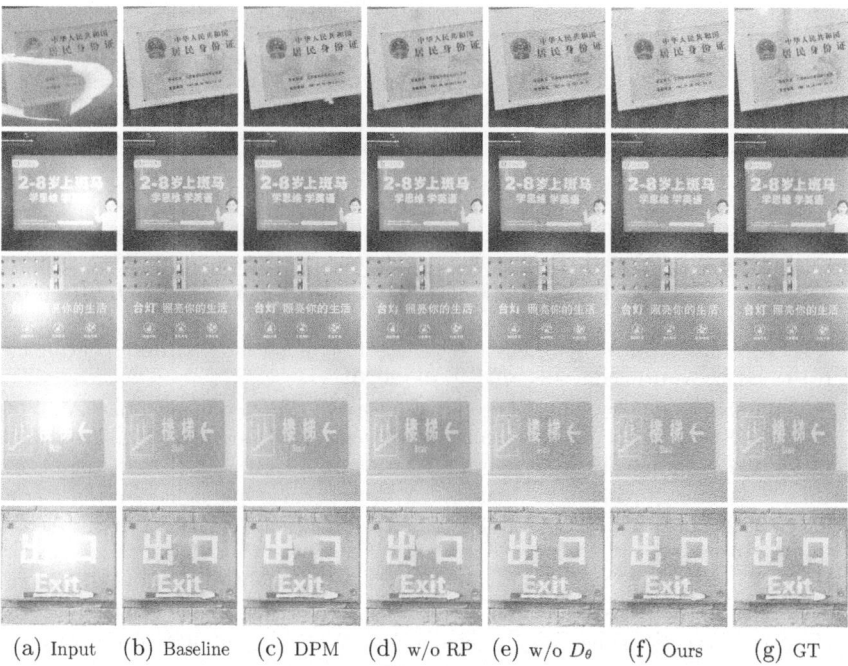

(a) Input   (b) Baseline   (c) DPM   (d) w/o RP   (e) w/o D_θ (f) Ours   (g) GT

Fig. 5. Visual examples from ablation resutls of our proposed components.

5   Discussion

This study presents a novel coarse-to-fine framework for addressing the complex task of removing specular highlights in a single document image without prior highlight detection. The coarse predictor incorporates a transformer mechanism to capture global highlight and background information for highlight removal. Subsequently, the refinement predictor leverages a conditional and full convolutional denoising diffusion probabilistic algorithm to capture local details between the coarse prediction and Ground Truth. The efficacy of the proposed neural network is demonstrated through numerous experiments conducted on public benchmark datasets and challenging real-world images.

Acknowledgements. This work has been supported by the National Key Research and Development Program Grant 2020AAA0109700, the National Natural Science Foundation of China (NSFC) Grant U23B2029, and the National Natural Science Foundation of China (NSFC) Grant No. 62276258.

References

1. Son, M., Lee, Y., Chang, H.S.: Toward specular removal from natural images based on statistical reflection models. IEEE Trans. Image Process. **29**, 4204–4218 (2020)
2. Du, B., Zhang, M., Zhang, L., Hu, R., Tao, D.: PLTD: patch-based low-rank tensor decomposition for hyperspectral images. IEEE Trans. Multimedia **19**(1), 67–79 (2016)
3. Arbelaez, P., Maire, M., Fowlkes, C., Malik, J.: Contour detection and hierarchical image segmentation. IEEE Trans. Pattern Anal. Mach. Intell. **33**(5), 898–916 (2010)
4. Hou, S., Wang, C., Quan, W., Jiang, J., Yan, D.M.: Text-aware single image specular highlight removal. In: Pattern Recognition and Computer Vision: 4th Chinese Conference, PRCV 2021, Beijing, China, October 29–November 1, 2021, Proceedings, Part IV 4, pp. 115–127. Springer (2021). https://doi.org/10.1007/978-3-030-88013-2_10
5. Xue, M., et al.: Arbitrarily-oriented text detection in low light natural scene images. IEEE Trans. Multimedia **23**, 2706–2720 (2020)
6. Muhammad, S., Dailey, M.N., Farooq, M., Majeed, M.F., Ekpanyapong, M.: Spec-Net and Spec-CGAN: deep learning models for specularity removal from faces. Image Vis. Comput. **93**, 103823 (2020)
7. Guo, S., Wang, X., Zhou, J., Lian, Z.: A fast specular highlight removal method for smooth liquor bottle surface combined with u2-net and lama model. Sensors **22**(24), 9834 (2022)
8. Ha, J.W., Lee, K.K., Yoo, J.S., Kim, J.O.: Deep highlight removal using temporal dark prior in high-speed domain. IEEE Access **11**, 20136–20149 (2023)
9. Hao, J., Zhao, Y., Peng, Q.: A specular highlight removal algorithm for quality inspection of fresh fruits. Remote Sens. **14**(13), 3215 (2022)
10. Wu, Z., Zhuang, C., Shi, J., Xiao, J., Guo, J.: Deep specular highlight removal for single real-world image. In: SIGGRAPH Asia 2020 Posters, pp. 1–2 (2020)
11. Anwer, A., Ainouz, S., Saad, M.N.M., Ali, S.S.A., Meriaudeau, F.: SpecSeg network for specular highlight detection and segmentation in real-world images. Sensors **22**(17), 6552 (2022)
12. Wu, Z., et al.: Single-image specular highlight removal via real-world dataset construction. IEEE Trans. Multimedia **24**, 3782–3793 (2021)

13. Wu, Z., Guo, J., Zhuang, C., Xiao, J., Yan, D.M., Zhang, X.: Joint specular high-light detection and removal in single images via unet-transformer. Comput. Vis. Media **9**(1), 141–154 (2023)

14. Vaswani, A., et al.: Attention is all you need. Adv. Neural Inf. Proc. Syst. **30** (2017)

15. Shafer, S.A.: Using color to separate reflection components. Color Res. Appl. **10**(4), 210–218 (1985)

16. Lee, S.W., Bajcsy, R.: Detection of specularity using color and multiple views. In: Computer Vision-ECCV'92: Second European Conference on Computer Vision Santa Margherita Ligure, Italy, May 19–22, 1992 Proceedings 2. pp. 99–114. Springer (1992). https://doi.org/10.1007/3-540-55426-2_13

17. Lin, S., Li, Y., Kang, S.B., Tong, X., Shum, H.Y.: Diffuse-specular separation and depth recovery from image sequences. In: Computer Vision-ECCV 2002: 7th European Conference on Computer Vision Copenhagen, Denmark, May 28–31, 2002 Proceedings, Part III 7. pp. 210–224. Springer (2002. https://doi.org/10.1007/3-540-47977-5_14

18. Yu, B., Chen, W., Zhong, Q., Zhang, H.: Specular highlight detection based on color distribution for endoscopic images. Frontiers Phys. **8**, 616930 (2021)

19. Fu, G., Zhang, Q., Lin, Q., Zhu, L., Xiao, C.: Learning to detect specular highlights from real-world images. In: Proceedings of the 28th ACM International Conference on Multimedia, pp. 1873–1881 (2020)

20. Huang, Z., Hu, K., Wang, X.: M2-Net: multi-stages specular highlight detection and removal in multi-scenes. arXiv preprint arXiv:2207.09965 (2022)

21. Huang, G., Yao, J., Huang, P., Han, L.: A mutual enhancement framework for spec-ular highlight detection and removal. In: Chinese Conference on Pattern Recog-nition and Computer Vision (PRCV), pp. 457–468. Springer (2023). https://doi.org/10.1007/978-981-99-8552-4_36

22. Chen, H., Li, L., Yu, N.: Mask-guided joint single image specular highlight detec-tion and removal. In: Chinese Conference on Pattern Recognition and Computer Vision (PRCV), pp. 457–468. Springer (2023). https://doi.org/10.1007/978-981-99-8546-3_37

23. Jadhav, P., Sawal, M., Zagade, A., Kamble, P., Deshpande, P.: Pix2pix generative adversarial network with ResNet for document image denoising. In: 2022 4th Inter-national Conference on Inventive Research in Computing Applications (ICIRCA), pp. 1489–1494. IEEE (2022)

24. Chen, J., Chen, J., Chao, H., Yang, M.: Image blind denoising with generative adversarial network based noise modeling. In: Proceedings of the IEEE Conference on Computer Vision and Pattern Recognition, pp. 3155–3164 (2018)

25. Souibgui, M.A., Kessentini, Y.: DE-GAN: a conditional generative adversarial net-work for document enhancement. IEEE Trans. Pattern Anal. Mach. Intell. **44**(3), 1180–1191 (2020)

26. Gangeh, M.J., Plata, M., Nezhad, H.R.M., Duffy, N.P.: End-to-end unsupervised document image blind denoising. In: Proceedings of the IEEE/CVF International Conference on Computer Vision, pp. 7888–7897 (2021)

27. Ho, J., Jain, A., Abbeel, P.: Denoising diffusion probabilistic models. Adv. Neural. Inf. Process. Syst. **33**, 6840–6851 (2020)

28. Li, H., et al.: SRDiff: single image super-resolution with diffusion probabilistic models. Neurocomputing **479**, 47–59 (2022)

29. Niu, A., et al.: CDPMSR: conditional diffusion probabilistic models for single image super-resolution. arXiv preprint arXiv:2302.12831 (2023)

30. Ren, M.S., Zhang, Y.M., Wang, Q.F., Yin, F., Liu, C.L.: Diff-writer: a diffusion model-based stylized online handwritten Chinese character generator. In: International Conference on Neural Information Processing, pp. 86–100. Springer (2023)
31. Saharia, C., Ho, J., Chan, W., Salimans, T., Fleet, D.J., Norouzi, M.: Image super-resolution via iterative refinement. IEEE Trans. Pattern Anal. Mach. Intell. **45**(4), 4713–4726 (2022)
32. Saharia, C., et al.: Palette: image-to-image diffusion models. In: ACM SIGGRAPH 2022 Conference Proceedings, pp. 1–10 (2022)
33. Whang, J., Delbracio, M., Talebi, H., Saharia, C., Dimakis, A.G., Milanfar, P.: Deblurring via stochastic refinement. In: Proceedings of the IEEE/CVF Conference on Computer Vision and Pattern Recognition, pp. 16293–16303 (2022)
34. Yang, Z., Liu, B., Xxiong, Y., Yi, L., Wu, G., Tang, X., Liu, Z., Zhou, J., Zhang, X.: DocDiff: document enhancement via residual diffusion models. In: Proceedings of the 31st ACM International Conference on Multimedia, pp. 2795–2806 (2023)
35. Zamir, S.W., Arora, A., Khan, S., Hayat, M., Khan, F.S., Yang, M.H.: Restormer: efficient transformer for high-resolution image restoration. In: Proceedings of the IEEE/CVF Conference on Computer Vision and Pattern Recognition, pp. 5728–5739 (2022)
36. Fu, G., Zhang, Q., Zhu, L., Li, P., Xiao, C.: A multi-task network for joint specular highlight detection and removal. In: Proceedings of the IEEE/CVF Conference on Computer Vision and Pattern Recognition, pp. 7752–7761 (2021)
37. Lin, J., El Amine Seddik, M., Tamaazousti, M., Tamaazousti, Y., Bartoli, A.: Deep multi-class adversarial specularity removal. In: Image Analysis: 21st Scandinavian Conference, SCIA 2019, Norrköping, Sweden, June 11–13, 2019, Proceedings 21, pp. 3–15. Springer (2019). https://doi.org/10.1007/978-3-030-20205-7_1
38. Yu, J., Lin, Z., Yang, J., Shen, X., Lu, X., Huang, T.S.: Free-form image inpainting with gated convolution. In: Proceedings of the IEEE/CVF International Conference on Computer Vision, pp. 4471–4480 (2019)

SlideCraft: Synthetic Slides Generation for Robust Slide Analysis

Travis Seng[1,2](✉) ⓘ, Axel Carlier[1,2] ⓘ, Thomas Forgione[3] ⓘ,
Vincent Charvillat[1] ⓘ, and Wei Tsang Ooi[4] ⓘ

[1] IRIT Toulouse, Toulouse, France
travis.seng@irit.fr
[2] IPAL, IRL2955, Singapore, Singapore
[3] Polymny Studio, Toulouse, France
[4] National University of Singapore, Singapore, Singapore

Abstract. The increasing amount of slide presentations in various sectors has amplified the need for effective slide layout and semantic analysis. However, we found that current slide datasets contain inconsistencies, mislabels, and incomplete annotations. Using them as a basis for developing deep learning-based slide analysis models could lead to models that are not robust and suboptimal. Addressing these challenges, we introduce SlideCraft, a tool for creating synthetic slide datasets that imitate real-world presentations. This tool overcomes the drawbacks of existing datasets by allowing users to create balanced, diverse, and accurately annotated slide data. We demonstrate SlideCraft's efficacy in enhancing slide layout analysis algorithms, focusing on its capability to improve dataset quality and object detection performance.

Keywords: Slide datasets · Slide analysis · Open source tool · Synthetic dataset

1 Introduction

Presentations in sectors such as education and research are increasingly being recorded and shared online, particularly since the pandemic of the early 2020s. This trend fueled a renewed interest in innovative applications for processing and interacting with recorded digital presentations, such as summarization [30,31], indexing [2,16,23,32], and enhanced browsing [2,16,23,32]. Slides are important visual aids for a presentation that often drive the structure and key points of the content of a presentation. Extracting the semantics from the slides, therefore, is an important fundamental task that enables these downstream applications. Much of the slides available online, however, are in image or video format, making computational analysis and understanding of the slides a non-trivial problem.

Recent efforts in slide analysis have pivoted to data-driven, learning-based, methods [15,28]. For such methods to be effective, there needs to be a large corpus of data with high-quality annotations. However, existing slide datasets

E. H. Barney Smith et al. (Eds.): ICDAR 2024, LNCS 14804, pp. 79–96, 2024.
https://doi.org/10.1007/978-3-031-70533-5_6

[8,9,15,28] often either suffer from inconsistencies and inaccuracies, or lack the essential annotations [3] necessary for the development of effective analysis algorithms. Beyond these issues, there are additional concerns that slow down progress in this domain. Firstly, the process of accurately annotating slide datasets is labor-intensive and expensive, requiring significant expertise and time to ensure high-quality annotations. Secondly, this challenge is compounded by the scarcity of comprehensive slide datasets available in the public domain, limiting the diversity and scope of training data for analysis algorithms. Finally, there is often an imbalance in the type of elements present on the slides within these datasets. For example, some classes of visual or textual elements may be overrepresented while others are underrepresented, leading to skewed learning outcomes and limiting the algorithms' ability to generalize across different slide layouts and content types.

To address these challenges, we introduce SlideCraft, a tool developed for generating synthetic slide datasets. SlideCraft first tackles the issue of labor-intensive annotations. Generating content with annotation eliminates the costly and time-consuming nature of manual annotation, ensuring high-quality, consistent annotations across the dataset. Second, it mitigates the scarcity of comprehensive slide datasets by generating a large and diverse array of realistic slides. Finally, it addresses the imbalance in slide elements by allowing for customizable element distribution in generated slides. This ensures a balanced representation of both visual and textual elements. In summary, SlideCraft's design not only imitates real-world presentations but also methodically overcomes the existing limitations of slide datasets, thereby facilitating the development of slide analysis algorithms.

In this paper, we present the development, capabilities, and potential impact of SlideCraft. Our focus is on its contribution to overcoming the current limitations in slide dataset availability and quality, thereby paving the way for future advancements in digital presentation analysis. **We show that by augmenting the existing dataset with slides generated from our tool, we can improve the mAP50 up to 13% on object detection models.** The source code of SlideCraft will be made publicly available for research purposes, fostering further innovation and collaboration in the field.

2   Related Work

Studies in slide dataset creation have primarily focused on collating slides from existing sources, with an emphasis on real-world diversity. However, these datasets often face challenges in terms of size, annotation accuracy, consistency, and imbalance. WiSe [9] and SPaSe [8] datasets, while offering segmentation masks across 25 diverse classes, are limited by their small size, containing only 2000 images – a quantity insufficient for comprehensive analysis. FitVid [15] introduces a more subject-diverse object detection dataset over 12 classes, yet it comprises only 5527 images. In contrast, Slideshare-1M [3], despite being the largest slide collection, lacks annotations, because it was made for content retrieval. Meanwhile, SlideVQA [28] stands out as the most extensive annotated slide dataset available, with 52,480 slides across 9 classes designed for object detection, showcasing a step forward in annotated slide data availability. Despite this, it faces two notable issues: a disparity between text and visual content, and the presence of inconsistencies as shown in Fig. 1.

Document layout analysis has recently garnered increased attention in the machine learning community, a development largely fueled by the introduction of more popular benchmarks and high-quality datasets such as DocBank [18], DocLayNet [24], and PubLayNet [33]. These datasets have been instrumental in advancing the field, offering comprehensive and diverse collections of document layouts that facilitate the training and testing of sophisticated layout analysis algorithms. However, despite these advancements, a significant gap remains in the specific domain of slide layout analysis. Unlike standard documents, slides have distinct characteristics: they are typically more visual, presented in landscape format, and contain less structured text. Furthermore, the classification needs for slide elements differ markedly from those in traditional document layouts. For instance, slides often require specific classes for elements like bullet points, titles, and visual aids, which are not commonly accounted for in general document layout datasets. The existing datasets, while robust for general document analysis, fall short of addressing the unique demands of slide presentations and potential applications.

Specialized datasets focused on document elements like tables [17], charts [21], diagrams [14], plots [22], equations [6] and handwriting [20] have been developed to encourage better analysis and recognition. Given that slide presentations often incorporate such elements, leveraging these specific datasets within SlideCraft presents an excellent opportunity to refine and broaden the analysis capabilities of models trained using SlideCraft-generated slides.

Historically, synthetic datasets have been approached with caution in machine learning applications due to their tendency to diverge from real-world scenarios. This divergence often results in a performance gap when algorithms trained on synthetic data are applied to actual tasks. In contrast to this trend, some researchers have employed advanced techniques like Generative Adversarial Networks (GANs) to create more realistic synthetic datasets [4,5,7]. These methods involve training neural networks to generate data that closely mimic real-world scenarios, a technique that has shown promise in fields such as image and document generation but which cannot be easily used to create reliable annotations. Others have utilized photorealistic open-world computer games to generate large datasets, offering reliable pixel-level semantic annotations for enhanced accuracy in various applications [27].

Furthermore, another approach in the generation of synthetic datasets for document layout analysis has involved creating layouts first by using a generative model and then populating them with content [25]. This method targets article layouts, where the focus is on arranging text and graphical elements in a manner consistent with traditional document formats, thereby creating synthetic datasets that resemble the structure of real-world documents. However, these techniques are predominantly tailored to the needs of standard document

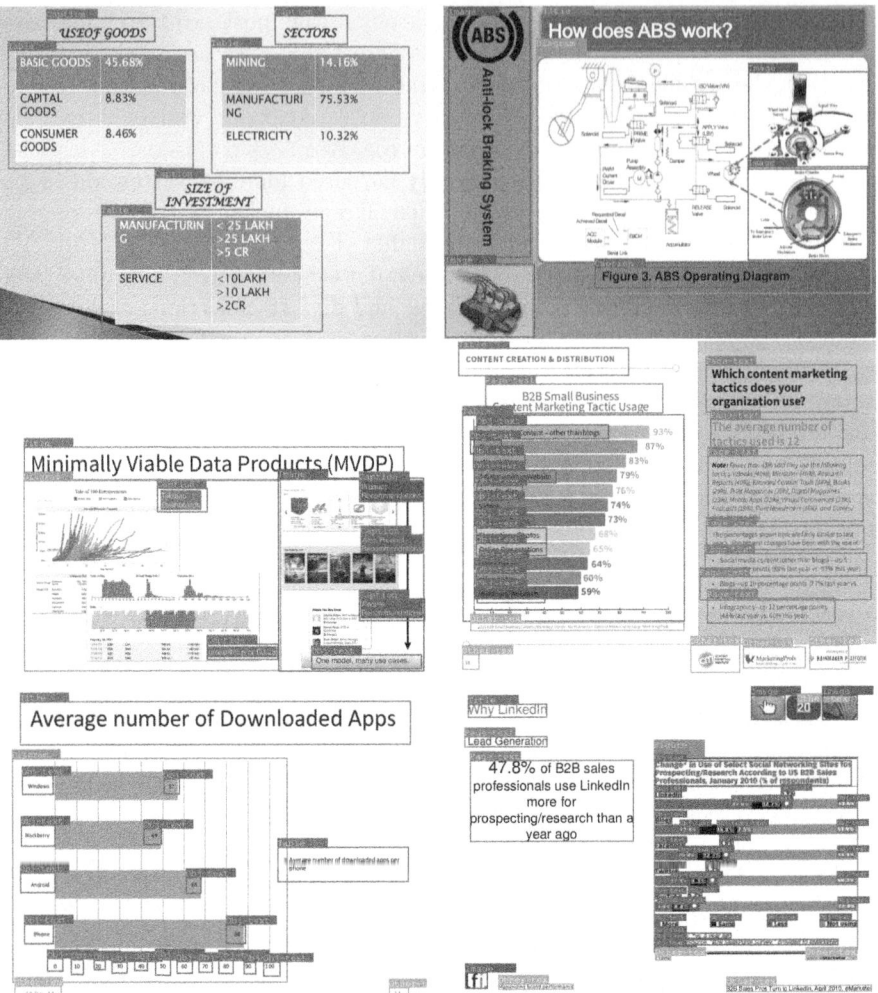

Fig. 1. SlideVQA contains some inconsistencies and errors. The Obj-Text class which represents all the text seen on any graphical elements (Image, Diagram, Table, Figure) is not always annotated. Moreover, there is some confusion between the class Figure and Diagram shown in the last two slides. Even though they are both horizontal charts, one is labeled as Diagram and in the other, as Figure. This affects training performance and evaluation reliability.

layouts, such as articles and reports, and do not directly address the unique requirements of slide presentations, such as having more visual content such as charts, graphs, and images, as well as the necessity to accommodate diverse layouts. The semantic organization within slides, how information is structured and prioritized, differs significantly from traditional documents, necessitating tools,

and datasets specifically designed to understand and generate these visually oriented formats.

We provide a detailed analysis of SlideCraft in the next section.

3   SlideCraft

SlideCraft is a tool designed for generating extensive, labeled datasets for training learning-based slide layout analysis methods. Its primary function is to facilitate the production of a wide array of annotated slide data. This tool operates through a set of flexible rules that ensure the generation of coherent, well-
structured, and readable presentations. These rules can be dynamically adjusted
by users, allowing for the creation of customized layouts and styles to suit specific
requirements.

The tool can be broken into four components, each responsible for generating
a different element: content, layout, style, and annotation. Figure 2 shows the
whole pipeline of SlideCraft.

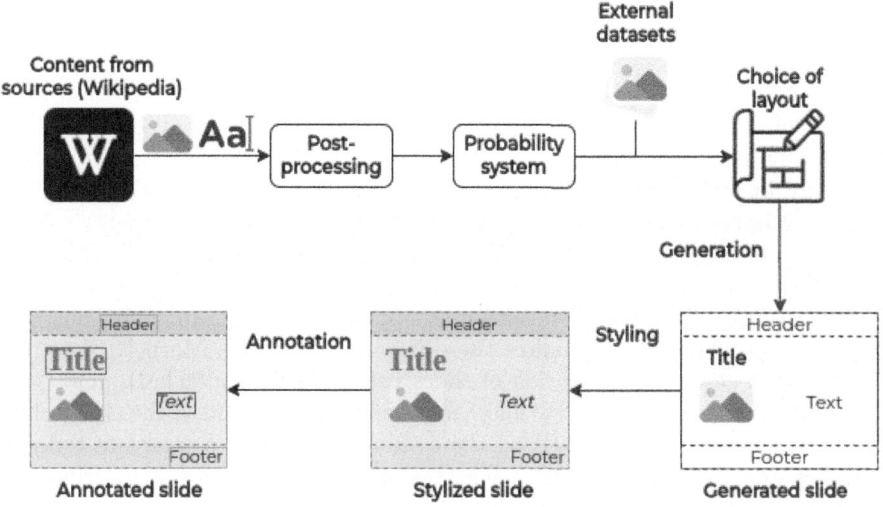

Fig. 2. Pipeline of SlideCraft

3.1   Content

The content component of SlideCraft is integral to generating the textual and
visual elements of the slides, with primary reliance on Wikipedia articles for
source material in our current implementation (the source material could come
from anywhere). Each section and subsection of an article is adeptly converted
into an individual slide, aligning with the original content's structure. To adapt
the typically lengthy and detailed text from Wikipedia articles to a slide-friendly

format, we employed Mistral-7B [11], a Large Language Model, for summarization. This LLM was tasked with distilling the textual elements into bullet points and short sentences, which are more suitable for the concise and direct nature of slide presentations.

However, given that Wikipedia articles are predominantly text-based and slides necessitate a strong visual component, we introduce the possibility of supplementing them with various graphical elements from other public datasets. This addition includes charts (ChartQA [21]), plots (PlotQA [22]), diagrams (AI2D [14]), tables (TableBank [17]), equations (im2latex-100k [6]), handwriting (IAM-OnDB [20]), logos (Logodet-3k [29]) and more. While incorporating these diverse visual elements could introduce a degree of incoherence with the original text, it is a deliberate choice aimed at enhancing the overall visual diversity of the slides. This step is beneficial for the primary objective of the tool - to improve the performance of slide layout analysis algorithms. We show in our experiments that the use of this additional material helps to improve object detection performance in several classes.

By expanding the range of visual content in the slides, we provide a richer and more challenging dataset for training these algorithms, ultimately contributing to their enhanced accuracy and effectiveness in real-world applications.

3.2   Layout

The layout component is crucial in determining the arrangement and presentation of content within each slide. This component is designed to emulate traditional layouts commonly found in popular presentation tools like Google Slides and PowerPoint. It assesses the quantity and type of content elements - be it text, images, or graphical data - and selects the most suitable layout for each slide. The decision-making process is guided by principles of design and readability, ensuring that the final output is not only visually appealing but also easy to comprehend. To actualize these layouts, we utilize Marp [1], a versatile tool that enables the creation of slides using Markdown, HTML, and CSS. Marp's flexibility and simplicity allow for efficient translation of the chosen layouts into polished slides. Based on the content, we choose a template tailored to the content, and we generate a markdown file that contains instructions for Marp to translate into a slide in HTML format. Templates are written in HTML and CSS and contain code to create layouts with columns, headers, footers, and different positions for elements. This approach ensures that SlideCraft can produce slides that are aesthetically consistent with standard presentation formats, thereby making the generated datasets ideal for enhancing slide layout analysis algorithms. Figure 3 shows some examples of layouts we can generate.

A probability system also plays a part in determining the composition of each slide. This system allows for control over the occurrence probabilities of various elements within the slides. Users can adjust the generation process by specifying the weights for different classes and determining their inclusion in a slide in a greedy way. The system plays a role both in the content phase, where it determines the likelihood of incorporating external content, and in the layout phase, deciding whether to include specific classes like titles, headers, and footers. This feature is particularly beneficial when aiming to balance an existing dataset or when focusing on enhancing specific classes that may be underrepresented or inadequately portrayed in the dataset. By adjusting these probabilities, we can tailor the layout of the slides to address specific needs or deficiencies in the dataset. For example, if the initial analysis indicates that slides with diagrams are less frequent, the system can be configured to generate more slides with diagrams. This level of customization in the layout component not only adds versatility to SlideCraft, but also ensures that the resulting dataset is well-rounded and accurately reflective of diverse presentation scenarios. Such targeted adjustments are instrumental in fine-tuning the dataset for more effective training and evaluation of slide layout analysis models.

3.3   Style

The style component is responsible for the aesthetic diversity of the slides. This component operates by randomly selecting various stylistic parameters to ensure that each slide is distinct from the others, but can also be modified to target certain styles of slides. These parameters include a wide range of elements such as colors for different text sections (like paragraphs, titles, headers, and footers), font types, text sizes, the degree of text boldness, and more. Additionally, it also encompasses choices for background styles and colors. This is done through the modification of the CSS in the markdown files read by Marp.

Randomization in selecting these stylistic features is key to producing a dataset with high variability. This diversity is not merely cosmetic; it is needed for the robustness of the slide layout analysis algorithms. By exposing these algorithms to diverse styles, we challenge and enhance their ability to accurately analyze and interpret slides across a broad spectrum of designs. In doing so, the style component contributes to the generation of a comprehensive and varied dataset. Figure 4 shows some outputs of SlideCraft with different layouts and styles, and Fig. 5 shows the same but for the same source material.

While SlideCraft represents the first attempt at creating a synthetic slide dataset for slide layout analysis, it is important to acknowledge its limitations in replicating the full spectrum of slide styles. The tool is not designed to mimic every possible type of slide, as the variety and complexity of presentation styles are vast and often context-specific. However, our goal with SlideCraft is to cover the most common and widely used slide styles. By focusing on these prevalent formats, we aim to provide a comprehensive and representative dataset that reflects the majority of real-world scenarios. This strategic approach allows us to optimize the tool's effectiveness in training and improving slide layout analysis algorithms, while realistically addressing the practical constraints of such a generative system.

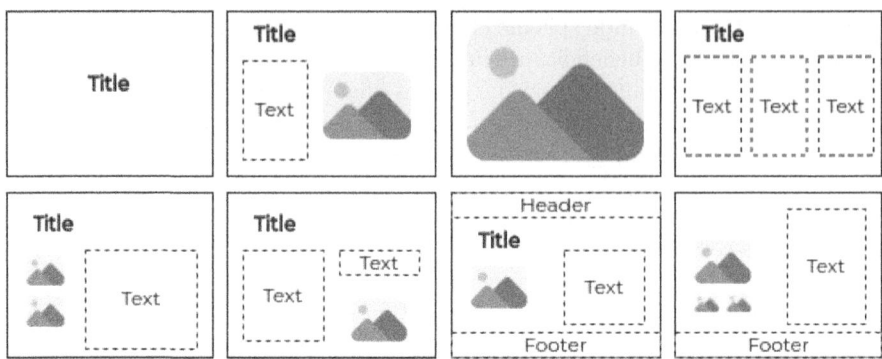

Fig. 3. Examples of different layouts: different number of columns, display of header and footer, display of the title. The number of elements taken from the content can be adjusted to choose between more visual or more textual output.

3.4   Annotations

SlideCraft generates annotations through the processing of the HTML generated by Marp for each slide. We employ JavaScript to access each text element within the Document Object Model (DOM) and obtain precise bounding boxes for these elements. To enhance accuracy, we modify the HTML by adding span tags around paragraphs and words, enabling us to capture more detailed bounding boxes around the text at a paragraph and word level. This approach ensures precise demarcation of text elements, critical for effective layout analysis.

In parallel, segmentation masks are created by isolating individual elements on the slides with CSS and computing the image difference. This method is particularly effective for identifying and delineating graphical elements within the slides.

For the classification of text elements, we use two information. Primarily, we rely on HTML tags to determine their classes (e.g., 'p' for general text, 'li' for bullet points, 'h1' for titles, etc.). Additionally, we incorporate class information derived during the layout computation phase, further refining the accuracy of our text annotations.

Graphical elements, on the other hand, are annotated based on the classifications available from the labeled images we use from existing datasets. By incorporating these pre-labeled images into our slides, we ensure that each graphical element is accurately categorized, significantly enhancing the reliability of our annotations. We also apply OCR to graphical elements to extract bounding boxes for any text they contain, enhancing text recognition across diverse slide components.

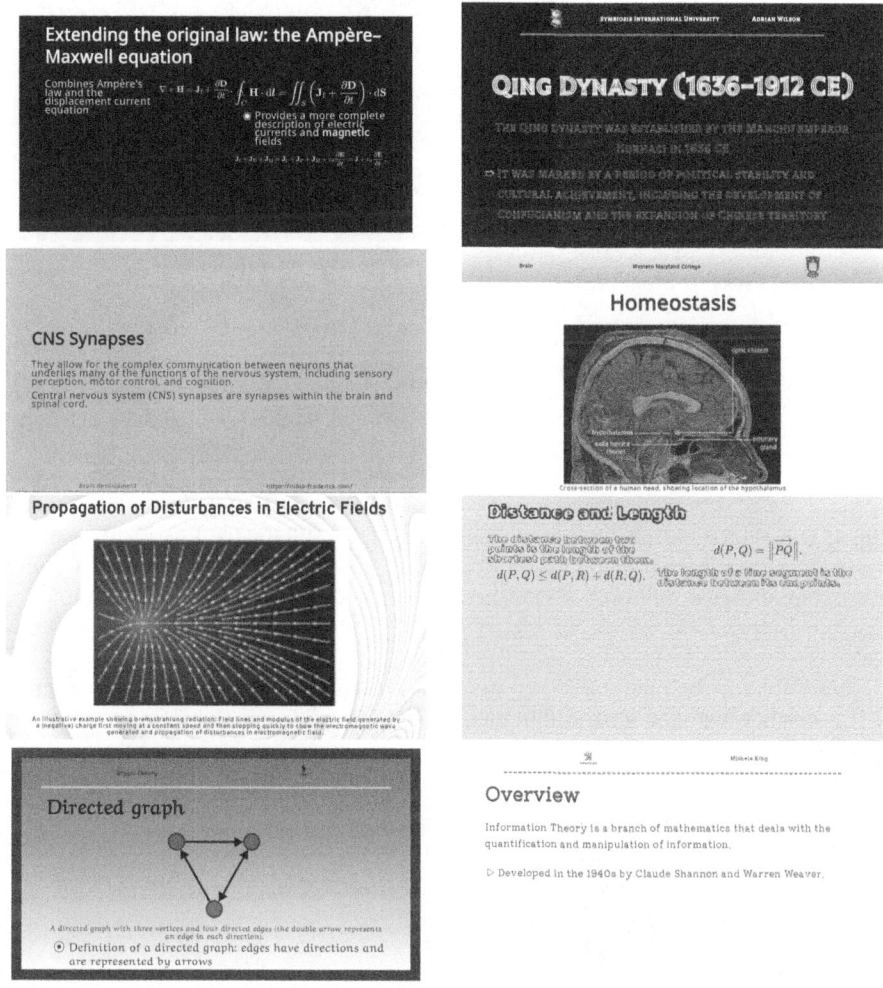

Fig. 4. Examples of SlideCraft's generated slides. Different styles and layouts are shown. Colors, backgrounds, font size, and font style are chosen randomly to increase the diversity of the generation.

This comprehensive approach to annotation, combining DOM manipulation, image processing, and existing dataset labels, ensures that SlideCraft not only creates visually varied slides but also provides richly annotated data.

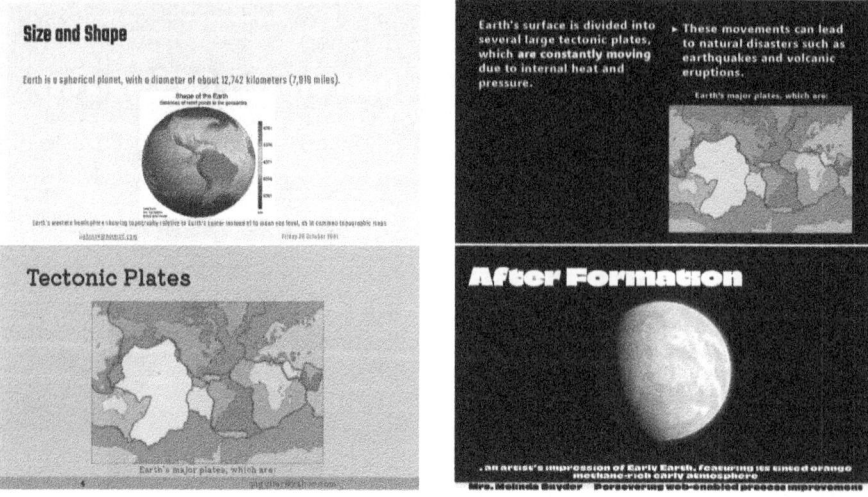

Fig. 5. Different styles of slides from the same source (Wikipedia: Earth). We can see diverse fonts and backgrounds, various font sizes and colors, the presence of a footer or not, and different numbers of elements.

The tool can be customized to generate custom classes as needed as shown in Fig. 6 where we generate different classes depending on the dataset we want to extend.

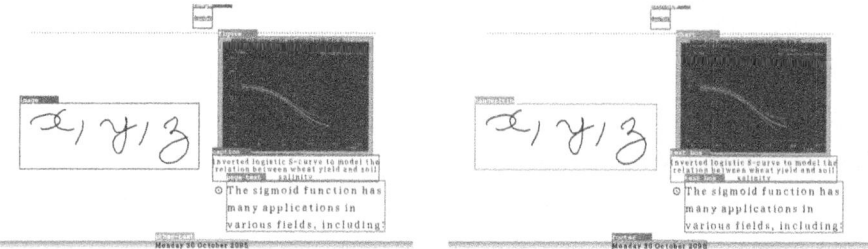

Fig. 6. Example of generated slides for SlideVQA annotations (Left) and Fitvid annotations (Right). Handwriting, header, footer, and equation, called Figure in FitVid, do not exist in SlideVQA.

4   Experiments

To demonstrate SlideCraft's effectiveness, we evaluate object detection performance using two models and use our generated synthetic dataset to extend two existing datasets.

4.1   Datasets

We selected the two largest annotated datasets available. The FitVid dataset, featuring 5,527 images, covers a wide array of 12 classes, including Title, Text Box, Picture, Chart, Figure, Diagram, Table, Schematic Diagram, Header, Footer, Handwriting, and Instructor. Given that there is no test dataset, we create a split by carefully separating slides from the same presentation. We obtain a train set with 4,421 slides and a test set with 1,106 slides. Similarly, the SlideVQA dataset brings together a vast collection of 52,480 slides, categorized into 9 distinct classes: Title, Page-Text, Obj-Text (text on graphical elements), Caption, Other-Text, Diagram, Table, Image, and Figure. The authors provide a train of 37,023 images, a validation set of 5,839 images, and a test set of 7,727 images.

4.2   Models

We selected FasterRCNN [26] with a ResNet50FPN [10, 19] backbone and YoloV8 [12] as our models, noting that FasterRCNN tends to be more sensitive to class imbalance, while YoloV8 appears to show less sensitivity, probably due to the use of focal loss. We trained FasterRCNN with a batch size of 16 using the SGD optimizer, starting with a learning rate of 0.005 and a momentum of 0.95, coupled with a Cosine Annealing Learning Rate scheduler. Data augmentation techniques included color jittering, random gamma adjustments, and brightness and contrast modulation, alongside blurring. For YoloV8, we used a batch size of 64 and adhered to standard settings but changed the input resolution to 1280 and omitted mosaic and flip augmentations. We trained both models for 50 epochs with an early stopping (patience of 5 epochs).

4.3   Experiments

Dataset Completion. Utilizing SlideCraft, we generate 25,000 synthetic slides for each dataset, enriching them with classes underrepresented in existing data, such as Diagram, Table, Plot, Handwriting, and Equation, by incorporating additional datasets for balance. We specifically target the underrepresented classes with the SlideCraft probability system while trying to fit the original dataset by adjusting the layout and style components. For example, in SlideVQA, to imitate the Obj-Text class, we run publicly available OCR Tesseract [13] on top of all the graphical elements. To imitate captions, we include a randomly generated smaller text close to the top or the bottom of a graphical element. We now compare the performance of each model trained on each dataset with and without added synthetic slides, evaluating their respective test datasets.

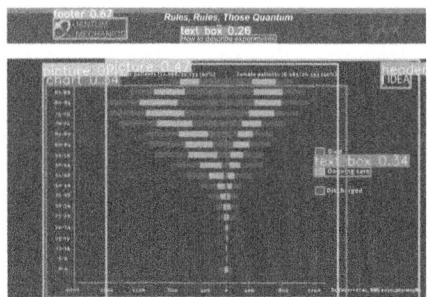 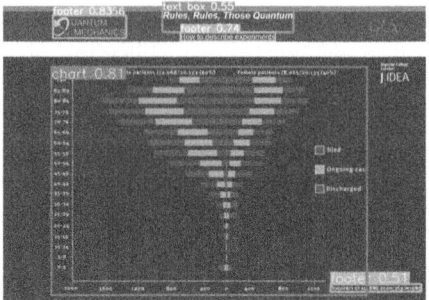

Fig. 7. Predictions on FitVid test dataset with our models. **Left**: Inference with YOLOv8 trained without added synthetic slides. **Right**: Inference with YOLOv8 trained with added synthetic slides. The model trained with synthetic data from Slide-Craft demonstrates higher precision in box predictions, correlating with the marked improvement in mAP scores.

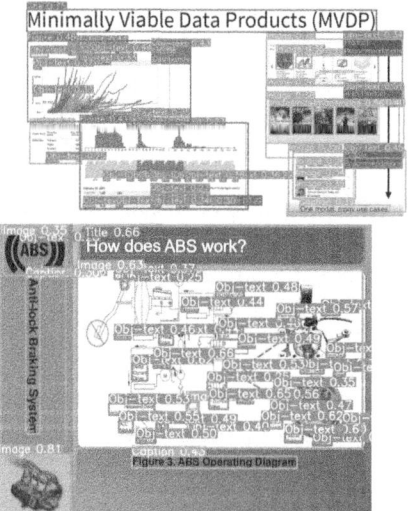 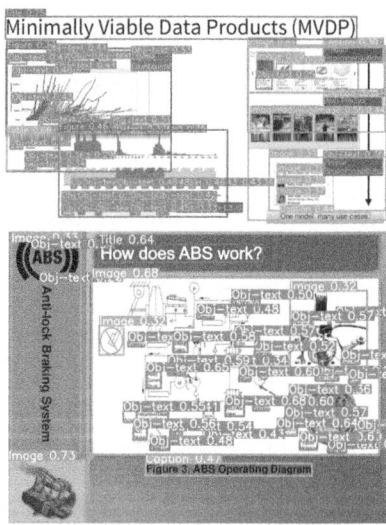

Fig. 8. Predictions on SlideVQA test dataset with our models. **Left**: Inference with YOLOv8 trained without added synthetic slides. **Right**: Inference with YOLOv8 trained with added synthetic slides. We can see that the model trained with synthetic data contradicts the ground-truth, despite looking accurate. For example, here, we predict more Obj-Text with added synthetic slides, but the ground truth in Fig. 1 does not contain these elements.

Table 1. mAP50 of FasterRCNN and YoloV8 on SlideVQA test set and FitVid test set, training with and without the slides generated by SlideCraft.

Dataset	FasterRCNN			Yolov8		
	Original mAP50	Mix mAP50	Improvement	Original mAP50	Mix mAP50	Improvement
SlideVQA	0.641	**0.645**	0.59%	0.682	**0.685**	0.44%
FitVid	0.485	**0.530**	9.22%	0.513	**0.581**	13.26%

Our first results, displayed in Table 1, reveal that mixing SlideCraft-generated data into training can lead to superior performance. With synthetic slides, Slide-VQA observes a slight improvement of the mAP50 by 0.59% and 0.44% with FasterRCNN and YoloV8 respectively. Conversely, FitVid benefits significantly from the synthetic slides, showing an improvement of the mAP50 by 9.22% and 13.26% for FasterRCNN and YoloV8, respectively. This gain is illustrated in Fig. 7. These outcomes underscore the importance of large datasets for achieving optimal performance, as evidenced by the more pronounced improvements in FitVid compared to SlideVQA when training with an extended and larger dataset.

The marginal gains on SlideVQA suggest its extensive size may inherently limit performance improvements. However, we think that the dataset inconsistencies likely skew results. For example, we get a lower mAP50 (-0.3%) for the Obj-Text class on our model trained with SlideCraft's slides than the one trained solely with the original dataset. Despite this, our predictions, as highlighted in the images from Fig. 8, show that we predict Obj-Text elements that should be correct based on how this class was defined. We also predict two Figures and a Table, which seem correct. However, as showcased in Fig. 1, these Obj-Text elements do not exist in the ground truth, and the whole left part of the slide is labeled as Diagram. Such cases happen in other images, as highlighted by Fig. 1. **Our first experiment shows that extending a dataset with SlideCraft-generated data can improve performance, especially on smaller datasets.**

Ground Truth Data Ratio. Furthermore, we conduct an additional experiment where we vary the ratio of real to synthetic slides from SlideVQA to see the impact of SlideCraft on different dataset sizes.

The findings from our second experiment, illustrated in Fig. 9, reveal that smaller datasets experience more significant enhancements when trained with synthetic slides. Our performance only decreases by 2% when we train with 50% of the SlideVQA dataset augmented with synthetic slides, compared to training with the entire SlideVQA dataset.

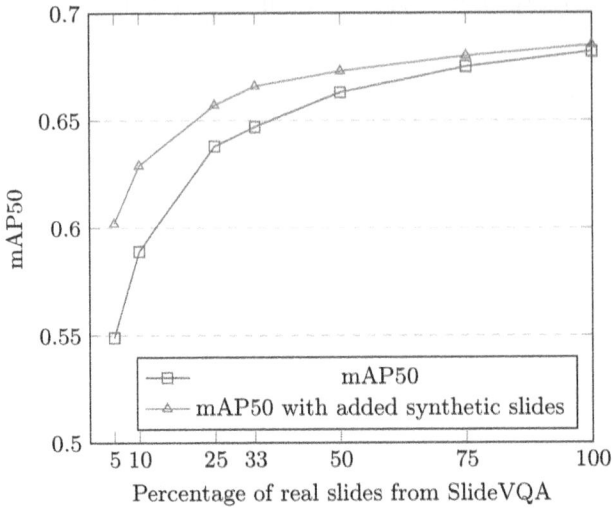

Fig. 9. Comparison of mAP50 obtained after training YoloV8 with and without added 25,000 SlideCraft's generated slides. Testing on SlideVQA test dataset. The results show a direct correlation between dataset size and the impact from adding synthetic slides: smaller datasets see greater benefits from the inclusion of synthetic slides.

This highlights SlideCraft's potential to significantly reduce the necessity for manually labeled slides in training object detection models, while potentially improving performance. This efficiency suggests a strategic approach in dataset annotation, emphasizing quality and consistency over sheer volume.

Balancing Dataset. We also study whether balancing the datasets affects the performance of FasterRCNN. To achieve this, we created additional synthetic datasets comprising 25,000 slides each, deliberately designed to match the imbalance of the original datasets. The new sets were then combined with the original SlideVQA and FitVid datasets to observe the effects of skewed class distribution. We then train a FasterRCNN on the imbalanced datasets.

Table 2. Comparison of mAP50 obtained after training FasterRCNN on original dataset, balanced dataset and imbalanced dataset. The balanced and imbalanced dataset were created after adding SlideCraft's generated slides with different distribution. The distribution was chosen to balance or not the original dataset.

Dataset	Original	Imbalanced	Balanced
SlideVQA	0.641	0.637	**0,645**
FitVid	0.511	0.522	**0.530**

The results shown in Table 2 reveal that training on balanced datasets give a slightly better performance on FasterRCNN. For SlideVQA, the model trained on the imbalanced dataset has a mAP50 of 0.637 which is slightly lower than training only on the original SlideVQA dataset. The FasterRCNN trained on the more balanced dataset, however, displays a mAP50 of 0.645 which is slightly better. For FitVid, FasterRCNN gets better performance with added synthetic slides but the balanced dataset gets the best performance. This suggests that balancing the dataset has a positive impact on performance, while an imbalanced dataset can lead to worse performance.

This result underscores the value of SlideCraft's probability system in enabling dataset balancing.

5   Discussion and Future Work

We demonstrated that integrating SlideCraft's synthetic data generation enhances existing datasets while minimizing the need for extensive manual labeling. Prioritizing high-quality annotations and supplementing them with SlideCraft-generated data emerges as a strategic approach to boost model performance. However, challenges persist, notably the inconsistencies within and between datasets, particularly in how ambiguous classes such as diagrams are annotated, and the lack of standardized classes for dataset creation. SlideCraft offers a solution by accommodating these discrepancies, enabling researchers to tailor datasets to specific needs and applications, marking a step toward more standardized and effective dataset creation for object detection models. This adaptability paves the way for more focused and customized research, enhancing the overall quality and applicability of object detection within diverse contexts. In order to further enhance SlideCraft's capabilities, we see potential in refining the generation of specific classes, such as captions, headers, and footers and expanding the diversity of layouts and styles to better mirror real-world presentations. Although not extensively studied yet, SlideCraft also possesses the ability to generate other types of annotations such as pixel-level segmentation masks and slide to Markdown code, broadening its utility for visual analysis. Another potential use of SlideCraft is to enhance presentations' clarity and adaptability, aiming to improve accessibility for diverse audiences.

References

1. marp-team/marp (2024). https://github.com/marp-team/marp. Original-date: 2018-03-25T12:47:38Z
2. Adcock, J., Cooper, M., Denoue, L., Pirsiavash, H., Rowe, L.A.: TalkMiner: a search engine for online lecture video. In: Proceedings of the International Conference on Multimedia - MM 2010, Firenze, Italy, p. 1507. ACM Press (2010). https://doi.org/10.1145/1873951.1874263. http://dl.acm.org/citation.cfm?doid=1873951.1874263
3. Araujo, A., Chaves, J., Lakshman, H., Angst, R., Girod, B.: Large-Scale Query-by-Image Video Retrieval Using Bloom Filters. arXiv arXiv:1604.07939 (2015)

4. Blanc-Beyne, T., Carlier, A., Mouysset, S., Charvillat, V.: Unsupervised human pose estimation on depth images. In: Dong, Y., Mladenić, D., Saunders, C. (eds.) ECML PKDD 2020. LNCS (LNAI), vol. 12460, pp. 358–373. Springer, Cham (2021). https://doi.org/10.1007/978-3-030-67667-4_22
5. Capobianco, S., Marinai, S.: DocEmul: A Toolkit to Generate Structured Historical Documents, pp. 1186–1191 (2017). https://doi.org/10.1109/ICDAR.2017.196
6. Deng, Y., Kanervisto, A., Ling, J., Rush, A.M.: Image-to-markup generation with coarse-to-fine attention (2017)
7. Ferreira, A., Nowroozi, E., Barni, M.: VIPPrint: validating synthetic image detection and source linking methods on a large scale dataset of printed documents. J. Imaging **7**(3), 50 (2021). https://doi.org/10.3390/jimaging7030050. https://www.mdpi.com/2313-433X/7/3/50
8. Haurilet, M., Al-Halah, Z., Stiefelhagen, R.: SPaSe - multi-label page segmentation for presentation slides. In: 2019 IEEE Winter Conference on Applications of Computer Vision (WACV), Waikoloa Village, HI, USA, pp. 726–734. IEEE (2019). https://doi.org/10.1109/WACV.2019.00082. https://ieeexplore.ieee.org/document/8659181/
9. Haurilet, M., Roitberg, A., Martinez, M., Stiefelhagen, R.: WiSe - slide segmentation in the wild. In: 2019 International Conference on Document Analysis and Recognition (ICDAR), Sydney, Australia, pp. 343–348. IEEE (2019). https://doi.org/10.1109/ICDAR.2019.00062. https://ieeexplore.ieee.org/document/8978089/
10. He, K., Zhang, X., Ren, S., Sun, J.: Deep Residual Learning for Image Recognition (2015). https://doi.org/10.48550/arXiv.1512.03385. http://arxiv.org/abs/1512.03385. arXiv:1512.03385
11. Jiang, A.Q., et al.: Mistral 7B (2023)
12. Jocher, G., Chaurasia, A., Qiu, J.: Ultralytics YOLO (2023). https://github.com/ultralytics/ultralytics
13. Kay, A.: Tesseract: an open-source optical character recognition engine. Linux J. **2007**(159), 2 (2007)
14. Kembhavi, A., Salvato, M., Kolve, E., Seo, M., Hajishirzi, H., Farhadi, A.: A diagram is worth a dozen images. In: Leibe, B., Matas, J., Sebe, N., Welling, M. (eds.) ECCV 2016. LNCS, vol. 9908, pp. 235–251. Springer, Cham (2016). https://doi.org/10.1007/978-3-319-46493-0_15
15. Kim, J., Choi, Y., Kahng, M., Kim, J.: FitVid: responsive and flexible video content adaptation. In: CHI Conference on Human Factors in Computing Systems, New Orleans, LA, USA, pp. 1–16. ACM (2022). https://doi.org/10.1145/3491102.3501948. https://dl.acm.org/doi/10.1145/3491102.3501948
16. Kim, J., Guo, P.J., Cai, C.J., Li, S.W.D., Gajos, K.Z., Miller, R.C.: Data-driven interaction techniques for improving navigation of educational videos. In: Proceedings of the 27th Annual ACM Symposium on User Interface Software and Technology, UIST 2014, pp. 563–572. Association for Computing Machinery, New York (2014). https://doi.org/10.1145/2642918.2647389
17. Li, M., Cui, L., Huang, S., Wei, F., Zhou, M., Li, Z.: TableBank: table benchmark for image-based table detection and recognition. In: Calzolari, N., et al. (eds.) Proceedings of the Twelfth Language Resources and Evaluation Conference, Marseille, France, pp. 1918–1925. European Language Resources Association (2020). https://aclanthology.org/2020.lrec-1.236

18. Li, M., Xu, Y., Cui, L., Huang, S., Wei, F., Li, Z., Zhou, M.: DocBank: a benchmark dataset for document layout analysis. In: Scott, D., Bel, N., Zong, C. (eds.) Proceedings of the 28th International Conference on Computational Linguistics, Barcelona, Spain, pp. 949–960. International Committee on Computational Linguistics (2020). https://doi.org/10.18653/v1/2020.coling-main.82. https://aclanthology.org/2020.coling-main.82

19. Lin, T.Y., Dollar, P., Girshick, R., He, K., Hariharan, B., Belongie, S.: Feature pyramid networks for object detection. In: 2017 IEEE Conference on Computer Vision and Pattern Recognition (CVPR), Honolulu, HI, pp. 936–944. IEEE (2017). https://doi.org/10.1109/CVPR.2017.106. http://ieeexplore.ieee.org/document/8099589/

20. Liwicki, M., Bunke, H.: IAM-OnDB-an on-line english sentence database acquired from handwritten text on a whiteboard. In: Eighth International Conference on Document Analysis and Recognition (ICDAR 2005), pp. 956–961. IEEE (2005)

21. Masry, A., Do, X.L., Tan, J.Q., Joty, S., Hoque, E.: ChartQA: a benchmark for question answering about charts with visual and logical reasoning. In: Muresan, S., Nakov, P., Villavicencio, A. (eds.) Findings of the Association for Computational Linguistics: ACL 2022, Dublin, Ireland, pp. 2263–2279. Association for Computational Linguistics (2022). https://doi.org/10.18653/v1/2022.findings-acl. 177. https://aclanthology.org/2022.findings-acl.177

22. Methani, N., Ganguly, P., Khapra, M.M., Kumar, P.: PlotQA: Reasoning over Scientific Plots (2020). http://arxiv.org/abs/1909.00997. arXiv:1909.00997

23. Mukhopadhyay, S., Smith, B.: Passive capture and structuring of lectures. In: Proceedings of the seventh ACM International Conference on Multimedia (Part 1) - MULTIMEDIA 1999, Orlando, Florida, United States, pp. 477–487. ACM Press (1999). https://doi.org/10.1145/319463.319690. http://portal.acm.org/citation.cfm?doid=319463.319690

24. Pfitzmann, B., Auer, C., Dolfi, M., Nassar, A.S., Staar, P.: DocLayNet: a large human-annotated dataset for document-layout segmentation. In: Proceedings of the 28th ACM SIGKDD Conference on Knowledge Discovery and Data Mining, Washington DC, USA, pp. 3743–3751. ACM (2022). https://doi.org/10.1145/3534678.3539043. https://dl.acm.org/doi/10.1145/3534678.3539043

25. Pisaneschi, L., Gemelli, A., Marinai, S.: Automatic generation of scientific papers for data augmentation in document layout analysis. Pattern Recogn. Lett. **167**, 38–44 (2023). https://doi.org/10.1016/j.patrec.2023.01.018. https://www.sciencedirect.com/science/article/pii/S0167865523000247

26. Ren, S., He, K., Girshick, R., Sun, J.: Faster R-CNN: towards real-time object detection with region proposal networks. In: Cortes, C., Lawrence, N., Lee, D., Sugiyama, M., Garnett, R. (eds.) Advances in Neural Information Processing Systems, vol. 28. Curran Associates, Inc. (2015). https://proceedings.neurips.cc/paper_files/paper/2015/file/14bfa6bb14875e45bba028a21ed38046-Paper.pdf

27. Richter, S.R., Vineet, V., Roth, S., Koltun, V.: Playing for data: ground truth from computer games. In: Leibe, B., Matas, J., Sebe, N., Welling, M. (eds.) ECCV 2016. LNCS, vol. 9906, pp. 102–118. Springer, Cham (2016). https://doi.org/10.1007/978-3-319-46475-6_7

28. Tanaka, R., Nishida, K., Nishida, K., Hasegawa, T., Saito, I., Saito, K.: SlideVQA: A Dataset for Document Visual Question Answering on Multiple Images (2023). https://doi.org/10.48550/arXiv.2301.04883. http://arxiv.org/abs/2301.04883. arXiv:2301.04883

29. Wang, J., Min, W., Hou, S., Ma, S., Zheng, Y., Jiang, S.: Logodet-3k: a large-scale image dataset for logo detection. ACM Trans. Multimedia Comput. Commun. Appl. (TOMM) **18**(1), 1–19 (2022)

30. Xu, C., et al.: Lecture2Note: automatic generation of lecture notes from slide-based educational videos. In: 2019 IEEE International Conference on Multimedia and Expo (ICME), pp. 898–903 (2019). https://doi.org/10.1109/ICME.2019. 00159. ISSN: 1945-788X

31. Yoo, T., Jeong, H., Lee, D., Jung, H.: LectYS: a system for summarizing lecture videos on YouTube. In: 26th International Conference on Intelligent User Interfaces, College Station, TX, USA, pp. 90–92. ACM (2021). https://doi.org/10.1145/ 3397482.3450722. https://dl.acm.org/doi/10.1145/3397482.3450722

32. Zhao, B., Xu, S., Lin, S., Wang, R., Luo, X.: A new visual interface for searching and navigating slide-based lecture videos. In: 2019 IEEE International Conference on Multimedia and Expo (ICME), pp. 928–933 (2019). https://doi.org/10.1109/ ICME.2019.00164. ISSN: 1945-788X

33. Zhong, X., Tang, J., Jimeno-Yepes, A.: PubLayNet: Largest Dataset Ever for Document Layout Analysis (2019). https://doi.org/10.1109/ICDAR.2019.00166

KVP10k : A Comprehensive Dataset for Key-Value Pair Extraction in Business Documents

Oshri Naparstek[1]([✉]), Ophir Azulai[1], Inbar Shapira[1], Elad Amrani[1], Yevgeny Yaroker[1], Yevgeny Burshtein[1], Roi Pony[1], Nadav Rubinstein[1], Foad Abo Dahood[1], Orit Prince[1], Idan Friedman[1], Christoph Auer[2], Nikolaos Livathinos[2], Maksym Lysak[2], Ahmed Nassar[2], Peter Staar[2], and Udi Barzelay[1]

[1] IBM Research Israel, University of Haifa Campus, Mount Carmel, Haifa 3498825, Israel
oshri.naparstek@ibm.com
[2] IBM Research Zurich, Säumerstrasse 4, 8803 Rüschlikon, Switzerland

Abstract. In recent years, the challenge of extracting information from business documents has emerged as a critical task, finding applications across numerous domains. This effort has attracted substantial interest from both industry and academy, highlighting its significance in the current technological landscape. Most datasets in this area are primarily focused on Key Information Extraction (KIE), where the extraction process revolves around extracting information using a specific, predefined set of keys. Unlike most existing datasets and benchmarks, our focus is on discovering key-value pairs (KVPs) without relying on predefined keys, navigating through an array of diverse templates and complex layouts. This task presents unique challenges, primarily due to the absence of comprehensive datasets and benchmarks tailored for non-predetermined KVP extraction. To address this gap, we introduce KVP10k , a new dataset and benchmark specifically designed for KVP extraction. The dataset contains 10707 richly annotated images. In our benchmark, we also introduce a new challenging task that combines elements of KIE as well as KVP in a single task. KVP10k sets itself apart with its extensive diversity in data and richly detailed annotations, paving the way for advancements in the field of information extraction from complex business documents.

1 Introduction

Extracting KVPs from business documents is a critical task that holds significant importance for businesses today. In an increasingly data-driven world, organizations generate and receive vast amounts of unstructured textual data in the form of invoices, contracts, reports, and other documents. The ability to efficiently extract relevant information in the form of key-value pairs from these documents can greatly benefit businesses. It not only streamlines data entry processes but also enables quick and accurate access to essential information,

E. H. Barney Smith et al. (Eds.): ICDAR 2024, LNCS 14804, pp. 97–116, 2024.
https://doi.org/10.1007/978-3-031-70533-5_7

leading to improved decision-making, enhanced efficiency, and better overall business operations. In this context, KVP extraction plays an important role in transforming unstructured data into actionable insights, helping companies stay competitive in their respective industries.

In exploring the landscape of information extraction from documents, it is essential to distinguish among KIE, Document Question Answering (DQA), and KVP extraction. These tasks, while related, diverge significantly in their objectives and methodologies.

KIE, as the most established of the trio, focuses on categorizing text snippets into a predefined set of classes. This process often involves the aggregation of related textual entities under unified labels, making it a task of entity recognition and classification at its core. The simplicity of KIE, relative to the other tasks, stems from its reliance on a known set of classes, reducing the complexity to the identification and classification of text according to these categories. Representative datasets in this domain include CORD [14], SROIE [5], Kleister-NDA [17], VRDU [20], Kleister-Charity [17], and EPHOIE [18].

Another related task is Document Question Answering. This task introduces a different paradigm, where the task is not to classify text into predefined categories but to locate and extract answers to user-posed questions directly from the text. This task eliminates the need for a fixed set of labels, instead requiring the model to understand the question's intent and retrieve relevant information from the document. The dynamic nature of the questions introduces variability and complexity, as the model must adapt to the diverse range of inquiries. DocVQA [11] is a notable dataset in this field, challenging models with a wide array of question-answer scenarios.

KVP extraction presents challenges akin to those encountered in document-based question answering, largely due to the absence of predefined keys. In question answering, complexity often arises when synthesizing an answer necessitates aggregating information from various sections within a document. Despite this, the objective remains to distill a singular, precise answer. In contrast, KVP extraction demands a comprehensive retrieval of all pertinent keys and values scattered throughout a document, expanding the scope beyond seeking a specific answer to encompass a broader extraction task. This task demands an understanding of document structure and content to discern the relationships between different pieces of information, often dealing with hierarchical key-value structures. Datasets such as FUNSD [7] and XFUND [21] are examples of the complexities of KVP extraction, presenting diverse documents where models must infer and extract a broad spectrum of information without relying on a fixed schema.

Despite the comprehensive and demanding nature of KVP extraction, a notable challenge within this domain is the current state of available datasets. The existing datasets for KVP extraction, such as FUNSD [7] and XFUND [21], are somewhat limited in scope and diversity. They tend to be smaller in size, which can restrict the depth of training and the robustness of models developed using these resources. Even newer datasets such as Form-NLU [2] and SIBR [22] are only 857 and 1,000 pages respectively. Furthermore, these datasets often lack variety in their sources, presenting a narrow view of the potential applications and scenarios where KVP extraction could be applied. This limitation in dataset quality and diversity poses significant challenges for researchers and practitioners aiming to develop models that are capable of performing well across a wide range of real-world documents and contexts. The need for larger, more varied datasets is crucial in pushing the boundaries of what KVP extraction models can achieve, ensuring they are versatile and effective across diverse document types and industries.

In recent years, there has been a growing interest in the domain of key information extraction and key-value pair extraction from various types of documents and many models were developed for the task of document understanding [3,4,6,9,12,15,19]. This growing enthusiasm is reflected in both academic research and industry applications, where the need to automate the extraction of critical data from documents such as legal contracts and medical records is increasingly apparent. However, despite this surge in interest and the evident practical importance of this task, a noticeable gap remains in the field: the absence of a comprehensive and high-quality dataset tailored specifically for key-value pair extraction from documents. This notable void has underscored the necessity for collaborative efforts within the research community to address this deficiency and create a resource that can significantly advance state-of-the-art document analysis and information extraction, ultimately benefiting a wide array of businesses and organizations across diverse domains.

In response to the gap in Key-Value Pair (KVP) extraction from documents, we introduce KVP10k . This dataset is distinguished by its comprehensive scope and focus on KVP extraction. Our contributions through this work are threefold:

1. **New Dataset** – KVP10k includes 10707 pages, making it the largest dataset available for KVP extraction. It features a broad array of keys and precise annotations, with text labeled as keys or values, providing a solid basis for training and evaluation.
2. **New Benchmark with Metrics** – We present a benchmark for KVP extraction, offering a framework for model comparison and performance evaluation. This facilitates a clearer understanding of model capabilities in the field.
3. **Baseline Results** – Initial baseline results are shared to establish a foundation for subsequent research, aiming to enhance KVP extraction methods.

KVP10k aims to support the advancement of document processing technologies by providing high-quality data and a platform for rigorous research.

2   Related Work

In this work, we have considered a wide range of existing datasets that contribute to the fields of KVP extraction and KIE. Notably, FUNSD [7] and XFUND [21] are prominent datasets for KVP tasks. Additionally, the SIBR [22] dataset is relevant for KVP tasks, particularly focusing on camera-captured image scenarios with real-world complexities such as blur, noise, and uneven illumination (in the wild). However, it is important to recognize that these datasets are relatively small and typically classify entities of KVPs as either questions or answers, without annotating common class types like dates or names. This limitation restricts their applicability for more complex or nuanced applications. In our endeavor, we have enhanced this foundation by annotating both KVPs and KIE elements, introducing a more detailed classification system with 17 distinct classes.

Turning our attention to KIE, datasets such as CORD [14], SROIE [5], Kleister-NDA [17], VRDU [20], Kleister-Charity [17], and EPHOIE [18] have provided valuable insights and benchmarks. Yet, a common limitation among these resources is their size, which often restricts the depth and breadth of research and application development. Moreover, many of these datasets are derived from specific, homogeneous sources or templates, which may not fully represent the diversity and variability encountered in real-world data.

In contrast, our dataset has been meticulously compiled to ensure a broad representation of sources and templates. This diversity is crucial for developing robust models capable of handling the wide range of formats and contexts in which KVP and KIE tasks are applied. Furthermore, unlike XFUND, which relies on synthetic data, our dataset is composed entirely of real-world documents. This choice underlines our commitment to authenticity and applicability, ensuring that the insights and models derived from our dataset are useful in real-world scenarios. Detailed comparison is provided in Fig 1.

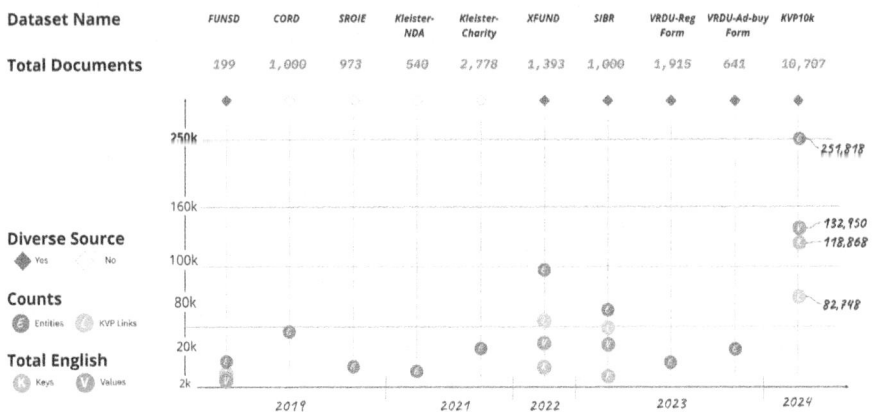

Fig. 1. Comparative overview of KVP10k versus other datasets: Comparing the Number of Documents, Entities, Keys, Values, and Links

In summary, while existing datasets have laid important groundwork in the fields of KVP and KIE, our dataset seeks to address some of their key limitations by offering a larger, more detailed and diverse collection of real-world data.

We hope that this contribution will not only support the advancement of current research but also inspire new directions in the extraction and understanding of key information from complex documents.

3   Data Acquisition

Our data acquisition process leveraged two primary sources to ensure a diverse and comprehensive dataset: extensive web data from Common Crawl and a collection of images from publicfiles.fcc.gov.

From Common Crawl, we employed a systematic approach to download indices and identify URLs of PDFs, focusing on over 40,000 targeted domains. These domains encompassed a broad spectrum of sources, including various companies, government bodies, and educational institutions. Applying filters to these URLs based on a pre-defined list of relevant and reliable domains, we efficiently collected a vast dataset pertinent to our research needs.

In the data filtering phase, we chose a subset of 8 million documents from the initially extracted web data. For categorizing these documents, we created a classifier to distinguish between documents suitable for KVP extraction and the rest. This classifier was developed employing a Longformer model [1], combined with a Roberta tokenizer(2019) [10]. Utilizing the 'allenai/longformer-base-4096' architecture as a foundation. The classifier was trained for binary classification over 20 epochs with a learning rate of 3e-4, employing the Adam optimizer. This training process used a dataset comprising 2,378 documents suitable for KVP extraction and 12,610 documents classified as non-KVP from the 8-million-document subset. The classifier's performance was notable, achieving a precision of 0.97, a recall of 0.55, and an F1-score of 0.7, culminating in an overall accuracy of 0.92. We also engineered two distinct rule-based classifiers. The primary classifier targets documents containing more than a specified number of words, N, which include pre-established substrings frequently observed in business documents such as "bill," "ship," "total," "sub," etc. Our secondary classifier is tailored to identify documents featuring more than K independent clauses, concluding with a colon. It then scans for an ensuing independent clause either directly after the colon or in the subsequent line. A schematic describing the data acquisition process for the common crawl data is shown in Fig. 2.

Besides the 7,000 images, we also acquired additional pages from public-files.fcc.gov by retrieving the initial 44,000 PDF links. Out of these, we randomly chose 5,000 PDFs, from which we then randomly selected between 1 to 5 pages per PDF.

To maintain diversity in the data and mitigate potential biases, a deduplication filter was employed, leveraging OCR [13] and sentence outputs. A criterion was set where documents were deemed alike if they had a minimum overlap of six sentences, with each sentence comprising at least three words. This deduplication method was applied to the document batch processed in the second phase, organizing them into clusters. Within each cluster, a single document was chosen at random.

In total, we gathered 3,524 pages from publicfiles.fcc.gov and 7,183 pages from Common Crawl.

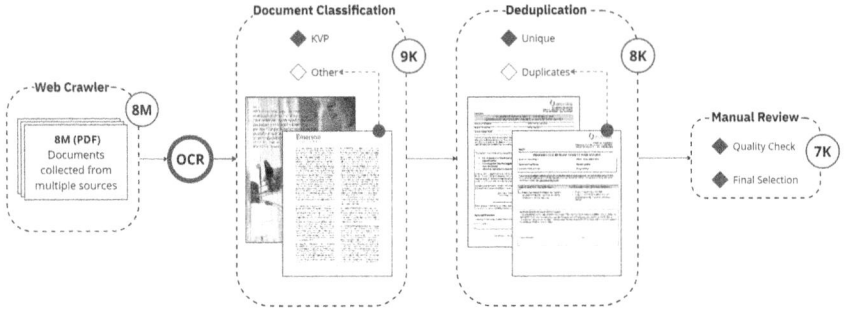

Fig. 2. A schematic describing the data collection process using web crawling.

4   Data Annotation

In the process of data annotation, a set of specific guidelines was established to ensure consistency and accuracy. These guidelines cover various annotation types, each with its own defined characteristics and purpose. The guidelines were designed to cater to different elements within the documents. The annotation process is structured to categorize and link textual elements within a given task. This involves enclosing relevant text within a defined area and assigning an appropriate label from a pre-determined set of label types. These labels include 'Text', and 'Handwriting-text'. Subsequently, a linkage is established from a value to its corresponding key.

In addition we introduce two label types without linking, 'unvalued-key' and unkeyed-values. Moreover, for unkeyed-values, a range of label types is available to categorize them appropriately. These types encompass 'Floating Document Type', 'Floating Document Title', 'Floating Year', 'Floating Date', 'Floating Name', 'Floating Address', 'Floating Phone', 'Floating Email', 'Floating Website', 'Floating Amount', and 'Floating Text' for instances that do not align with the aforementioned categories. This systematic approach ensures clarity and coherence in the annotation of the document, facilitating a structured and comprehensible representation of the data. An example for an annotated page is given in Fig 3. Examples of the annotation format are given in appendix A.

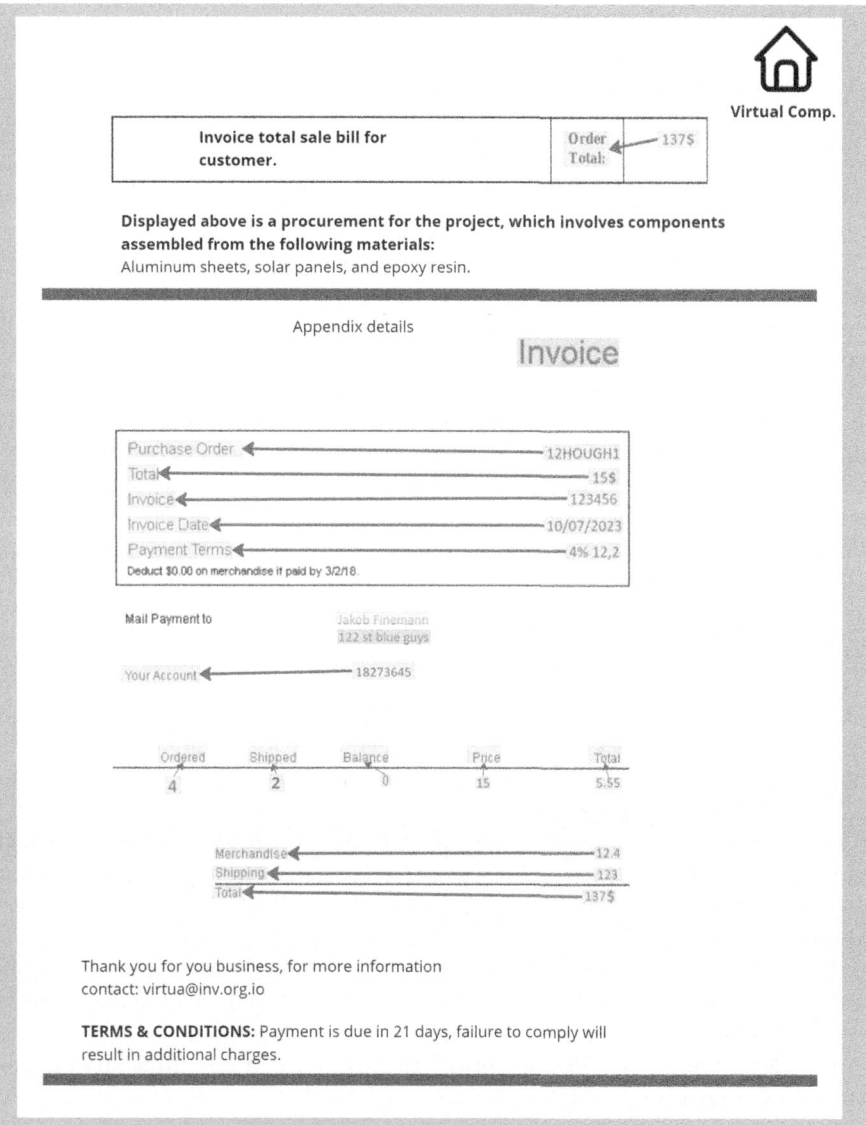

Fig. 3. Example of an annotated page

5 Dataset Characteristics

5.1 Diverse Source of Images

Our dataset goes beyond mere visual richness, providing a deep understanding of the text and how things relate to each other, setting new standards for how we understand documents, particularly in key-value pair extraction. This richness in text is not just about having different kinds of documents or complex designs. It's about exploring the detailed layers of text and the intricate ways text parts interact across various documents. Covering a wide range of document types, from business papers to scientific studies, KVP10k includes a variety of words and specialized terms. This range of documents helps in studying complex text patterns, meanings, and hints that help train models to understand not just what documents look like but also the following two properties: 1) what the information means, 2) why it is important. Figure 4 presents a selection of images from different types of documents. These examples showcase the visual and text variety in KVP10k . In addition, they showcase the complexity of document designs and the thorough assignment of labels that help train deep learning models.

5.2 Dataset Statistics

In this section, we present some dataset statistics to provide an understanding of the characteristics of KVP10k . The dataset is diverse and richly annotated, covering various document types and layouts. In Fig. 5, we present the distribution of entities per page in the dataset.

Figure 6 provides an overview of the distribution of entity labels in the benchmark dataset. This pie chart shows the relative proportions of different entity labels present in the dataset, including labels such as floating name, text, phone, date, key/value, etc.

Again, Fig. 1 presents a comparative analysis of KVP10k against existing ones, highlighting our dataset's superior quantity of documents, entities, links, keys, and values.

Together, these statistics illustrate the diversity and complexity of this dataset, providing insights into its composition and structure.

6 Benchmark

To support the community in evaluating KVP extraction systems, we have developed a comprehensive benchmarking tool. This utility is crafted to assess the performance of various KVP extraction models, providing a uniform and reliable method for researchers and developers to gauge the effectiveness of their solutions.

Our benchmarking code includes implementations of the metrics focusing on the location of an entity, the textual meaning of an entity, and a combined

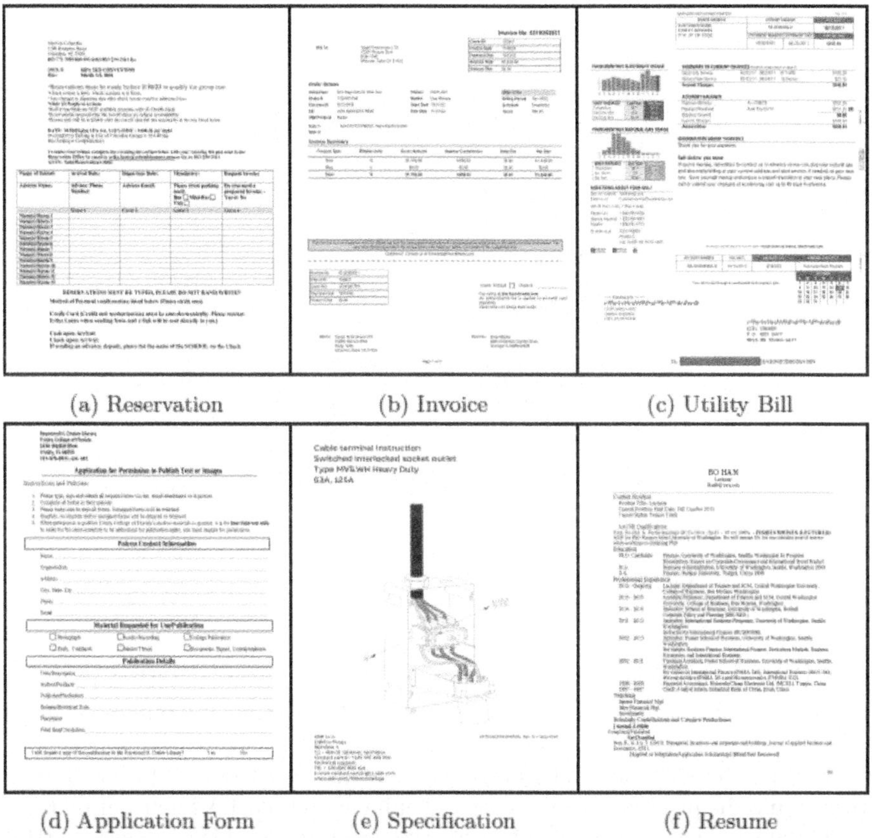

(a) Reservation   (b) Invoice   (c) Utility Bill

(d) Application Form   (e) Specification   (f) Resume

Fig. 4. Exemplifying Versatility: A collage of diverse document categories from KVP10k Dataset

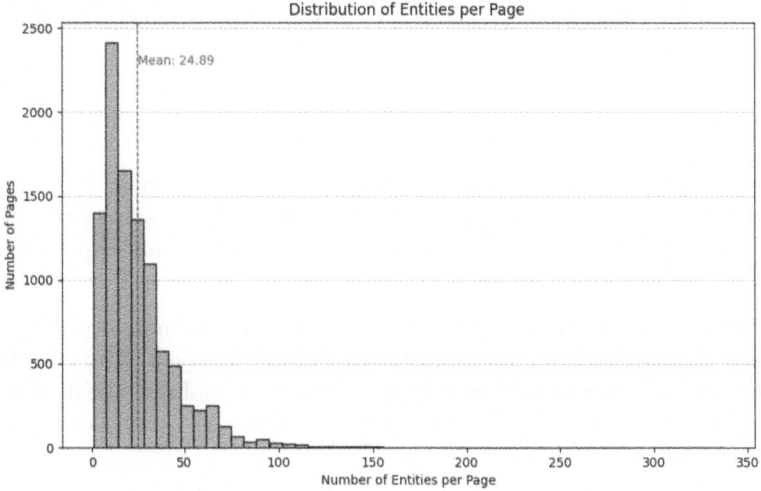

Fig. 5. Distribution of entities per page in KVP10k .

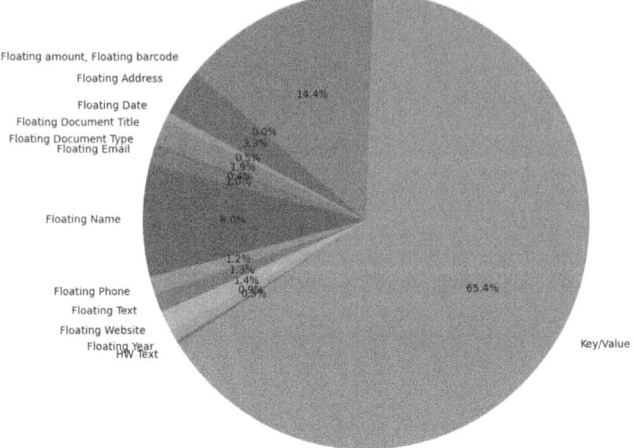

Fig. 6. Distribution of entity labels in KVP10k .

approach. It facilitates a nuanced assessment, shedding light on different facets of a model's performance. We have designed the code with user-friendliness in mind, ensuring it can be easily integrated with diverse KVP extraction models. This feature is particularly beneficial for researchers and practitioners in the domain, streamlining the process of evaluating and enhancing their KVP extraction tools.

The benchmark code is openly accessible and can be found at our GitHub repository[1]. We invite the community to use this resource to propel forward the development of more precise and efficient KVP extraction technologies.

Our goal with this benchmarking code is to establish a standardized approach for evaluating KVP extraction systems, thus promoting continuous progress and innovation in this area.

6.1   Tasks

This benchmark is designed to rigorously evaluate the performance of algorithms in the extraction of key-value pairs from documents. It consists of two distinct tasks, each tailored to assess specific aspects of the algorithms' capabilities.

Entity Recognition Task. Similar to FUNSD [7], the primary objective of this task is to identify key and value entities within a document. The effectiveness of an algorithm is quantified through two metrics: normalized edit distance and Intersection Over Union (IOU). An entity is considered a successful 'hit' if it satisfies the following criteria:

---

[1] https://github.com/IBM/KVP10k.

– The normalized edit distance from the ground truth is below 0.2.
– The IOU with the ground truth exceeds 0.3.

The overall performance of the algorithm is assessed using the F1 score.

Key-Value Pair Detection Task. The second task extends the challenge by focusing on the detection of key-value pairs in the document, as well as identifying unkeyed values and unvalued keys. An unkeyed value refers to a value present without an explicit key, such as a date without a preceding "Date:" label. An unvalued key refers to a key present with empty value, such as in empty fillable form. This task also encompasses the detection of values with no associated keys.

We have three metrics for this task. Location only, text only and a combination of the two.

– Location only metric: a key-value pair is considered a 'hit' if both the key and value have IOU above 0.3.
– Text only metric: a key-value pair is considered a 'hit' if both the key and value has normalized edit distance below 0.2.
– Combined metric: a key-value pair is considered a 'hit' if both the key and the value have normalized edit distances below 0.2and have IOU above 0.3.

Unkeyed values are handled as key-value pairs where the value is subject to the previously defined criteria, and the key is identified through an exact text match. Unlike the standard key-value pairs, the detection of the key in these instances does not require a threshold for Intersection over Union (IOU) to be considered a successful hit. Unvalued keys are processed as key-value pairs where the key meets the established criteria and the value is explicitly recognized as empty. For these instances, the assessment of the value does not necessitate the Intersection over Union (IOU) threshold.

The evaluation metric remains consistent with the first task, utilizing the F1 score. The precision and recall are calculated based on two steps: 1) number of correctly identified pairs, unkeyed values, and unvalued keys in relation to the total number of such entities in each image, and 2) then averaging the scores per image over the total number of images.

This task presents a novel integration of elements from both KIE and KVP tasks, offering a unique approach to document analysis. For unkeyed values, the task aligns with KIE tasks, akin to those seen in datasets like CORD [14]. Conversely, when addressing key-value pairs, the methodology parallels entity detection and linking tasks, similar to what is observed in datasets such as FUNSD [7]. Combining the approaches allows for appropriate handling of unpredictable and diverse document information.

6.2 Output Format

The format we have adopted for presenting the results of our benchmark analysis is intentionally straightforward, designed for ease of interpretation and use.

Listing 1.1. Illustrative example of the output format

```
{
    "kvps_list": [
        {
            "type": "unkeyed",
            "key": {
                "text": "Example Key Text 0"
            },
            "value": {
                "text": "Associated Value Text 0",
                "bbox": [x1, y1, x2, y2]
            }
        },
        {
            "type": "unvalued",
            "key": {
                "text": "Example key text 1",
                "bbox": [x1, y1, x2, y2]
            },

        },
        {
            "type": "kvp",
            "key": {
                "text": "Example Key Text 2",
                "bbox": [x1, y1, x2, y2]
            },
            "value": {
                "text": "Associated Value Text 2",
                "bbox": [x1, y1, x2, y2]
            }
        }
    ]
}
```

Specifically, the data output is organized as a sequence of dictionaries, with each dictionary entry comprising a pair of key-value elements. This structure captures not only the textual content but also the spatial positioning of each element, facilitating a comprehensive understanding of the data's layout.

In Listing 1.1, we provide a sample of the output format. This example illustrates the structure of the data, including the categorization of key-value pairs ('unkeyed', 'unvalued' or as a specific 'kvp') and the associated spatial information ('bbox') for each text element, represented as coordinates on the document.

In this format, each 'key' and 'value' is delineated by their respective textual content and a bounding box ('bbox'), the latter specifying the coordinates of the text's location on the page, thus offering a dual perspective on the data: textual and spatial. The 'type' field distinguishes between 'unkeyed' entries, where the value's text is provided and the key's text is set as the appropriate label from the unkeyed-values range of label types, 'unvalued' entries, where only the key's text is provided, and 'kvp' entries, where both key and value include spatial data. This nuanced approach ensures a richer dataset, conducive to more insightful analysis.

6.3 Baselines

In this baseline section, we delve into the preliminary outcomes of our exploration into the tasks outlined earlier, employing a strategy influenced by the LMDX framework [15]. Our initial step involves processing the OCR-derived text from each document, converting it into a format amenable to integration with a large language model. This process entails arranging the OCR [16] text in a systematic top-to-bottom and left-to-right order, where each text line is tagged with its bounding box coordinates, denoted as [x1,y1,x2,y2].

While considering various models for this task, we recognized certain limitations in encoder-based models like LayoutLMv3 [4]. These models primarily function as token classifiers, which inherently restricts their capability in performing tasks that require key-value pair linking, as they lack the mechanism to generate new tokens necessary for establishing these links. Moreover, their performance is significantly influenced by the order in which the input is read, which can be a drawback in processing complex document layouts.

This realization steered us towards adopting an LMDX [15]-like method. Our choice of using an LMDX-like method was motivated by its generative capabilities, which we believe offer a more flexible and effective approach for the tasks at hand. We proceeded to fine-tune the Mistral-7B model [8] to produce text outputs that align closely with the ground truth data. This fine-tuning process was executed on a single A100 GPU and spanned over a 24-hour timeframe. The insights gleaned from this exercise are encapsulated in the accompanying table, providing a clear overview of our findings. Our initial baseline results for the key-value pair detection task are given in Table 1

Table 1. Baseline Results for the key-value pair detection task

	Text Only			Location Only			Text + Location (All)		
	Precision	Recall	F1	Precision	Recall	F1	Precision	Recall	F1
Regular	0.678	0.641	0.659	0.670	0.631	0.650	0.627	0.595	0.611
Unkeyed	0.584	0.620	0.601	0.635	0.672	0.653	0.568	0.601	0.584
Unvalued	0.617	0.586	0.601	0.634	0.604	0.618	0.603	0.573	0.588
All	0.645	0.640	0.643	0.665	0.657	0.661	0.615	0.608	0.612

7   Conclusions

In conclusion, the task of extracting information from business documents, especially without the crutch of predefined keys, presents a formidable challenge that spans across various domains. Our work sheds light on the critical gap within current datasets and benchmarks that are largely tailored for KIE with predetermined keys. By introducing KVP10k , not only do we provide a resource that caters to the nuanced demands of non-predetermined KVP extraction, we also set a new precedent for the depth of diversity and annotation detail required for meaningful progress in this field.

The significance of our contribution lies not just in the dataset itself but also in the potential it unlocks for future research and applications. By offering a platform that is both challenging and reflective of real-world complexities, KVP10k invites a broader exploration of methodologies and technologies in the realm of information extraction. This, in turn, could catalyze a wave of innovations that enhance the efficiency and accuracy of processing complex business documents.

As the community engages with KVP10k , we anticipate the emergence of novel approaches that not only excel in the context of our benchmark but also inspire the development of more adaptive, robust solutions for information extraction at large. Thus, our work not only addresses an immediate need within the field but also lays the groundwork for ongoing advancements that could redefine the boundaries of what's possible in information extraction from business documents.

A   Annotation Format

An exaple for flat KVP annotation is given in listing 1.2. An exaple for unkeyed value annotation is given in listing 1.3 and an example with a section is given in 1.4. Note that in the annotation the x any y coordinates are relative to the page size and take values between 0 and 1

B   Additional Statistics

Here we provide mode detailed statistics of the data. Figure 7 is a histogram that specifically focuses on documents sourced from public files. Figure 8, on the other hand, is a histogram that specifically focuses on documents sourced from web crawl.

These plots provides an overview of the variability in the number of entities in total, including key-value pairs and other types of entities, across different documents and within specific sources. The aim is to offer a comprehensive view of the dataset's characteristics, highlighting variations in document complexity across different sources.

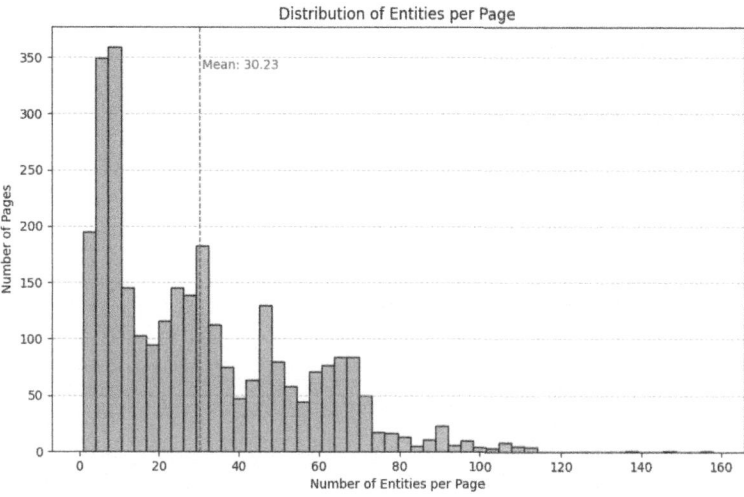

Fig. 7. Distribution of entities per page in KVP10k sourced from Public files.

In addition to the overall distribution, we have generated two more pie charts to examine the distribution of entity labels within specific subsets of the dataset:

– Figure 9 focuses on documents sourced from public files.
– Figure 10 specifically examines documents sourced from web crawl.

The following pie charts offer a more intuitive representation of the distribution of entity labels in total across different documents and within specific sources, facilitating a better understanding of the characteristics of the dataset.

Listing 1.2. Example of Annotation Format

```
{
"rectangles": [
{
  "_id": "3ba88dfc-aee5-433b-a68d-a2616033cbd3",
  "annotationQCEvaluations": null,
  "annotationQcEvals": null,
  "attributes": {},
  "color": "rgb(253, 255, 0)",
  "comments": [],
  "coordinates": [
    {
       "x": 0.657255,
       "y": 0.4894
    },
    {
       "x": 0.66902,
```

```
         "y":  0.4894
      },
      {
         "x":  0.66902,
         "y":  0.542701
      },
      {
         "x":  0.657255,
         "y":  0.542701
      }
   ],
   "evaluations":  null,
   "label":  "Text",
   "label_id":  null,
   "origin":  "manual",
   "state":  "editable",
   "type":  ""
},

}
```

Listing 1.3. Example of Annotation Format

```
{
"rectangles": [
{
  "_id":  "deaed54b-ef4c-4152-a916-94715312ad11",
  "annotationQCEvaluations":  null,
  "annotationQcEvals":  null,
  "attributes":  {},
  "color":  "rgb(142, 165, 183)",
  "comments":  [],
  "coordinates":  [
     {
        "x":  0.171097,
        "y":  0.393928
     },
     {
        "x":  0.312631,
        "y":  0.393928
     },
     {
        "x":  0.312631,
        "y":  0.408258
     },
     {
```

```
      "x": 0.171097,
      "y": 0.408258
  }
],
"evaluations": null,
"label": "Floating Address",
"label_id": null,
"origin": "manual",
"state": "editable",
"type": ""
},
]
}
```

Listing 1.4. Example of Annotation Format

```
{
"rectangles": [
 {
   "_id": "5a331350-94d0-4dad-b5db-06e992d0a902",
   "annotationQCEvaluations": null,
   "annotationQcEvals": null,
   "attributes": {},
   "color": "rgb(216, 82, 82)",
   "comments": [],
   "coordinates": [
     {
       "x": 0.084706,
       "y": 0.66808
     },
     {
       "x": 0.213333,
       "y": 0.66808
     },
     {
       "x": 0.213333,
       "y": 0.685645
     },
     {
       "x": 0.084706,
       "y": 0.685645
     }
   ],
   "evaluations": null,
   "label": "Section",
   "label_id": null,
```

```
    "origin":  "manual",
    "state":  "editable",
    "type":  ""
  },

}
```

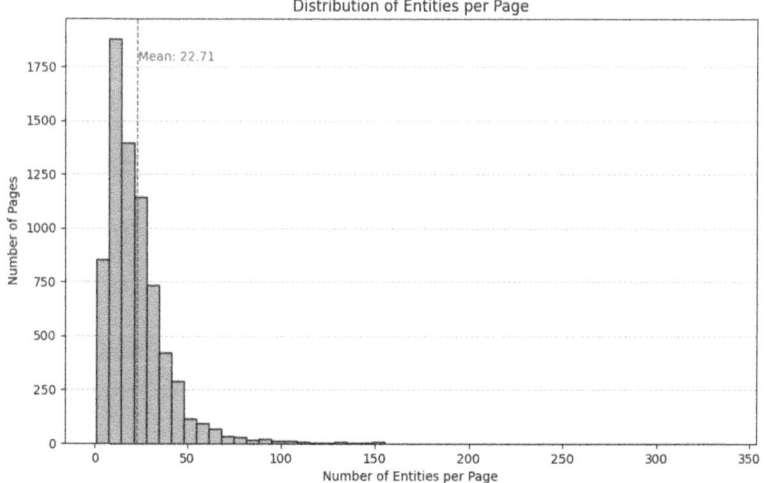

Fig. 8. Distribution of entities per page in KVP10k sourced from web crawl.

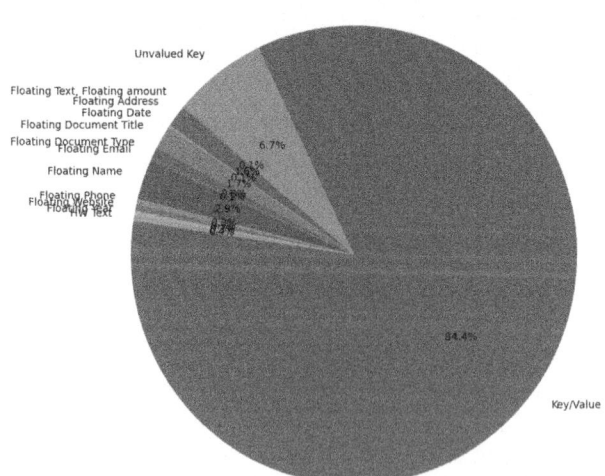

Fig. 9. Distribution of entity labels in KVP10k sourced from Public files.

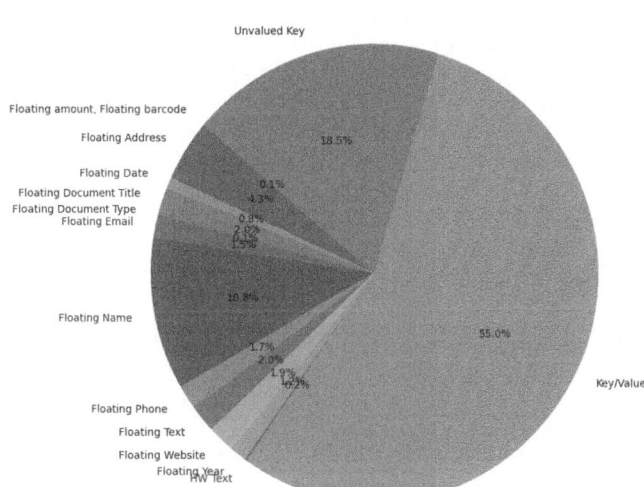

Fig. 10. Distribution of entity labels in KVP10k sourced from web crawl.

References

1. Beltagy,I., Peters, M.E., Cohan, A.: Longformer: the longdocument transformer. arXiv preprint arXiv:2004.05150 (2020)
2. Ding, Y., et al.: Form-NLU: dataset for the form natural language understanding. In: Proceedings of the 46th International ACM SIGIR Conference on Research and Development in Information Retrieval, pp. 2807–2816 (2023)
3. Hong, T., et al.: Bros: A pre-trained language model for understanding texts in document. (2020)
4. Huang, Y., et al.: Layoutlmv3: Pre-training for document AI with unified text and image masking. In: Proceedings of the 30th ACM International Conference on Multimedia, pp. 4083–4091 (2022)
5. Huang, Z., et al.: Icdar2019 competition on scanned receipt OCR and information extraction. In: 2019 International Conference on Document Analysis and Recognition (ICDAR). IEEE, pp. 1516–1520 (2019)
6. Hwang, W., et al.: Spatial dependency parsing for semi-structured document information extraction. arXiv preprint arXiv:2005.00642 (2020)
7. Jaume, G., Ekenel, H.K., Thiran, J.P.: Funsd: a dataset for form understanding in noisy scanned documents. In: 2019 International Conference on Document Analysis and Recognition Workshops (ICDARW), Vol. 2, pp. 1–6. IEEE (2019)
8. Jiang, A.Q., et al.: Mistral 7B. arXiv preprint arXiv:2310.06825 (2023)
9. Lee, C.Y., et al.: FormNetV2: multimodal Graph Contrastive Learning for Form Document Information Extraction. arXiv preprint arXiv:2305.02549 (2023)
10. Liu, Y., et al.: RoBERTa: a robustly optimized BERT pretraining approach. arXiv preprint arXiv:1907.11692 (2019)
11. Mathew, M., Karatzas, D., Jawahar, C.V.: Docvqa: a dataset for VQA on document images. In: Proceedings of the IEEE/CVF Winter Conference on Applications of Computer Vision, pp. 2200–2209(2021)

12. Mathur, P., et al.: LayerDoc: layer-wise extraction of spatial hierarchical structure in visually-rich documents. In: Proceedings of the IEEE/CVF Winter Conference on Applications of Computer Vision, pp. 3610–3620 (2023)
13. Naparstek, O., et al.: BusiNet-a light and fast text detection network for business documents. I arXiv preprint arXiv:2207.01220 (2022)
14. Park, S., et al.: CORD: a consolidated receipt dataset for post- OCR parsing. In: Workshop on Document Intelligence at NeurIPS 2019 (2019)
15. Perot, V., et al.: LMDX: Language Model-based Document Information Extraction and Localization. arXiv preprint arXiv:2309.10952 (2023)
16. Smith, R.: An overview of the Tesseract OCR engine. In: Ninth International Conference on Document Analysis and Recognition (ICDAR 2007), Vol. 2, pp. 629–633. IEEE (2007)
17. Stanisławek, T., et al.: Kleister: key information extraction datasets involving long documents with complex layouts. In: International Conference on Document Analysis and Recognition, pp. 564–579. Springer (2021). https://doi.org/10.1007/978-3-030-86549-8_36
18. Wang, J., et al.: Towards robust visual information extraction in real world: new dataset and novel solution. In: Proceedings of the AAAI Conference on Artificial Intelligence, Vol. 35 no. 4 , pp. 2738–2745 (2021)
19. Wang, Z., et al.: DocStruct: a multimodal method to extract hierarchy structure in document for general form understanding. arXiv preprint arXiv:2010.11685 (2020)
20. Wang, Z., et al.: VRDU: a benchmark for visually-rich document understanding. In: Proceedings of the 29th ACM SIGKDD Conference on Knowledge Discovery and Data Mining, pp. 5184–5193 (2023)
21. Xu, Y., et al.: XFUND: a benchmark dataset for multilingual visually rich form understanding. In: Findings of the Association for Computational Linguistics: ACL 2022, pp. 3214–3224 (2022)
22. Yang, Z., et al.: Modeling entities as semantic points for visual information extraction in the wild. In: Proceedings of the IEEE/CVF Conference on Computer Vision and Pattern Recognition, pp. 15358–15367 (2023)

Chinese Text Recognition

Visual Prompt Learning for Chinese Handwriting Recognition

Gang Yao[1,2], Ning Ding[1,2], Tianqi Zhao[1,2], Kemeng Zhao[1,2],
Pei Tang[1,2], Yao Tao[3], and Liangrui Peng[1,2]

[1] Department of Electronic Engineering, Tsinghua University, Beijing, China
[2] Beijing National Research Center for Information Science and Technology,
Tsinghua University, Beijing, China
{yg19,dn22,zhaotq20,zkm23,tp21}@mails.tsinghua.edu.cn,
penglr@tsinghua.edu.cn
[3] Noah's Ark Lab, Huawei Technologies, Shenzhen, China
taoyao2@huawei.com

Abstract. Recognizing Chinese handwriting in unconstrained scenarios remains a challenging task due to wide variations in writing styles and imaging conditions. Recently, prompt learning in natural language processing has shown success in leveraging context awareness in various domains. This paper proposes to incorporate visual prompt learning into an encoder-decoder model for handwriting recognition. For the encoder, multi-scale meta prompts are incorporated to utilize contextual information in internal feature representations. For the decoder, additional character-level visual prompts are used along with the embeddings of previously predicted text to guide the decoding process. Experiments conducted on the SCUT-HCCDoc, SCUT-EPT and CASIA-HWDB Chinese handwriting datasets validate the effectiveness of the proposed methods.

Keywords: handwriting recognition · visual prompt learning · encoder-decoder model

1 Introduction

In recent years, there have been increasing demands to recognize Chinese handwriting in low-quality images captured by mobile phones in unconstrained scenarios. In addition to large character set and variations of writing styles, different illumination, viewpoints, and degradation in imaging process bring more challenges for handwriting recognition.

Deep learning based handwriting recognition methods generally fall into two categories, connectionist temporal classification (CTC) based methods [18] and attention mechanism based methods [2]. Recently, Transformer-based methods with self-attention mechanism [11,12,15,24,32] have shown promising performance for handwriting tasks.

Deep learning based handwriting recognition models usually require a large amount of labeled samples for training. For Chinese handwriting images captured

© The Author(s), under exclusive license to Springer Nature Switzerland AG 2024
E. H. Barney Smith et al. (Eds.): ICDAR 2024, LNCS 14804, pp. 119–133, 2024.
https://doi.org/10.1007/978-3-031-70533-5_8

in unconstrained scenarios, it is hard to collect and annotate enough training samples. Moreover, samples from real applications in the inference stage may not obey the same statistical distribution as that of training samples, thus affecting the generalization ability of the recognition model.

To alleviate this problem, it is feasible to learn adaptive representations with deep learning frameworks. Inspired by the prompt learning originated in natural language processing (NLP), we investigate learning adaptive representations via visual prompts in the Transformer-based encoder-decoder model for Chinese handwriting recognition. Usually, convolutional neural networks (CNNs) are used as the backbone to extract feature maps. Recurrent neural network (RNN) [22] or Transformer [24] is used in an encoder-decoder architecture to model the contextual information for recognition.

Recently, prompt learning has been broadly used to adapt pre-trained models to NLP downstream tasks and different domains [5,10,29,30]. Instead of manually designing a prompt [3,4,17], Zhou et al. [30] introduce automatic prompts that can be optimized through minimizing the classification loss. Zhou et al. [29] further propose conditional context optimization (CoCoOp) to achieve better model generalization ability by employing a lightweight neural network to generate an input-conditional token (vector) for each image. However, CoCoOp [29] only learns a global visual prompt for the whole image, whereas for handwriting recognition task, both local and global structures are important.

To better learn local and global contextual information for handwriting recognition, learnable meta prompts at the patch level of feature maps are introduced in this paper. For each input image, the feature maps are first divided into patches. Cross attention between meta prompts and patch-level features generates adaptive prompts that focus on local style information. To learn global contextual information, self-attention mechanism among prompts of different patches is used to update prompts. The updated prompts are utilized to further update patch-level features via cross attention.

The above prompt learning scheme can be added after one or more blocks of a CNN-based backbone network to learn contextual information, which can be used as an encoder instead of using an RNN encoder for handwriting recognition.

In order to guide the learning of visual prompts to focus on the geometric structure information of handwriting images, additional loss functions are designed at local and global levels to maximize the mutual information between the learned prompts and the edge images.

In the previous Transformer-based decoder for text recognition [12,20], only text embedding or visual embedding is used. We propose to use character-level visual prompts along with predicted text embeddings to guide the decoding process for handwriting images with diverse styles. To generate the character-level visual prompts, the location of each character in the image is required. Therefore, we propose a method to estimate the location of each character by incorporating an additional CTC (Connectionist Temporal Classification) decoding branch in the decoder, which does not require extra annotations of character positions in the training process.

In summary, we propose a novel visual prompt learning method for Chinese handwriting recognition. The main contributions of our work are as follows:

- For the encoder, multi-scale learnable meta prompts are introduced to learn adaptive representations from text images with diverse styles. The training process is guided by maximizing mutual information between the learned prompts and the edge information of an input image.
- For the decoder, character-level visual prompts are integrated with text embedding to guide the decoding process.
- The proposed method achieves state-of-the-art performance on Chinese handwriting datasets including the SCUT-HCCDoc and SCUT-EPT datasets with unconstrained scenarios.

2   Method

We propose to incorporate visual prompt learning into an encoder-decoder based handwriting recognition model in order to adapt to variations of handwriting samples acquired in unconstrained environments. For the encoder, we focus on enriching contextual information in internal feature representations. Inspired by the idea of generating learnable vectors to model contextual information for each image in CoCoOp [30], meta prompts are introduced in the encoder. For the decoder, we also propose to generate character-level visual prompts for the decoding process. The system framework of our proposed visual prompt learning method for Chinese handwriting recognition is shown in Fig. 1.

The research work in this paper is built on an encoder-decoder based handwriting recognition model in our previous work [19]. It comprises a ResNet backbone, a bi-directional LSTM encoder and a Transformer decoder. The ResNet backbone together with the LSTM encoder can be viewed broadly as one encoder to learn both spatial and temporal contextual information. Our research results in Sect. 3 show the LSTM network can be removed after introducing visual prompts in the ResNet backbone to learn contextual information. We adopt an improved Transformer decoder proposed by Li et al. [12] as the decoder in our baseline model.

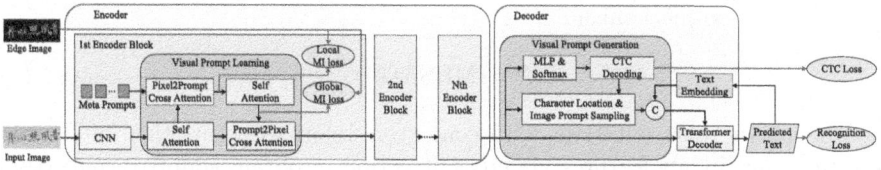

Fig. 1. The architecture of our Chinese handwriting recognition network. It consists of an encoder with multi-scale adaptive prompts to learn contextual representations and a character-level visual prompt-guided decoder to generate recognition results.

2.1 Encoder with Multi-scale Visual Prompts

To generate adaptive style representations from samples collected in uncon-strained environments, we integrate adaptive prompts into the ResNet backbone, serving as an encoder.

At the i-th block of the encoder, the image features $X^{(i)} \in \mathbb{R}^{H \times W \times C}$ are first evenly partitioned into non-overlapping patches $X_p^{(i)} \in \mathbb{R}^{N \times P \times P \times C}$, where (H, W) is the height and width of feature maps, C is the channel number of feature maps, (P, P) is the size of patches and $N = HW/P^2$ is the number of patches.

For simplicity, the index of the encoder block i is omitted in the following expressions. To retain positional information, 2D sinusoidal position embeddings PE_p are added to the patch embedding of images.

For a single patch $x_p \in \mathbb{R}^{P \times P \times C}$, we perform multi-head self-attention (MHSA) to learn local features:

$$x_p = \text{MHSA}(x_p). \tag{1}$$

To extract local style features of the images, M learnable meta prompts $z \in \mathcal{R}^{M \times C}$ are introduced and shared among different patches. To distinguish different patches, the prompts with positional embedding are obtained by adding the average position embeddings of all pixels in the patch to the meta prompts. The local prompts are obtained via cross attention between the prompts with positional embedding and local features. This process can be formulated as:

$$z^l = z_p + \text{CA}(q = z_p, k = x_p, v = x_p), \tag{2}$$

where $z_p = (z + \text{Mean}(PE_p))$ is the prompts with positional embedding and CA is the cross attention formulated by:

$$\text{CA}(q, k, v) = \text{softmax}(q W_q W_k^T k^T / \sqrt{C}) v W_v, \tag{3}$$

where $W_q, W_k, W_v \in \mathcal{R}^{C \times C}$ are linear projection matrices.

The aggregated local prompts $Z^l \in \mathbb{R}^{N \times M \times C}$ are obtained by stacking the adaptive prompts z^l of all patches. In order to represent contextual style informa-tion between different patches, the global prompts $Z^g \in \mathbb{R}^{N \times M \times C}$ are computed via self-attention mechanism:

$$Z^g = \text{MHSA}(Z^l). \tag{4}$$

These style representations with contextual information are then integrated into the input image features via cross attention to obtain adaptive contextual representations $X_p' \in \mathbb{R}^{N \times P \times P \times C}$:

$$X_p' = X_p + CA(q = X_p, k = Z^g, v = Z^g). \tag{5}$$

The final output $X' \in \mathbb{R}^{H \times W \times C}$ are obtained by reshaping X_p'.

To acquire multi-scale style information, we hierarchically apply the above process to various blocks of the ResNet, yielding an encoder with multi-scale adaptive prompts. The output of the final block is fed into the decoder for text recognition.

Our method is different from other vision Transformer methods such as ViT in two ways: First, our method introduces meta prompts to learn local prompts via cross-attention mechanism. Second, our method updates local features at pixel level instead of patch level in feature maps with global contextual information.

2.2 Objective Function in the Encoder

To guide the learning of prompts, we propose to use edge information of input images to present local and global geometric information of handwriting styles. Local and global loss functions are added in the encoder to maximize the mutual information (MI) between the learned prompts and the edge information of input images. The mutual information is estimated by using Jensen-Shannon mutual information estimator [7].

The edge images are generated via the Canny operator [1]. The feature maps E of edge images extracted by a CNN are partitioned into patches $\{e_p\}$. In order to guide local prompts z^l to learn the geometric information of each patch, we propose to maximize the local mutual information I_l between z^l and the corresponding patch of feature maps of edge images.

$$\mathcal{I}_l = \sum_{z^l \in Z^l} \left(-\mathbb{E}\left[softplus(-T(e_p, z^l))\right] - \mathbb{E}\left[softplus(T(e'_p, z^l))\right] \right) \qquad (6)$$

where $T(\cdot, \cdot) = MLP(Concate(\cdot, \cdot))$ is a light-weight neural net, and the softplus function $softplus(z) = \log(1 + e^z)$ is the adopted nonlinear activation function to discriminate whether the two inputs originate from the same sample. The inputs e_p, z^l are generated from the same sample, while e'_p, z^l are generated from different samples.

To guide global prompts Z^g to learn contextual geometric information through self-attention, we propose to maximize the global mutual information I_g between Z^g and the whole feature maps of edge images.

$$\mathcal{I}_g = -\mathbb{E}\left[softplus(-T(E, Z^g))\right] - \mathbb{E}\left[softplus(T(E', Z^g))\right] \qquad (7)$$

where E, Z^g are generated from the same sample and E', Z^g are generated from different samples.

Since the low-level feature maps of the encoder preserve finer-grained geometric information, mutual information maximization is only applied after the first encoder block.

2.3 Decoder with Character-Level Visual Prompts

We propose to incorporate character-level visual prompts into Transformer decoder. The purpose of character-level visual prompts is to focus on the related local feature in the encoder output.

The visual prompts are obtained after estimating the position of each character in the feature representation by using the CTC decoding results. The position of the next character is estimated by using the CTC classification scores and the locations of previous characters. To generate the visual prompt for a character to be decoded, a window centered at the estimated character position $\hat{p}_t$ is used to sample image features in the output of the encoder X_{out}, then the sampled features are fed into a multi-layer perceptron (MLP) with two layers to generate visual prompts:

$$Prompt_t = \text{MLP}(X_{out}[\hat{p}_t - w/2 : \hat{p}_t + w/2]), \tag{8}$$

where w is the width of the window, whose value is set according to the estimated average character width.

The visual prompts are concatenated with the embeddings of previously decoded text to guide the decoding process for the current character to be decoded.

2.4   Total Loss Function

The total loss function in the training stage is calculated as follows:

$$\mathcal{L} = \mathcal{L}_{rec} + \beta_1 \mathcal{L}_{ctc} - \beta_2 \mathcal{I}_l - \beta_3 \mathcal{I}_g, \tag{9}$$

where L_{rec} is the recognition loss of the transformer decoder, L_{ctc} is the CTC loss, I_l is the local mutual information, I_g is the global mutual information, and $\beta_1, \beta_2, \beta_3$ are coefficients. The recognition loss of the transformer decoder is calculated by the cross-entropy loss between the predicted text and its corresponding ground truth. The CTC loss can be formulated as follows:

$$L_{ctc} = -\log P(Y|C) \tag{10}$$

where C is the predicted probability distribution obtained by an MLP layer and softmax function over the final output of the encoder and Y is the corresponding ground truth. The probability of Y over C is:

$$P(Y|C) = \sum_{\pi \in M^{-1}(Y)} \Pi_{t=1}^{T} C_{t,\pi_t} \tag{11}$$

where M denotes the CTC decoding process and $M^{-1}(Y)$ denotes all paths whose decoding results are Y. C_{t,π_t} denotes the predicted probability of character π_t at position t and T is the length of the predicted results.

3   Experimental Results

In this section, the proposed method is compared with other methods on the SCUT-HCCDoc, SCUT-EPT and CASIA-HWDB Chinese handwriting datasets. Ablation experiments are carried out on the SCUT-HCCDoc dataset. Visualization for the prompts in the encoder and decoder are described. Error analysis is also presented.

3.1   Experimental Setup

Datasets and Evaluation Metrics. The SCUT-HCCDoc, SCUT-EPT and CASIA-HWDB Chinese handwriting datasets are used in our experiments.

The SCUT-HCCDoc [27] dataset is a large-scale dataset consisting of camera-captured Chinese handwriting images obtained by Internet search. There are both horizontal and vertical text images in the dataset. It contains 93,254 training images and 23,389 test images.

The SCUT-EPT [31] dataset is collected from examination papers with numerous challenging cases including character erasure, etc. It contains 40,000 training images and 10,000 test images.

The databases CASIA-HWDB 2.0-2.2 [14] comprise handwriting samples collected for research purpose from 1,019 distinct writers, encompassing a total of 5,091 pages, 52,230 lines, and 1,349,414 characters.

We employ the Accurate Rate (AR) and Correct Rate (CR) as the evaluation metrics to assess the effectiveness of our methods. These metrics are calculated by using edit distance between the recognition results and the ground truth text as follows:

$$AR = (N_t - D_e - S_e - I_e)/N_t$$
$$CR = (N_t - D_e - S_e)/N_t, \qquad (12)$$

where D_e, S_e, and I_e represent the total number of deletion, substitution, and insertion errors, respectively, and N_t is the total number of characters in ground truth text.

Implementation Details. In our experiment, we employ a variant of ResNet [6] as the backbone. The image height is normalized to 64 pixels while retaining the same respect ratio as that of the original image. For samples with an aspect ratio greater than a preset threshold, e.g. 4, they are treated as vertical text line images, and a 90° counterclockwise rotation is performed. Furthermore, data augmentation techniques are used, including color and geometric transformations such as rotation and color jitter. The loss weights $\beta_1, \beta_2, \beta_3$ are set to 1, 0.1, and 1 according to empirical values respectively. The ADADELTA [26] optimizer with an initial learning rate of 0.3 is adopted. The learning rate is decayed by a factor of 0.1 at the 30-th, 40-th, 60-th, and 80-th epoch. Training is performed on four 32GB V100 GPUs with batch size set to 32.

3.2   Comparison with Other Methods

The comparisons of our method with other methods on the SCUT-HCCDoc, SCUT-EPT and CASIA-HWDB datasets are shown in Tables 1, 2 and 3 respectively. The comprehensive experiments conducted on these datasets validate the effectiveness of our proposed method.

Examples of recognition results on the SCUT-HCCDoc dataset are shown in Fig. 2. These examples show that our method can support both horizontal and vertical text images, and can adapt to different levels of background interference in text images captured in unconstrained scenarios.

Table 1. Results on the SCUT-HCCDoc dataset. Extra training data are from the CASIA-HWDB [14] and SCUT-EPT [31] datasets.

Method	AR(%)	CR(%)
CTC-based [18,27]	87.46	88.83
Attention-based [2,27]	83.30	84.81
Liu *et al.* [13]	89.06	90.12
Peng *et al.* [16]	90.71	92.01
Wu *et al.* [21]	91.25	92.75
SegCTC [9]	92.08	93.40
Li *et al.* [12]	92.72	94.02
Baseline	92.25	93.80
Ours	92.78	94.05
Ours (Use extra training data)	**93.18**	**94.32**

Table 2. Results on the SCUT-EPT dataset.

Method	AR(%)	CR(%)
CTC [31]	75.97	80.26
SAN [25]	76.42	80.83
Baseline	78.16	82.33
Ours	**80.03**	**83.71**

3.3   Ablation Studies

To demonstrate the effectiveness of the components in our model, we conduct a series of ablation experiments on the SCUT-HCCDoc dataset. The default settings in the ablation studies use the number of meta prompts as 4.

Visual Prompt Learning in Encoder and Decoder. The effect of multiscale visual prompt learning in encoder and decoder is shown in Table 4. The baseline model built from our previous work comprises a ResNet backbone, an

Table 3. Results on the CASIA-HWDB 2.0-2.2 test set and ICDAR 2013 dataset.

Method	HWDB test set		ICDAR 2013	
	AR(%)	CR(%)	AR(%)	CR(%)
Hu *et al.* [8]	94.29	94.95	95.48	96.13
Wu *et al.* [23]	95.88	95.95	96.20	96.32
Baseline (with language model)	97.04	97.35	93.05	94.15
Ours	97.66	98.45	95.26	96.02

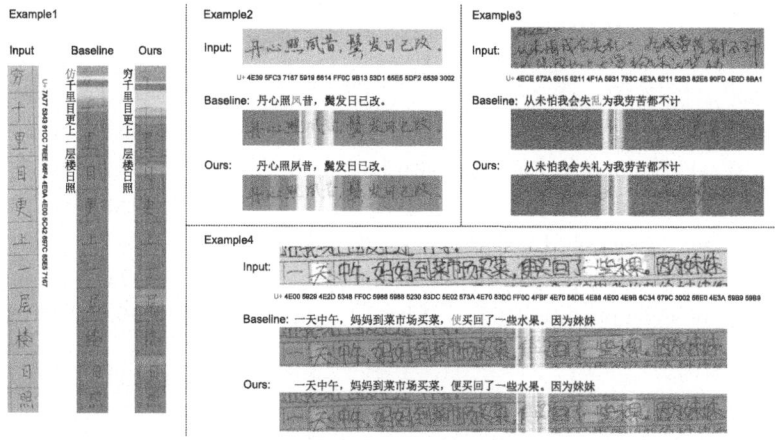

Fig. 2. Examples of recognition results on the SCUT-HCCDoc dataset. For each sample, the input image, recognition results (in which errors are marked in red) of the baseline and our method. The attention coefficients of the last cross-attention layer for decoding the corresponding character in the Transformer decoder are shown in the images. (Color figure online)

LSTM encoder, and a modified Transformer decoder. For the encoder, introducing single scale meta prompt at the last convolutional block in the ResNet can improve the recognition performance and removing the LSTM encoder can achieve better results. It indicates that the prompts can learn contextual information better than LSTM encoder, thus the LSTM encoder is removed in the subsequent experiments. Then, for the multi-scale prompts, the meta prompts are added to four blocks of ResNet to capture multi-scale contextual information and the result demonstrates that the utilization of multi-scale prompts in encode has improved performance. For the decoder, simply introducing additional CTC loss impairs the performance of the model slightly, while the proposed method with character-level visual prompts in decoder can achieve improved recognition performance.

We further show the effect of visual prompts in encoder by analyzing the mutual information between the input images and the feature representations output by the encoder.

The input image is denoted as X and the feature map extracted by the encoder as Y. The mutual information between X and Y is

$$I(X;Y) = H(Y) - H(Y|X). \tag{13}$$

From Zhao, et al. [28], we get the estimation of the lower bound of mutual information between X and Y, neglecting $H(Y)$,

$$\hat{I}(X;Y) = -\frac{1}{2}\log\left(\det\left(M\right)\right) = -\frac{1}{2}\mathrm{Tr}\left(\log(M)\right) \tag{14}$$

Table 4. The effects of visual prompt learning in encoder and decoder on the SCUT-HCCDoc dataset.

Method	AR(%)	CR(%)
Baseline	92.25	93.80
Single-Scale Prompts in Encoder (with LSTM)	92.28	93.76
Single-Scale Prompts in Encoder (w/o LSTM)	92.35	93.79
Multi-Scale Prompts in Encoder	92.53	93.85
Multi-Scale Prompts in Encoder + Additional CTC Loss	92.03	93.46
Multi-Scale Prompts in Encoder + Visual Prompts in Decoder	**92.58**	**93.91**

where

$$M = \Sigma_Y - \text{Cov}(Y, X)(\Sigma_X^{-1})^T \text{Cov}(Y, X)^T, \tag{15}$$

Σ_Y and Σ_X are the covariance matrix of Y and X. The mutual information between input images and extracted features before and after using prompts are recorded in Table 5, and we can find that the mutual information is increased after using adaptive prompts.

Table 5. The estimated mutual information before and after using prompt.

Method	Mutual Information
Before using prompts	1239.44
After using prompts	1316.02

Number of Prompts. The performance of our method with different numbers of meta prompts is shown in Table 6. The recognition accuracy has achieved the best with 10 meta prompts.

Table 6. Results on the SCUT-HCCDoc dataset with different number of prompts.

Nums	AR(%)	CR(%)	Model Params(M)
2	92.50	93.81	60.66
4	92.58	93.91	62.41
6	92.55	93.91	64.15
8	92.69	93.95	65.90
10	**92.71**	93.97	67.64
12	92.48	**93.99**	69.38

Mutual Information Loss. The effect of introducing local and global mutual information loss functions in the decoder at the training stage is shown in Table 7. We compare two ways of utilizing edge images that contain geometric structure information, including using edge images as input and using mutual information loss functions guided by edge images. Simply using edge images as input is less effective than using original images as input. Alternatively, using mutual information loss functions guided by edge images can better take advantage of both original images and edge images. The results show that the model has achieved the best recognition performance with both local and global mutual information objective functions.

Table 7. The effects of adding local mutual information (LMI) loss and global mutual information (GMI) loss on the SCUT-HCCDoc dataset.

	AR(%)	CR(%)
Before using MI losses (4 meta prompts)	92.58	93.91
Before using MI losses & using edge image as input(4 meta prompts)	91.19	92.79
LMI (4 meta prompts)	92.65	94.05
GMI (4 meta prompts)	92.61	94.04
LMI+GMI (4 meta prompts)	92.70	**94.10**
LMI+GMI (10 meta prompts)	**92.78**	94.05

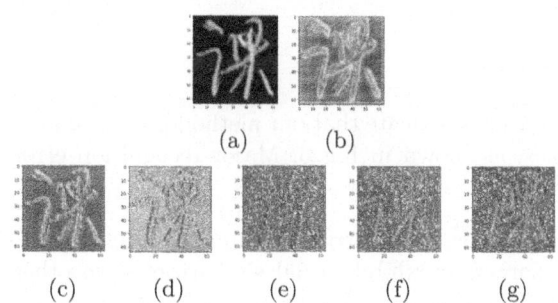

(a)   (b)

(c)   (d)   (e)   (f)   (g)

Fig. 3. Attention weights of different prompts.

3.4 Visualization

Visual Prompts in the Encoder. To better understand the influence of the adaptive prompts in the encoder, we conduct an additional isolated Chinese character recognition task on the CASIA-HWDB isolated character samples to visualize the cross-attention weights in the step of updating feature maps, as Fig. 3 shows.

Figure 3a shows the original character image. Figure 3b and Fig. 3c show the feature map before and after applying the cross-attention mechanism, while Fig. 3d, 3e, 3f and 3g show the attention weights generated by 4 different prompts. In the calculation of the cross-attention mechanism described by Eq. (5) in Sect. 3.1, feature maps are used as queries, and prompts are used as both keys and values. The attention weights are the output of the Softmax function defined in Eq. (3). The attention weights visualized in Fig. 3d, 3e, 3f and 3g exhibit higher responses on the left and right edges of the strokes respectively. As edges of strokes contain geometric information about handwriting styles, the learned prompts have revealed visual contextual information of style representations.

Visual Prompt Generation in the Decoder. The estimated location of each character reflects the quality of the generated image prompt. As shown in Fig. 4, the proposed method can estimate character locations in text lines with different lengths and character intervals.

Fig. 4. The visualization of the estimated locations (marked in green) of characters for generating visual prompts in decoder. (Color figure online)

3.5   Error Analysis

The experimental results indicate that our method produced identification errors on certain samples, as shown in Fig. 5. Major recognition errors can classified into two categories:

1. Similar character errors, as illustrated in images in Fig. 5a and Fig. 5b. Certain handwritten characters exhibit visual similarities with others, which might be improved by adopting semantic textual prompts via pre-trained language models.
2. Errors in cursive handwritings, as illustrated in Fig. 5b and Fig. 5c. Inter-character interference arising from cursive writing in handwriting images affects recognition, which might be alleviated by introducing a more precise character segmentation method to generate character-specific visual prompts in the decoder.

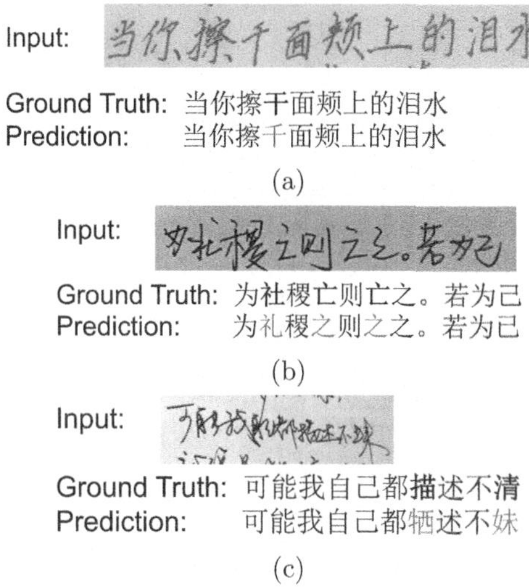

Input:

Ground Truth: 当你擦干面颊上的泪水
Prediction: 当你擦干面颊上的泪水

(a)

Input:

Ground Truth: 为社稷亡则亡之。若为己
Prediction: 为礼稷之则之之。若为己

(b)

Input:

Ground Truth: 可能我自己都**描述不清**
Prediction: 可能我自己都牺述不妹

(c)

Fig. 5. Examples of misrecognized handwriting images. For each example, the input image, ground truth and recognition results (errors marked in red) are presented. (Color figure online)

4   Conclusion

This paper investigates to incorporate visual prompt learning in an encoder-decoder model to utilize contextual features for Chinese handwriting recognition. For the encoder, learnable prompts are introduced for multi-scale feature maps to learn adaptive representations. Additional loss functions are designed to maximize the mutual information between the learned prompts and the edge images at local and global levels so as to focus on the geometric structure information of handwriting images. For the decoder, additional character-level vision prompts are used along with the embeddings of the predicted text to guide the decoding process. Ablation studies are carried out on the SCUT-HCCDoc dataset. Experiments results on the SCUT-HCCDoc, SCUT-EPT and CASIA-HWDB Chinese handwriting datasets validate the effectiveness of the proposed method. Future research work will further explore other forms of prompt learning including semantic textual prompts and more precise character-level visual prompts for handwriting recognition. More experiments will be carried out on other handwriting datasets.

Acknowledgements. This research work was supported by Huawei Noah's Ark Lab.

References

1. Canny, J.: A computational approach to edge detection. PAMI **8**(6), 679–698 (1986). https://doi.org/10.1109/TPAMI.1986.4767851
2. Chorowski, J.K., Bahdanau, D., Serdyuk, D., Cho, K., Bengio, Y.: Attention-based models for speech recognition. In: NIPS, vol. 28 (2015)
3. Daumé III, H., Brill, E.: Web search intent induction via automatic query reformulation. In: Proceedings of HLT-NAACL, pp. 49–52 (2004)
4. Devlin, J., Chang, M.W., Lee, K., Toutanova, K.: BERT: pre-training of deep bidirectional Transformers for language understanding. In: Proceedings of NAACL, vol. 1, pp. 4171–4186 (2019)
5. Gan, Y., Bai, Y., Lou, Y., Ma, X., Zhang, R., Shi, N., Luo, L.: Decorate the newcomers: visual domain prompt for continual test time adaptation. In: AAAI, pp. 7595–7603 (2023)
6. He, K., Zhang, X., Ren, S., Sun, J.: Deep residual learning for image recognition. In: CVPR, pp. 770–778 (2016)
7. Hjelm, R.D., Fedorov, A., Lavoie-Marchildon, S., et al.: Learning deep representations by mutual information estimation and maximization. In: ICLR (2018)
8. Hu, S., Wang, Q., Huang, K., Wen, M., Coenen, F.: Retrieval-based language model adaptation for handwritten Chinese text recognition. IJDAR **26**(2), 109–119 (2023)
9. Huang, J., Peng, D., Li, H., Ni, H., Jin, L.: SegCTC: offline handwritten Chinese text recognition via better fusion between explicit and implicit segmentation. In: Fink, G.A., Jain, R., Kise, K., Zanibbi, R. (eds.) ICDAR 2023. LNCS, vol. 14190, pp. 332–349. Springer, Cham (2023). https://doi.org/10.1007/978-3-031-41685-9_21
10. Jia, M., Tang, L., Chen, B.C., et al.: Visual prompt tuning. In: ECCV, pp. 709–727 (2022)
11. Kang, L., Riba, P., Rusiñol, M., Fornés, A., Villegas, M.: Pay attention to what you read: non-recurrent handwritten text-line recognition. Pattern Recogn. **129**, 108766 (2022)
12. Li, T., Wu, S., Wang, Z.: Mask guided selective context decoding for handwritten Chinese text recognition. In: ICASSP, pp. 1–5 (2023)
13. Liu, B., Sun, W., Kang, W., Xu, X.: Searching from the prediction of visual and language model for handwritten Chinese text recognition. In: ICDAR, pp. 274–288 (2021)
14. Liu, C.L., Yin, F., Wang, D.H., Wang, Q.F.: CASIA online and offline Chinese handwriting databases. In: ICDAR, pp. 37–41 (2011)
15. Lu, N., Yu, W., Qi, X., et al.: Master: multi-aspect non-local network for scene text recognition. Pattern Recogn. **117**, 107980 (2021)
16. Peng, D., Jin, L., Ma, W., et al.: Recognition of handwritten Chinese text by segmentation: a segment-annotation-free approach. IEEE Trans. Multimedia **25**, 2368–2381 (2023)
17. Ponti, E.M., Glavaš, G., Majewska, O., et al.: XCOPA: a multilingual dataset for causal commonsense reasoning. arXiv preprint arXiv:2005.00333 (2020)
18. Shi, B., Bai, X., Yao, C.: An end-to-end trainable neural network for image-based sequence recognition and its application to scene text recognition. PAMI **39**(11), 2298–2304 (2016)
19. Tang, P., Peng, L., Yan, R., et al.: Domain adaptation via mutual information maximization for handwriting recognition. In: ICASSP, pp. 2300–2304 (2022)

20. Wang, T., Zhu, Y., Jin, L., et al.: Decoupled attention network for text recognition. In: AAAI, pp. 12216–12224 (2020)
21. Wu, S., Li, Y., Wang, Z.: Improving CTC-based handwritten Chinese text recognition with cross-modality knowledge distillation and feature aggregation. In: ICME, pp. 792–797 (2023)
22. Wu, Y.C., Yin, F., Chen, Z., Liu, C.L.: Handwritten Chinese text recognition using separable multi-dimensional recurrent neural network. In: ICDAR, pp. 79–84 (2017)
23. Wu, Y.C., Yin, F., Liu, C.L.: Improving handwritten Chinese text recognition using neural network language models and convolutional neural network shape models. Pattern Recogn. **65**, 251–264 (2017)
24. Xie, X., Fu, L., Zhang, Z., Wang, Z., Bai, X.: Toward understanding wordart: corner-guided transformer for scene text recognition. In: ECCV, pp. 303–321 (2022)
25. Yan, S., Wu, J.W., Yin, F., Liu, C.L.: Recognizing handwritten Chinese texts with insertion and swapping using a structural attention network. In: ICDAR, pp. 557–571 (2021)
26. Zeiler, M.D.: Adadelta: an adaptive learning rate method. arXiv preprint arXiv:1212.5701 (2012)
27. Zhang, H., Liang, L., Jin, L.: SCUT-HCCDoc: a new benchmark dataset of handwritten Chinese text in unconstrained camera-captured documents. Pattern Recogn. **108**, 107559 (2020)
28. Zhao, S., Wang, Y., Yang, Z., Cai, D.: Region mutual information loss for semantic segmentation. In: NeurIPS, vol. 32 (2019)
29. Zhou, K., Yang, J., Loy, C.C., Liu, Z.: Conditional prompt learning for vision-language models. In: CVPR, pp. 16816–16825 (2022)
30. Zhou, K., Yang, J., Loy, C.C., Liu, Z.: Learning to prompt for vision-language models. IJCV **130**(9), 2337–2348 (2022)
31. Zhu, Y., Xie, Z., Jin, L., Chen, X., Huang, Y., Zhang, M.: SCUT-EPT: new dataset and benchmark for offline Chinese text recognition in examination paper. IEEE Access **7**, 370–382 (2018)
32. Zhu, Z.Y., Yin, F., Wang, D.H.: Attention combination of sequence models for handwritten Chinese text recognition. In: ICFHR, pp. 288–294 (2020)

Context-Aware Confidence Estimation for Rejection in Handwritten Chinese Text Recognition

Yangyang Liu[1,2(✉)], Yi Chen[1,2], Fei Yin[1,2], and Cheng-Lin Liu[1,2]

[1] State Key Laboratory of Multimodal Artificial Intelligence Systems (MAIS),
Institution of Automation, Chinese Academy of Sciences,
Beijing 100190, China
liuyangyang2021@ia.ac.cn, liucl@nlpr.ia.ac.cn
[2] School of Artificial Intelligence, University of Chinese Academy of Sciences, Beijing
100049, China

Abstract. Handwritten Chinese Text Recognition (HCTR) has been advanced largely by deep learning in recent years. However, the remaining recognition errors still hinder reliability-critical applications where zero-error is desired. Rejecting low-confidence patterns can help reduce the error rate but the increased rejection rate is also harmful. In this paper, we propose a character confidence estimation method incorporating contexts for character rejection in HCTR. Based on a text line recognizer outputting character segmentation and classification results, the confidence of each segmented character is estimated by combining the scores of a re-trained character classifier, the linguistic and geometric contexts. We introduce a probabilistic formula for estimating the confidence by combining the classifier and contextual scores, and an improved approach for scoring the geometric context using unary and binary geometric features. Experimental evaluations on the CASIA-HWDB and ICDAR2013 datasets demonstrate that our method can significantly improve the rejection performance in respect of low error rate at moderate rejection rate. The re-trained classifier, the linguistic context and the geometric context are all justified effective to improve the confidence.

Keywords: Handwritten Chinese Text Recognition · Confidence Estimation · Geometric Context · Bayesian probability formula

1 Introduction

In recent years, Handwritten Chinese Text Recognition (HCTR) has experienced significant advancements due to extensive training data and sophisticated deep network architectures. Despite the impressive performance of deep learning-based HCTR systems, they remain prone to recognition errors due to a range of challenging factors inherent in handwritten Chinese texts. These include the presence of extremely small punctuation marks, the overlapping and variable spacing of character areas, and the similarity between different characters (see Fig. 1). Such errors pose substantial challenges, especially in high-stake application scenarios like archive digitization, finance, courts and hospitals, where the accuracy and reliability of character recognition are critical. In these applications, manual verification is usually needed to guarantee high accuracy. Considering that current HCTR systems are not perfect, it is helpful to estimate the character confidence and reject those samples of low confidence, and the manual verification cost can be largely reduced by focusing on these low-confidence (rejected) samples only. Confidence estimation [8,10,24,44], the process of translating a model's output into the **Actual Probability** of the predicted class, plays a pivotal role in this scenario. It aids in making decisions about which samples to reject, thereby enhancing the reliability of the recognized text.

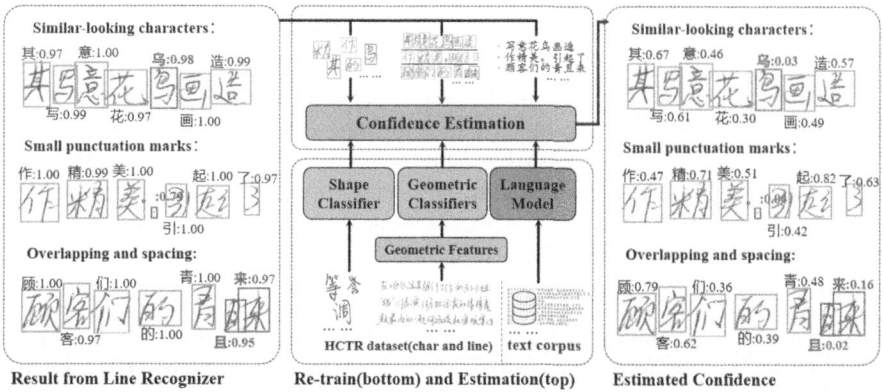

Fig. 1. Overview of our HCTR confidence estimation method. The green bounding boxes represent correctly predicted samples, while the red ones indicate incorrect ones. The left panel shows the MSP (maximum softmax probability) output by the text line recognizer. The middle panel shows the proposed context-aware confidence estimation process, and the right panel shows the confidence values produced by the proposed method. (Color figure online)

Confidence estimation was initially proposed to convert the output of a traditional machine learning algorithm, such as SVM, into a posterior probability [24]. In neural network classifiers, the largest class probability output by softmax layer can be directly used as the final confidence. This is known as MSP (maximum softmax probability) [8]. However, experiments have demonstrated that this approach suffers from severe overconfidence (Fig. 1 Left). Therefore, the confidence estimation on model output remains an important issue in the era of deep learning.

Confidence estimation has also been concerned in handwriting recognition and automatic speech recognition [4,12,15,16]. In handwriting recognition (usually taking a text line as the object of a sequence recognition problem), the text line recognizer's output of each character can serve as a confidence. However, the classifier output confidence may not be reliable, and the contexts of text (linguistic and geometric contexts) are ignored. Existing approaches primarily calibrate the confidence at the text line level, rejecting samples based on the entire line, without modelling each character individually [9,23,27]. While this approach is straightforward and intuitive, it lacks reliability in determining the confidence of each output character, and the confidence of the entire line lacks interpretability. In practical applications of HCTR, the text is usually quite lengthy, and the following manual verification requires the confidence of each individual character. Consequently, character-level confidence estimation and rejection is a more appropriate and effective strategy for the HCTR task.

In this paper, we propose a character confidence estimation method incorporating contexts for character rejection in HCTR. A Bayesian probability formula models the confidence of each predicted character, integrating the posterior probabilities of a re-trained character classifier and the linguistic and geometric contexts, finally outputting a calibrated confidence. We also proposed an improved approach for scoring geometric features of characters, from binary geometric features [42] to Bilateral Binary Geometric Features, which allow the classifier to simultaneously consider the geometric features of both the preceding and subsequent characters. We estimate character confidence on a recently proposed high accuracy HCTR system [39]. Experiments on the CASIA-HWDB dataset [19] and ICDAR2013 dataset [38] demonstrate that our method significantly enhances the rejection performance in terms of error-rejection tradeoff.

The main contributions in this paper are summarized as follows:

- We propose a method for character-level confidence estimation in Handwritten Chinese Text Line Recognition, by combining the probabilities of a re-trained character and linguistic and geometric contexts.
- We proposed an improved approach for scoring geometric context from Bilateral Binary Geometric ($BiBG$) Features.
- Our extensive experiments on a high accuracy text line recognizer show that the proposed method can significantly improve the rejection performance in terms of error-rejection tradeoff.

The rest of this paper is organized as follows: Sect. 2 introduces some related work. Section 3 presents our proposed confidence estimation method. Section 4 introduces geometric context and $BiBG$. Section 5 presents experimental results and Sect. 6 concludes our work.

2   Related Work

2.1   Confidence Estimation for Single Classifier

Confidence estimation for single object classification task under closed set is aimed to support tasks of misclassification detection (MisD) and confidence calibration.

Misclassification detection is to determine whether a test sample from a known category is incorrectly classified or not according to class probability. In deep neural networks, the output of softmax layer sums to 1, so it can be considered as a posterior probability in closed-set setting. Hendrycks et al. [8] regarded the highest softmax output probability as the output's confidence, known as MSP (maximum softmax probability), which is commonly treated as baseline method for misclassification detection.

Some works focus on misclassification detection during the training process. Experiments have shown that simple samples are often classified correctly earlier than difficult ones in training process. Moon et al. [29] proposed Confidence Ranking Loss (CRL), utilising the frequency with which each sample is correctly classified during training. Rabanser et al. [26] saved intermediate states of the model during training and inspect the consistency of samples during testing. Zhu et al. [43] combined two different flat minima techniques to boost misclassification detection. Moreover, regularization methods such as mixup [28], label smoothing [22], focal loss [21], have been demonstrated to be effective for confidence estimation. Other works treat misclassification detection as a binary classification task. Corbiere et al. [2] introduced an auxiliary model (ConfidNet) to learn confidence. This method first trains a main classification model with a softmax layer and uses the True Class Probability (TCP) as the standard confidence for each sample. Qu et al. [25] divided the training set into different parts before each iteration of training, employing a meta-learning strategy for the auxiliary model.

In the context of handwritten character recognition, confidence evaluation has been used to boost accuracy or combining multiple classifiers. Liu et al. [16] proposed a precise candidate selection method for large character set recognition by confidence evaluation of distance-based classifiers. Furthermore, they investigated a number of confidence evaluation methods for measurement-level combination of classifiers [6,17,18]. He et al. [7] analyzed the recognition confidence of handwritten Chinese character, based on the softmax output of convolutional neural network (CNN).

2.2   Context-Aware Confidence Estimation

Confidence estimation on text line recognition usually involves the use of contextual information.

Some recent works have focused on confidence estimation for scene-text recognition. Huang et al. [9] extended label smoothing to every class by leveraging the contextual dependency in sequences. Slossberg et al. [27] conducted an in-depth analysis of the current status of calibration in scene-text recognition models. Peng et al. [23] proposed a Perception and Semantic aware Sequence Regularization framework, which explores perceptively and semantically correlated sequences as regularization. These works all perform confidence calibration at the sequence-level, lacking modeling for individual characters, so cannot be used for character rejection (misclassification detection).

In the field of handwritten Chinese text recognition, Wang et al. [33] proposed a candidate segmentation-recognition path scoring method by confidence transformation. Wang et al. [30] proposed two regularized class-dependent confidence transformation methods and a string-level confidence learning method under the Minimum Classification Error (MCE) criterion. They further investigated into different confidence-learning methods for HCTR [31]. In the context of keyword spotting, Zhang et al. [40,41] proposed to directly estimate the posterior probability of candidate characters based on the N-best paths from the candidate segmentation-recognition lattice. Li et al. [15] proposed a new CNN based confidence estimation method featured with softmax. However, most of these works did not consider the problem of rejection. Li et al. [15] rejected recognition results whose confidence is lower than the threshold, but did not fully utilize contextual information, especially geometric information.

Confidence estimation has received intensive attention in Automatic Speech Recognition (ASR), which is a sequence recognition problem similar to text line recognition. These methods can be classified into three categories: a) methods based on posterior probabilities; b) utterance verification; and c) binary classification. The first approach is non-parametric and mainly focuses on post-processing of output probabilities. Wessel et al. [34] showed that posterior word probabilities computed on word graphs and N-best lists clearly outperform non-probabilistic confidence measures. Kemp et al. [11] exploited the use of word lattices as information sources for the measure of confidence. The utterance verification approach employs statistical hypothesis testing to train two models: one for the null hypothesis and another for the alternative hypothesis. Rahim et al. [14] described a HMM-based utterance verification system using the framework of statistical hypothesis testing. Sukkar et al. [13] presented a framework for vocabulary independent utterance verification in subword-based ASR. The binary classification approach uses a classifier to classify a word as correct or incorrect based on a combination of speech features. Del et al. [4] presented a comprehensive study of confidence measures using deep bidirectional recurrent neural network (DBRNN) and deep directional long short-term memory (DBLSTM) models. Kumar et al. [12] computed confidence on the predictions made by an end-to-end speech recognition model via sequence posterior probability-based and binary classification-based methods.

3   Proposed Method

3.1   Problem Formulation

Given a set of input images of handwritten Chinese text line $\mathcal{X} = \{\mathcal{X}_i\}_{i=1}^{N}$ as the training set and image $\mathcal{X}_i$ has a sequence label $\{(C_{ij}, B_{ij})\}_{j=1}^{L_i}$, where C_{ij} and B_{ij} are the class index and the bounding box of the j-th character in i-th image, respectively.

Suppose we already have a well pre-trained Chinese text recognizer f. Through f we get the prediction results $\mathcal{R}$:

$$\mathcal{R} = \{\mathcal{R}_i \mid \mathcal{R}_i = f(\mathcal{X}_i)\}_{i=1}^{N}, \quad f(\mathcal{X}_i) = \left\{\left(\hat{C}_{ij}, \hat{B}_{ij}\right)\right\}_{j=1}^{\hat{L}_i}. \tag{1}$$

Among these prediction results, we define a prediction as correctly predicted if its bounding box overlaps with any bounding box in the ground truth, and the character categories are the same. As is expressed as:

$$R_{ij} \text{ predicts correctly} \iff \exists k \in \{0, 1, \ldots, L_i\} : O(\hat{B}_{ij}, B_{ik}) \text{ and } \hat{C}_{ij} = C_{ik}, \tag{2}$$

where O denotes overlapping in bounding box. Our objective is to create a character-level confidence estimation function, denoted as κ, and $\hat{p} := \kappa(\mathcal{X}, \mathcal{R}, f)$ provides the **Actual Probability**. The lower the $\hat{p}$, the more likely to reject the character prediction.

3.2   Probabilistic Model

We try to address this problem from a probabilistic perspective. After text line recognition, we have obtained the pseudo-labels of the text lines $\left\{\left(\hat{C}_{ij}, \hat{B}_{ij}\right)\right\}_{j=1}^{\hat{L}_i}$. We consider the confidence of a character as the posterior probability with respect to the pseudo-label of the character category:

$$\hat{p}(\mathcal{R}_j) := P(\hat{C}_j \mid \mathcal{X}, \hat{B}_j) = P(\hat{C}_j \mid \mathcal{X}, \hat{\mathcal{X}}_j). \tag{3}$$

Here, for simplicity, we omit subscript i, assuming text line $\mathcal{X}_i$. j denotes the j-th prediction result in a text line image. $\hat{\mathcal{X}}_j$ denotes the image patch cut out from the text line image using the bounding box $\hat{B}_j$. Using the Bayesian probability formula, we can decompose the posterior probability:

$$
\begin{aligned}
P(\hat{C}_j \mid \mathcal{X}, \hat{\mathcal{X}}_j) &= \frac{P_{LM}(\hat{C}_j)P(\mathcal{X}, \hat{\mathcal{X}}_j \mid \hat{C}_j)}{P(\mathcal{X}, \hat{\mathcal{X}}_j)} \\
&= \frac{P_{LM}(\hat{C}_j)P(\mathcal{X} \mid \hat{C}_j)P(\hat{\mathcal{X}}_j \mid \hat{C}_j)}{P(\mathcal{X})P(\hat{\mathcal{X}}_j)} \\
&= \frac{P(\mathcal{X} \mid \hat{C}_j)}{P(\mathcal{X})} \frac{P_{LM}(\hat{C}_j)P(\hat{\mathcal{X}}_j \mid \hat{C}_j)}{P(\hat{\mathcal{X}}_j)}.
\end{aligned}
\tag{4}
$$

This can be viewed as the fusion of line-based probability and character-based probability. Following previous work [32], the prior probability of $\hat{C}_j$, which is a probability prediction made solely from the language model, without considering any visual information, is denoted as $P_{LM}(\hat{C}_j)$.

The first factor in Eq. (4) is equivalent to the posterior probability output during the text line recognition process:

$$\frac{P(\mathcal{X} \mid \hat{C}_j)}{P(\mathcal{X})} = P(\hat{C}_j \mid \mathcal{X}) = P_{recog}(\hat{C}_j). \tag{5}$$

In Eq. (4), $P(\hat{\mathcal{X}}_j)$ denotes the marginal likelihood of image patch $\hat{\mathcal{X}}_j$. With geometric context features [42], it can be decomposed into:

$$P(\hat{\mathcal{X}}_j) = P(x_j)P(ug_j)P(bg_j), \tag{6}$$

x, ug, bg denote shape features, unary geometric features, binary geometric features respectively (see Sect. 4). Similarly, $P(\hat{\mathcal{X}}_j \mid \hat{C}_j)$ can be decomposed in the same way, and we can make the following deduction:

$$
\begin{aligned}
\frac{P(\hat{\mathcal{X}}_j \mid \hat{C}_j)}{P(\hat{\mathcal{X}}_j)} &= \frac{P(x_j \mid \hat{C}_j)P(ug_j \mid \hat{C}_j)P(bg_j \mid \hat{C}_j)}{P(x_j)P(ug_j)P(bg_j)} \\
&= \frac{P(\hat{C}_j \mid x_j)P(\hat{C}_j \mid ug_j)P(\hat{C}_j \mid bg_j)}{P_1(\hat{C}_j)P_2(\hat{C}_j)P_3(\hat{C}_j)},
\end{aligned}
\tag{7}
$$

where $P(\hat{C}_j \mid x_j)$, $P(\hat{C}_j \mid ug_j)$, $P(\hat{C}_j \mid bg_j)$ are the posterior probability of character shape classifier, unary geometric [42] classifier, binary geometric classifier, which are all trained from single-character datasets. $P_1(\hat{C}_j)$, $P_2(\hat{C}_j)$, $P_3(\hat{C}_j)$ are the prior probabilities of the three classifiers.

Combining (4), (5), (7), Eq. (3) can be further transformed as:

$$\hat{p}(\mathcal{R}_j) = P_{recog}(\hat{C}_j)P_{LM}(\hat{C}_j)\frac{P(\hat{C}_j \mid x_j)P(\hat{C}_j \mid ug_j)P(\hat{C}_j \mid bg_j)}{P_1(\hat{C}_j)P_2(\hat{C}_j)P_3(\hat{C}_j)}, \tag{8}$$

Eq. (8) represents the result of the confidence estimation of prediction $\mathcal{R}_j$. It indicates the combination of text line recognition, language model, and the classifiers on character shape and geometric features to obtain a confidence that more accurately reflects the reliability of prediction results.

To normalize the multiplicative effect of these components, it is optional to take the Mth root of $\hat{p}(\mathcal{R}_j)$. Considering that the degree of Eq. (8) is 2, M is set to 2 by default. M does not affect the metrics in this paper.

3.3   Language Model

The language model we use in confidence estimation is a BERT-like model [5]. This model masks the middle token at each timestep and predicts the masked content based on the tokens before and after it. In Chinese text line recognition, the token of the language model is a single character. We take the results of text recognition as the input to the language model, outputting the probability of each character. The higher the probability, the more semantically reliable the character is.

3.4   Character Feature Classifiers

As is inferred in Eq. (8), we utilize three character feature classifiers to boost the confidence estimation, which are character shape classifier, unary geometric classifier, binary geometric classifier. Each of these classifiers has been individually trained on segmented character samples.

Character Shape Classifier. The character shape classifier is a CNN-based neural network [35] with a prototype classifier [36] substituting the last linear layer. On inputting a normalized isolated-character images, it outputs the probabilities of 7356 classes (labeled classes in our dataset). We use a large number of isolated-character images for re-training the character classifier, rather than using the existing classifier in the text line recognizer. Re-training is aimed to design a classifier generating more reliable confidence measures.

Character Geometric Classifiers. The geometric classifiers utilized in this study are Multi-Layer Perceptrons (MLPs) with a prototype classifier substituting the last linear layer as in character shape classifier. The inputs are unary and binary geometric features, which are represented as 42-dimensional and 24-dimensional vectors, respectively, extracted from text line training data with character-level annotations. The MLPs are employed to classify these unary and binary geometric features into some discrete classes.

The classes are obtained by clustering the character classes via K-means, based on the prototypes from unary geometric classifier. The number of classes for unary geometric features is supposed to be N, and that for binary geometric features is N^2, considering that binary geometric features taking a character-pair as input.

The geometric features will be further explained in following section.

4   Geometric Context

Geometric Context was first proposed for online handwritten Japanese character string recognition [42]. A statistical method was proposed for modeling both unary and binary geometric features, which were proven to be efficient in the performance of string recognition. Geometric features are particularly effective to distinguish between the punctuation marks and normal characters, which have large difference in character size and relative position [1]. Such features are useful in text line recognition, via, e.g., normalizing characters according to the vertical position in text line.

We follow the modeling method in [37] and proposed an improved approach for modeling binary geometry. Unary and binary geometric features are extracted as in Table 1 and Table 2. Unary geometric features reflect the position and distribution of individual characters in a text line, while binary geometric features reflect the relative positional relationships between characters.

Bilateral Binary Geometric ($BiBG$) Features. The conventional binary geometric features [42] solely address the geometric relationship between the current character and its predecessor. This is inadequate for fully capturing the geometric relationships among characters. For instance, the proximity of two punctuation marks can significantly skew the distribution of binary geometric features. To address this limitation, we introduce the Bilateral Binary Geometric ($BiBG$) Features. $BiBG$ enhances binary geometric features by concatenating the binary geometric features of the current character with the previous and the next character. This modification yields a more precise representation of the local geometric context of the current character within the text line, thereby enhancing robustness compared to traditional binary geometric features. The *BiBG* classifier takes the binary geometric features of two character-pairs as input, so its number of classes is N^3, the cube of that of unary geometric features (two intersecting character-pairs have three characters).

Table 1. Unary geometric features (The last column denotes whether normalized w.r.t. the text line height or not).

No.	Feature	Norm
1–2	Height and width of bounding box	Y
3	Sum of inner gaps	Y
4–5	Distances of horizontal/vertical gravity center to left/upper bound	Y
6	Logarithm of aspect ratio	N
7	Square root of bounding box area	Y
8	Diagonal length of bounding box	Y
9–10	Distances of horizontal/vertical gravity center to horizontal/vertical geometric center	Y
11–12	Distances of vertical gravity/geometric center to text line vertical gravity/geometric	Y
13–14	Distances of upper/lower bound to text line vertical geometric center	Y
15–16	Means of horizontal/vertical projection profiles	Y
17–22	Normalized amplitude deviations, coefficients of skewness and kurtosis of the horizontal and vertical projection profiles	N
23–30	Means and deviations of the upper, lower, left and right outline profiles	Y
31–42	Normalized amplitude deviations, coefficients of skewness and kurtosis of the upper, lower, left and right outline profiles	N

Table 2. Binary geometric features.

No.	Feature	Norm
1–6	Distances between the upper bounds, lower bounds, upper-lower bounds, lower-upper bounds, left bounds and right bounds	Y
7–8	Distances between the horizontal gravity centers and the vertical gravity centers	Y
9–10	Distances between the horizontal geometric centers and the vertical geometric centers	Y
11–12	Height and width of the box enclosing two consecutive characters	Y
13	Gap between the bounding boxes	Y
14	Ratio of heights of the bounding boxes	N
15	Ratio of widths of the bounding boxes	N
16	Square root of the common area of the bounding boxes	Y
17–20	Differences between the mean of upper, lower, left and right outline profiles	Y
21-24	Differences between the deviation of upper, lower, left and right outline profiles	Y

5   Experiments

5.1   Datasets

We trained the text line recognition model on the CASIA-HWDB 2.0-2.2 text line dataset and a synthetic dataset [39] as baseline. Subsequently, we re-train the character shape classifier on the CASIA-HWDB 1.0-1.2 isolated character dataset and cropped characters from CASIA-HWDB 2.0-2.2. We re-train the unary geometric classifier, and $BiBG$ classifier on the CASIA-HWDB 2.0-2.2 text line dataset. Our method is tested on the ICDAR2013 test set.

CASIA-HWDB. The CASIA-HWDB dataset consists of two distinct sections. CASIA-HWDB 1.0-1.2 contains 3,895,135 isolated character samples of 7,356 classes from 1,020 writers. CASIA-HWDB 2.0-2.2 has 52,230 text lines segmented from 5,091 handwritten pages. We use CASIA-HWDB dataset for training shape classifier and geometric classifiers.

ICDAR2013. ICDAR2013 is a competition dataset with 3,342 text lines segmented from 300 handwritten pages.

5.2   Implementation Details

For the text line recognition model, training was conducted on four Titan RTX GPUs, with a batch size of 8 per GPU. We utilized the Adam optimizer, starting with an initial learning rate of 1×10^{-3}, and multiplied it by 0.1 at 50% and 80% of the total training iterations [39]. The character shape classifier was trained on a single Titan RTX GPU, with a batch size of 256, using the SGD optimizer. Its initial learning rate was set at 0.1, which was reduced by a factor of 0.1 at 50% of the total training iterations. The training process for the geometric classifier was similar to that of the shape classifier, but with a batch size of 32.

For character geometric classifiers, N is set to be 10, so unary geometric features, binary geometric features, and $BiBG$ features are classified into 10, 100, 1000 classes respectively.

5.3   Performance Metrics

To evaluate the performance of our method in character rejection (misclassification detection), we use AUROC [3], AUPR-Success, AUPR-Error [20], and FPR at 95% TPR [43] as our performance metrics. Here, "Success" means treating correctly classified samples as positive, while "Error" means the opposite.

Additionally, we use a more intuitive metric for measuring the tradeoff between rejection rate and error rate.: **RP99 (R**ejecting **R**ate at **P**recision of **99**%), which denotes the proportion of samples rejected while the accuracy of remaining samples is 99%. This metric is simpler in definition and intuitively reflects the need of human verification. Our experimental results show that this new metric is also consistent with other widely used metrics. For more comprehensive measure of error-rejection tradeoff, we also give the receiving characteristic curve of error-rejection rates.

5.4   Ablation Study

To verify the effectiveness of each part in confidence estimation, we conducted ablation experiments on the ICDAR2013 testing dataset. The baseline is MSP (treating the maximum softmax probability as confidence). These experiments demonstrate that each part in our method is effective in terms of misclassification detection performance, even in the absence of language model, as is shown in Table 3 and Table 4. The experiments in Table 4 are conducted on the results of text line recognition with language model (LM), while those of Table 3 are based on the results of text line recognition without LM (LM is used in post confidence estimation, however).

Table 3. Effectiveness of each part in confidence estimation. (*LM* denotes Language Model used in confidence estimation. *ug* denotes unary geometric features. *Bg* denotes Bilateral Binary Geometric (*BiBG*) Features. The same in Table 4)

	LM	ug	Bg	AUROC↑	AUPR-S↑	AUPR-E↑	FPR95↓	RP99↓
Baseline[a]				88.76	99.26	38.50	45.97	25.27
Ours				90.30	99.39	43.94	41.79	20.50
Ours	✓			91.94	99.51	45.95	39.36	16.46
Ours	✓	✓		92.28	99.53	47.21	38.06	15.54
Ours		✓	✓	91.55	99.49	44.62	40.60	17.42
Ours	✓	✓	✓	**93.08**	**99.58**	**50.95**	**34.09**	**13.66**

[a]The precision of baseline without rejection is 95.39%

Table 4. Effectiveness of each part in confidence estimation on outputs with language model.

	LM	ug	Bg	AUROC↑	AUPR-S↑	AUPR-E↑	FPR95↓	RP99↓
Baseline[a]	✓			81.65	99.49	7.29	70.17	11.47
Ours				87.08	99.70	8.37	69.74	10.36
Ours	✓			89.29	99.75	10.14	64.77	8.09
Ours	✓	✓		89.42	99.75	10.32	64.58	7.82
Ours		✓	✓	88.94	99.74	9.71	65.99	8.65
Ours	✓	✓	✓	**89.95**	**99.77**	**11.25**	**59.77**	**6.82**

[a]The precision of baseline without rejection is 98.16%

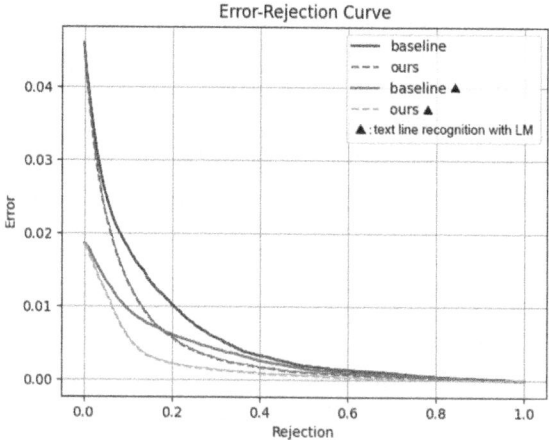

Fig. 2. Error-Rejection Curve. The horizontal axis represents the percentage of rejected samples, while the vertical axis indicates the error rate of the remaining samples. The solid line represents using the output of the line recognizer as the confidence score, while the dashed line is our confidence estimation method. (Color figure online)

The ablation study proves the effectiveness of our method's framework and the geometric context in confidence estimation. The language model, unary geometric features, and binary geometric features classifiers all contribute to performance gain. Combining them yields the best performance. Furthermore, the results without language model in confidence estimation shows that even if our method relies solely on the visual features of characters, it can still significantly enhance the performance of misclassification detection.

Effectiveness of Bilateral Binary Geometric Features. We compare Bilateral Binary Geometric ($BiBG$) Features proposed in this paper with the conventional binary geometric features [37]. The results are shown in Tables 5 and 6.

It is evident that $BiBG$, upon integrating bidirectional geometric features, surpasses the conventional bg in terms of confidence performance. Furthermore, this enhancement is pronounced in both scenarios - with and without the use of a language model in text line recognition.

Tradeoff Between Rejection and Error. The Error-Rejection Curve is shown in Fig. 2, which measures the tradeoff between the rejection rate and the error rate. As can be seen from the curve, the curve descends more rapidly with the use of our confidence estimation method, indicating that fewer samples need to be rejected to achieve a low error rate. Notably, our confidence estimation method remains significantly effective in the case of text line recognition with language model. It is noteworthy that the curve for confidence estimation on outputs of text line recognition without language model (orange dashed line) descends more sharply compared to baseline with language model (blue solid line).

Table 5. Comparison of Binary Geometric Features on outputs of text line recognition without LM.

	AUROC↑	AUPR-S↑	AUPR-E↑	FPR95↓	RP99↓
Ours (w/ *bg*)	92.84	99.56	49.74	35.21	14.29
Ours (w/ *BiBG*)	**93.08**	**99.58**	**50.95**	**34.09**	**13.66**

Table 6. Comparison of Binary Geometric Features on outputs of text line recognition with LM.

	AUROC↑	AUPR-S↑	AUPR-E↑	FPR95↓	RP99↓
Ours (w/ *bg*)	89.82	99.76	11.01	61.76	7.12
Ours (w/ *BiBG*)	**89.95**	**99.77**	**11.25**	**59.77**	**6.82**

5.5   Comparison with Other Methods

We also compare our method with other confidence calibration approaches. The experimental result is presented in Table 7.

Table 7. Comparison with other methods on outputs of text line recognition without LM.

	AUROC↑	AUPR-S↑	AUPR-E↑	FPR95↓	RP99↓
Baseline	88.76	99.26	38.50	45.97	25.27
Top1-Top2	88.58	99.27	37.09	46.31	25.36
LS [22]	88.85	98.56	41.22	48.23	26.74
Focal [21]	90.36	98.54	42.92	47.89	26.28
FMFP [43]	89.05	99.28	35.79	48.48	24.89
CASLS [9]	90.01	99.30	40.57	48.92	21.82
Ours (shape only)	90.30	99.39	43.94	41.79	20.50
Ours	**93.08**	**99.58**	**50.95**	**34.09**	**13.66**

According to the experiment results, traditional methods of confidence estimation (Top1-Top2, using the difference between the probability of the highest and the second-highest class as confidence) are ineffective for HCTR. Methods based on regularization (Label Smoothing, Focal Loss, FMFP) aim to create a smoother distribution in feature space by introducing regularization terms and disturbances during training. This paradigm offers a slight improvement in misclassification detection performance, but it is not significant, as it fails to utilize confidence from aspects like the semantics or shape of the text. CASLS, which relies on contextual information for label smoothing, can be considered as implicitly using a language model, thus showing a noticeable improvement in performance.

In contrast, our method, designed specifically for HCTR, combines the output of the recognizer with multiple classifiers. Our method has led to a significant improvement in misclassification detection performance, surpassing all the compared methods. Especially, when only the character shape classifier is used, our method still shows better performance compared to others. The character shape classifier reevaluates the recognition results of text lines, which is trained on a large-scale single-character dataset.

5.6 Visualization and Analysis

We visualize some sample the results of baseline and our method in Fig. 3. It is observable that recognition errors are caused by shape similarity to error class, small punctuation marks, and overlapping and variable spacing in character

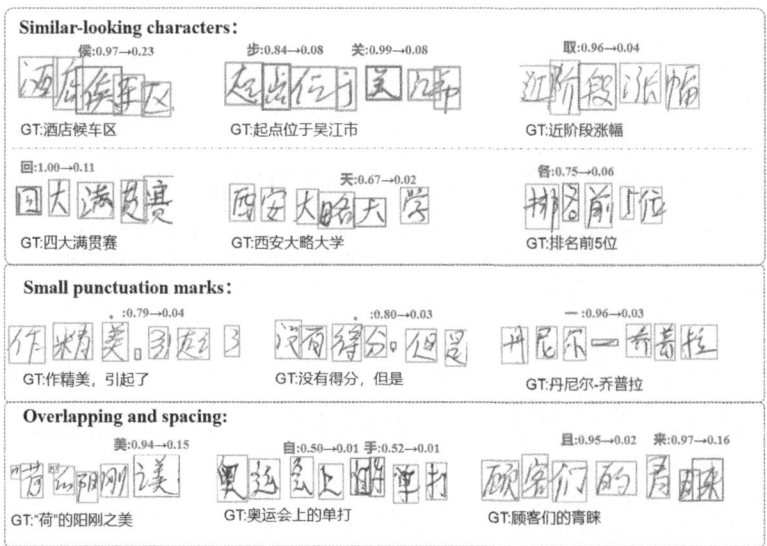

Fig. 3. Visualization of confidence estimation. The red bounding boxes denote misclassified samples. Red characters indicate the predicted character classes, with the right arrow showing the change in our confidence estimation compared to the baseline method MSP. GT denotes the ground truth. (Color figure online)

areas. In the results of MSP (maximum softmax probability), the phenomenon of overconfidence is obvious, with the confidence of misclassified samples being very high and tending towards 1. After applying our confidence estimation method, the confidence of misclassified samples have been generally corrected to a very low value.

6   Conclusion

In this paper, we present a character-level confidence estimation method for Handwritten Chinese Text Recognition (HCTR) by combining re-trained character classifier and linguistic and geometric contexts. From a probabilistic perspective, we derive a formula for calculating the confidence of each predicted character. An improved geometric context scoring method is also proposed to leverage bilateral binary geometric features. Extensive ablation and comparative experiments were conducted on the ICDAR2013 test set to demonstrate the effectiveness of our method. Our method outperforms all other confidence calibration and estimation methods in terms of misclassification detection and error-rejection tradeoff. The method can be applied to more handwriting datasets and scripts in the future.

Acknowledgements. This work has been supported by the National Key Research and Development Program Grant 2020AAA0109700, the National Natural Science Foundation of China (NSFC) Grants U23B2029 and U20A20223.

References

1. Chen, Y., Zhang, H., Liu, C.L.: Improved learning for online handwritten Chinese text recognition with convolutional prototype network. In: Fink, G.A., Jain, R., Kise, K., Zanibbi, R. (eds.) ICDAR 2023. LNCS, vol. 14190, pp. 38–53. Springer, Cham (2023). https://doi.org/10.1007/978-3-031-41685-9_3
2. Corbière, C., Thome, N., Bar-Hen, A., Cord, M., Pérez, P.: Addressing failure prediction by learning model confidence. In: Advances in Neural Information Processing Systems, vol. 32 (2019)
3. Davis, J., Goadrich, M.: The relationship between precision-recall and roc curves. In: Proceedings of the 23rd International Conference on Machine Learning (ICML), pp. 233–240 (2006)
4. Del-Agua, M.A., Gimenez, A., Sanchis, A., Civera, J., Juan, A.: Speaker-adapted confidence measures for ASR using deep bidirectional recurrent neural networks. IEEE/ACM Trans. Audio Speech Lang. Process. **26**(7), 1198–1206 (2018)
5. Devlin, J., Chang, M.W., Lee, K., Toutanova, K.: Bert: pre-training of deep bidirectional transformers for language understanding. arXiv preprint arXiv:1810.04805 (2018)

6. Hao, H., Liu, C.L., Sako, H., et al.: Confidence evaluation for combining diverse classifiers. In: Proceedings of the 7th International Conference on Document Analysis and Recognition (ICDAR), vol. 3, pp. 760–765 (2003)

7. He, M., Zhang, S., Mao, H., Jin, L.: Recognition confidence analysis of handwritten Chinese character with CNN. In: Proceedings of the 13th International Conference on Document Analysis and Recognition (ICDAR), pp. 61–65 (2015)

8. Hendrycks, D., Gimpel, K.: A baseline for detecting misclassified and out-of-distribution examples in neural networks. arXiv preprint arXiv:1610.02136 (2016)

9. Huang, S., Luo, Y., Zhuang, Z., Yu, J.G., He, M., Wang, Y.: Context-aware selective label smoothing for calibrating sequence recognition model. In: Proceedings of the 29th ACM International Conference on Multimedia, pp. 4591–4599 (2021)

10. Jaeger, P.F., Lüth, C.T., Klein, L., Bungert, T.J.: A call to reflect on evaluation practices for failure detection in image classification. arXiv preprint arXiv:2211.15259 (2022)

11. Kemp, T., Schaaf, T., et al.: Estimating confidence using word lattices. In: Proceedings of the 5th European Conference on Speech Communication and Technology (EUROSPEECH), pp. 827–830 (1997)

12. Kumar, A., Singh, S., Gowda, D., Garg, A., Singh, S., Kim, C.: Utterance confidence measure for end-to-end speech recognition with applications to distributed speech recognition scenarios. In: Proceedings of the 21th Conference of the International Speech Communication Association (INTERSPEECH), vol. 2020, pp. 4357–4361 (2020)

13. Lee, C.H.: Vocabulary independent discriminative utterance verification for non-keyword rejection in subword based speech recognition. IEEE Trans. Speech Audio Process. 4(6), 420–429 (1996)

14. Lee, C.H., Juang, B.H.: Discriminative utterance verification for connected digits recognition. IEEE Trans. Speech Audio Process. 5(3), 266–277 (1997)

15. Li, P., Peng, L., Wen, J.: Rejecting character recognition errors using CNN based confidence estimation. Chin. J. Electron. 25(3), 520–526 (2016)

16. Liu, C.L.: Precise candidate selection for large character set recognition by confidence evaluation. IEEE Trans. Pattern Anal. Mach. Intell. 22(6), 636–641 (2000)

17. Liu, C.L.: Classifier combination based on confidence transformation. Pattern Recogn. 38(1), 11–28 (2005)

18. Liu, C.L., Hao, H., Sako, H.: Confidence transformation for combining classifiers. Pattern Anal. Appl. 7, 2–17 (2004)

19. Liu, C.L., Yin, F., Wang, D.H., Wang, Q.F.: CASIA online and offline Chinese handwriting databases. In: Proceedings of the 11th International Conference on Document Analysis and Recognition (ICDAR), pp. 37–41 (2011)

20. Manning, C., Schutze, H.: Foundations of Statistical Natural Language Processing. MIT Press, Cambridge (1999)

21. Mukhoti, J., Kulharia, V., Sanyal, A., Golodetz, S., Torr, P., Dokania, P.: Calibrating deep neural networks using focal loss. In: Advances in Neural Information Processing Systems, vol. 33, pp. 15288–15299 (2020)

22. Müller, R., Kornblith, S., Hinton, G.E.: When does label smoothing help? In: Advances in Neural Information Processing Systems, vol. 32, pp. 4694–4703 (2019)

23. Peng, Z., Luo, Y., Chen, T., Xu, K., Huang, S.: Perception and semantic aware regularization for sequential confidence calibration. In: Proceedings of the IEEE/CVF Conference on Computer Vision and Pattern Recognition, pp. 10658–10668 (2023)

24. Platt, J., et al.: Probabilistic outputs for support vector machines and comparisons to regularized likelihood methods. In: Advances in Large-Margin Classifiers, pp. 61–74 (1999)

25. Qu, H., Li, Y., Foo, L.G., Kuen, J., Gu, J., Liu, J.: Improving the reliability for confidence estimation. In: Avidan, S., Brostow, G., Cissé, M., Farinella, G.M., Hassner, T. (eds.) ECCV 2022. LNCS, vol. 13687, pp. 391–408. Springer, Cham (2022). https://doi.org/10.1007/978-3-031-19812-0_23
26. Rabanser, S., Thudi, A., Hamidieh, K., Dziedzic, A., Papernot, N.: Selective classification via neural network training dynamics. arXiv preprint arXiv:2205.13532 (2022)
27. Slossberg, R., et al.: On calibration of scene-text recognition models. In: Karlinsky, L., Michaeli, T., Nishino, K. (eds.) ECCV 2022. LNCS, vol. 13804, pp. 263–279. Springer, Cham (2022). https://doi.org/10.1007/978-3-031-25069-9_18
28. Thulasidasan, S., Chennupati, G., Bilmes, J.A., Bhattacharya, T., Michalak, S.: On mixup training: improved calibration and predictive uncertainty for deep neural networks. In: Advances in Neural Information Processing Systems, vol. 32 (2019)
29. Toneva, M., Sordoni, A., Combes, R.T.D., Trischler, A., Bengio, Y., Gordon, G.J.: An empirical study of example forgetting during deep neural network learning. arXiv preprint arXiv:1812.05159 (2018)
30. Wang, D.H., Liu, C.L.: String-level learning of confidence transformation for Chinese handwritten text recognition. In: Proceedings of the 21st International Conference on Pattern Recognition (ICPR), pp. 3208–3211 (2012)
31. Wang, D.H., Liu, C.L.: Learning confidence transformation for handwritten Chinese text recognition. Int. J. Doc. Anal. Recognit. (IJDAR) 17, 205–219 (2014)
32. Wang, Q.F., Yin, F., Liu, C.L.: Integrating language model in handwritten Chinese text recognition. In: Proceedings of the 10th International Conference on Document Analysis and Recognition (ICDAR), pp. 1036–1040 (2009)
33. Wang, Q.F., Yin, F., Liu, C.L.: Improving handwritten Chinese text recognition by confidence transformation. In: Proceedings of the 11th International Conference on Document Analysis and Recognition (ICDAR), pp. 518–522 (2011)
34. Wessel, F., Macherey, K., Ney, H.: A comparison of word graph and n-best list based confidence measures. In: Proceedings of the 6th European Conference on Speech Communication and Technology (EUROSPEECH), pp. 315–318 (1999)
35. Xiao, X., Jin, L., Yang, Y., Yang, W., Sun, J., Chang, T.: Building fast and compact convolutional neural networks for offline handwritten Chinese character recognition. Pattern Recogn. 72, 72–81 (2017)
36. Yang, H.M., Zhang, X.Y., Yin, F., Yang, Q., Liu, C.L.: Convolutional prototype network for open set recognition. IEEE Trans. Pattern Anal. Mach. Intell. 44(5), 2358–2370 (2022)
37. Yin, F., Wang, Q.F., Liu, C.L.: Transcript mapping for handwritten Chinese documents by integrating character recognition model and geometric context. Pattern Recogn. 46(10), 2807–2818 (2013)
38. Yin, F., Wang, Q.F., Zhang, X.Y., Liu, C.L.: ICDAR 2013 Chinese handwriting recognition competition. In: Proceedings of the 12th International Conference on Document Analysis and Recognition (ICDAR), pp. 1464–1470 (2013)
39. Yu, M.M., Zhang, H., Yin, F., Liu, C.L.: An approach for handwritten Chinese text recognition unifying character segmentation and recognition. Pattern Recognit. 110373 (2024)
40. Zhang, H., Wang, D.H., Liu, C.L.: A confidence-based method for keyword spotting in online Chinese handwritten documents. In: Proceedings of the 21st International Conference on Pattern Recognition (ICPR), pp. 525–528 (2012)
41. Zhang, H., Wang, D.H., Liu, C.L.: Character confidence based on n-best list for keyword spotting in online Chinese handwritten documents. Pattern Recogn. 47(5), 1880–1890 (2014)

42. Zhou, X.D., Yu, J.L., Liu, C.L., Nagasaki, T., Marukawa, K.: Online handwritten Japanese character string recognition incorporating geometric context. In: Proceedings of the 9th International Conference on Document Analysis and Recognition (ICDAR), vol. 1, pp. 48–52 (2007)
43. Zhu, F., Cheng, Z., Zhang, X.Y., Liu, C.L.: Rethinking confidence calibration for failure prediction. In: Avidan, S., Brostow, G., Cissé, M., Farinella, G.M., Hassner, T. (eds.) ECCV 2022. LNCS, vol. 13685, pp. 518–536. Springer, Cham (2022). https://doi.org/10.1007/978-3-031-19806-9_30
44. Zhu, F., Zhang, X.Y., Wang, R.Q., Liu, C.L.: Learning by seeing more classes. IEEE Trans. Pattern Anal. Mach. Intell. **45**(6), 7477–7493 (2022)

Radical Similarity Based Model Optimization and Post-correction for Chinese Character Recognition

Zhongyuan Han⬛, Jun Du^(✉)⬛, Mobai Xue⬛, Jiefeng Ma⬛, Pengfei Hu⬛, and Zhenrong Zhang⬛

University of Science and Technology of China, Hefei, China
{hanzhongy,xmb15,jfma,pengfeihu,zzr666}@mail.ustc.edu.cn,
jundu@ustc.edu.cn

Abstract. Radical-based methods for Chinese character recognition (CCR) have been proven effective and offer substantial advantages. Different from character-based methods, Chinese characters are described as combinations of structures and radicals, and character recognition is achieved by the proper identifications of these components. However, there are visual similarities among radicals, leading to the ambiguity problem for CCR, which is not fully utilized in previous work. Accordingly, in this study, we first employ the stroke order information of Chinese radicals to establish a radical similarity metric. Then we improve the radical-based CCR in two ways. During the training stage, we propose a new loss function called minimum Bayesian risk (MBR) based on the radical similarity metric to yield better performance. During the recognition stage, the radical similarity is adopted to post-correct the potential error recognition results, offering a low-cost yet effective solution. Experimental results on different radical-based CCR models and datasets demonstrate the effectiveness and robustness of our proposed method.

Keywords: Radical similarity · Chinese character recognition · Bayesian risk

1 Introduction

Chinese character recognition has been under extensive investigation for decades yet remains challenging due to the large-scale vocabulary and complex structure of Chinese characters. From a compositional perspective, Chinese characters can be represented by the combinations of structures and radicals, which are the fundamental components of Chinese characters. Compared with numerous categories of characters, the number of radicals is limited. Only about 500 radicals are adequate to describe over 20,000 Chinese characters [27]. Consequently, it is natural to decompose Chinese characters to the spatial structures shown in Fig. 1(a) and radicals, through which the objective for CCR can be transformed into the recognition of structures and radicals. As illustrated in Fig. 1(b), Each Chinese character has a tree representation. In this tree, intermediate nodes represent structural components, while the leaf nodes are specific radicals. By performing the depth-first traversal of the tree, we can obtain an Ideographic Description Sequence (IDS) [18], which is a description grammar of Chinese character structure, defined under the Unicode standard.

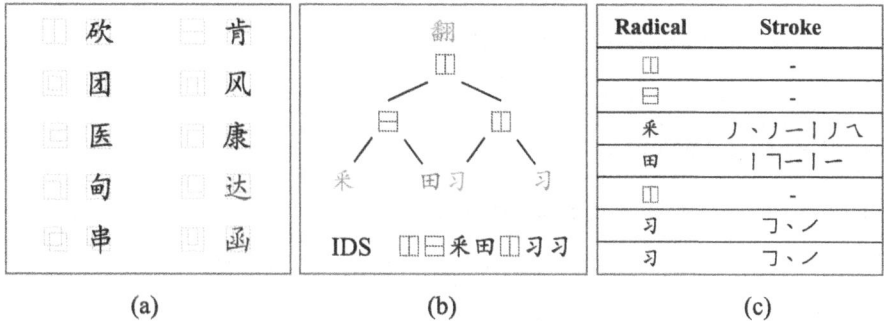

Fig. 1. (a) Ten structures in Chinese character. (b) An example of radical level representation of a Chinese character. (c) Stroke sequence of radicals in (b).

Moreover, Chinese radicals can be further broken down into strokes which are the smallest indivisible units of Chinese characters, as shown in Fig. 1(c). Since radicals are composed by combining strokes on a plane, it's hard to precisely describe such two-dimensional relations to correspond to a radical with its strokes. Radicals can only be approximately represented by the writing order of strokes. Although it is inadequate to denote each radical uniquely, it can still illustrate the diversities among most radicals and preserve valuable information. Consequently, we decompose a radical into stroke sequence representation according to its writing order and leverage this information to describe the similarities between different radicals.

In this paper, we first introduce a metric to measure the similarity between different radicals, which is defined by the normalized Levenshtein distance [31] between the stroke sequences of different radicals. During the training stage, the similarity information is utilized to calculate the MBR loss function, aiding in optimizing the model. In the test stage, we leverage the radical similarity to propose a post-correction module to rectify erroneous radical sequences through syntactic constraints. To validate the effectiveness of our proposed method, we conduct experiments on various radical-based CCR models. The experimental results of artistic and handwritten datasets demonstrate that the MBR loss function and the post-correction module can help improve performance across various models. The contributions of this paper are summarized as follows:

– We first propose a similarity metric for Chinese radicals, which is defined according to the stroke sequence of radicals, allowing for quantifiably evaluating the similarity between different radicals.

- We incorporate the similarity metric into the training stage, and the corresponding MBR loss function based on our metric can help optimize the model. Additionally, a post-correction module is proposed to rectify the decoded radical sequence through syntactic constraints with the assistance of the similarity metric.
- Our experimental results, obtained from various models and datasets, incontrovertibly demonstrate the effectiveness of our approach for radical-based CCR methods. We further analyze the experimental results, which indicate that our approach has exerted a favorable impact on models, thereby contributing to enhanced recognition efficiency.

2 Related Works

2.1 Chinese Character Recognition

Chinese character recognition technology has undergone over four decades of development since Casey and Nagy published the pioneering paper in 1966 [3]. Early approaches [11,21] relied on morphology-based observations to obtain hand-crafted features for CCR tasks. With the development of deep learning, this manual feature extraction step is gradually being replaced by automatic feature extraction through neural networks. DirectMap [35] combines the traditional normalization-cooperated direction decomposed feature map with the deep convolution neural network. [26] proposes the template-instance loss functions to alleviate the imbalance problem between easy and difficult character instances [10]. Radical-based methods are proposed to address the issue of zero-shot learning [20] problems in CCR tasks. Chinese characters are represented by structures and radicals, which transforms the task of character classification into accurate recognition of these components. Compared to the character-based method, the number of required categories for recognition has been significantly reduced, thereby injecting renewed vitality into the domain of CCR. RAN [34] utilizes the aforementioned IDS to represent a Chinese character, employing an encoder-decoder architecture as the basic framework. FewShotRAN [23] follows the Encoder-Decoder framework, adopts prototype learning [29] to learn robust radical features, and proposes a character analysis decoder to avoid inflexible match decoding. HDE [2] leverages the radical tree to encode a Chinese character to an embedding and learns the compatibility between the image feature and the embedding. RTN [28] employs Transformer [22], decoding structures and radicals via the self-attention mechanism. These methods can be applied to recognize unseen Chinese characters because the necessary radicals and structures have been learned from characters observed during training [34]. Despite the remarkable success achieved by radical-based methods in CCR tasks, regrettably, they regard all radicals as mutually independent and unrelated entities without considering the inherent similarity among various radicals. The utilization of this additional information is often neglected. Recently, some stroke-based methods [4,33] have also been proposed. Resembling radical-based methods, they utilize a stroke sequence to represent a Chinese character. Compared to radicals, strokes involve even fewer types, which offers a notable advantage in zero-shot learning CCR tasks.

2.2   Minimum Bayesian Risk

Minimum Bayesian Risk (MBR) stems from statistical decision theory from the principle of maximization of expected utility [1], which has been applied to parsing [8] and speech recognition [7]. A large number of experiments have verified its effectiveness. The same idea was later applied to bilingual word alignment [14] and machine translation [15]. In this paper, we extend the application of this concept to the field of Chinese character recognition.

The definition of MBR originates from the principle of Bayesian decision theory. Considering data points $\boldsymbol{X}$ taking values in sample space $\mathcal{X}$, $\boldsymbol{\theta}$ is the model parameter to be estimated according to $\boldsymbol{X}$. The estimator is denoted by $\delta(\boldsymbol{X})$ and $\delta : \mathcal{X} \rightarrow \Theta$ is a mapping from the sample space to the parameter space. If $\boldsymbol{\theta}$ is sampled from an underlying prior distribution w, the minimum Bayesian risk R_{Bayes} associated with this estimator is:

$$R_{\text{Bayes}} := \inf_{\delta} \int_{\Theta} \mathbb{E}_{\boldsymbol{\theta}} L(\boldsymbol{\theta}, \delta(\boldsymbol{X})) w(\mathrm{d}\theta) \tag{1}$$

where L is a non-negative loss function.

3   Methodology

Our method originates from the desire to formulate a quantitative description of the relationship between different radicals. In Sect. 3.1, we elaborate on the definition of the similarity metric. In Sect. 3.2, following previous work [6,12, 13], we demonstrate how to leverage the radical similarity to calculate MBR loss function during the training stage. In Sect. 3.3, we introduce the principle and procedure of our post-correction module. The architecture of our proposed method is shown in Fig. 2.

3.1   Radical Similarity Metric

A widely-used notion of string similarity is the edit distance [19], the minimum number of insertions, deletions, and substitutions required to transform one string into the other. Levenshtein distance [31] is a kind of common edit distance calculated by the dynamic programming algorithm. In our research, we regard the stroke sequence as a string. Therefore, we can utilize the Levenshtein distance of stroke sequences between two radicals to quantifiably evaluate their similarity. We use $\text{Lev}(\boldsymbol{p}, \boldsymbol{q})$ to denote the Levenshtein distance between stroke sequence $\boldsymbol{p}$ and $\boldsymbol{q}$ with size of m and n respectively, then:

$$\text{Lev}(\boldsymbol{p}, \boldsymbol{q}) = \min \begin{cases} \text{Lev}(\boldsymbol{p}_{<m}, \boldsymbol{q}) + 1 \\ \text{Lev}(\boldsymbol{p}, \boldsymbol{q}_{<n}) + 1 \\ \text{Lev}(\boldsymbol{p}_{<m}, \boldsymbol{q}_{<n}) + \boldsymbol{I}_{(p_m, q_n)} \end{cases} \tag{2}$$

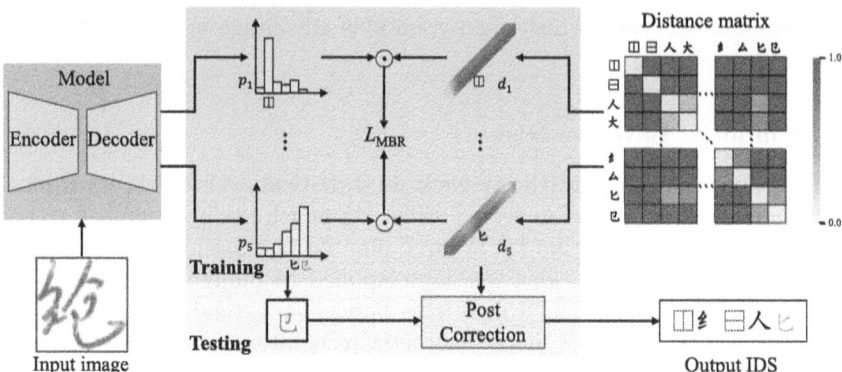

Fig. 2. The flowchart of our proposed method. In this framework, a Chinese character image undergoes processing within a radical-based CCR model, yielding a series of output probability vectors indicating predicted radicals. Subsequently, these predicted probabilities are then multiplied with the corresponding target radical distance vector to calculate L_{MBR}. Afterwards, the predicted radical results are passed through a post-processing module to obtain the output IDS.

where the subscript $< m$ and $< n$ denote the first $m-1$ and $n-1$ elements of $\boldsymbol{p}$ and $\boldsymbol{q}$. p_m and q_n are the mth and n^{th} strokes in $\boldsymbol{p}$ and $\boldsymbol{q}$ respectively. $\boldsymbol{I}_{(p_m, q_n)}$ is the indicator function and its value is 1 only when p_m is equal to q_n, otherwise 0.

Subsequently, normalizing the Levenshtein distance between the stroke sequences of the two radicals by the length of the longer stroke sequence, we acquire a symmetric distance matrix $\boldsymbol{D}$ that stores the distance between each pair of radicals. Its dimension is the same as the number of radicals involved in the CCR task. During the experiment, it suffices to extract the required distance vector from this matrix without redundant calculations. Each element $d(r_p, r_q)$ in $\boldsymbol{D}$ represents the distance between radical r_p and r_q with stroke sequences $\boldsymbol{p}$ and $\boldsymbol{q}$. The radical similarity measure for radicals r_p and r_q is derived by subtracting $d(r_p, r_q)$ from 1.

$$d(r_p, r_q) = \frac{\text{Lev}(\boldsymbol{p}, \boldsymbol{q})}{\max(m, n)} \tag{3}$$

3.2 Minimum Bayesian Risk Loss Function

In the task of the radical-based method of CCR, the Bayesian risk R_{Bayes} hinges on two essential components: the recognition model and a utility metric. Recognition model $\boldsymbol{P_\theta}(\boldsymbol{H}|\boldsymbol{O})$ estimates the radical dimension probability distribution $\boldsymbol{H}$ given source visual feature segment $\boldsymbol{O}$; the utility metric $\text{u}(\boldsymbol{H}, \boldsymbol{G})$ estimates risk of hypothesis probability distribution $\boldsymbol{H}$ with reference $\boldsymbol{G}$. Maximum likelihood estimation serves as a widely employed criterion for parameter estimation, equivalent to minimizing the cross-entropy between $\boldsymbol{H}$ and $\boldsymbol{G}$. Now, in light of the radical distance, our objective is to select the optimal hypothesis radical probability vector distribution based on its expected utility within the recognition model. The radical level minimum Bayesian risk loss function L_{MBR} can be denoted as:

$$L_{\mathrm{MBR}}(\boldsymbol{\theta}) = \mathbb{E}_{\boldsymbol{H} \sim P_{\boldsymbol{\theta}}(\cdot|\boldsymbol{O})}\mathrm{u}(\boldsymbol{H}, \boldsymbol{G}) \tag{4}$$

$$= \frac{1}{|\mathcal{O}|}\sum_{o \in \mathcal{O}}\sum_{r_h \in \mathcal{R}} P_{\boldsymbol{\theta}}(r_h|\boldsymbol{o}) \cdot d(r_h, r_{g(o)}) \tag{5}$$

where $\mathcal{R} = \{r_1, r_2, \cdots, r_K\}$ is the radical set which is constant, K is the number of radicals. $\mathcal{O}$ is the set of all visual features obtained by the encoder to be decoded as radicals, $|\mathcal{O}|$ is the number of elements in $\mathcal{O}$. For a given feature $\boldsymbol{o}$, $r_g(\boldsymbol{o})$ represents the corresponding ground truth radical label. $d(r_h, r_{g(o)})$ denotes the distance between the ground truth radical and a hypothesized radical r_h. Finally, to emphasize the impact of the target radical, the model is optimized by the sum of cross-entropy loss function [5] L_{CE} and MBR loss function L_{MBR}:

$$L = L_{\mathrm{CE}} + \lambda L_{\mathrm{MBR}} \tag{6}$$

where λ is the hyperparameter.

Here, we provide a brief analysis of the impact of the proposed MBR loss on the probability distribution. In the training stage, when the network incorporates a softmax output nonlinearity, the input to the softmax layer is denoted by $\boldsymbol{a}$, and its output, denoted as $\hat{\boldsymbol{p}}$, corresponds to the predicted probability vector. Let $\boldsymbol{y}$ represent the one-hot label, we know that:

$$\frac{\partial L_{\mathrm{CE}}}{\partial \boldsymbol{a}} = \hat{\boldsymbol{p}} - \boldsymbol{y} \tag{7}$$

$\boldsymbol{y}_d$ is the distance label with dimension K, and:

$$\boldsymbol{y}_d[i] = d(r_i, r_{g(o)}), i = 1, \cdots, K \tag{8}$$

It can explained as the distance between other radicals and the target radical, then:

$$\frac{\partial L_{\mathrm{MBR}}}{\partial \boldsymbol{a}} = (\boldsymbol{y}_d - L_{\mathrm{MBR}} \cdot \mathbf{1}^K) \odot \hat{\boldsymbol{p}} \tag{9}$$

where $\mathbf{1}^K$ is a K dimensional vector with all elements are 1, $\odot$ represents the operation of element-wise multiplication.

The definition of the MBR loss implies that it imposes a larger penalty on radicals that exhibit significant differences at the stroke level compared to the target radical, potentially resulting in low visual similarity. As a consequence, the model tends to allocate more attention to similar radicals, i.e., assigns greater weights to these probability dimensions, that's how the MBR loss plays the role. Additionally, the gradients also indicate that the MBR loss function can be interpreted as a weighted sum of the distances between the target radical and other radicals, where the weight is determined by the model's output probability $\hat{p}$. Whenever the distance between the target radical and another radical surpasses this threshold, a penalty is applied to the corresponding dimension of that radical. Conversely, if a radical's distance is sufficiently close to the target radical, the MBR loss may have a negligible impact.

3.3   Post-correction Module

When evaluating the accuracy of a radical-based CCR model, the optimal approach is to directly compare the output radical sequence with its ground truth label. Only when they are perfectly identical will the output be considered correct. Otherwise, even minor differences will result in rejection. Some methods like RAN [34] calculate the Levenshtein distance between the output radical sequence and IDS in the candidate label set $\mathcal{L}$, then select the closest label as the final predicted result. However, this approach fails to yield remarkable improvement in most scenarios because it treats the weights of insertion, deletion, and substitute as uniform. FewShotRAN [23] incorporates an additional module specifically designed to determine the potential corresponding Chinese character based on the decoded radical sequence. HDE [2] calculates the distance between the encoded feature with all predefined character embedding and chooses the nearest label as the predicted result.

For a well-trained recognition model, although the output radical sequence may not always be consistent with the labels, they may resemble each other. For this reason, we refine the weight of the substitute operation according to the pre-defined radical distance matrix $\boldsymbol{D}$. This refinement allows us to calculate the weighted Levenshtein distance $\text{wLev}(\boldsymbol{I}_c, \boldsymbol{I}_l)$ between radical sequence $\boldsymbol{I}_c$ and $\boldsymbol{I}_l$. Furthermore, due to the decomposability of Chinese character radicals, a single radical may be equivalent to a combination of multiple structures and radicals. To incorporate this decomposability into our similarity calculation, we decompose the radical sequence into the stroke sequence representation and also calculate their normalized Levenshtein distance as a weighting factor. The new measure for comparing the two radical sequences is the product of the weighting factor and the weighted Levenshtein distance. The procedure for the post-correction module is outlined in Algorithm 1 and we exhibit an example in Fig. 3.

Algorithm 1. Main algorithm of post-correction module

1: Input radical sequence $\boldsymbol{I}_c$, candidate label set $\mathcal{L}$
2: **if** $\boldsymbol{I}_c \in \mathcal{L}$ **then**
3:     **output** $\boldsymbol{I}_c$
4: **else**
5:     Initialize output set $\mathcal{W} \leftarrow \{\}, d_{min} \leftarrow \infty$
6:     Decompose $\boldsymbol{I}_c$ into stroke representation $\boldsymbol{S}_c$
7:     **for** $\boldsymbol{I}_l \in \mathcal{L}$ **do**
8:         Decompose $\boldsymbol{I}_l$ into stroke representation $\boldsymbol{S}_l$
9:         $w_{c,l} = \frac{\text{Lev}(\boldsymbol{S}_c, \boldsymbol{S}_l)}{\max(\boldsymbol{S}_c.\text{size}, \boldsymbol{S}_l.\text{size})}$
10:         $d(\boldsymbol{I}_c, \boldsymbol{I}_l) = \text{wLev}(\boldsymbol{I}_c, \boldsymbol{I}_l) \cdot w_{c,l}$
11:         **if** $d_{min} == d(\boldsymbol{I}_c, \boldsymbol{I}_l)$ **then**
12:             add $\boldsymbol{I}_l$ to $\mathcal{W}$
13:         **else if** $d_{min} > d(\boldsymbol{I}_c, \boldsymbol{I}_l)$ **then**
14:             $\mathcal{W} \leftarrow \{\}, d_{min} \leftarrow d(\boldsymbol{I}_c, \boldsymbol{I}_l)$
15:             add $\boldsymbol{I}_l$ to $\mathcal{W}$
16:         **end if**
17:     **end for**
18:     **output** the element in $\mathcal{W}$
19: **end if**

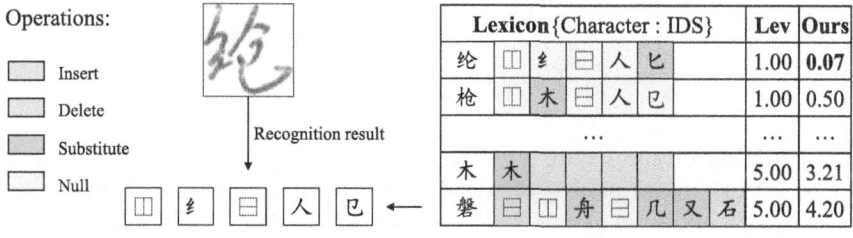

Fig. 3. Example for the Levenshtein distance and our proposed distance measure for rectifying erroneous radical sequence.

4 Experiment

In this section, we provide an overview of the experimental setup, which includes details about the datasets, baseline models, and implementation specifications. Subsequently, we conduct a series of comprehensive experiments on benchmark datasets to evaluate the effectiveness of our proposed method from both qualitative and quantitative perspectives.

4.1 Datasets and Metrics

We apply our method to both artistic and handwritten Chinese characters. The printed Artistic Dataset [4] is employed, which contains 3755 characters in 105 printed artistic fonts in zero-shot format. For handwritten Chinese character recognition, we use the HWDB1.0-1.1 Dataset [16], consisting of 2.73 million offline handwritten Chinese character images from 720 writers serving as the training and validation set with 3881 characters. The ICDAR2013 handwritten Chinese character Dataset [30] is selected as the testing set, comprising 224,419 offline handwritten Chinese characters from 60 writers with 3755 characters. The accuracy of Chinese character image recognition serves as a metric for evaluating the effectiveness of the methodology.

4.2   Implementation Details

To emphasize the effectiveness of our proposed method, we have constructed two radical-based baseline models for CCR, including RAN [34] and RTN [28]. Both of them utilize IDS to represent a Chinese character label. In the experiments, the artistic images are sized to 32×32; the handwritten Chinese character images are sized to 64×64. λ in Eq. 6 is set to 1. We set the batch size to 64 for training. The optimizer is Adadelta [32] with a weight decay of 1×10^{-4} and the learning rate is set to 0.01. All experiments are performed on a single NVIDIA GeForce RTX 3090 GPU with 24GB RAM. Note that the experimental results of baseline and MBR loss models are obtained through greedy search without any post-processing techniques.

4.3   Printed Artistic Benchmark Comparison in Zero-Shot Setting

We analyze the ability of our method on the printed artistic characters. Experiments are set in the zero-shot [24] format, where the data is divided into subsets containing 500, 1000, 1500, 2000, and 2755 categories following previous work [4]. The recognition results are presented in Table 1. Compared with handwritten Chinese characters, printed artistic characters exhibit clearer strokes and consistent writing styles relatively, which are easier to correctly recognize [10]. The experiment result in Table 1 demonstrates that both models benefit from the proposed MBR loss with the improvement in recognition accuracy. As the size of the training set increases, the model substantially enhances its capacity to recognize Chinese radicals, even though not correctly recognized but bear a resemblance to them. Such a characteristic is effectively captured by our proposed post-correction module, resulting in the improvement of the overall recognition performance.

4.4   Handwritten Benchmark Comparison in Zero-Shot Setting

Besides the printed artistic characters, experimental results on handwritten Chinese characters in the zero-shot setting are summarized in Table 2. We fine-tune the RAN [34] and RTN [28] baseline models. For the training set, we select the first m classes of 3755 characters from HWDB1.0-1.1, where m ranges in 500, 1000, 1500, 2000, 2755. The test set consists of samples with labels from the last 1000 classes of the ICDAR2013 dataset. Our partition method is the same as that used in [4] for a fair comparison. The result shows that our method is able to improve the performance of RAN [34] and RTN [28] baselines across all partition settings.

Table 1. Performance comparison in the character zero-shot setting on the Printed Artistic benchmark (%).

Printed Artistic	Strategy	Character Zero-Shot Setting				
		500	1000	1500	2000	2755
SLD [4]	Baseline	7.03	26.22	48.42	54.86	65.44
HDE [2]	Baseline	7.48	21.13	31.75	40.43	51.41
RAN [34]	Baseline	2.89	12.07	25.69	35.36	46.74
	+MBR	3.21	12.82	27.50	36.98	50.34
	+Post	5.28	18.35	36.93	47.39	59.66
	+MBR+Post	**5.98**	**19.42**	**38.02**	**50.23**	**62.06**
RTN [28]	Baseline	1.21	6.93	14.11	20.09	32.70
	+MBR	1.23	6.95	15.27	22.06	36.39
	+Post	2.23	9.87	18.97	25.45	43.14
	+MBR+Post	**2.24**	**10.06**	**20.98**	**28.41**	**46.14**

Table 2. Performance comparison in the character zero-shot setting on the handwritten benchmark (%).

Handwritten	Strategy	Character Zero-Shot Setting				
		500	1000	1500	2000	2755
SLD [4]	Baseline	5.60	13.85	22.88	25.73	37.91
ACPM [37]	Baseline	9.72	18.50	27.74	34.00	42.43
RAN [34]	Baseline	2.65	10.10	16.92	21.56	31.78
	+MBR	3.95	11.62	17.91	23.72	34.51
	+Post	5.59	19.01	25.27	31.49	42.24
	+MBR+Post	**7.41**	**21.33**	**28.79**	**35.73**	**47.05**
RTN [28]	Baseline	0.91	4.56	8.67	12.75	17.51
	+MBR	0.97	5.10	9.29	13.58	18.67
	+Post	1.50	6.65	11.77	16.71	22.98
	+MBR+Post	**1.54**	**8.86**	**12.62**	**16.94**	**23.88**

4.5   Handwritten Benchmark Comparison in Seen Setting

To evaluate the impact of our approach on the entire handwritten dataset, we use the ICDAR2013 dataset as the testing set, where all labels have appeared in the training and validation dataset HWDB1.0-1.1. Certainly, all the radicals defined in the dictionary are accessible during the training process. The experimental results under the seen character setting are presented in Table 3, which indicates that both the RAN [34] and RTN [28] benefit from the MBR loss function and the post-correction module. The term "Decomposition" in Table 3 represents the modeling approach of the CCR model as mentioned in Sect. 2. Different

Table 3. The results in seen character setting on ICDAR2013 (%).

Method	Decomposition	Accuracy
HCCR-GoogLeNet [36]	Character	96.35
DirectMap+ConvNet+Adaptation [35]	Character	97.37
template+instance [26]	Character	97.45
DenseRAN [25]	Radical	96.66
FewShotRAN [23]	Radical	96.97
SLD [4]	Stroke	96.74
STAR [33]	Stroke&Radical	97.14
ACPM [37]	All	97.80
RAN [34]	Radical	96.51
RAN+MBR		96.74
RAN+MBR+post		**97.01**
RTN [28]	Radical	96.60
RTN+MBR		96.93
RTN+MBR+post		**97.05**
HDE [2]	Radical	97.14
HDE+MBR		**97.22**
Dense Classifier [10]	Character	97.23
Dense Classifier+MBR		**97.50**

from caption generation methods, HDE [2] compares the predicted results with all candidate characters' embedding instead of creating radical captions step by step, which is more like a character-based method. We also design a Dense classifier [10] which combines a DenseNet [9] and a linear classifier. The MBR loss on HDE [2] and Dense classifier [10] is defined on character level distance, calculated by the Levenshtein distance between different characters' IDS to extend our method on character-based CCR methods. ACPM [37] model is the temporary SOTA method, but considering its strategy, which incorporates character level, radical level, and stroke level information, our method only focuses on radical level information. The experimental results illustrate the effectiveness of our method and achieve the highest score among radical-based CCR methods.

To examine the influence of the solely radical distance-based MBR loss function on the model without cross-entropy loss function, we conduct experiments on the ICDAR2013 handwritten dataset and find that there is a decrease in the performance relative to the original model with an accuracy of 95.69% for RAN [34] and 94.95% for RTN [28]. From a computational perspective, the cross-entropy loss function is commonly employed to measure the disparity between one-hot labels and predicted probabilities, aiming to maximize the probability associated with the target radical dimension while neglecting the probability distribution across other dimensions. In contrast, our proposed MBR loss function for Chinese radicals assigns a distance between 0 and 1 to the target radical and other radicals. However, when certain radicals closely resemble the target radical, assigning higher probability values to their dimensions does not increase the MBR loss. Through experimental analysis, we find that when the distance between two radicals is close, solely relying on the MBR loss function substantially reduces the discriminative capacity of models on the target radical, contributing to a decrease in the model's overall recognition accuracy. Consequently, we opt for a hybrid loss function that integrates both cross-entropy loss and MBR loss. This strategic choice serves to magnify the impact of the target radical on the model, simultaneously offering flexibility to accommodate similar radicals and penalize dissimilar ones. Such a strategy contributes to the effective optimization of the model, aligning with established human habits and conventions.

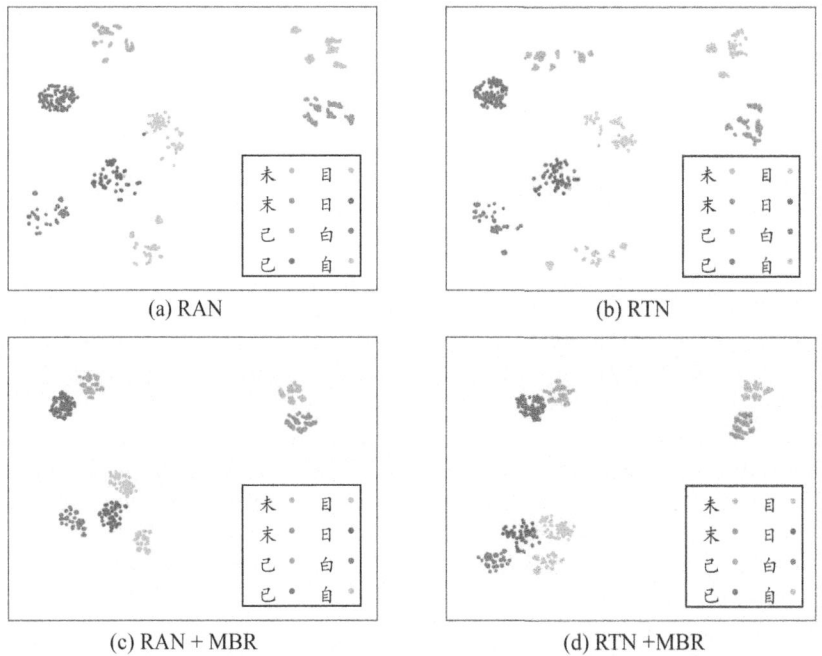

(a) RAN

(b) RTN

(c) RAN + MBR

(d) RTN +MBR

Fig. 4. T-SNE visualization of features of the same radicals from baseline and the MBR loss function

To verify the impact of the additional MBR loss on radical-based CCR models, we visualize the features of several radicals produced by RAN [34] and RTN [28] with/without the inclusion of MBR loss, i.e., "RAN" and "RAN+MBR", "RTN" and "RTN+MBR". For a fair comparison, the features are obtained from the same image but under various optimization strategies. They are embedded into a 2-D space using t-SNE visualization [17] and the results are shown in

164   Z. Han et al.

Fig. 4. Each color represents a radical and each point represents a feature of the corresponding radical. Obviously, with the assistance of the MBR loss function, the features associated with the same radical category exhibit heightened compactness, which indicates a higher degree of similarity and coherence. Moreover, the features of visually similar radicals display relatively proximate inter-cluster distances, reflecting their resemblance. Despite this, the model still effectively maintains distinctiveness between different clusters for clear classification. The MBR loss function induces more precise and valuable feature representations.

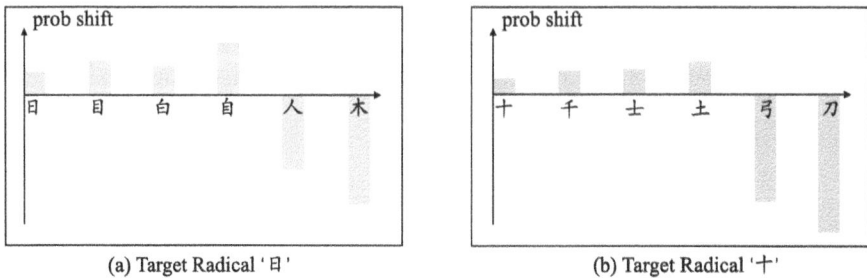

(a) Target Radical ' 日 '   (b) Target Radical '十'

Fig. 5. Probability shift after applying the MBR loss function for the same radicals. Predicted probabilities increase for visually similar radicals while decreasing for dissimilar ones.

We also observe the output predicted probability changes to verify the impact of the MBR loss. When the prediction is correct, apart from the maximum probability value in the target radical dimension, there is also a distribution of probability weights across other dimensions. Figure 5 illustrates the changes in the probabilities across some radical dimensions after introducing the MBR loss function. This observation indicates that the output probability of visually similar radicals slightly increases. In contrast, the probability of the distinct radical is reduced, which supports the conclusion of the visualization results and is in line with the analysis in Sect. 3.2.

Different parameter λ in Eq. 6 may result in different model efficiency. To verify its impact on the model, we conducted a series of extended experiments on the RAN [34] and RTN [28] models, following the settings outlined in Table 2, with recognition accuracy as the ultimate metric. We set $\lambda = 0.5, 1, 2$ and 5. The experimental results are presented in Table 4. From these results can see that when $\lambda = 1$, the addition of the proposed MBR loss performs best.

Table 4. Performance comparisons on ICDAR2013 dataset with different parameters λ (%).

λ	0.5	1	2	5
RAN [34]	96.56	96.74	96.71	96.64
RTN [28]	96.62	96.95	96.88	96.79

Finally, we list several instances of recognition results that violate semantic constraints, which were rectified to the correct Chinese characters by our post-processing module as shown in Fig. 6. This serves as an exemplar of the efficacy of our post-processing module in disambiguation. The method of calculating the edit distances between radical sequences and all IDS may match multiple results. However, our post-processing module selects the most plausible IDS from these options. According to the above results, it is evident that the introduction of radical similarity-based MBR loss function draws the feature distances of similar radicals closer, while still ensuring distinguishability between categories. Moreover, even if there are slight discrepancies between the output results and the actual labels, the post-processing module is capable of correcting these results under the constraints of Chinese character rules, thereby enhancing the recognition accuracy of the model.

Image				
Char	肥	杖	役	贼
Predict	⿰舟巴	⿰木大	⿰彳⿱日⿵几又	⿰火⿱日戈⿸ナ
Lev	⿰口巴 ⿰革巴 ⋮	⿰木果 ⿰木木 ⋮	⿰氵⿱日⿵几又 ⿰月⿱日⿵几又 ⋮	⿰纟⿱日戈ナ ⿰犭⿱日戈ナ ⋮
Ours	⿰月巴	⿰木丈	⿰彳⿱日⿵几又	⿰贝⿱日戈ナ

Fig. 6. Several examples of irrational recognition result corrected by the post-processing module. The post-processing module corrects certain typos in writing by substituting them with correct Chinese characters based on the similarity of their radicals.

5 Conclusion and Future Work

In this paper, we propose a radical similarity metric for Chinese radicals according to their normalized Levenshtein distance of stroke sequences. Based on this criterion, we can calculate the radical-level MBR loss function and integrate it into the training stage to help optimize model parameters. In addition, we also leverage the similarity metric to propose a post-correction module to recall some erroneous recognition results. Experimental results demonstrate the effectiveness of the addition of the MBR loss and the post-processing module. However, our current method only takes into account the similarity of radicals, thereby neglecting the impact of spatial structures. Furthermore, the defined similarity between radicals still has limitations, the stroke sequence of radicals is not the optimal way to represent radicals, as it cannot perfectly capture the two-dimensional plane structure of radicals. In future work, we aim to propose a more generalized measurement to automatically calculate the similarity of Chinese radicals and characters, broadening the application scope of our approach to other languages, such as Japanese and Korean.

References

1. Bickel, P.J., Doksum, K.A.: Mathematical statistics: basic ideas and selected topics, volumes I-II package. CRC Press (2015)
2. Cao, Z., Lu, J., Cui, S., Zhang, C.: Zero-shot handwritten Chinese character recognition with hierarchical decomposition embedding. Pattern Recogn. **107**, 107488 (2020)
3. Casey, R., Nagy, G.: Recognition of printed Chinese characters. IEEE Trans. Electron. Comput. **1**, 91–101 (1966)
4. Chen, J., Li, B., Xue, X.: Zero-shot Chinese character recognition with stroke-level decomposition (2021). arXiv preprint arXiv:2106.11613
5. De Boer, P.T., Kroese, D.P., Mannor, S., Rubinstein, R.Y.: A tutorial on the cross-entropy method. Ann. Oper. Res. **134**, 19–67 (2005)
6. Freitag, M., Grangier, D., Tan, Q., Liang, B.: High quality rather than high model probability: minimum bayes risk decoding with neural metrics. Trans. Assoc. Comput. Linguist. **10**, 811–825 (2022)
7. Goel, V., Byrne, W.J.: Minimum bayes-risk automatic speech recognition. Comput. Speech Lang. **14**(2), 115–135 (2000)
8. Goodman, J.: Parsing algorithms and metrics (1996). arXiv preprint cmp-lg/9605036
9. Iandola, F., Moskewicz, M., Karayev, S., Girshick, R., Darrell, T., Keutzer, K.: DenseNet: Implementing efficient convnet descriptor pyramids (2014). arXiv preprint arXiv:1404.1869
10. Jiang, X., et al.: Group, contrast and recognize. a self-supervised method for chinese character recognition. In: Fink, G.A., Jain, R., Kise, K., Zanibbi, R. (eds.) Document Analysis and Recognition - ICDAR 2023. ICDAR 2023. LNCS, vol. 14190. Springer, Cham (2023). https://doi.org/10.1007/978-3-031-41685-9_26
11. Jin, L.W., Yin, J.X., Gao, X., Huang, J.C.: Study of several directional feature extraction methods with local elastic meshing technology for HCCR. In: Proceedings of the Sixth International Conference for Young Computer Scientist, pp. 232–236 (2001)
12. Kaiser, J., Horvat, B., Kacic, Z.: A novel loss function for the overall risk criterion based discriminative training of hmm models. In: INTERSPEECH, pp. 887–890 (2000)
13. Kingsbury, B.: Lattice-based optimization of sequence classification criteria for neural-network acoustic modeling. In: 2009 IEEE International Conference on Acoustics, Speech and Signal Processing, pp. 3761–3764. IEEE (2009)
14. Kumar, S., Byrne, B.: Minimum bayes-risk word alignments of bilingual texts. In: Proceedings of the 2002 Conference on Empirical Methods in Natural Language Processing (EMNLP 2002), pp. 140–147 (2002)
15. Kumar, S., Byrne, B.: Minimum bayes-risk decoding for statistical machine translation. In: Proceedings of the Human Language Technology Conference of the North American Chapter of the Association for Computational Linguistics: HLT-NAACL 2004, pp. 169–176 (2004)

16. Liu, C.L., Yin, F., Wang, D.H., Wang, Q.F.: Online and offline handwritten Chinese character recognition: benchmarking on new databases. Pattern Recogn. **46**(1), 155–162 (2013)
17. Van der Maaten, L., Hinton, G.: Visualizing data using t-SNE. J. Mach. Learn. Res. **9**(11), 2579–2605 (2008)
18. Morioka, T.: Integration of a Chinese character ontology and historical glyph examples. In: Proceedings of the 9th International Conference of Digital Archives and Digital Humanities (DADH2018), pp. 287–300. Organizer of 9th International Conference of Digital Archives and Digital (2018)
19. Ristad, E.S., Yianilos, P.N.: Learning string-edit distance. IEEE Trans. Pattern Anal. Mach. Intell. **20**(5), 522–532 (1998)
20. Romera-Paredes, B., Torr, P.: An embarrassingly simple approach to zero-shot learning. In: International Conference on Machine Learning, pp. 2152–2161. PMLR (2015)
21. Su, Y.M., Wang, J.F.: A novel stroke extraction method for Chinese characters using Gabor filters. Pattern Recogn. **36**(3), 635–647 (2003)
22. Vaswani, A., et al.: Attention is all you need. In: Advances in Neural Information Processing Systems, vol. 30 (2017)
23. Wang, T., Xie, Z., Li, Z., Jin, L., Chen, X.: Radical aggregation network for few-shot offline handwritten Chinese character recognition. Pattern Recogn. Lett. **125**, 821–827 (2019)
24. Wang, W., Zheng, V.W., Yu, H., Miao, C.: A survey of zero-shot learning: settings, methods, and applications. ACM Trans. Intell. Syst. Technol. (TIST) **10**(2), 1–37 (2019)
25. Wang, W., Zhang, J., Du, J., Wang, Z.R., Zhu, Y.: DenseRAN for offline handwritten Chinese character recognition. In: 2018 16th International Conference on Frontiers in Handwriting Recognition (ICFHR), pp. 104–109. IEEE (2018)
26. Xiao, Y., Meng, D., Lu, C., Tang, C.K.: Template-instance loss for offline handwritten Chinese character recognition. In: 2019 International Conference on Document Analysis and Recognition (ICDAR), pp. 315–322. IEEE (2019)
27. Xue, M., Du, J., Zhang, J., Wang, Z.-R., Wang, B., Ren, B.: Radical composition network for Chinese character generation. In: Lladós, J., Lopresti, D., Uchida, S. (eds.) ICDAR 2021. LNCS, vol. 12821, pp. 252–267. Springer, Cham (2021). https://doi.org/10.1007/978-3-030-86549-8_17
28. Yang, C., Wang, Q., Du, J., Zhang, J., Wu, C., Wang, J.: A transformer-based radical analysis network for Chinese character recognition. In: 2020 25th International Conference on Pattern Recognition (ICPR), pp. 3714–3719. IEEE (2021)
29. Yang, H.M., Zhang, X.Y., Yin, F., Liu, C.L.: Robust classification with convolutional prototype learning. In: Proceedings of the IEEE Conference on Computer Vision and Pattern Recognition, pp. 3474–3482 (2018)
30. Yin, F., Wang, Q.F., Zhang, X.Y., Liu, C.L.: ICDAR 2013 Chinese handwriting recognition competition. In: 2013 12th International Conference on Document Analysis and Recognition, pp. 1464–1470. IEEE (2013)
31. Yujian, L., Bo, L.: A normalized Levenshtein distance metric. IEEE Trans. Pattern Anal. Mach. Intell. **29**(6), 1091–1095 (2007)
32. Zeiler, M.D.: ADADELTA: an adaptive learning rate method (2012). arXiv preprint arXiv:1212.5701
33. Zeng, J., Xu, R., Wu, Y., Li, H., Lu, J.: STAR: Zero-shot Chinese character recognition with stroke-and radical-level decompositions (2022). arXiv preprint arXiv:2210.08490

34. Zhang, J., Du, J., Dai, L.: Radical analysis network for learning hierarchies of Chinese characters. Pattern Recogn. **103**, 107305 (2020)
35. Zhang, X.Y., Bengio, Y., Liu, C.L.: Online and offline handwritten Chinese character recognition: a comprehensive study and new benchmark. Pattern Recogn. **61**, 348–360 (2017)
36. Zhong, Z., Jin, L., Xie, Z.: High performance offline handwritten Chinese character recognition using GoogLeNet and directional feature maps. In: 2015 13th International Conference on Document Analysis and Recognition (ICDAR), pp. 846–850. IEEE (2015)
37. Zu, X., Yu, H., Li, B., Xue, X.: Chinese character recognition with augmented character profile matching. In: Proceedings of the 30th ACM International Conference on Multimedia, pp. 6094–6102 (2022)

Puzzle Pieces Picker: Deciphering Ancient Chinese Characters with Radical Reconstruction

Pengjie Wang[1], Kaile Zhang[1], Xinyu Wang[2(✉)], Shengwei Han[3], Yongge Liu[3], Lianwen Jin[4], Xiang Bai[1], and Yuliang Liu[1]

[1] Huazhong University of Science and Technology, Wuhan 430074, China
[2] The University of Adelaide, Adelaide, SA 5005, Australia
xinyu.wang02@adelaide.edu.au
[3] Anyang Normal University, Anyang 450000, China
[4] South China University of Technology, Guangzhou 510641, China

Abstract. Oracle Bone Inscriptions is one of the oldest existing forms of writing in the world. However, due to the great antiquity of the era, a large number of Oracle Bone Inscriptions (OBI) remain undeciphered, making it one of the global challenges in the field of paleography today. This paper introduces a novel approach, namely Puzzle Pieces Picker (P^3), to decipher these enigmatic characters through radical reconstruction. We deconstruct OBI into foundational strokes and radicals, then employ a Transformer model to reconstruct them into their modern counterparts, offering a groundbreaking solution to ancient script analysis. To further this endeavor, a new Ancient Chinese Character Puzzles (ACCP) dataset was developed, comprising an extensive collection of character images from seven key historical stages, annotated with detailed radical sequences. The experiments have showcased considerable promising insights, underscoring the potential and effectiveness of our approach in deciphering the intricacies of ancient Chinese scripts. Through this novel dataset and methodology, we aim to bridge the gap between traditional philology and modern document analysis techniques, offering new insights into the rich history of Chinese linguistic heritage.

Keywords: Historic Chinese Characters · Oracle Bone Characters · Optical Character Recognition · Radical Recognition

1 Introduction

Chinese characters represent one of the few writing systems in the world with a documented history of continuous evolution. Originating from the Oracle Bone Inscriptions (OBI) of the Shang Dynasty [3], they have evolved over more than 3,000 years into modern standard Chinese characters, undergoing significant changes in both quantity and form. As a result, many ancient characters have lost their meanings, making it difficult to find their modern equivalents and

E. H. Barney Smith et al. (Eds.): ICDAR 2024, LNCS 14804, pp. 169–187, 2024.
https://doi.org/10.1007/978-3-031-70533-5_11

Fig. 1. The consistency and evolution of the roof-like component "宀" in Chinese characters. The first row demonstrates the uniformity in the representation of the radical among various characters during the Oracle Bone Inscriptions period. The second row depicts the evolutionary path of the character "安", highlighting the changes in the roof-like component from 1700 BC through subsequent historical stages to the standardized form in contemporary Chinese script.

interpretations. These ancient scripts, particularly the OBI, contain valuable information about the human and geographic landscape of millennia ago, offering crucial insights for archaeology and history, thus attracting the interest of many scholars. However, more than two-thirds of the over 4,500 discovered OBI characters still have unknown meanings, presenting a significant challenge to interpreting complete oracle bone documents [1].

In recent years, with the advancement of Artificial Intelligence (AI), particularly in document analysis and natural language processing, scholars have begun to digitize ancient characters [27] and employ machine learning models to recognize them. However, these approaches often rely on existing Optical Character Recognition (OCR) or object detection algorithms and depend on training data with known labels, which limits their ability to generalize to texts with unknown labels, thus hindering their capacity to decipher ancient texts with unknown meanings. Pioneering approaches have also made initial attempts to decipher unknown ancient Chinese scripts using AI algorithms. For instance, Diao et al. [6] and Lin et al. [19] have explored the potential of training models on manually annotated radicals in OBIs, enabling these models to recognize and infer meanings for characters they have never seen before. This method's advantage lies in its utilization of the models' generalization abilities, extending beyond the constraints of training solely on data with known labels to interpret previously unlabeled characters. While these innovative strategies have shown promise, they are not without their challenges. The intensive labor required for manual annotation restricts these studies to relatively small datasets, often limited to a few hundred characters. Additionally, integrating cross-era character evolution to enhance deciphering efforts poses another complex challenge.

In this paper, inspired by the concept of a jigsaw puzzle, we innovatively propose a method for deciphering ancient Chinese characters named **P**uzzle **P**ieces **P**icker (P³). This approach is fundamentally grounded in two key observations:

the *shape consistency* in the forms of radicals from the same era, and the *structured consistency* across different epochs in their evolutionary trajectory. Specifically, as illustrated in Fig. 1, the first row displays various OBI composed of the roof-like component "宀", showcasing a remarkable uniformity in their written forms. The second row delineates the evolutionary journey of the Chinese character "安" from 1700 BC to its contemporary form. It is evident that despite the variations in the portrayal of the "宀" radical over time, the structural configuration of the character has remained consistent, maintaining an upper-lower structure where the "宀" is positioned above the "女" character.

Building upon the above findings, and taking into account that Chinese characters from different periods are composed of similar elements such as radicals, it becomes feasible to deduce the manifestation of a radical in other periods once its meaning in a specific era is understood. Consequently, as depicted in Fig. 2, we treat the decipherment of ancient Chinese texts as a puzzle game, wherein ancient characters from various periods are deconstructed into distinct puzzle pieces according to their radicals and other components, with the original textual structure sequences as recipes for reassembly. Subsequently, our proposed P^3 framework employs a Transformer-based [28] sequence prediction model trained to examine the evolutionary patterns of components across different epochs. Thus, during the inference phase, when introduced to new, previously unseen ancient texts, it is capable of selecting the appropriate puzzle pieces and providing the reassembly recipe, thereby unveiling the decipherment results. Specifically, the main contributions of this paper are as follows:

Different Period Characters   Puzzle Pieces   Radical Reconstruction

Fig. 2. The decipherment of ancient Chinese characters is treated as a puzzle-solving game, where characters from various periods are first broken down into pieces according to radical strokes. These pieces are then analyzed by the proposed P^3 that examines potential evolutionary patterns. In the inference stage, the model predicts a reconstruction recipe when presented with new, unseen samples, thus aiding in decipherment.

- We have constructed a large-scale dataset named Ancient Chinese Character Puzzles (ACCP), comprising nearly 90,000 categories of Chinese characters from 7 distinct historical periods, totaling over 340,000 images. For each character, we provide detailed annotations, including its category and radical sequence.
- We introduce a novel approach, Puzzle Pieces Picker (P^3), a sequence prediction model based on a Transformer architecture. To the best of our knowledge,

this represents the first attempt to decipher ancient texts through a method of deconstructing and reconstructing characters from different eras.

- We conducted extensive experiments to evaluate the performance of the proposed P^3 model in deciphering texts from various periods and assess the impact of training on cross-era data. The results demonstrate that P^3 shows promising effectiveness in simulated deciphering tasks, highlighting its potential in advancing the field of ancient language analysis.

2   Related Work

The incorporation of AI, particularly OCR and other document analysis techniques, into the study of ancient Chinese characters, marks a significant shift in the way historical documents are analyzed.

Ancient Text Detection and Recognition. The challenge lies in accurately detecting [7, 20, 23, 24, 32, 34] and recognizing [11, 13, 19, 29, 38] ancient characters from scanned images of historical documents. This task shares considerable similarities with OCR and general object detection, leading much of the existing research to focus on adapting established models to meet the unique challenges of working with ancient languages. For instance, Meng et al. [24], Liu et al. [23], and Yang et al. [23] have applied well-known general object detectors like SSD [21], Faster R-CNN [9], and YOLO-v4 [2] to the detection of ancient Chinese characters. Their efforts were primarily aimed at tackling the intricate challenges posed by the dense text arrangement in ancient manuscripts.

In the realm of recognition, OCR technology has already achieved remarkable success in processing modern documents and scene texts. However, the challenges associated with ancient languages are notably more complex due to the limited availability of training data, which is often further complicated by issues such as missing pieces, damage, and noise in historical documents. For example, datasets such as OBI-100 [8], OBI-125 [35], Oracle-20k [12], and OBC306 [16] each contain only a few hundred Oracle Bone Inscription characters, and some of these characters are affected by significant noise. Consequently, research efforts have increasingly shifted towards the creation of innovative data augmentation strategies and the implementation of few-shot learning techniques. For example, Han et al. [13] introduced the ORC-BERT Augmentor, a novel data augmentation approach pre-trained through self-supervised learning, specifically designed for the recognition of ancient characters with limited training data. Lin et al. [19] developed a radical extraction and recognition framework, which approaches ancient Chinese characters as combinations of radicals and identifies them by detecting these constituent radicals. Wang et al. [29] proposed the structure-texture separation network, which serves as an end-to-end learning framework designed to perform disentanglement, transformation, adaption, and recognition tasks on ancient characters in a unified process.

Zero-Shot and Few-Shot Learning of Chinese Characters. Despite these methods' success in employing CV for ancient text detection and recognition, their reliance on data from known categories limits their applicability to deciphering tasks, which require testing on unlabeled data. In response, some studies [4–6,18,31,36,37] have turned to zero-shot or few-shot learning, aiming to apply methods trained on known data to new, unseen texts, thus facilitate the decipherment of unknown characters. For modern Chinese characters, some research [4,18,31,37] has concentrated on the analysis of strokes and radical structures within modern Chinese characters to facilitate their interpretation. Li et al. [18], proposed the Radical Counter Network, designed to detect and quantify radicals and their spatial configurations within characters. Similarly, Cao et al. [4] delved into the hierarchical decomposition of modern Chinese characters, devising an algorithm that translates characters into vectors representative of their structural elements. However, due to the significant differences between ancient and modern Chinese characters in terms of radicals and structures, these previous methods cannot be directly applied to decryption tasks of ancient Chinese characters. For ancient Chinese characters, Chang et al. [5] developed a four-stage cascade GAN, trained using images of ancient Chinese scripts from four distinct historical epochs. This innovative model can simulate the evolutionary trajectory of specific characters, allowing for the generation of potential decipherment clues when presented with images of unknown texts. Diao et al. [6] conducted zero-shot learning on Oracle Bone Inscriptions by manually annotating radicals, employing common object detection methods. However, these prior approaches still exhibit shortcomings, including excessive dependence on manually annotated radical information and a lack of validation on large-scale datasets.

Fig. 3. Examples of the evolution of Chinese characters across seven historical periods in our ACCP dataset. Each row showcases a category of characters, while each column corresponds to a specific period, illustrating the developmental trajectory within the same character category.

By contrast, the proposed model can extract and learn from the evolutionary patterns of radicals in nearly 90,000 Chinese characters across seven periods. It can automatically decompose more radical pieces for the puzzle, increasing the likelihood of reconstructing the ancient characters that need to be deciphered.

3   The ACCP Dataset

Chinese characters have undergone a millennia-long evolution, with their iconic forms broadly categorized into seven historical stages: Oracle Bone Inscriptions, Bronze Inscriptions, Warring States Script, Seal Script, Clerical Script, the script of the Kangxi Dictionary, and Modern Regular Chinese Characters. Despite the vast transformation from the original OBI to contemporary characters, there are notable similarities between scripts from successive periods. For example, as illustrated in Fig. 3, the temporal gap between OBI and Bronze Inscriptions is merely about 300 years, leading to substantial similarities in their forms. Following the unification of the six states by the Qin dynasty in 200 BC, Clerical Script emerged as the predominant script, closely resembling modern Chinese characters. This clear evolutionary trajectory facilitates the cross-validation of character forms across different eras, a common approach in the study of ancient Chinese scripts. However, existing datasets rarely provide a comprehensive collection of characters across these varied historical stages. Among the larger datasets, EVOBC stands out, yet it only encompasses characters from four distinct periods: Bronze Inscriptions, Warring States Script, Seal Script, and Clerical Script. To better leverage the evolutionary characteristics of Chinese characters from deciphering ancient scripts, we expanded upon one of the largest OBI datasets, HUST-OBC [30], and the evolutionary dataset EVOBC [10]. As demonstrated in Table 1, our extension incorporates additional data from the Kangxi Dictionary, a comprehensive work from the Qing Dynasty known for its detailed compilation of character forms, and modern Regular Characters. This expanded

Table 1. The ACCP dataset spans over 3,000 years, covering seven script types from various historical periods, and comprises nearly 90k categories with more than 340k image samples.

Source	Script Type	Time Period	#Category	#Sample
HUST-OBC [30]	Oracle Bone Inscriptions	1,700BC - 1100BC	1,781	77,064
EVOBC [10]	Bronze Inscription	1,300BC - 200BC	4,801	42,573
	Warring States script	475BC - 211BC	5,343	80,119
	Seal Script	221BC - 420AD	14,158	29,751
	Clerical Script	200BC - 220AD	2,890	3,568
Extension	Kangxi Dictionary	1,662AD - 1722AD	9,354	9,354
	Regular Character	Now	88,899	103,915
Total		1,700BC - Now	88,901	346,344

collection forms the basis of our dataset, named Ancient Chinese Character Puzzles **(ACCP)**, designed specifically to advance the study and decipherment of ancient Chinese texts. ACCP will be available at https://github.com/Pengjie-W/Puzzle-Pieces-Picker

4   Method

In the introduction, we briefly outlined the concept of the Puzzle Piece Picker (P^3) model, which is inspired by jigsaw puzzles and primarily involves two main steps: *deconstruction* and *reconstruction*. In this section, we will delve into the specific implementation details of the P^3 model.

4.1   Radical Decomposition

The P^3 model facilitates the deciphering of unknown ancient Chinese characters by deconstructing them according to radicals and then reconstructing them based on the rules of radicals in modern Chinese characters. This process necessitates the use of training data annotated at the radical level, which aids the model in learning the local structures of radical components. However, as outlined in Sect. 3, the ACCP dataset, which is expanded from HUST-OBC [30] and EVOBC [10], lacks such annotations. Manually annotating using image editing software like Photoshop is prohibitively expensive due to the high costs involved.

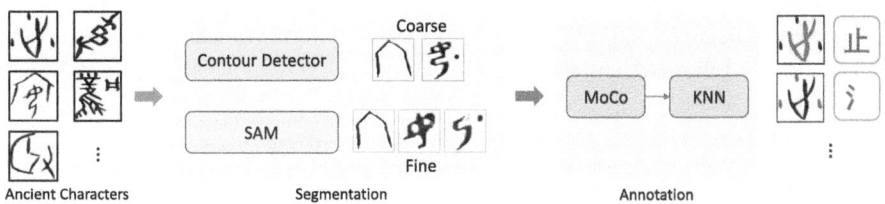

Fig. 4. Flowchart of radical deconstruction.

Segmentation. To acquire annotations at the radical level, a straightforward approach involves utilizing segmentation models to obtain segmentation masks for radicals. However, there isn't an off-the-shelf model capable of performing this task independently. Consequently, we devised a radical decomposition process that operates on both coarse and fine levels, as illustrated in Fig. 4. Specifically, we employed a contour detector for coarse-grained segmentation and the Segment Anything Model (SAM) [17], a large-scale, general-purpose semantic segmentation model, for fine-grained segmentation.

Contour-Based: For contour-based decomposition, we developed a heuristic method that leverages OpenCV to detect the edges of ancient Chinese character images. By setting a distance threshold, we merged components within a certain range to delineate the radical components. This method is simple yet effective for characters with dispersed structures. However, its limitations become apparent with characters that are closely adjacent or overlapping, where it struggles to accurately separate individual elements.

SAM-Based: Regarding the SAM-based approach, SAM, has been pre-trained on a vast dataset and is adept at zero-shot generalization for segmenting virtually any object. Thus we also employed it to obtain radical masks for ancient character images. Compared to the contour-based method, SAM is more proficient at distinguishing radicals that are stuck together. However, its drawback lies in over-segmenting, where it might divide a single component into multiple fragments.

As a result, each ancient character image yields two distinct sets of segmentation outcomes, one from the contour-based method and the other from the SAM-based approach, as depicted in Fig. 5. The effectiveness of each method can vary; for some characters, the SAM-based results may prove superior, while for others, the contour-based might be more effective.

Annotation. Merely possessing segmentation masks is insufficient. It is also essential to acquire labels corresponding to the radicals of modern Chinese characters associated with these segmented components. To address this, we employed MoCo [14] to learn the representations of different components and then grouped them using KNN.

Specifically, *MoCo* [14] takes all of the radical pieces segmented by both contour-based and SAM-based methods as training samples, and maps them into a feature space, facilitating the measurement of their similarities. During each iteration, an image of a radical serves as the positive sample, while others are treated as negative samples. This setup aims to minimize the contrastive loss defined by the following equation:

$$\mathcal{L}_q = -\log \frac{\exp\left(q \cdot k_+ / \tau\right)}{\sum_{i=1}^{K} \exp\left(q \cdot k_i / \tau\right)} \tag{1}$$

where τ represents the temperature parameter, which aids in controlling the separation of distribution, K denotes the number of selected samples for comparison, q is the feature vector derived from the enhanced positive sample, and k_+ is the feature vector of the actual positive sample. By iteratively optimizing this loss, our model effectively discriminates between the positive samples and a multitude of negative samples, thereby learning robust representations of radicals in ancient Chinese characters.

Following the completion of the training process, we utilized ResNet-50 [15] to transform images of each radical into a 128-dimensional feature vector, denoted as $f \in \mathbb{R}^{128}$. This encoding facilitates the subsequent application of a *KNN* algorithm for clustering and annotating these radicals. As illustrated in Fig. 6, the procedure for an unlabeled radical component begins with identifying its K nearest previously annotated neighbors within the feature space, utilizing Euclidean distances for this determination. These annotated neighbors are either components that have already been labeled or are part of an ancient Chinese character

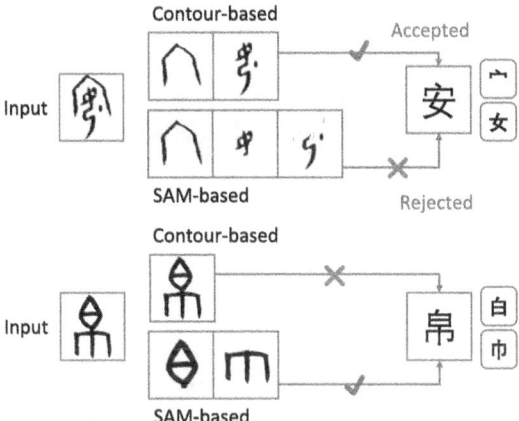

Fig. 5. Contour-based and SAM-based methods yield different sets of masks for radicals. These segmented components are tested for their ability to reconstruct the specified modern Chinese characters. Successful reconstructions are retained as final annotations, while failed attempts indicate which segmentation masks to discard.

whose interpretation has been established. Each annotated component is associated with a dictionary where the dictionary keys represent potential radicals, and the values indicate confidence levels. This dictionary is initially populated based on the Ideographic Description Sequence (IDS) of the corresponding Chinese character, with keys derived from modern Chinese radical components. The confidence level assigned to each key is set to $\frac{1}{n}$, where n, typically ranging from 2 to 3, is the total number of radicals comprising the Chinese character. Upon identifying the marked neighbors and their associated radical dictionaries, we proceeded to refine the dictionary corresponding to the target ancient character radical:

$$F_c = softmax(\sum_{i=1}^{K} r_i f_c) \tag{2}$$

In this equation, F_c represents the updated confidence level, r_i signifies the Euclidean distance between the target component and a marked neighbor within the feature space, and f_c denotes the confidence levels associated with the marked neighbor, with a value of 0 assigned if the key is absent from the dictionary. The final step involves determining the label for the target ancient character radical by selecting the dictionary key that exhibits the highest value. Ultimately, as shown in Fig. 5, the labeled radical components are utilized in attempts to reconstruct corresponding modern Chinese characters. Successful reconstructions are accepted as final annotations, while failures lead to rejection. This process effectively filters out inferior segmentation masks produced during the initial coarse and fine segmentation phases.

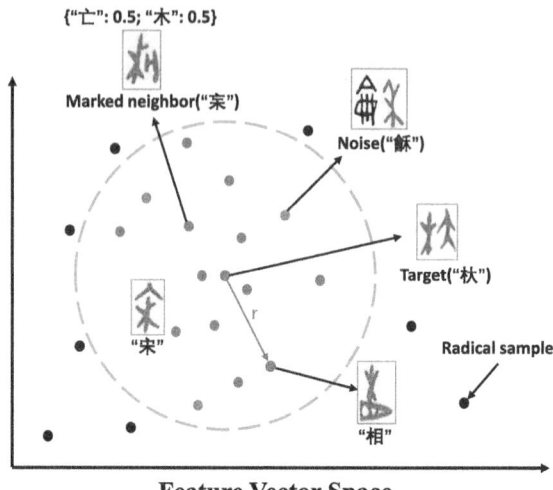

Feature Vector Space

Fig. 6. Illustration of the KNN clustering in the feature space. The target radical component is marked in red, surrounded by its 15 nearest neighbors within the delineated boundary. Green points represent similar, yet non-matching radicals, while blue points correspond to radicals sharing the same radical as the target, aiding in the accurate labeling process. (Color figure online)

Given that OBIs represent a historical phase with the most undeciphered characters and suffer from a limited number of training samples, we primarily employ radical decomposition techniques to synthesize supplementary radical data, thereby enriching the training dataset. This augmentation aids the model in examining the structures and shapes of radicals more effectively. It is important to note, however, that the proposed method for radical decomposition is adaptable not only to ancient Chinese characters from any period but also to other radical-based languages, such as Korean and Japanese. Figure 7 showcases examples of radical-level annotations produced by our methods, demonstrating their effectiveness in segmenting ancient Chinese characters by radicals and in automatically assigning accurate labels.

4.2   Radical Reconstruction

Preliminary. In the endeavor to decipher ancient Chinese characters, AI-assisted models play a pivotal role in suggesting possible modern equivalents for undeciphered characters. Our innovative P^3 model introduces a novel pipeline that first deconstructs ancient characters into their radical components and then reconstructs these components into modern Chinese characters. This reconstruction process is guided by an understanding of the structural principles of modern Chinese writing. As demonstrated in Fig. 8, the framework of modern Chinese

Radicals   Samples

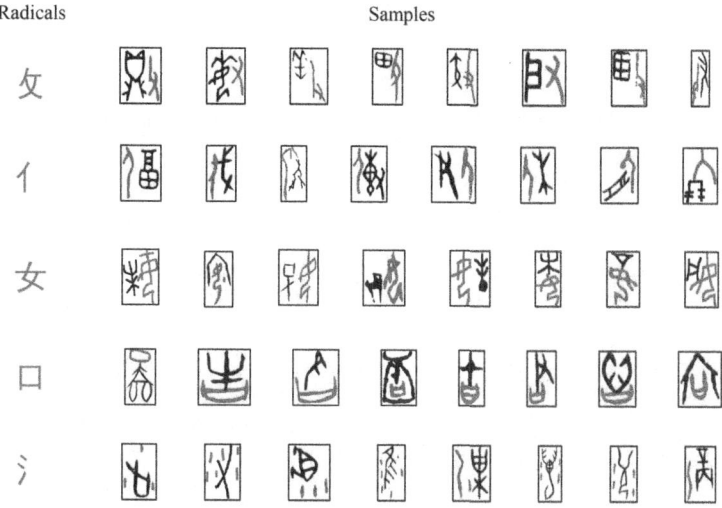

Fig. 7. Examples of segmented radical components by radical decomposition.

characters can be articulated through Ideographic Description Sequence (IDS)[1], which highlight 12 common structural motifs, such as left-right, top-bottom, and left-middle-right. These motifs provide insights into the spatial organization of radicals within a character. Armed with the deconstructed radicals from ancient texts, we apply the corresponding IDS to reassemble these radicals into the organized structure of modern Chinese characters, thus fulfilling the deciphering objective.

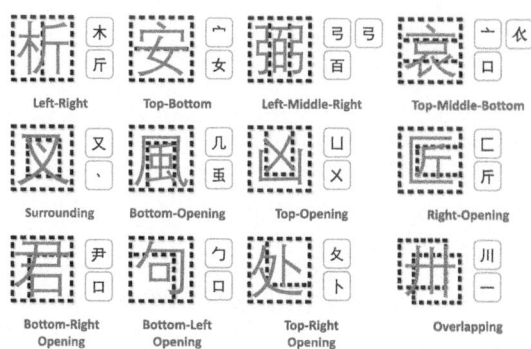

Fig. 8. 12 common structures of Chinese characters in IDS.

---

[1] https://github.com/cjkvi/cjkvi-ids.

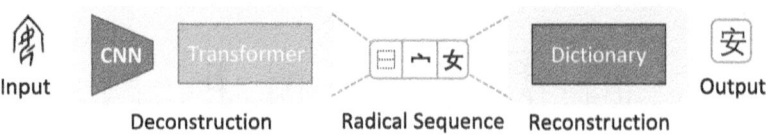

Fig. 9. Overview of the network architecture of the proposed methods.

Network Architecture. As depicted in Fig. 9, the network architecture of our method is elegantly simple. Drawing inspirations from the scene text spotter method SPTS [22,25], we have eliminated the coordination information typically required in text spotting tasks and have instead integrated IDS to form the predictive sequence. This means that for any given input image of an ancient Chinese character, the model is designed to output the deconstructed radical strokes along with the corresponding IDS. These elements are then utilized to reconstruct the modern standard Chinese characters, thus achieving the desired decipherment. To elaborate, given an input image $X \in \mathbb{R}^{3 \times H_0 \times W_0}$, the Deconstruction first employs a CNN to extract image features, producing a feature representation of $X^{1024 \times 8 \times 8}$. Subsequently, a Transformer module processes these features to deconstruct the ancient script into radical strokes and the possible IDS. Following this decomposition, the Reconstruction reassembles the identified radicals into modern Chinese characters. This reassembly is guided by predetermined IDS frameworks, ensuring each radical sequence is accurately matched to the correct character. This streamlined approach not only accelerates the reconstruction process but also enhances the accuracy of deciphering ancient scripts, making efficient use of the structural nuances of Chinese characters.

5   Experiments

5.1   Implementation Details

In our implementation, the ResNet-50 serves as the foundational backbone network. For the Transformer components, both the encoder and decoder are structured with six layers and feature eight attention heads, utilizing the Pre-Layer Normalization Transformer configuration as outlined by Xiong et al. [33]. Optimization is achieved through the AdamW optimizer, starting with an initial learning rate of $5e^{-4}$ which is linearly decreased to a minimum of $1e^{-6}$. The model is trained with a batch size of 256 for 100 epochs. Our demo will be available at https://github.com/Pengjie-W/Puzzle-Pieces-Picker

5.2   Deciphering Experiments

Given the intricate nature of deciphering ancient scripts, which often demands extensive validation by experts and presents challenges in quantitatively assessing the performance of proposed models, we opted for an alternative evaluation approach. We leveraged already deciphered ancient Chinese characters as test data to quantitatively analyze the performance of our model. Specifically, we assessed the deciphering capabilities of our proposed P^3 model across characters from seven distinct periods. To this end, we adopted a methodology similar to cross-validation, training seven different models for deciphering ancient Chinese characters, each tailored to a specific historical period.

Table 2. Quantitative results of the proposed P^3 on ancient character decipherment task. 'OBIs*' represents the experiments of deciphering OBIs after incorporating Radical Decomposition

	OBIs	OBIs*	Bronze	Warring	Seal	Clerical	Kangxi	Regular
Sample-wise Acc.	17.5%	19.2%	25.6%	14.1%	29.4%	49.7%	96.4%	93.6%
#Undeciphered	1,000	1000	682	592	935	491	848	999
#Success	394	414	246	197	361	278	814	963
Category-wise Acc.	39.4%	41.4%	36.1%	33.3%	38.6%	56.6%	96.0%	96.4%

In detail, we commenced by randomly selecting 1,000 character categories from the dataset with the smallest sample size, referred to the OBI period, and designated these as *undeciphered characters*. We then sought corresponding character classes in other historical periods to serve as the *undeciphered characters* for those periods. However, due to the potential non-existence of certain characters across different eras, the number of *undeciphered characters* in other periods might be less than 1,000. The total number of categories deemed as *undeciphered* in each era is summarized in Table 2, illustrating variations across periods, with Bronze inscriptions featuring 682 classes while Seal script encompassing 935 classes. Subsequently, for each period's deciphering model, we formulated the training datasets by integrating two distinct components, while excluding the aforementioned categories of undeciphered characters. The first component comprised all other *deciphered* characters from the same period within the ACCP dataset, aimed at aiding the model in learning the typographic features of that era. The second component consisted of data from all other periods, intended to facilitate the model's understanding of the evolutionary patterns of characters across different epochs. This setup led to the training of a total of seven distinct models, each evaluated for its effectiveness in deciphering characters from the respective periods. Furthermore, for the model of OBIs deciphering, due to its fewer categories, we conducted Radical Decomposition on the deciphered samples of 781 classes of OBIs, generating 9,787 radical samples, to further experiment with deciphering OBIs*. To evaluate the performance, we employed two primary metrics: sample-wise accuracy and category-wise accuracy. Sample-wise accuracy is defined as the ratio of correctly deciphered samples to the total number of samples, providing a granular view of the model's deciphering capability on an individual character basis. The latter metric deems a category successfully deciphered if any single sample within it is correctly interpreted, reflecting the practical value of the real-world decipherment process.

Table 3. Ablation study results showing the impact of including data from various historical periods on the decipherment accuracy of the OBI decipherment model. The '√' symbol represents the usage of data corresponding to the respective period in the experiment, while the symbol "-" represents non-usage.

OBI	Bronze	Warring	Seal	Clerical	Kangxi	Regular	OBIs Acc.	OBIs* Acc.
√	-	-	-	-	-	-	0.02%	4.55%
√	-	-	-	-	-	√	0.42%	4.88%
√	-	-	-	-	√	√	0.45%	4.54%
√	-	-	-	√	√	√	0.80%	4.96%
√	-	-	√	√	√	√	2.53%	6.64%
√	-	√	√	√	√	√	6.23%	9.24%
√	√	√	√	√	√	√	17.47%	19.16%

The experimental results, as presented in Table 2, highlight the varying degrees of deciphering success across different historical periods. Notably, the Kangxi Period exhibits the highest sample-wise accuracy at 96.4% and a similarly high category-wise accuracy of 96.0%. This outstanding performance can likely be attributed to the closer proximity of the Kangxi period to modern times, which entails a more consistent writing standard and an abundance of available data. In contrast, the OBIs period shows a very low sample-wise accuracy at 17.5% and a category-wise accuracy of 39.4%. After incorporating Radical Decomposition, the sample-wise accuracy increased to 19.2%, and the category-wise accuracy increased to 41.4%. This can be attributed to the significant age of the OBI period, which not only limits the availability of data but also introduces a greater intra-category variability. Characters from this era are more stylized and less standardized, posing a substantial challenge for deciphering efforts.

5.3   Ablation Study

To further explore the efficacy of our model, particularly its capacity to learn the evolutionary patterns of ancient Chinese characters across different epochs, we selected the OBI decipherment model (including OBIs and OBIs*) as a case study for an in-depth ablation analysis.

In our ablation analysis presented in Table 3, we meticulously examined the impact of systematically reintroducing samples from various historical periods into the training set of our OBI decipherment model. For the OBI model, starting with an initial setup that exclusively used OBI data, we observed a decipherment accuracy close to 0, highlighting the inherent difficulty of deciphering ancient scripts without a broader historical context.

Upon sequentially adding data from more recent periods, specifically the Regular and Kangxi scripts, we noted only a marginal improvement in accuracy. This modest increase, with accuracy reaching 0.42% upon the inclusion of Regular script data and slightly fluctuating with the addition of Kangxi script data, suggests that the considerable temporal distance between these scripts and the

Table 4. Results of the extended ablation experiments. Each row represents one deciphering experiment, and the symbol '✓' indicates whether samples from the corresponding period were included in the training set.

OBIs*	Bronze	Warring	Seal	Clerical	Kangxi	Regular
4.78%	2.62%	0.96%	14.16%	10.91%	67.43%	76.76%
4.71%	2.95%	1.66%	16.25%	17.87%	92.63%	✓
4.75%	2.67%	1.48%	16.51%	21.01%	✓	✓
4.97%	3.65%	2.31%	19.26%	✓	✓	✓
5.59%	5.47%	2.93%	✓	✓	✓	✓
7.70%	15.06%	✓	✓	✓	✓	✓
19.16%	✓	✓	✓	✓	✓	✓

OBI period limits their effectiveness in providing useful contextual information for decipherment. The narrative changed significantly with the inclusion of Clerical, Seal, and Warring States period data, where we observed a steady increase in accuracy, culminating at 6.23%. This increment attests to the model's capability to extract and learn potential evolutionary patterns from cross-era data, thereby enhancing its decipherment performance.

The most pronounced surge in performance was witnessed with the incorporation of Bronze script data, which led to an 11.24% jump in accuracy, bringing it to a notable 17.47%. This substantial improvement underscores a strong correlation between the OBI and Bronze scripts, hinting at a direct lineage or significant shared characteristics that facilitate decipherment.

For the OBIs* model, as we sequentially introduced training data by more recent periods, the trend in decipherment accuracy was similar to that of the OBIs model. When data from the same period were added, the accuracy increased by approximately 2% to 4% compared to the OBIs model. This indicates that our method of decomposing components of Oracle Bone Inscription samples is effective.

Building upon the above insights, we further extended our investigation to a broader range of historical scripts, as detailed in Table 4. This comprehensive examination reaffirmed the efficacy of our model in deciphering ancient Chinese characters by leveraging data from multiple eras. The consistent improvement in decipherment performance across various settings underscored the model's capacity to extract and learn from the evolutionary patterns of Chinese characters. This finding not only validates our approach but also highlights the importance of incorporating a diverse historical corpus to enhance the decipherment of ancient scripts.

5.4   Visualization

To gain a more profound insight into our methodology, we employed Grad-CAM [26] for the visualization of the radical deconstruction and reconstruction process. As illustrated in Fig. 10, we demonstrate how the model deciphers

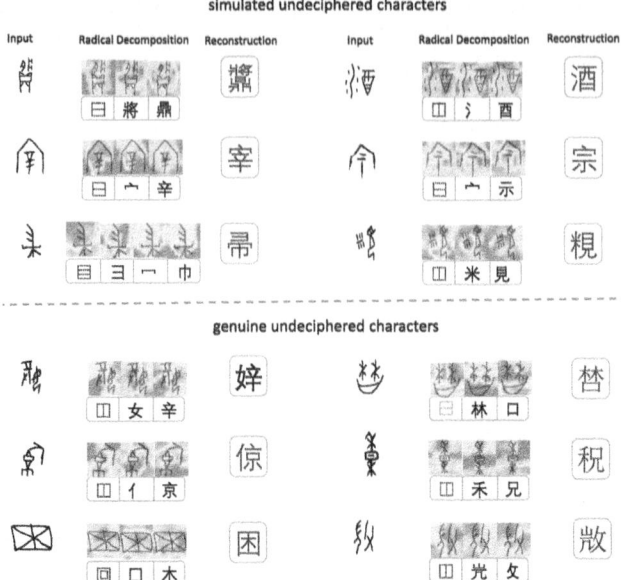

Fig. 10. Inputting the oracle bone inscription images into the model yields the radical sequence as output, along with the corresponding visualization images for each component. In the visualization images, the intensity of red indicates the areas where the component is more relevant. In the upper half of the image, *simulated undeciphered characters* represent images from a test set containing 1000 classes of OBI that have not been seen before. In the lower half, *genuine undeciphered characters* represent truly undeciphered OBI images from the HUST-OBC dataset. (Color figure online)

unseen categories of OBI images. The model discerns radicals in a fashion similar to human intuition while also considering the proximity of components to predict their structural arrangement.

6   Conclusion

In this paper, we have introduced the Puzzle Pieces Picker, an innovative method inspired by jigsaw puzzles for deciphering ancient Chinese characters. This approach, leveraging the deconstruction and reassembly of radicals, was supported by the expansion of existing datasets to form a new ACCP dataset. Our tests on the ACCP dataset demonstrated the P^3 model's promising accuracy across different historical periods. Furthermore, our ablation study confirmed the model's effectiveness in identifying evolutionary patterns in ancient characters. While our focus was on ancient Chinese characters, the versatility of the P^3 model suggests its applicability to other radical-based writing systems, offering new avenues for linguistic research and historical document decipherment.

Acknowledgments. This work was supported by the National Natural Science Foundation of China (No. 62225603, No. 62206104). This research is supported in part by National Natural Science Foundation of China (Grant No.: 62441604).

References

1. Bazerman, C.: Handbook of research on writing: History, society, school, individual, text. Routledge (2009)
2. Bochkovskiy, A., Wang, C.Y., Liao, H.Y.M.: Yolov4: optimal speed and accuracy of object detection. arXiv preprint arXiv:2004.10934 (2020)
3. Boltz, W.G.: Early Chinese writing. World Archaeol. **17**(3), 420–436 (1986)
4. Cao, Z., Lu, J., Cui, S., Zhang, C.: Zero-shot handwritten Chinese character recognition with hierarchical decomposition embedding. Pattern Recogn. **107**, 107488 (2020)
5. Chang, X., Chao, F., Shang, C., Shen, Q.: Sundial-gan: a cascade generative adversarial networks framework for deciphering oracle bone inscriptions. In: Proc. ACM Int. Conf. Multimedia, pp. 1195–1203 (2022)
6. Diao, X., et al.: Toward zero-shot character recognition: a gold standard dataset with radical-level annotations. In: Proc. ACM Int. Conf. Multimedia, pp. 6869–6877 (2023)
7. Fang, L., Huabiao, L., Jin, M., Sheng, Y., Peiran, J.: Automatic detection and recognition of oracle rubbings based on mask r-cnn. Data Analysis and Knowledge Discovery **5**(12), 88–97 (2022)
8. Fu, X., Yang, Z., Zeng, Z., Zhang, Y., Zhou, Q.: Improvement of oracle bone inscription recognition accuracy: a deep learning perspective. ISPRS Int. J. Geo-Information **11**(1) (2022). https://doi.org/10.3390/ijgi11010045, https://www.mdpi.com/2220-9964/11/1/45
9. Girshick, R.: Fast r-cnn. In: Proc. IEEE Int. Conf. Comp. Vis., pp. 1440–1448 (2015)
10. Guan, H., et al.: An open dataset for the evolution of oracle bone characters: Evobc. arXiv preprint arXiv:2401.12467 (2024)
11. Guo, J., Wang, C., Roman-Rangel, E., Chao, H., Rui, Y.: Building hierarchical representations for oracle character and sketch recognition. IEEE Trans. Image Process. **25**(1), 104–118 (2016)
12. Guo, J., Wang, C., Roman-Rangel, E., Chao, H., Rui, Y.: Building hierarchical representations for oracle character and sketch recognition. IEEE Trans. Image Process. **25**(1), 104–118 (2016). https://doi.org/10.1109/TIP.2015.2500019
13. Han, W., Ren, X., Lin, H., Fu, Y., Xue, X.: Self-supervised learning of orc-bert augmentator for recognizing few-shot oracle characters. In: Proc. Asian Conf. Comp. Vis. (2020)
14. He, K., Fan, H., Wu, Y., Xie, S., Girshick, R.: Momentum contrast for unsupervised visual representation learning. In: Proc. IEEE Conf. Comp. Vis. Patt. Recogn., pp. 9729–9738 (2020)
15. He, K., Zhang, X., Ren, S., Sun, J.: Deep residual learning for image recognition. In: Proc. IEEE Conf. Comp. Vis. Patt. Recogn., pp. 770–778 (2016)
16. Huang, S., Wang, H., Liu, Y., Shi, X., Jin, L.: Obc306: a large-scale oracle bone character recognition dataset. In: 2019 International Conference on Document Analysis and Recognition (ICDAR), pp. 681–688 (2019). https://doi.org/10.1109/ICDAR.2019.00114

17. Kirillov, A., et al.: Segment anything. arXiv preprint arXiv:2304.02643 (2023)
18. Li, Y., Zhu, Y., Du, J., Wu, C., Zhang, J.: Radical counter network for robust Chinese character recognition. In: Proc. Int. Conf. Patt. Recogn., pp. 4191–4197 (2021)
19. Lin, X., Chen, S., Zhao, F., Qiu, X.: Radical-based extract and recognition networks for oracle character recognition. Int. J. Doc. Anal. Recognit. **25**(3), 219–235 (2022)
20. Liu, G., Xing, J., Xiong, J.: Spatial pyramid block for oracle bone inscription detection. In: Proceedings of the 2020 9th International Conference on Software and Computer Applications, pp. 133–140. Association for Computing Machinery (2020)
21. Liu, W., et al.: SSD: single shot MultiBox detector. In: Leibe, B., Matas, J., Sebe, N., Welling, M. (eds.) ECCV 2016. LNCS, vol. 9905, pp. 21–37. Springer, Cham (2016). https://doi.org/10.1007/978-3-319-46448-0_2
22. Liu, Y., Zhang, J., Peng, D., Huang, M., Wang, X., Tang, J., Huang, C., Lin, D., Shen, C., Bai, X., et al.: Spts v2: single-point scene text spotting. IEEE Trans. Pattern Anal. Mach, Intell (2023)
23. Liu, Z., et al.: Oracle character detection based on improved faster r-cnn. In: 2021 International Conference on Intelligent Transportation, Big Data Smart City (ICITBS), pp. 697–700 (2021)
24. Meng, L., Lyu, B., Zhang, Z., Aravinda, C.V., Kamitoku, N., Yamazaki, K.: Oracle bone inscription detector based on SSD. In: Cristani, M., Prati, A., Lanz, O., Messelodi, S., Sebe, N. (eds.) ICIAP 2019. LNCS, vol. 11808, pp. 126–136. Springer, Cham (2019). https://doi.org/10.1007/978-3-030-30754-7_13
25. Peng, D., et al.: Spts: single-point text spotting. In: Proc. ACM Int. Conf. Multimedia, pp. 4272–4281 (2022)
26. Selvaraju, R.R., Cogswell, M., Das, A., Vedantam, R., Parikh, D., Batra, D.: Gradcam: visual explanations from deep networks via gradient-based localization. In: Proc. IEEE Int. Conf. Comp. Vis., pp. 618–626 (2017)
27. Suzuki, A., Suzuki, T.: A survey on the achievements of the oracle bone digitization projects and prior definitions of the targets. J. Inf. Process. Soc. Japan. DD (Digital Document) (6), 1–8 (2013)
28. Vaswani, A., et al.: Attention is all you need. Proc. Advances in Neural Inf. Process, Syst. (2017)
29. Wang, M., Deng, W., Liu, C.L.: Unsupervised structure-texture separation network for oracle character recognition. IEEE Trans. Image Process. **31**, 3137–3150 (2022)
30. Wang, P., et al.: An open dataset for oracle bone script recognition and decipherment. arXiv preprint arXiv:2401.15365 (2024)
31. Wang, W., Zhang, J., Du, J., Wang, Z.R., Zhu, Y.: Denseran for offline handwritten Chinese character recognition. In: 2018 16th International Conference on Frontiers in Handwriting Recognition (ICFHR), pp. 104–109. IEEE (2018)
32. Xing, J., Liu, G., Xiong, J.: Oracle bone inscription detection: a survey of oracle bone inscription detection based on deep learning algorithm. In: Proceedings of the International Conference on Artificial Intelligence, Information Processing and Cloud Computing, pp. 1–8 (2019)
33. Xiong, R., et al.: On layer normalization in the transformer architecture. In: Proc. Int. Conf. Mach. Learn., pp. 10524–10533. PMLR (2020)
34. Yang, Z., Fu, T.: Oracle detection and recognition based on improved tiny-yolov4. In: Proceedings of the 2020 4th International Conference on Video and Image Processing, pp. 128–133. Association for Computing Machinery (2021)

35. Yue, X., Li, H., Fujikawa, Y., Meng, L.: Dynamic dataset augmentation for deep learning-based oracle bone inscriptions recognition. J. Comput. Cult. Herit. **15**(4), December 2022. https://doi.org/10.1145/3532868. https://doi.org/10.1145/3532868

36. Zhang, G., Liu, D., Smyth, B., Dong, R.: Deciphering ancient chinese oracle bone inscriptions using case-based reasoning. In: Sánchez-Ruiz, A.A., Floyd, M.W. (eds.) Case-Based Reasoning Research and Development. pp. 309–324. Springer International Publishing (2021)

37. Zhang, J., Zhu, Y., Du, J., Dai, L.: Radical analysis network for zero-shot learning in printed Chinese character recognition. In: Proc. IEEE Int. Conf. Multimedia & Expo., pp. 1–6. IEEE (2018)

38. Zhang, Y.K., Zhang, H., Liu, Y.G., Yang, Q., Liu, C.L.: Oracle character recognition by nearest neighbor classification with deep metric learning. In: Proc. Int. Conf. Doc. Anal. and Recognit., pp. 309–314 (2019)

Document Understanding and NLP

Light-Weight Multi-modality Feature Fusion Network for Visually-Rich Document Understanding

Jeff Yang[ID], Huynh Vu The[(✉)][ID], and Hai Luu Tuan[ID]

Cinnamon AI, Tokyo, Japan
vuthe-huynh@yahoo.com
https://cinnamon.is

Abstract. Entity extraction (EE) is an important task in visually-rich document understanding (VrDU) which leverages multi-modal features of text, layout, and image. Recent transformer-based architectures enable an effective fusion of these features, showing great performance on the EE task. However, these models are heavy, leading to substantially high training cost and low inference speed. Thus, we propose a light-weight transformer-based model (named LMFFN) with a novel layout-self-attention layout-aware multi-modal fusion mechanism that allows an efficient entity extraction. Specifically, the proposed framework uses just a simple pre-training objective coupled with an effective batch implementation. In addition, no constraints are required with regard to the input sequence length or the reading order. This relaxation gives our model an advantage when it comes to camera and skewed documents, as we observed a 7% F1-score improvement when we compared our model to previous SOTA models on camera data. Evaluation results of three public datasets (CORD, SROIE, and XFUND) show that our proposed architecture achieves competitive performance compared to recent SOTA models while having 5 to 10 times fewer parameters.

1 Introduction

Visually-rich documents (VrDs) are tightly connected to our daily life from insurance, logistic, finance to travel and leisure. As they contain lots of useful information, there are demands for archiving, indexing or structuring document information in form of digital format. In visually-rich documents understanding (VrDU), documents of various formats (PDF, images, etc.) are digitized, structured and organized automatically, which plays a crucial role in developing digital transformation processes in daily business operation.

The VrDU tasks such as semantic entity extraction and entity linking [18] are built upon the inherent complexities of co-existed multi-modal features. Hence, an effective feature learning and fusion of multi-modal data play an important role to achieve good performance. Another challenge of VrDU tasks is the lack of data. While there is a variety of document templates, some of them are in limited

E. H. Barney Smith et al. (Eds.): ICDAR 2024, LNCS 14804, pp. 191–207, 2024.
https://doi.org/10.1007/978-3-031-70533-5_12

amounts, which can affect a VrDU model performance on limited-data templates. In some VrDU approaches [27,39,44,45], when a new document template is introduced, a VrDU model may need to be re-trained from scratch, preventing knowledge transfer among different document templates. To improve this limitation, visual-language pre-training (VLP) has been exploring for cross-template knowledge transfer. The recent *LayoutLM* [41] has obtained a strong pre-training using Masked Visual Language Modelling (MVLM) objective given a combination of textual, visual, and spatial input features. The following *LayoutLMv2* [40] further enhances the pre-training by introducing two-new training objectives to improve the correlation of images and text features.

While *LayoutLMv3* [15] has achieved good performance in VrDU tasks, it is costly to pre-train the model on a large pre-training dataset. On the other hand, the effectiveness of applying this pre-training weight in low-resource language dataset (e.g. Japanase) is open to question. We propose a new light-weight transformer-based architecture to address this issue. Our model contains 21M parameters compared to 133M of $LayoutLMv3_{BASE}$ [15] and 368M of $LayoutLMv3_{LARGE}$ [15], which is 6 times more compact than $LayoutLMv3_{BASE}$. The light-weight backbone is obtained due to the efficient design of text encoders and transformer layers. Instead of using heavy BERT-based language model [9], we use just two bidirectional gated recurrent units (GRU) [6] for text encoders. Experimental results on private datasets show that simple GRU with segment-level aligned text encoders achieve a similar F1-score compared to the heavy language model with token-level based encoders. However, for under-segmented layout data, our segment-level based model does not perform as well as token-level based approaches [15,40]. In addition to the efficient design, our transformer layer just contain 0.7M parameters compared to 7M of *LayoutLMv3*, which is 10 times smaller. It is noted that, the current count of 21M parameters does not account for the image encoder (24M), as its parameters are frozen during training. Additionally, during inference, the image features extracted by the image encoder are derived from a previous layout step, where a sequence of 2D locations is generated as input to our LMFFN model. The efficient model architecture design contributes to reducing the inference speed and lowering the pre-training cost substantially.

In addition to using a lightweight backbone, we introduce a novel fusion mechanism for textual, visual, and layout features that enables the lightweight model to achieve competitive performance. Inspired by the layout-aware implementation of [41], we explore different multi-modal fusion techniques (see Fig. 3). In contrast to the linear layer encoding of location information used in [41], our design leverages self-attention for layout using Transformer1 (shown in Fig. 1), which has been proven to be highly effective, improving the F1-score by 2

We also demonstrate an advantage of our model over cameras images and skewed images due to the relaxed constraint in the optimal order of 1D position encoding. In previous approaches [14,40,41,44], an input sequence is required to be sorted in reading order. However, in skewed and camera images, texts are often not aligned well which might cause a sorted sequence to be incorrect.

On the other hand, our model design does not require that the input sequences are sorted in reading order. We observed a 7% F1-score improvement when we compared our model to SOTA approaches on internally collected camera data. Our model also imposes no condition on the length of the input sequences. On the contrary, previous approaches [14,24,40,41,44] apply a threshold for the maximum length of input sequences. This constraint reduces the generalization of a model on dense-text documents in which a whole input sequence might not be fully observed by the model. On the other hand, an arbitrary length of input sequence contributes to improving the efficiency of batch training. Our proposed padding scheme relies on the longest sequence in a batch, which greatly reduces the number of paddings, causing the batch training to be more efficient.

The contributions of this paper are summarized as follow:

- A generic light-weight transformer-based backbone for VrDU tasks.
- A novel layout-self-attention layout-aware attention design for multi-modal feature fusion. Different multi-modal feature fusions are proposed and analyzed.
- A transformer-based framework with a simple pre-training objective, and the relaxed constraints with regard to the input sequence length and the optimal reading order.
- A new benchmark test set (the ct_SROIE) that includes 49 camera images which are built from the test set of SROIE dataset [16]. These camera images are generated by a mobile device with various conditions of light, geometry, camera and background

2 Related Works

Early VrDU approaches [1,4,8,10,22] are based on a set of heuristics rules and empirical features to extract pairs of key-value from VrDs. Rules range from simple regular expression [1] to complicated patterns designed by several seniors in a long period of time [22]. Although the rule-based approaches worked well in some cases, they often require considerable effort and expertise to develop and configure given a wide range of document templates.

Limitations in rule-based approaches have motivated researchers for learning-based algorithms. Initial attempts in this direction focus on plain text parsing and left the visual features as well as layout information untapped. Some plain-text frameworks [29,34] formulated the entity extraction (EE) as a sequential labeling problem by serializing text segments in optimal orders before a subsequent BiLSTM-CRF layer [23]. The plain text framework provides a strong baseline for a couple of academic datasets but shows limitations in real world applications, in which text alone is not representative enough to provide comprehensive semantic information. In fact, there are several numeric text segments that represent different amount entities in invoice documents, and without spatial cues it's barely possible to make precise prediction under different layouts.

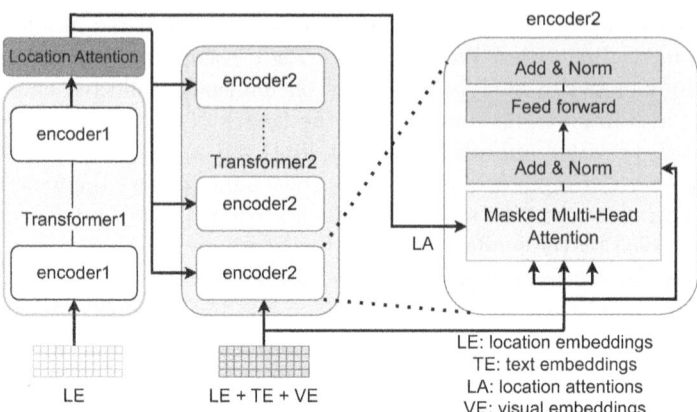

Fig. 1. The illustration of layout-self-attention layout-aware fusion block. The block contains Transformer1 for self-attention on location embeddings (LE) and Transformer2 for feature fusion of multi-modal data (LE+TE+VE).

Learning-based approaches that model both text and spatial features have been proposed by [5,17,20,26] using graph neural networks [11]. In graph-based approaches for VrDs [31], the whole document is modelled as a graph, text features as nodes, and spatial relationships among the text segments (e.g. left, right, top, bottom, text width, text height) as edges. This modelling can help preserve the spatial relation for a document content. However, the propagation of relational information among edges is constraint by the shallowness of the network, which hinders the spatial modeling of far-distance nodes. To overcome this drawback, [5,12,26] apply graph-based attention [36] for VrDU tasks to model non-local entities in a document [36]. A similar approach to model complex spatial relationship using graph is proposed by [17], in which the EE task is formulated as a spatial dependency parsing problem, being able to model full spatial hierarchy in a document.

Recent works further explore vision-based transformer architecture [21] to capture full semantic structure of VrDs by the cross-modality integration of textual, visual and layout information [14,24,37,40,41]. In this direction, LayoutLM [41] and LayoutLMV2 [40] extract image features (following the bottom-up approach) and incorporates them alongside textual and layout features. In the bottom-up approach, image features are taken from a set of salient image regions proposed by a detection model (e.g., Faster-CNN [33]). In contrast, our work extracts image features following the top-down mechanism whereby output features are leveraged from an image encoder [19] instead of from proposed image regions. We find that the top-down mechanism enables a simple concatenation or summation of multi-modal features.

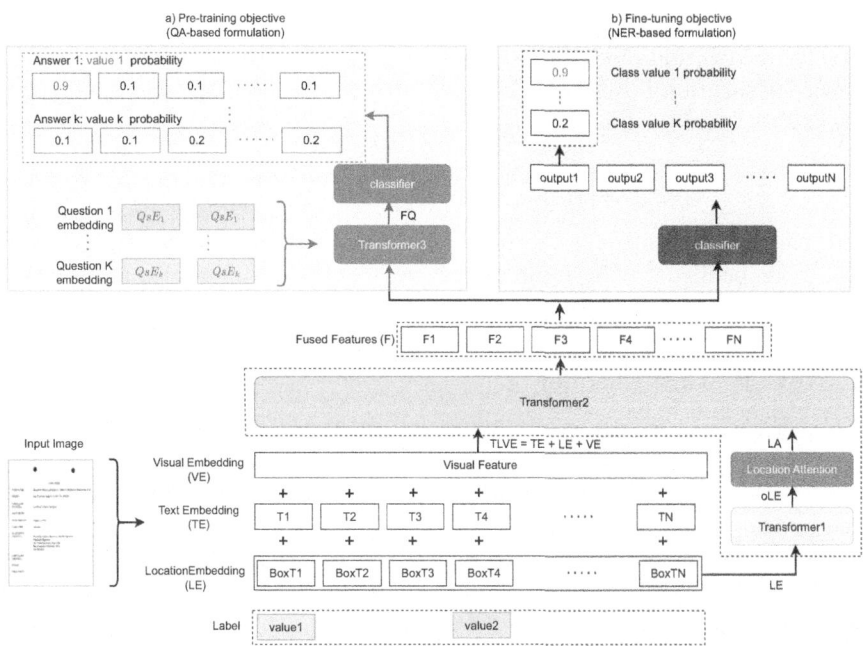

Fig. 2. The LMFFN architecture includes three types of feature encoders: one for visual features, one for text features, and one for location features. These encoders extract visual embeddings (VE), text embeddings (TE), and location embeddings (LE), respectively. The location embeddings (LE_i) corresponds to a location box Box_{Ti} while the text embedding (TE_i) corresponds to a text segment T_i. The fusion block comes after the encoders and has two transformers. It produces fused features (F) that can be supervised using QA-based or NER-based approaches.

3 Model Architecture

Figure 2 provides an overview of the proposed LMFFN model. The model utilizes separate encoders for different modalities, including text, location, and visual features. The location encoder is followed by a self-attention module, called Transformer1, to further encode location embeddings (LE). A multi-modal fusion block (as shown in both Fig. 2 and Fig. 1) is formed by the combination of Transformer1 and Transformer2, which takes the feature embeddings of the encoder outputs for feature fusion and outputs the fused feature (F). The final layer of the model is designed based on downstream tasks, such as QA and NER formulations. The QA pre-training strategy allows the model to learn rich representation under weak-supervised datasets. Our proposed LMFFN does not impose constraints in the optimal reading order of input sequences, and thus, sequence tagging [9] and 1D position encoding are not required in the segment-level aligned encoders.

For each input image I, using a text segmentation model and an optical character recognition model, we obtain a sequence of text segments S_{text} and a sequence of location boxes S_{loc}

$$S_{text} = \{T_1, T_2, ..., T_i, T_{pad}\} \tag{1}$$

$$S_{loc} = \{Box_{T1}, Box_{T2}, ..., Box_{Ti}, Box_{pad}\} \tag{2}$$

where each text segment T_i corresponds to a location box Box_{Ti}. T_{pad} and Box_{pad} are padded to the input sequences to obtain similar sequence length with regard to the longest sequence in a batch.

3.1   Text Feature Encoder

Different from most recent works [24, 40, 41] which adopt pre-trained Bert-based language model [9], we deploy a light-weight sequence model for text embeddings. The text embedding model (Emb_{text}) contains learnable embedding layers followed by two bidirectional gated recurrent units (GRU) [6].

Before the embedding step, each text segment T_i is tokenized to a number of tokens $\{tk_1^i, tk_2^i, tk_j^i\}$. Then, Pad is added so that each text segment has the same number of elements. Finally, text embeddings (TE) is obtained by the formula,

$$TE = Emb_{text}(S_{text}) \tag{3}$$

3.2   Location Feature Encoder

We leverage four corner points of each text bounding box $Box_i = \{v_i^{tl}, v_i^{tr}, v_i^{bl}, v_i^{br}\}$ where each point $v_i = (x, y)$. The x, y coordinates are divided by the image width (w) and height (h) to be normalized between 0 and 1. The normalized features are concatenated before being embedded by the location embedding layer Emb_{loc}, which consists of some linear layers. The location embeddings (LE) is obtained by the following formula,

$$LE = Linear(Concat(Norm((v_i^{tl}, v_i^{tr}, v_i^{bl}, v_i^{br}))) \tag{4}$$

3.3   Visual Feature Encoder

We use ResNet50 [13] with deformable convolution [7, 25] as our encoder backbone and reuse its pre-train weights for visual feature extraction. The deformable convolution provides a flexible receptive field, which is beneficial to the arbitrary shape of document objects such as text-lines and stamps. All input images are first resized to (768, 768) to unify the output dimensions. We take the grid features from the last layer and flatten the height and width dimensions to align the transformer input feature shape. It is worth noting that our visual feature encoder was stacked with decoder architecture to form a segmentation backbone and pretrained with our private dataset. The dataset consists of thousands of document images with four common document objects including text-lines, stamps, tables, and checkmarks.

3.4   Layout-Self-Attention Layout-Aware Fusion Block

The fusion block, depicted in Fig. 1 and Fig. 2, is comprised of two transformer modules, each with a hidden size of 256, which contributes to the transformer's lightweight design. Transformer1 has two layers of an 8-head encoder that encodes the location embeddings $LE \in R^{N \times D}$, where D represents the feature dimension and N denotes the length of the input sequence. The Transformer1 output, referred to as oLE, is multiplied by its transpose to obtain the location attention (LA). If $Q_{l,i}, K_{l,i}$ and $V_{l,i}$, are queries, keys and values of the input of the ith Transformer1 encoder, d as feature dimension of the queries and keys, then the attention function of this encoder is defined as

$$softmax(\frac{Q_{l,i}K_{l,i}^T}{\sqrt{d}})V_{l,i} \tag{5}$$

Transformer2 contains 8 layers of an 8-head encoder. It creates fused features (F) given the location attention (LA) and the combined feature $TLVE$ of text, image and location ($TLVE = TE + LE + VE \in R^{N \times D}$). If $Q_{f,i}, K_{f,i}, V_{f,i}$, are queries, keys and values of the $TLVE$ input of the ith Transformer2 encoder, the attention function of the encoder is described as

$$softmax(\frac{Q_{f,i}K_{f,i}^T}{\sqrt{d}} + LA)V_{f,i} \tag{6}$$

Fine-Tuning Objective : We formulate the fine-tuning task as a name entity extraction task (NER-based task), similar to [14,40,41].

Pre-training Objective: Question-Answering Based Formulation (QA-Based Formulation) . Unlike previous approaches that utilize complex pre-training objectives (such as text-image matching, text-image alignment and visual language modelling) on a large, unlabelled English dataset of 11 million images, we use a QA-based formulation (inspired by [38]) for the pre-training objective.

Corresponding to a set of K entities is a set of questions Qs = $[qs_1, ...qs_i, ...qs_K]$. For each question qs_i, we obtain a corresponding embedding QsE_i by a sentence transformer [32] followed by some linear layers.

We use 2 layers of Transformer3 to obtain the attention fusion FQ_i of the output feature F and question embedding QsE_i. Denote K_i and V_i as keys and values of the input embedding QsE_i and Q_i as queries of the output feature F. The attention function of the ith Transformer3 encoder is given as follow

$$softmax(\frac{Q_iK_i^T}{\sqrt{d}})V_i \tag{7}$$

$$loss_{QA} = \frac{\sum_{i=i}^{n1} loss(f_1(FQ_i), y_i))}{n1} \tag{8}$$

4   Experiment

4.1   Public Datasets

CORD. [30] stands for a Consolidated Receipt Dataset for post-OCR parsing. The dataset consists of all receipt-typed forms in which 800 forms are used for training, 100 for validation, and 100 for testing. There are more than thirty semantic entities (e.g., menu price, menu name, etc.) annotated for the entity extraction task.

SROIE [16] consists of 953 receipt documents. The dataset is divided into 626 forms for training and 327 for testing. Digit and English characters held the majority of the dataset. Annotations are provided for four semantic entities including: company, address, date and total.

ct_SROIE consists of 49 camera images which are built from the test set of SROIE dataset [16]. The test set of SROIE are manually classified into groups and images are sampled from these groups to obtain a collection of 49 scan images (named st_SROIE). These images are then captured by a mobile device with various conditions of light, geometry, camera and background. The ct_SROIE dataset will be publicly available.

4.2   Private Dataset

Japanese Medical Camera Document (JMCD) contains 809 of both camera and scan images with 48 target entities. Camera and scan images are mixed up for both training and testing. Training data contains 269 camera images and 257 scan images while testing data includes 109 images in camera format and 104 in scan format.

Japanese Pre-training Document (JPD) is a private dataset containing approximately 18000 business documents of various types such as invoices, medical receipts, health checkups, insurance contracts, etc. Documents are provided with more than 100.000 labeled entities. The dataset is used for the pre-training task and more than 100 distinct entities are included.

4.3   Experiment Setup

In our experiments, pre-training and fine-tuning are performed end-to-end using a single NVIDIA Tesla T4 with 16G memory. The model is first pre-trained on the JPD dataset for the entity extraction task formulated as the QA-based approach. The model is then fine-tuned on private datasets (JMCD) as well as public datasets including CORD [30] and SROIE [16]. The fine-tuning is performed for the entity extraction task, similar to that of pre-training. We compare results using the entity-level F1-score, commonly used in previous approaches [14,17].

For both pre-training and fine-tuning tasks, input images are resized to 768×768 and words in each text segments are padded with reference to the longest text segment in a document. For batch implementation, text segments are also padded to obtain the same number of segments for all images of the same batch. Contrary to previous approaches [14, 24, 40], no constraint is imposed on the length of input text segments. We use three transformers to encode and fuse multi-modal features. Transformer1 and Transformer3 contain 2 layers while 8 layers are applied for Transformer2.

About hyperparameters, all transformers are configured with 256 hidden sizes, and 8 attention heads. The weight parameters of the image feature module are initialized from a ResNet-based [13] segmentation model trained on internal datasets. The rest of the parameters (e.g., transformers, text feature extraction) are initialized randomly. The training applies KLDivLoss using AdamW as the optimizer with an initial learning rate of 10^{-4} in the first epoch. In the following epochs, the learning rate follows the CosineAnnealingLR scheduler.

4.4   Experiment Results

Our model LMFFN is benchmarked on three public datasets: CORD [30], SROIE [16] and XFUND [42].

As shown in Table 1, our model LMFFN obtains a competitive F1-score of 96.47%, which is slightly lower than the 96.56% of $LayoutLMv3_{BASE}$ [15] on the CORD dataset, while containing only 21M parameters, just one sixth of the 133M parameters of $LayoutLMv3_{BASE}$ [15]. For large model configurations, our model achieves 1% F1-score lower than $LayoutLMv3_{LARGE}$ [15] (96.47% vs 97.46%) while having fewer parameters (21M vs 368M) and being pre-trained in a small dataset of 18K images compared to 11M used by $LayoutLMv3_{LARGE}$. It is worth noting that we use a Japanese pre-training dataset to fine-tune on the English public datasets. An English pre-training dataset with a large number of images will likely improve our model's performance. On the SROIE dataset, our model LMFFN obtains a SOTA F1-score of 97.51% (Table 3). Compared to StrucText [24], our model obtains a 0.6% F1-score higher with fewer parameters (21M vs 107M). On Japanese documents of XFUND dataset, our model outperforms LayoutXLM_BASE [42] by 2% (81.04% vs 79.21%) and LayoutXLM_LARGE [42] by 1% (80.33% vs 81.04%) (Table 2).

5   Ablation Study

We conducted several experiments to evaluate the effectiveness of some components (e.g. pre-training, image feature as well as fusion mechanism) on the overall performance. Experiments are evaluated on private and public datasets.

5.1   Camera Data

In practical applications, it is often necessary to work with both scanned and camera-captured data. However, the majority of publicly available datasets consist of scanned data. To assess the performance of our model on camera-captured

Table 1. Results of entity extraction on CORD dataset.

Method	F1	Params.
$BERT_{BASE}$ [9]	89.68	110M
$UniLMv_{BASE}$ [3]	90.92	125M
$LayoutLM_{BASE}$ [41]	94.72	113M
$LayoutLMv2_{BASE}$ [40]	94.95	200M
SPADE [17]	91.50	-
$DocFormer_{BASE}$ [2]	96.33	183M
$LayoutLMv3_{BASE}$ [15]	**96.56**	133M
$BERT_{LARGE}$ [9]	90.25	340M
$UniLMv2_{LARGE}$ [3]	92.05	355M
$LayoutLM_{LARGE}$ [41]	94.93	343M
BROS [14]	95.36	-
$LayoutLMv2_{LARGE}$ [40]	96.01	426M
$DocFormer_{LARGE}$ [2]	96.99	536M
$LayoutLMv3_{LARGE}$ [15]	**97.46**	368M
Our LMFFN	96.47	21M

Table 2. Results of entity extraction on SROIE dataset

Method	F1	Params.
LAYOUTLM_BASE [41]	94.38	113M
LAYOUTLM_LARGE [41]	95.24	343M
PICK [44]	96.12	-
VIES [37]	96.12	-
TRIE [45]	96.18	-
LayoutLMv2_BASE [40]	96.25	200M
MatchVIE [35]	96.57	-
LayoutLMV2_LARGE [40]	96.61	426M
StrucText [24]	96.88	107M
Our LMFFN	**97.51**	21M

Table 3. Results of entity extraction on Japanese documents of the XFUND dataset.

Method	F1	Params.
LayoutXLM_BASE [42]	79.21	345M
LayoutXLM_LARGE [42]	80.33	625M
Our LMFFN	**81.04**	**21M**

data, we use JP-LayoutLMv2 [28], a version of LayoutLMv2 that has been pre-trained on Japanese documents, and evaluate it on both public camera dataset (ct_SROIE) and private camera dataset (JMCD). The results on Table 4 show that our model outperforms JP-LayoutLMv2 by 7% on the JMCD (79.27% vs 72.64%) and by 9% on the ct_SROIE (91.37% vs 82.03%). It is noted that the evaluation on ct_SROIE and st_SROIE uses the same train set of SROIE. While LMFFN model on ct_SROIE achieves a lower 2% F1-score compared to the evaluation on st_SROIE (91.37% vs 93.93%), JP-LayoutLMv2 accuracy drops 9% from 91.2% on st_SROIE to 82.03% on ct_SROIE. We observe that the low F1-score of JP-LayoutLMv2 is attributed to skewed images where text segments are rotated and not aligned well. As a result, an optimal reading order cannot be obtained, leading to decreased extraction results.

Table 4. Ablation study on camera data JMCD and ct_SROIE.

Models	JMCD	st_SROIE	ct_SROIE
JP-LayoutLMv2 [28]	72.64	91.2	82.03
our LMFFN	79.27	93.93	91.37

5.2   Training and Inference Memory

In Table 5, we compare the computation resources required for LayoutLMv3 [15] and our LMFFN model. Both models are trained and run using a batch size of 2 on an NVIDIA Tesla T4. Our model uses only half the GPU memory during training compared to LayoutLMv3 [15] (2.8G versus 6.5G), and during inference, our LMFFN model consumes just 55% of the GPU resources required by LayoutLMv3 (1.5G vs 2.7G). In addition, the model weight size of LMFFN is just approximately 170M while that of LayoutLMv3 is more than 1.4G. The large size of the model weight may cause problems for resource optimization in the deployment process.

Table 5. Ablation study on the computation resources among *LayoutLMv3_base* [15] and our LMFFN.

Resource	LayoutLMv3 [15]	Our LMFFN
Training memory	6.5G	2.8G
Inference memory	2.7G	1.5G

5.3   Processing Time

We conducted experiments to compare the processing time of LayoutLMv3_base [15], LayoutXLM_base [42], and our LMFFN model on the CORD [30] and XFUND-ja [43] datasets. Using NVIDIA Tesla T4, we ran the experiments with a batch size of 2 and measured the average run-time on the test set of these two datasets. We then divided the number of samples in the test set by the average run-time to obtain the number of processed samples per second, as shown in Tables 6 and 7.

Our LMFFN model has shown impressive results on the CORD dataset, processing double the number of samples in just one second compared to LayoutLMv3_base (12.5 vs 6.7). Similarly, on the XFUND-ja dataset, our model was able to process triple the number of samples compared to LayoutXLM_base. The long processing time of LayoutXLM_base may be due to the large number of model parameters compared to our LMFFN (345M vs 21M). This difference results in LayoutXLM_base generating a large model checkpoint of 4.2G, while the size of the model weight of our LMFFN is just 170M, making it much smaller than that of LayoutXLM_base.

5.4   Fusion Mechanism

Table 8 displays different fusion configurations used to evaluate the impact of various combinations of text and location embeddings. These configurations incorporate location attention (LA), text attention (TA), and location-text attention (LTA), in addition to location embeddings (LE), text embeddings (TE), and

Table 6. The number of processed samples per second on the test set of the CORD [30] dataset.

Method	No. samples per second
LayoutLMv3_base [15]	6.7
Our LMFFN	**12.5**

Table 7. The number of processed samples per second on the test set of the XFUND-ja [43] dataset.

Method	No. samples per second
LayoutXLM_base [42]	3.3
Our LMFFN	**12.8**

their combination (TLE). We define reference co-attention (Eq. 13, 15, 17) as the co-attention mechanism that maintains a constant value in the reference input across all layers of the same transformer. In contrast, the inputs of the layer-by-layer attention (Eq. 10, 11) are updated across transformer layers.

Denote $Q_{l,i}, K_{l,i}$ and $V_{l,i}$, as queries, keys and values of an $encode_i$ with location embeddings (LE) as input; $Q_{tx,i}, K_{tx,i}$ and $V_{tx,i}$, with text embeddings (TE) as input and $Q_{f,i}, K_{f,i}$ and $V_{f,i}$, with the fused feature (TLE) as input. We also define f_a as an attention function described by $fa(X) = XX^T$ (Fig. 3), and TA, LA, and LTA as text attention, location attention and location-text attention respectively.

$$Atten1 = f_s(\frac{Q_{f,i}K_{f,i}^T}{\sqrt{d}})V_{f,i} \tag{9}$$

$$Atten2 = f_s(\frac{Q_{f,i}K_{f,i}^T}{\sqrt{d}} + \frac{Q_{l,i}K_{l,i}^T}{\sqrt{d}})V_{f,i} \tag{10}$$

$$Atten3 = f_s(\frac{Q_{f,i}K_{f,i}^T}{\sqrt{d}} + \frac{Q_{l,i}K_{l,i}^T}{\sqrt{d}} + \frac{Q_{f,i}K_{l,i}^T}{\sqrt{d}})V_{f,i} \tag{11}$$

$$Atten4 = f_s(\frac{Q_{l,i}K_{l,i}^T}{\sqrt{d}})V_{l,i} \tag{12}$$

$$Atten5 = f_s(\frac{Q_{f,i}K_{f,i}^T}{\sqrt{d}} + LA)V_{f,i} \tag{13}$$

$$Atten6 = f_s(\frac{Q_{l,i}K_{tx,i}^T}{\sqrt{d}})V_{tx,i} \tag{14}$$

$$Atten7 = f_s(\frac{Q_{f,i}K_{f,i}^T}{\sqrt{d}} + LA + LTA)V_{f,i} \tag{15}$$

$$Atten8 = f_s(\frac{Q_{tx,i}K_{tx,i}^T}{\sqrt{d}})V_{tx,i} \qquad (16)$$

$$Atten9 = f_s(\frac{Q_{f,i}K_{f,i}^T}{\sqrt{d}} + LA + TA)V_{f,i} \qquad (17)$$

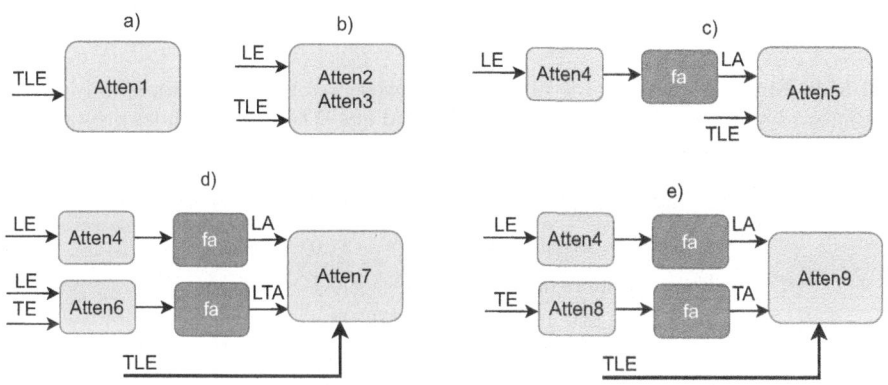

Fig. 3. A diagram of configurations for different fusion mechanisms.

Table 8. Fusion configurations and F1-score on public datasets $CORD$, $CORD^*$ and $SROIE$. $CORD^*$ denotes the $CORD$ dataset trained from scratch. All fusion configurations do not take into account visual embeddings (VE).

Configurations	Reference	$CORD^*$	$CORD$	$SROIE$
Fusion1	Fig. 3 a), Eq. 9	93.53	93.85	95.86
Fusion2	Fig. 3 b), Eq. 10	93.29	93.98	-
Fusion3	Fig. 3 b), Eq. 11	93.81	94.55	-
Fusion4	Fig. 3 c), Eq. 13	**95.26**	**95.72**	**97.36**
Fusion5	Fig. 3 d), Eq. 15	91.97	93.95	-
Fusion6	Fig. 3 e), Eq. 17	90.69	93.52	-

As shown in Table 8, the Fusion4 configuration with location-reference attention achieves the highest F1-score across Fusion configurations on the public datasets. We observe that the inclusion of text-reference attention (in Fusion6, Fusion7 configuration) decreases the F1-score. Similarly, the layer-by-layer attention configurations (Fusion2, Fusion3) do not performed as well as the location-reference attention (Fusion4).

5.5 Layout-Self-Attention

We compare the effect of using self-attention versus feedforward for layout on the overall accuracy of the XFUND and SROIE datasets. The Transformer1 module in Fig. 1 illustrates the use of layout-self-attention for location embeddings. Table 9 shows that without the pre-training model, self-attention performs better than feedforward by 1% on both datasets (74.8% vs 73.9% on XFUND and 95.6% vs 94.8% on SROIE). Especially, with the pre-training model, self-attention achieves a higher F1-score than the feedforward configuration by 3% (81.04% vs 77.7%) on the XFUND dataset and 1% (97.51% vs 96.15%) on the SROIE dataset.

Table 9. Ablation study on the differences between layout-self-attention and layout-feedforward methods for creating location embeddings. The study includes pre-training (PT) and compares the effectiveness of the two methods.

Configurations	XFUND	SROIE
layout-self-attention (with PT)	81.04	97.51
layout-feedforward (with PT)	77.7	96.15
layout-self-attention (without PT)	74.8	95.6
layout-feedforward (without PT)	73.9	94.8

6 Conclusion

We propose a light-weight transformer-based backbone with a novel layout-aware fusion mechanism for VrDU tasks. The proposed model has 5 to 10 times fewer parameters than recent SOTA models. Furthermore, our model does not apply constraints on the length and the optimal order of an input sequence. The relaxed constraints give our model an advantage when it comes to camera and skewed images. However, for under-segmented layout data, our segment-level based model does not perform as well as token-level based approaches. On public datasets, our LMFFN model achieves competitive performance compared to SOTA models although we use just a simple pre-training objective and a small Japanese-based pre-training dataset. We would further explore complex pre-training objectives on large English-based datasets to observe and analyze our model performance.

Acknowledgments. We acknowledge Cinnamon AI for supporting this study.

References

1. Adali, S., Sonmez, A.C., Gokturk, M.: An integrated architecture for processing business documents in Turkish. In: Gelbukh, A. (ed.) CICLing 2009. LNCS, vol. 5449, pp. 394–405. Springer, Heidelberg (2009). https://doi.org/10.1007/978-3-642-00382-0_32
2. Appalaraju, S., Jasani, B., Kota, B.U., Xie, Y., Manmatha, R.: Docformer: end-to-end transformer for document understanding. In: Proceedings of the IEEE/CVF International Conference on Computer Vision, pp. 993–1003 (2021)

3. Bao, H., et al.: Unilmv2: pseudo-masked language models for unified language model pre-training. In: International Conference on Machine Learning, pp. 642–652. PMLR (2020)

4. Belaïd, Y., Belaïd, A.: Morphological tagging approach in document analysis of invoices. In: Proceedings of the 17th International Conference on Pattern Recognition, 2004. ICPR 2004, vol. 1, pp. 469–472. IEEE (2004)

5. Carbonell, M., Riba, P., Villegas, M., Fornés, A., Lladós, J.: Named entity recognition and relation extraction with graph neural networks in semi structured documents. In: 2020 25th International Conference on Pattern Recognition (ICPR), pp. 9622–9627. IEEE (2021)

6. Chung, J., Gulcehre, C., Cho, K., Bengio, Y.: Empirical evaluation of gated recurrent neural networks on sequence modeling. arXiv preprint arXiv:1412.3555 (2014)

7. Dai, J., Qi, H., Xiong, Y., Li, Y., Zhang, G., Hu, H., Wei, Y.: Deformable convolutional networks. In: Proceedings of the IEEE International Conference on Computer Vision, pp. 764–773 (2017)

8. Dengel, A.R., Klein, B.: *smartFIX*: a requirements-driven system for document analysis and understanding. In: Lopresti, D., Hu, J., Kashi, R. (eds.) DAS 2002. LNCS, vol. 2423, pp. 433–444. Springer, Heidelberg (2002). https://doi.org/10.1007/3-540-45869-7_47

9. Devlin, J., Chang, M.W., Lee, K., Toutanova, K.: Bert: pre-training of deep bidirectional transformers for language understanding. arXiv preprint arXiv:1810.04805 (2018)

10. Esser, D., Schuster, D., Muthmann, K., Berger, M., Schill, A.: Automatic indexing of scanned documents: a layout-based approach. In: Document recognition and retrieval XIX, vol. 8297, pp. 118–125. SPIE (2012)

11. Gori, M., Monfardini, G., Scarselli, F.: A new model for learning in graph domains. In: Proceedings. 2005 IEEE International Joint Conference on Neural Networks, vol. 2, pp. 729–734 (2005)

12. Gui, T., Zou, Y., Zhang, Q., Peng, M., Fu, J., Wei, Z., Huang, X.J.: A lexicon-based graph neural network for chinese ner. In: Proceedings of the 2019 Conference on Empirical Methods in Natural Language Processing and the 9th International Joint Conference on Natural Language Processing (EMNLP-IJCNLP). pp. 1040–1050 (2019)

13. He, K., Zhang, X., Ren, S., Sun, J.: Deep residual learning for image recognition. In: Proceedings of the IEEE Conference on Computer Vision and Pattern Recognition, pp. 770–778 (2016)

14. Hong, T., Kim, D., Ji, M., Hwang, W., Nam, D., Park, S.: Bros: a pre-trained language model for understanding texts in document (2020)

15. Huang, Y., Lv, T., Cui, L., Lu, Y., Wei, F.: Layoutlmv3: pre-training for document ai with unified text and image masking. arXiv preprint arXiv:2204.08387 (2022)

16. Huang, Z., Chen, K., He, J., Bai, X., Karatzas, D., Lu, S., Jawahar, C.: Icdar2019 competition on scanned receipt ocr and information extraction. In: 2019 International Conference on Document Analysis and Recognition (ICDAR), pp. 1516–1520. IEEE (2019)

17. Hwang, W., Yim, J., Park, S., Yang, S., Seo, M.: Spatial dependency parsing for semi-structured document information extraction. arXiv preprint arXiv:2005.00642 (2020)

18. Jaume, G., Ekenel, H.K., Thiran, J.P.: Funsd: a dataset for form understanding in noisy scanned documents. In: 2019 International Conference on Document Analysis and Recognition Workshops (ICDARW), vol. 2, pp. 1–6. IEEE (2019)

19. Jiang, H., Misra, I., Rohrbach, M., Learned-Miller, E., Chen, X.: In defense of grid features for visual question answering. In: Proceedings of the IEEE/CVF Conference on Computer Vision and Pattern Recognition, pp. 10267–10276 (2020)

20. Katti, A.R., Reisswig, C., Guder, C., Brarda, S., Bickel, S., Höhne, J., Faddoul, J.B.: Chargrid: Towards understanding 2d documents. arXiv preprint arXiv:1809.08799 (2018)

21. Khan, S., Naseer, M., Hayat, M., Zamir, S.W., Khan, F.S., Shah, M.: Transformers in vision: a survey. ACM Computing Surveys (CSUR) (2021)

22. Klein, B., Agne, S., Dengel, A.: Results of a Study on invoice-reading systems in Germany. In: Marinai, S., Dengel, A.R. (eds.) DAS 2004. LNCS, vol. 3163, pp. 451–462. Springer, Heidelberg (2004). https://doi.org/10.1007/978-3-540-28640-0_43

23. Lample, G., Ballesteros, M., Subramanian, S., Kawakami, K., Dyer, C.: Neural architectures for named entity recognition. arXiv preprint arXiv:1603.01360 (2016)

24. Li, Y., Qian, Y., Yu, Y., Qin, X., Zhang, C., Liu, Y., Yao, K., Han, J., Liu, J., Ding, E.: Structext: Structured text understanding with multi-modal transformers. In: Proceedings of the 29th ACM International Conference on Multimedia. pp. 1912–1920 (2021)

25. Liao, M., Wan, Z., Yao, C., Chen, K., Bai, X.: Real-time scene text detection with differentiable binarization. In: Proceedings of the AAAI Conference on Artificial Intelligence, vol. 34, pp. 11474–11481 (2020)

26. Liu, X., Gao, F., Zhang, Q., Zhao, H.: Graph convolution for multimodal information extraction from visually rich documents. arXiv preprint arXiv:1903.11279 (2019)

27. Majumder, B., Potti, N., Tata, S., Wendt, J.B., Zhao, Q., Najork, M.: Representation learning for information extraction from form-like documents (2020)

28. Nguyen, T.-A.D., Vu, H.M., Son, N.H., Nguyen, M.-T.: A span extraction approach for information extraction on visually-rich documents. In: Barney Smith, E.H., Pal, U. (eds.) ICDAR 2021. LNCS, vol. 12917, pp. 353–363. Springer, Cham (2021). https://doi.org/10.1007/978-3-030-86159-9_25

29. Palm, R.B., Winther, O., Laws, F.: Cloudscan-a configuration-free invoice analysis system using recurrent neural networks. In: 2017 14th IAPR International Conference on Document Analysis and Recognition (ICDAR), vol. 1, pp. 406–413. IEEE (2017)

30. Park, S., Shin, S., Lee, B., Lee, J., Surh, J., Seo, M., Lee, H.: Cord: a consolidated receipt dataset for post-ocr parsing. In: Workshop on Document Intelligence at NeurIPS 2019 (2019)

31. Qian, Y., Santus, E., Jin, Z., Guo, J., Barzilay, R.: Graphie: a graph-based framework for information extraction. arXiv preprint arXiv:1810.13083 (2018)

32. Reimers, N., Gurevych, I.: Sentence-bert: sentence embeddings using siamese bert-networks. arXiv preprint arXiv:1908.10084 (2019)

33. Ren, S., He, K., Girshick, R., Sun, J.: Faster R-CNN: towards real-time object detection with region proposal networks. Advances in neural information processing systems 28 (2015)

34. Sage, C., Aussem, A., Elghazel, H., Eglin, V., Espinas, J.: Recurrent neural network approach for table field extraction in business documents. In: 2019 International Conference on Document Analysis and Recognition (ICDAR), pp. 1308–1313. IEEE (2019)

35. Tang, G., et al.: Matchvie: exploiting match relevancy between entities for visual information extraction. arXiv preprint arXiv:2106.12940 (2021)

36. Veličković, P., Cucurull, G., Casanova, A., Romero, A., Lio, P., Bengio, Y.: Graph attention networks. arXiv preprint arXiv:1710.10903 (2017)
37. Wang, J., et al.: Towards robust visual information extraction in real world: new dataset and novel solution. In: Proceedings of the AAAI Conference on Artificial Intelligence, vol. 35, pp. 2738–2745 (2021)
38. Wang, Z., Shang, J.: Towards few-shot entity recognition in document images: a label-aware sequence-to-sequence framework. arXiv preprint arXiv:2204.05819 (2022)
39. Wei, M., He, Y., Zhang, Q.: Robust layout-aware IE for visually rich documents with pre-trained language models. In: Proceedings of the 43rd International ACM SIGIR Conference on Research and Development in Information Retrieval, pp. 2367–2376 (2020)
40. Xu, Y., et al.: Layoutlmv2: Multi-modal pre-training for visually-rich document understanding. arXiv preprint arXiv:2012.14740 (2020)
41. Xu, Y., Li, M., Cui, L., Huang, S., Wei, F., Zhou, M.: Layoutlm: pre-training of text and layout for document image understanding. In: Proceedings of the 26th ACM SIGKDD International Conference on Knowledge Discovery & Data Mining, pp. 1192–1200 (2020)
42. Xu, Y., et al.: Layoutxlm: multimodal pre-training for multilingual visually-rich document understanding. arXiv preprint arXiv:2104.08836 (2021)
43. Xu, Y., et al.: Xfund: a benchmark dataset for multilingual visually rich form understanding. In: Findings of the Association for Computational Linguistics: ACL 2022, pp. 3214–3224 (2022)
44. Yu, W., Lu, N., Qi, X., Gong, P., Xiao, R.: Pick: processing key information extraction from documents using improved graph learning-convolutional networks. In: 2020 25th International Conference on Pattern Recognition (ICPR), pp. 4363–4370. IEEE (2021)
45. Zhang, P., et al.: Trie: end-to-end text reading and information extraction for document understanding. In: Proceedings of the 28th ACM International Conference on Multimedia, pp. 1413–1422 (2020)

GDP: Generic Document Pretraining to Improve Document Understanding

Akkshita Trivedi[1]([✉]) [iD], Akarsh Upadhyay[1] [iD], Rudrabha Mukhopadhyay[2] [iD], and Santanu Chaudhury[1] [iD]

[1] Indian Institute of Technology Jodhpur, Karwar, India
{trivedi.2,upadhyay.4}@iitj.ac.in
[2] International Institute of Information Technology, Hyderabad, India
radrabha.m@research.iiit.ac.in

Abstract. In this paper, we propose a novel pretraining approach for document analysis that advances beyond conventional methods. The approach, called the GDPerformer, trains a suite of unique architectures to predict both masked OCR tokens and masked OCR bounding boxes, fostering the network to learn document semantics such as structure and language. Our experiments with GDPerformerv1 and GDPerformerv2 show enhanced performance on various downstream tasks, including Semantic Entity Recognition and Extraction and Multi-Modal Document Classification with minimal task-specific data and generalization to a wide range of documents. Furthermore, our pretrained features exhibit robustness in handling noisy documents and can be easily extended to multiple languages. Our experiments indicate that the proposed pretraining strategy requires only $50K$ document images, making it particularly beneficial for low-resource languages.

Keywords: Pretraining · Document Analysis · Masked Language Modeling · Document Transformers · Document Imaging · Deep Learning · Masked Layout Modeling

1 Introduction

Printed documentation has been critical to human civilization for centuries, serving as a means of recording and conveying information. From ancient clay tablets to today's books, printed documents have played a vital role in preserving and disseminating knowledge. With the advent of the printing press, mass production of printed materials became possible, leading to an explosion of information. Printed documents continue to play a crucial role in daily life, providing news, knowledge, and entertainment. But with the increasing digitization of data, the ability to automatically extract information from document images has become more important. Using AI and machine learning algorithms, document analysis plays a critical role in this, enabling the understanding of even unstructured document images, such as handwritten notes. It is, therefore, essential to improve document analysis algorithms.

© The Author(s), under exclusive license to Springer Nature Switzerland AG 2024
E. H. Barney Smith et al. (Eds.): ICDAR 2024, LNCS 14804, pp. 208–226, 2024.
https://doi.org/10.1007/978-3-031-70533-5_13

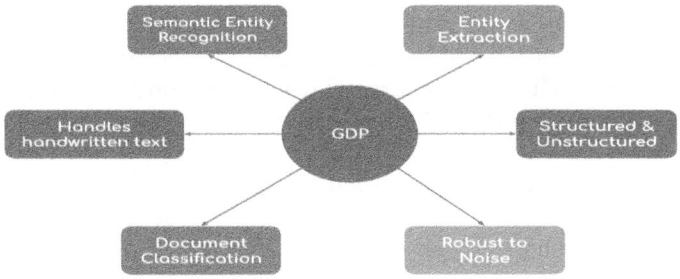

Fig. 1. We propose GDP: Generic Document Pretraining for solving a host of tasks related to document imaging.

Deep learning has greatly improved the performance of document analysis algorithms for efficient information retrieval. This is especially crucial in fields like healthcare for quick and accurate information retrieval from a large amount of medical literature. Document analysis can extract structured data from unstructured documents, like handwritten text, names, dates, and addresses. Tasks in document imaging include OCR (text extraction), layout analysis, information extraction, document image enhancement, and document classification. Traditionally, document imaging tasks have been approached using supervised learning, where specialized datasets are collected and annotated to solve specific tasks. While these efforts have achieved great progress in their designated tasks, the datasets require significant human effort in collection and annotation. The networks trained on them don't easily transfer [25,27,33] to other tasks because of differences between them. Additionally, the risk of overfitting to dataset biases and undesirable patterns increases, resulting in expensive training and difficulty scaling with growing research in the field. Recently, self-supervised learning and pretraining [1,14,32] have become more popular for document imaging tasks as they allow deep learning models to learn from vast amounts of unlabeled data. This reduces the cost and effort required for human annotation and improves generalization. In self-supervised learning, models learn from a large amount of unlabeled data, which is often more readily available as a proxy task. The models are then fine-tuned with a small amount of labeled task-specific data, resulting in improved performance. Pretraining models lead to better generalization [38] by learning useful features across different tasks instead of overfitting to a single task.

This paper presents a new approach for document analysis tasks that utilizes a novel pretraining strategy. We propose a couple of new models, namely, GDPerformerv1 and GDPerformerv2, and train them on a proxy task of predicting masked OCR tokens and masked OCR bounding boxes from multiple masked input streams. This allows the networks to learn high-level semantics in the document images, including the document's structure and language. We demonstrate that our model can effectively understand structured and unstructured documents. Our proposed approach contributes significantly to document analysis as it provides a new way to effectively understand and extract information from documents. Additionally, the ability of the pretrained embedding space to generalize well to a wide array of documents and perform well on noisy documents makes it a valuable tool for researchers and practitioners in the field. Our approach is a promising way for the community to address the challenges in document analysis and contribute to developing more robust models for natural language understanding and information retrieval tasks. We summarize our contributions below and in Fig. 1.

1.1 Our Contributions

1. A novel approach for document analysis tasks is proposed that utilizes pretraining of novel umbrella architectures called GDPerformer to predict both masked OCR tokens and corresponding masked OCR bounding boxes. In contrast to the current approaches, we provide strong empirical evidence that predicting the masked OCR bounding boxes and tokens together helps the network learn a better document structure.
2. The pretrained embedding space generated by this approach is used for various downstream tasks such as semantic entity recognition [2,11] and extraction [12,26], document classification [7,16–18], and more. The proposed method improves performance on various downstream tasks and handles noisy documents with high accuracy. Furthermore, our pretrained models extend directly to unseen languages like French without any additional training requirement.
3. Our pretraining approach is designed to be highly scalable and efficient, allowing for easy implementation in low-resource language setups. Despite using only a limited number of document images, our approach still effectively trains deep neural networks to learn high-level semantics in document images, including their structure and language. This demonstrates the versatility and robustness of our approach, as it can produce results even with limited training data.

2 Related Work

Pretraining transformer models [30] on structured and unstructured documents has been effective in improving performance on various document analysis tasks. End-to-end Transformers, such as DocFormer [1] and UniDoc [9], comprehend entire documents and handle a wide range of document types. BERT-based models, LAMBERT [8], and VL-BERT [31] are specialized models pre-trained on vast amounts of text data. Cross-modal pretraining methods, such as VSR [39], TRIE [40], LayoutLMv2 [35], LayoutXLM [37], and LAMPERT [34], combine text and image information for improved document understanding. Models for specific tasks, such as X-LXMERT [5], Look, Read and Ask [15], and Graph Convolution for Multimodal Information Extraction [22], perform question answering and leverage rich information from both modalities. Models for document analysis can be grouped into three categories: text-only, text and layout, and text, layout, and image features. The text-only category includes BERT [6], RoBERTa [23], and UniLMv2 [3]. The text and layout category includes BROS [13] and LiLT [32]. The text, layout, and image features category includes DocFormer [1], SelfDoc [20], XYlayout [10], StructuralLM [19], StrucTexT [21], and LayoutLMv3 [14], which aim to use all available information to perform various downstream tasks. The introduction of image features led to masked image modeling, where a separate image tokenizer, such as from DALL-E [28], is trained and used to extract masked language and visual features.

2.1 A Generic Weakness of Current Masked Image Modeling Techniques for Document Analysis

Mask Image Modeling is a demanding task that involves predicting masked image tokens, which requires a separate image tokenizer to be trained, as demonstrated by the LayoutLMv3 [14] approach, where the authors employed a tokenizer from DALL-E [28]. However, the masking of images can result in a challenge of white pixels, as most document backgrounds are typically white. This can lead to the model simply predicting the images as white pixels, potentially not capturing the complete information present in the document. We hypothesize that simply masking image tokens and then training the model to predict them does not guarantee that the model will learn the layout information effectively. While providing the layout tokens as input to the model offers a rudimentary understanding of the layout structure, conditioning the final output on such layout information remains challenging.

3 Generic Document Pretraining

To improve the model's understanding of the document layout, we propose a novel pretraining method that involves masking the layout tokens and forcing the model to predict the corresponding bounding boxes. This approach allows the model to learn the layout structure and improve its understanding of the document. We present two new models, GDPerformerv1 and GDPerformerv2, which are trained on predicting masked OCR tokens and masked OCR bounding boxes. While GDPerformerv1 uses masked OCR tokens and bounding boxes as input, GDPerformerv2 extends this by incorporating additional non-masked image features to provide a deeper understanding of the high-level semantic elements in document images. Our method offers a fresh perspective on effectively extracting information from documents and has great potential for overcoming challenges in document analysis.

3.1 OCR Tokens and Bounding Boxes

We utilize the text annotations and their corresponding bounding boxes from the annotated ground truth available in Industrial Digital Library [4] for pretraining

GDPerformerv1 and GDPerformerv2. We tokenize the given text annotations (OCR Tokens) in an image as $W_{1...n}$ and concat them in a sequence denoted by E_T. A learnable position encoding, P_1, is added to this sequence before finally being passed through a LayerNorm operation as given in Eq. 1.

$$E_N = \text{LN}(E_{\text{token}} + P_1) \qquad (1)$$

Here, $E_N \in \mathcal{R}^{n \times d_T}$, where $d_T = 768$ is the dimension of each token. We also utilize the corresponding bounding boxes for the OCR tokens, represented by $B_{1...n}$, as an additional input stream. Each of the boxes contains $(x_{min}, x_{max}, y_{min}, y_{max}, w, h)$ in a normalized fashion, wherein each value ranges between 0 to 1000 following the strategy presented in LayoutLM [36]. Please note that the w and h present here denote the height and width of each bounding box. A linear layer is used to encode the bounding boxes, and then a similar learnable positional encoding is added to the bounding box embeddings following Eq. 3.

$$B_{\text{embedding}} = \text{Linear}\left(\text{CAT}\left(E_{x_{\min}}, E_{x_{\max}} \; E_{y_{\min}}, E_{y_{\max}}, E_{\text{width}}, E_{\text{height}}\right)\right) \qquad (2)$$

$$B_N = \text{LN}(B_{\text{embedding}} + P_2) \qquad (3)$$

Here, P_2 is the learnable positional encoding used for the bounding box embeddings. We also ensure $B_N \in \mathcal{R}^{n \times d_T}$, to enable compatibility with the E_N. We mask a portion of both inputs following the masking strategy given below. The masked encodings serve as multi-modal inputs to the proposed GDPerformer networks and are denoted by E_N^m and B_N^m. During inference, we use a standard OCR like Tesseract [29] to extract both the OCR tokens and the bounding boxes on unseen documents.

3.2   Masking both OCR Tokens and Bounding Boxes

Unlike previous approaches, which either mask only the OCR tokens or uses masked image tokens along with OCR tokens, we mask both OCR tokens and bounding boxes to force the proposed models to learn the high-level semantics of document layout and content. Our masking process involved randomly applying the following operations to both the tokens and bounding boxes in the input data: (1) 80% of the time, selected OCR tokens and OCR bounding boxes were replaced with a masked token. (2) 10% of the time, an OCR token and the corresponding bounding box were replaced with random tokens. (3) The remaining 10% of the time, the OCR token, and the corresponding bounding box were left unchanged. The masked tokens in the input data are represented by W^m, with each masked word described as W_j^m, where j represents the indices of the masked word. Similarly, the corresponding masked bounding boxes are represented by B^m, with each masked bounding box represented as B_j^m. Please note that the masked words and bounding boxes were selected randomly and are not necessarily consecutive.

3.3   GDPerformer

The utilization of transformers in document analysis has proven to improve various tasks, as exemplified by models like LiLT [32] and LayoutLMv3 [14]. The advantages of using transformers stem from their self-attention mechanism, which allows the model to assign different weights to each token in the input, thus effectively capturing long-range dependencies. To take advantage of recent advancements in masked language modeling for document analysis, we propose two novel architectures, GDPerformerv1 and GDPerformerv2, for predicting masked OCR tokens (E_N^m) and masked bounding boxes (B_N^m) as a pretraining task. In GDPerformerv1, we use only masked OCR tokens and bounding boxes as input streams for the multimodal transformer network that aims to predict the masked values. We extend this approach in GDPerformerv2 by incorporating image features as an additional input modality to the transformer-based architecture.

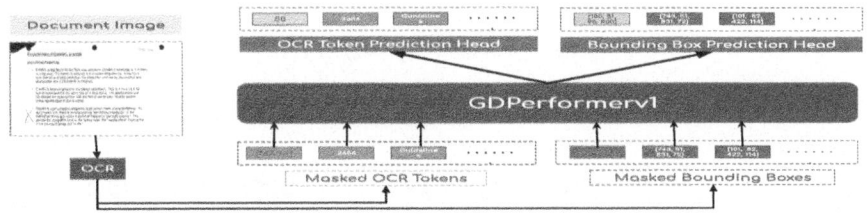

Fig. 2. Our proposed GDPerformerv1 architecture takes masked OCR tokens and their corresponding masked bounding boxes as input and trains a transformer network to predict both of the modalities as output.

GDPerformerv1. We modify the architecture proposed in [14] for our pretraining task. Our transformer architecture contains a multimodal setup and takes in sequences of multiple input modalities. We concatenate the sequences of the masked OCR tokens and the masked bounding boxes in the length dimension and create the final input embedding C_{2N}^m as given in Eq. 4.

$$C_{2N}^m = E_N^m \mid_c B_N^m \tag{4}$$

The input sequence C_{2N}^m is passed through standard transformer encoder-decoder architecture (inspired from [32]). We add two heads of linear layers as the final output from GDPerformerv1 to predict the masked OCR tokens and the bounding boxes. This dual prediction task forces the model to learn both the language component as well as the structure of the document, resulting in a highly semantic latent space. Having described the intuition behind the task and the masking procedure in the above section, we define the loss function, which focuses on penalizing the model, depending on how successful the model is, in retrieving the masked bounding box and the corresponding token. We calculate the loss between the masked OCR tokens and the ground truth using the standard cross entropy loss function and denote it as Masked Language Model (MLM) loss (Eq. 5). The masked bounding boxes are used to calculate the Masked Layout (MLL) loss (Eq. 6) using standard $L1$ distance between predicted and ground truth bounding box coordinates. The overall architectural pipeline is also depicted in Fig. 2.

$$L_{MLM}(\theta) = -\sum_{l \in j} \log p_\theta \left(\mathbf{E}'_l \mid \mathbf{E}^M, \mathbf{B}^M \right) \tag{5}$$

$$L_{MLL}(\theta) = \sum_{l \in j} \left(||\mathbf{B}'_l - \mathbf{B}_l||_1 \right) \tag{6}$$

Here θ depicts the trainable parameters in GDPerformerv1, j represents the indices of the masked OCR tokens, and $\mathbf{E}'_l$ denotes the predicted OCR tokens. The final loss function is presented in Eq. 7.

$$L(\theta) = L_{MLM}(\theta) + L_{MLL}(\theta) \tag{7}$$

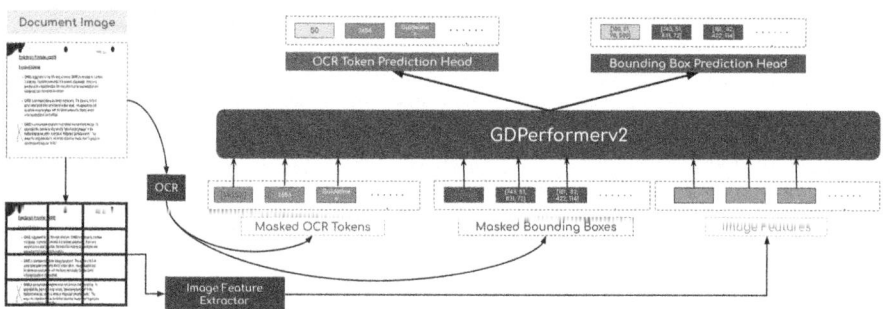

Fig. 3. We propose a modified variation of GDPerformerv1 architecture as GDPerformerv2. In this case, additional image features are concatenated with the sequence of masked OCR tokens and masked bounding boxes. Adding image features improves the network's overall understanding of the semantics of the document.

GDPerformerv2. The proposed GDPerformerv1 predicted masked OCR tokens and bounding box coordinates and was, in the process, forced to learn the document's structure successfully. The natural question was "will the performance improve if we provide an additional image embedding?" Therefore, we modify GDPerformerv1 to take an additional sequence of image features as input in the multimodal transformer network named GDPerformerv2. Following [14], we first resize a given document image to $H \times W$ and then divide the image into $k, i \times i$ patches. We pass each patch through a convolution layer before flattening them into a vector of d_I dimension. The sequence of image features is denoted by $I_{1...k} = I_K \in \mathcal{R}^{k \times d_I}$. As done for the previous encodings, we add a positional encoding P_3 with I_K. We concatenate I_K along with the sequences of the masked OCR tokens and the masked bounding boxes in the length dimension and create the final input embedding C^m_{2N+K} according to Eq. 8. Please note that we do not mask the image features like the bounding boxes and OCR tokens in the proposed formulation.

$$C^m_{2N+K} = E^m_N \mid_c B^m_N \mid_c I_K \tag{8}$$

We then pass C^m_{2N+K} through a similar transformer-based encoder-decoder architecture as shown in Fig. 3. The overall training strategy and loss functions follow that of GDPerformerv1 with certain modifications in the MLM loss function as given Eq. 9.

$$L_{MLM}(\theta) = -\sum_{l \in j} \log p_\theta \left(\mathbf{E}'_l \mid \mathbf{E}^M, \mathbf{B}^M, \mathbf{I}_K \right) \tag{9}$$

3.4 Training Dataset and Implementation Details

In this work, we utilize a transformer encoder with 12 heads in the self-attention mechanism and a hidden size of $d_t = 768$. Additionally, the model has 3072 intermediate feed-forward networks. Text input is pre-processed using Byte-Pair Encoding with a maximum sequence length of $L = 512$. We add the tokens [CLS] and [SEP] at the beginning and end of each text sequence and pad with [PAD] if the length is shorter than L. The bounding box coordinates for these special tokens are all set to zero. The image embedding parameters are defined as $H \times W = 224 \times 224$ and $k = 16$. Our final layers for the pretraining task consisted of a token prediction head with shape $(L, \text{Vocab Size})$, and for coordinates, it is $(L, 4)$. To improve efficiency, we employ distributed and mixed-precision training. More details about the training process can be found in [14]. Both the models were trained with a constant learning rate of 5×10^{-5}. We only use $50,000$ samples from the Industry Document Library [4] for training the GDPerformer networks, which are in stark contrast to millions of data samples used by comparable networks like DocFormer [1], etc. An ablation varying the number of training images is shown in Sect. 5.3. This potentially opens up avenues for similar pretraining techniques to work on low-resource languages like Hindi.

4 Experiments: Finetuning on Downstream Tasks

We perform several experiments with different downstream tasks by fine-tuning the pretrained GDPerformer models and evaluating them against multiple previous works on the same task-specific test sets. For fine-tuning, the task-specific classification head is attached to the pretrained model by removing the heads for token and coordinate prediction. To ensure a fair comparison in all experiments, the hyperparameters are kept constant to prevent any potential biases from affecting the results. This allows for effective utilization of the pre-trained weights while making necessary modifications for the target task. In this section, we investigate various downstream tasks and use standard metrics to fairly evaluate the performance of the proposed GDPerformerv1 and GDPerformerv2 against multiple other state-of-the-art works.

4.1   Metrics Used for Evaluation

We evaluate the model's performance using recall, precision, and F1, which are common performance metrics in machine learning. Recall measures the proportion of relevant items that were correctly classified. The recall is a measure of the proportion of relevant items that the model correctly classified. It gives an idea of how many relevant items the model can identify. A high recall value indicates that the model can identify the most relevant items. Precision measures the proportion of correct positive predictions made by the model. It gives an idea of how accurate the positive predictions made by the model are. A high precision value indicates that the positive predictions made by the model are accurate. F1 is the harmonic mean of precision and recall, balancing recall and precision. A high F1 value indicates that the model has both high recall and precision, i.e., it is able to identify most of the relevant items, and its positive predictions are accurate.

4.2   Comparison with Multiple Prior Approaches

We compare the performance of several models trained for solving different downstream tasks. These models are grouped based on the modality they used for the task: Text (T), layout (L), and image features (I). The models that only use text include BERT [6], RoBERTa [23], and UniLMv2 [3]. The models that use both text and layout include BROS [13] and LiLT [32]. The models that use text, layout, and image features include DocFormer [1], SelfDoc [20], XYlayout [10], StructuralLM [19], StrucTexT [21], and LayoutLMv3 [14]. We compared these models to showcase the effectiveness of our proposed models in different groups, GDPerformerv1 (T+L) and GDPerformerv2 (T+L+I).

4.3   Semantic Entity Recognition

The semantic entity recognition task aims to categorize each word in a given text into one of four pre-defined semantic entities: "Question", "Answer," "Header," or "Other." To tackle this task, we finetune the GDPerformer models described on the FUNSD dataset [11], adapting them for the entity recognition task by attaching a classification head. We finetune all the trainable parameters in both GDPerformerv1 and GDPerformerv2 while replacing the original heads with a classification head to predict one of the four classes. We finetune both models on the train set of FUNSD dataset [11] with a standard multi-class cross-entropy loss to predict the classes of each semantic entity. We compare the proposed

GDPerformer models with several other state-of-the-art approaches using the standard test set of the FUNSD dataset in Table 1. We observe a clear improvement in all the metrics achieved by both of our proposed models. Expectantly, the additional image feature information in GDPerformerv2 improves the overall performance compared to GDPerformerv1. We also observe significant gains by GDPerformerv2 compared to previous works with a similar setup like LayoutLMv3 [14]. For models that use text and layout information (T+L), the model LiLT [32] outperforms other models with a F1-Score of 88.41. Our proposed model GDPerformerv1 also performed well with a F1-Score of 87.81. For models that use text, layout, and image information (T+L+I), our proposed model GDPerformerv2 outperforms other models by a wide margin with a F1-Score of 91.41. This is also the highest performance among all models in all three modality categories. We also compare visual results achieved by different methods in Fig. 4. We visually compare [14,32] and both of the proposed models and note decipherable improvements.

Table 1. We compare the performance of different models trained on the FUNSD [11] dataset for the semantic entity recognition task. We observe that the performance of GDPerformerv2 outperforms all other methods by over a 1-point difference on metrics like the F1-Score.

Model	Modality	Precision	Recall	F1-Score	#Params(M)
BERT [6]	T	60.01	65.65	62.70	109
RoBERTa [23]	T	65.92	69.27	67.55	124
UniLMv2 [3]	T	67.80	73.91	70.72	125
BROS [13]	T+L	80.56	81.88	81.21	139
StrucTexT [21]	T+I	85.68	80.97	83.09	107
LiLT [32]	T+L	87.21	89.65	88.41	-
GDPerformerv1 (Ours)	T+L	86.21	89.45	87.81	130
DocFormer [1]	T+L+I	82.29	86.94	84.55	536
SelfDoc [20]	T+L+I	-	-	83.36	-
XYlayout [10]	T+L+I	-	-	83.35	-
StructuralLM [19]	T+L+I	83.52	86.81	85.14	355
LayoutLMv3 [14]	T+L+I	-	-	90.29	133
GDPerformerv2 (Ours)	**T+L+I**	**90.86**	**91.95**	**91.41**	**125**

4.4 Entity Extraction

The Entity extraction task focuses on labeling each word to its corresponding field. We use the CORD [26] dataset to finetune the GDPerformer models for extracting different entities. CORD [26] is an English receipt dataset that

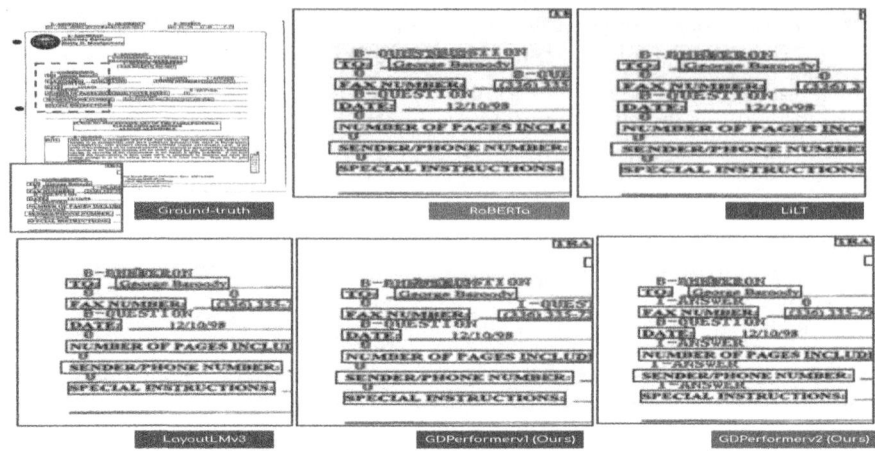

Fig. 4. We visually compare different algorithms on the Semantic Entity Recognition task. We plot the predicted labels from different methods and mark the correct entity predictions in green and the wrong predictions in red. A zoomed-in portion of the predictions is shown where the difference is most visible between different algorithms for demonstration purposes. As we observe, the outputs from GDPerformerv2 are the most accurate (highest share of green marked labels), indicating superior performance. (Color figure online)

includes 800 receipts for the training set, 100 for the validation set, and 100 for the test set. The dataset contains the photo and the OCR annotation corresponding to the receipt and has 30 fields under 4 categories. The task focuses on labeling each word to the right field. We use the official OCR annotations available. Similar to the previous task, we finetuned the proposed model in an end-to-end fashion. We report the results for entity extraction in Table 2. The table shows that the models with the combination of text and layout information (T+L) generally perform better than those with only text modality (T). Among the models with T+L, GDPerformerv1 achieves the highest performance. On the other hand, models incorporating text, layout, and image information (T+L+I) further improve the performance, as can be seen from the performance of Doc-Former and GDPerformerv2. Among these models, GDPerformerv2 outperforms other models by a small margin. Compared to the other models, the proposed GDPerformerv1 and GDPerformerv2 have relatively fewer parameters, suggesting that they are computationally more efficient. A visual comparison is also provided in Fig. 5. Overall, these results expectantly demonstrate the effectiveness of incorporating layout information in document analysis tasks.

4.5   Unstructured Document Classification

We utilize the Tobacco-3482 Dataset [18], which contains Gray-scale images corresponding to 10 classes: Advertisement, Email, Form, Letter, Memo, News,

Table 2. We present a comparison for Entity Extraction on the CORD [26] dataset.

Model	Modality	Precision	Recall	F1-Score	#Params(M)
BERT [6]	T	91.07	90.41	90.74	109
RoBERTa [23]	T	91.93	91.60	91.76	124
UniLMv2 [3]	T	91.23	92.89	92.05	125
BROS [13]	T+L	95.58	95.14	95.36	139
LiLT [32]	T+L	95.98	96.16	96.07	-
LayoutLMv3 [14]	T+L	-	-	96.56	133
GDPerformerv1	T+L	95.97	96.41	96.19	130
DocFormer [1]	T+L+I	96.52	96.14	96.33	536
GDPerformerv2	**T+L+I**	**96.41**	**96.63**	**96.62**	**125**

Note, Report, Resume, and Scientific Article, ranging in a total of 3482 images, in an unbalanced fashion. Due to the nature of the different classes, the document ranges from structured format to unstructured and free format. We split the dataset into a stratified split of 80% for training and 20% for test purposes. We adopt the finetuning approach discussed earlier with a similar classification head. In the comparison on the Tobacco dataset (Table 3), models using text (T) modality performed well, with BERT achieving an F1-Score of 82.21 and RoBERTa scoring 74.35. Models using both text and layout (T+L) showed improvement over the text-only models, with LiLT and GDPerformerv1 achieving F1-Scores of 75.66 and 85.60, respectively. The highest performance was achieved by models using all modalities (T+L+I), with GDPerformerv2 scoring the highest F1-Score of 88.42. We provide sample results of Unstructured Document Classification in Fig. 6.

Table 3. Comparison of Unstructured Document Classification on the Tobacco-3482 [18] dataset.

Model	Modality	Precision	Recall	F1-Score	#Params(M)
RoBERTa [23]	T	78.48	79.57	74.35	124
BERT [6]	T	84.56	81.36	82.21	109
LiLT [32]	T+L	80.22	76.17	75.66	-
GDPerformerv1	T+L	87.37	86.48	85.60	130
LayoutLMV3 [14]	T+L+I	63.28	68.00	62.21	133
GDPerformerv2	**T+L+I**	**89.67**	**88.87**	**88.42**	**125**

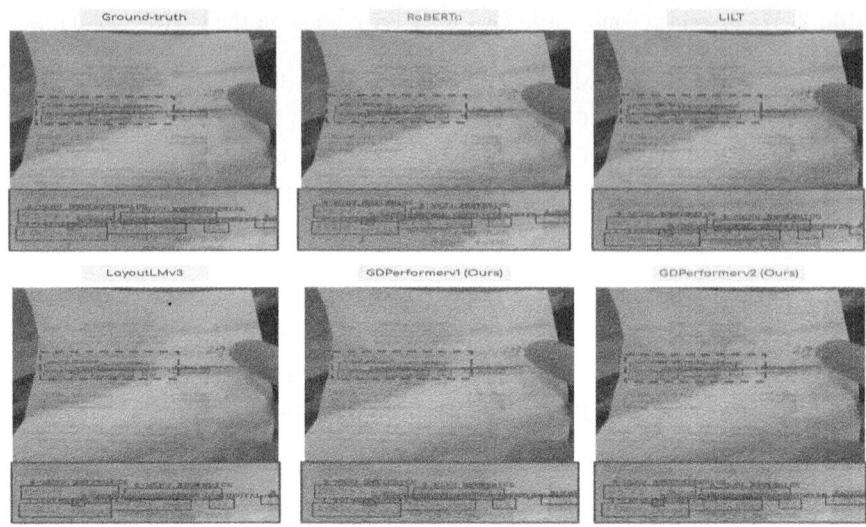

Fig. 5. We visually compare different algorithms on the Entity Extraction task. As done in the previous Figure, we plot the predicted labels from different methods and mark the correct entity predictions in green and the wrong predictions in red. The outputs from GDPerformerv2 are observed to be the most accurate (highest share of green-marked labels). A zoomed-in portion of the predictions is also shown. (Color figure online)

4.6   Document Visual Question Answering

For the task of Document Visual Question Answering, a model is tasked with receiving a document image and a corresponding question as input, and its objective is to output an answer. Following the similar approach to [14], we formulate our task as an extractive QA problem, where the model is asked to predict the start and end positions by classifying the last hidden state of each text token with a binary classifier.

We utilize the Doc-VQA [24] for exploring the capabilities of our proposed method for visual question answering on document images. The official partition of the DocVQA dataset contains 10,194 images and its corresponding 39,463 question/answer pairs in the training and 1,286 images and 5,349 question/answer pairs in validation dataset. We find that the top performance was achieved by models using all modalities (T+L+I), with GDPerformerv2 scoring the highest F1-Score of 68.99 (Table 4)

5   Ablation Studies

5.1   Extending GDP to French

We conducted an experiment on GDP by creating a synthetic French Document Classification task. The Tobacco3482 dataset was used and its text was translated into French using Google Lens. The evaluation was then repeated using the

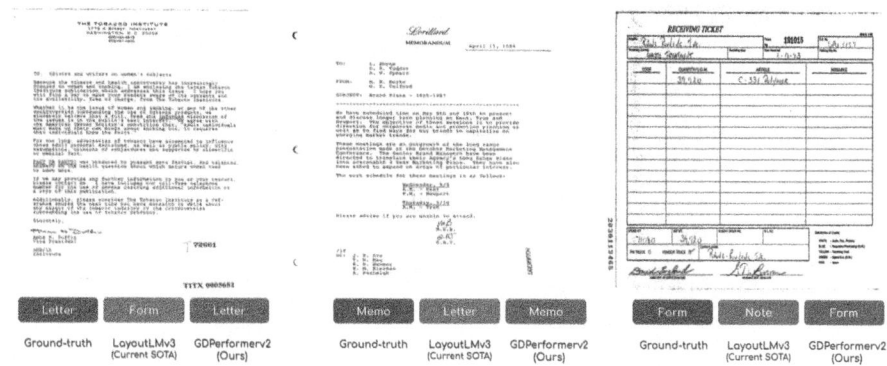

Fig. 6. We plot selected document classification results from GDPerformerv2 trained on the Tobacco-3482 dataset and compare it with the classification results from [14].

Table 4. Comparison of Document Visual Question Answering on DocVQA dataset [24] dataset.

Model	Modality	F1-Score	#Params(M)
BERT [6]	T	66.66	109
RoBERTa [23]	T	69.52	124
LiLT [32]	T+L	66.51	-
GDPerformerv1	T+L	67.98	130
LayoutLMV3 [14]	T+L+I	68.09	133
GDPerformerv2	**T+L+I**	**68.99**	**125**

synthetic dataset. The finetuning of the document classification model and other models like LiLT and LayoutLMv3 was performed for comparison purposes. The results showed that our proposed GDPerformer networks performed better than other approaches, as shown in Table 5. This suggests that our novel pretraining strategy has learned a more holistic linguistic representation that is less biased towards a specific language and can handle unknown languages effectively.

5.2 Performance of Unstructured Noisy Document Classification

Our algorithm is designed for document classification tasks, and we used the Tobacco3482 [18] dataset to demonstrate its robustness against various types of noise. In our experiments, different types of noises were added to the test set (used in Sect. 4.5) of the Tobacco-3482 dataset. The noises, including Gaussian noise with a specified mean and variance, Salt and Pepper noise, and Blur, were added equally to all the documents in the test set. Please note that none of the models were finetuned on noisy images but directly tested on the noisy documents. The results of different algorithms are presented in Table 6. This experiment demonstrates that our algorithm can handle real-world document

Table 5. Comparison on the synthetically created French dataset for Document Classification. We observe the least drop in performance from our proposed GDPerformerv2

Model	Modality	Precision	Recall	F1-Score
BERT [6]	T	32.38	42.86	33.19
LiLT [32]	T+L	42.86	23.81	42.86
GDPerformerv1	T+L	64.29	54.76	64.29
LayoutLMV3 [14]	T+L+I	17.14	35.71	23.13
GDPerformerv2	**T+L+I**	**68.57**	**68.84**	**68.70**

images where noise is expected and still achieve good results. This can be of great significance in practical applications, where documents often get distorted for various reasons, such as scanning errors, bad lighting, or camera shaking. The ability to classify such documents accurately can greatly improve the efficiency and effectiveness of document-based processes.

Table 6. Comparison of Unstructured Document Classification on the noisy Tobacco-3482 [18] test set. As we observe, our proposed models have the least drop in performance on noisy documents

Model	Modality	Precision	Recall	F1-Score	#Params(M)
BERT [6]	T	42.99	21.01	16.41	109
RoBERTa [23]	T	07.30	11.29	03.37	124
LiLT [32]	T+L	30.78	15.30	10.06	-
GDPerformerv1	T+L	58.26	21.61	18.17	130
LayoutLMV3 [14]	T+L+I	26.39	20.18	11.53	130
GDPerformerv2	**T+L+I**	**58.65**	**51.64**	**44.69**	**125**

5.3 Varying the Amount of Training Data

We experimented by varying the amount of training data used for pretraining the GDPerformer networks. We train additional versions of GDPerformerv1 and GDPerformerv2 using $100K$ and $150K$ training samples from the IDL [4] dataset. We then finetune the models for all three tasks following the strategy presented in Sect. 4 and note the F1-scores. We find the improvement in terms of the F1-score for all the tasks to be negligible with increasing training data. This is shown in Fig. 7 and thus, we stick to models trained on $50K$ images for all our other experiments. This phenomenon can benefit low-resource languages where large annotated datasets are not readily available. The model's performance can still be achieved by pretraining the models with limited data and improving document analysis in low-resource settings.

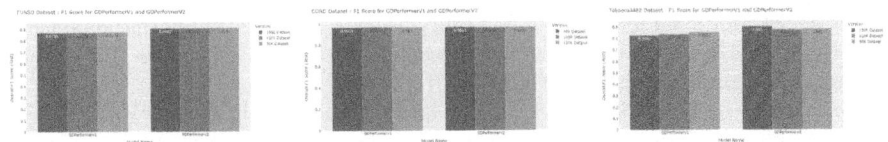

Fig. 7. We note that the performance does not improve with increasing training data. This can potentially lead to the extension of such pretraining techniques for low-resource languages.

6   Conclusion

In conclusion, this study presents a novel approach for document analysis tasks that leverages the power of pretrained transformer-based models, GDPerformerv1, and GDPerformerv2. Our approach enhances the understanding of both the document's structure and language, leading to improved accuracy and robustness in document analysis. The pretrained embedding space generated by our approach is utilized for various downstream tasks. It has shown improved performance on these tasks, with results showing its ability to generalize to a wide range of documents, including noisy documents. Furthermore, our approach has demonstrated its scalability, requiring a relatively small dataset of $50K$ to $150K$ documents for training. Additionally, our approach directly works on other languages, further highlighting its potential for widespread usage in document analysis.

References

1. Appalaraju, S., Jasani, B., Kota, B.U., Xie, Y., Manmatha, R.: Docformer: end-to-end transformer for document understanding. In: Proceedings of the IEEE/CVF International Conference on Computer Vision, pp. 993–1003 (2021)
2. Arshad, O., Gallo, I., Nawaz, S., Calefati, A.: Aiding intra-text representations with visual context for multimodal named entity recognition. In: 2019 International Conference on Document Analysis and Recognition (ICDAR), pp. 337–342 (2019)
3. Bao, H., et al.: Unilmv2: pseudo-masked language models for unified language model pre-training. In: International Conference on Machine Learning, pp. 642–652. PMLR (2020)
4. Biten, A.F., Tito, R., Gomez, L., Valveny, E., Karatzas, D.: Ocr-idl: Ocr annotations for industry document library dataset. arXiv preprint arXiv:2202.12985 (2022)
5. Cho, J., Lu, J., Schwenk, D., Hajishirzi, H., Kembhavi, A.: X-lxmert: aint, caption and answer questions with multi-modal transformers. arXiv preprint arXiv:2009.11278 (2020)
6. Devlin, J., Chang, M., Lee, K., Toutanova, K.: BERT: pre-training of deep bidirectional transformers for language understanding. CoRR **abs/1810.04805** (2018). http://arxiv.org/abs/1810.04805
7. Emanuela Boros, E., Toumi, A., Rouchet, E., Abadie, B., Stutzmann, D., Kermorvant, C.: Automatic page classification in a large collection of manuscripts based

on the international image interoperability framework. In: 2019 International Conference on Document Analysis and Recognition (ICDAR), pp. 756–762 (2019). https://doi.org/10.1109/ICDAR.2019.00126

8. Garncarek, Ł, Powalski, R., Stanisławek, T., Topolski, B., Halama, P., Turski, M., Graliński, F.: LAMBERT: layout-aware language modeling for information extraction. In: Lladós, J., Lopresti, D., Uchida, S. (eds.) ICDAR 2021. LNCS, vol. 12821, pp. 532–547. Springer, Cham (2021). https://doi.org/10.1007/978-3-030-86549-8_34

9. Gu, J., et al.: .: Unidoc: unified pretraining framework for document understanding. In: Ranzato, M., Beygelzimer, A., Dauphin, Y., Liang, P., Vaughan, J.W. (eds.) Advances in Neural Information Processing Systems. vol. 34, pp. 39–50. Curran Associates, Inc. (2021). https://proceedings.neurips.cc/paper/2021/file/0084ae4bc24c0795d1e6a4f58444d39b-Paper.pdf

10. Gu, Z., et al.: Xylayoutlm: towards layout-aware multimodal networks for visually-rich document understanding. In: Proceedings of the IEEE/CVF Conference on Computer Vision and Pattern Recognition, pp. 4583–4592 (2022)

11. Guillaume Jaume, Hazim Kemal Ekenel, J.P.T.: Funsd: a dataset for form understanding in noisy scanned documents. In: Accepted to ICDAR-OST (2019)

12. Guo, H., Qin, X., Liu, J., Han, J., Liu, J., Ding, E.: Eaten: entity-aware attention for single shot visual text extraction. In: 2019 International Conference on Document Analysis and Recognition (ICDAR), pp. 254–259 (2019)

13. Hong, T., Kim, D., Ji, M., Hwang, W., Nam, D., Park, S.: Bros: a pre-trained language model for understanding texts in document (2021). In: https://openreview.net/forum, p. 48

14. Huang, Y., Lv, T., Cui, L., Lu, Y., Wei, F.: Layoutlmv3: pre-training for document ai with unified text and image masking. In: Proceedings of the 30th ACM International Conference on Multimedia (2022)

15. Jahagirdar, S., Gangisetty, S., Mishra, A.: Look, read and ask: learning to ask questions by reading text in images. In: Lladós, J., Lopresti, D., Uchida, S. (eds.) ICDAR 2021. LNCS, vol. 12821, pp. 335–349. Springer, Cham (2021). https://doi.org/10.1007/978-3-030-86549-8_22

16. Jain, R., Wigington, C.: Multimodal document image classification. In: 2019 International Conference on Document Analysis and Recognition (ICDAR), pp. 71–77 (2019). https://doi.org/10.1109/ICDAR.2019.00021

17. Kosaraju, S.C., et al.: Dot-net: document layout classification using texture-based cnn. In: 2019 International Conference on Document Analysis and Recognition (ICDAR), pp. 1029–1034 (2019). https://doi.org/10.1109/ICDAR.2019.00168

18. Kumar, J., Ye, P., Doermann, D.: Learning document structure for retrieval and classification. In: Proceedings of the 21st International Conference on Pattern Recognition (ICPR2012), pp. 1558–1561 (2012)

19. Li, C., Bi, B., Yan, M., Wang, W., Huang, S., Huang, F., Si, L.: StructuralLM: Structural pre-training for form understanding. In: Proceedings of the 59th Annual Meeting of the Association for Computational Linguistics and the 11th International Joint Conference on Natural Language Processing (Volume 1: Long Papers), pp. 6309–6318. Association for Computational Linguistics, August 2021. https://doi.org/10.18653/v1/2021.acl-long.493, https://aclanthology.org/2021.acl-long.493

20. Li, P., Gu, J., Kuen, J., Morariu, V.I., Zhao, H., Jain, R., Manjunatha, V., Liu, H.: Selfdoc: self-supervised document representation learning. In: Proceedings of the IEEE/CVF Conference on Computer Vision and Pattern Recognition, pp. 5652–5660 (2021)

21. Li, Y., et al.: Structext: structured text understanding with multi-modal transformers. In: Proceedings of the 29th ACM International Conference on Multimedia, pp. 1912–1920 (2021)
22. Liu, X., Gao, F., Zhang, Q., Zhao, H.: Graph convolution for multimodal information extraction from visually rich documents. In: North American Chapter of the Association for Computational Linguistics (2019)
23. Liu, Y., et al.: Roberta: a robustly optimized bert pretraining approach. arXiv preprint arXiv:1907.11692 (2019)
24. Mathew, M., Karatzas, D., Jawahar, C.V.: Docvqa: a dataset for vqa on document images. In: 2021 IEEE Winter Conference on Applications of Computer Vision (WACV), pp. 2199–2208 (2021). https://doi.org/10.1109/WACV48630.2021.00225
25. Miranda, B., Wang, Y.X., Koyejo, O.: The curse of zero task diversity: on the failure of transfer learning to outperform maml and their empirical equivalence. ArXiv abs/2208.01545 (2021)
26. Park, S., et al.: Cord: a consolidated receipt dataset for post-ocr parsing (2019)
27. Raffel, C., et al.: Exploring the limits of transfer learning with a unified text-to-text transformer. ArXiv abs/1910.10683 (2019)
28. Ramesh, A., et al.: Zero-shot text-to-image generation. In: International Conference on Machine Learning, pp. 8821–8831. PMLR (2021)
29. Smith, R.: An overview of the tesseract ocr engine. In: Ninth International Conference on Document Analysis and Recognition (ICDAR 2007), vol. 2, pp. 629–633. IEEE (2007)
30. Studer, L., et al.: A comprehensive study of imagenet pre-training for historical document image analysis. In: 2019 International Conference on Document Analysis and Recognition (ICDAR), pp. 720–725 (2019)
31. Su, W., et al.: Vl-bert: pre-training of generic visual-linguistic representations. In: International Conference on Learning Representations (2020). https://openreview.net/forum?id=SygXPaEYvH
32. Wang, J., Jin, L., Ding, K.: Lilt: a simple yet effective language-independent layout transformer for structured document understanding. In: Annual Meeting of the Association for Computational Linguistics (2022)
33. Williams, J., Tadesse, A., Sam, T., Sun, H.M., Montañez, G.D.: Limits of transfer learning. In: International Conference on Machine Learning, Optimization, and Data Science (2020)
34. Wu, T.L., Li, C., Zhang, M., Chen, T., Hombaiah, S.A., Bendersky, M.: Lampret: layout-aware multimodal pretraining for document understanding. ArXiv abs/2104.08405 (2021)
35. Xu, Y., et al.: Layoutlmv2: multi-modal pre-training for visually-rich document understanding. In: ACL-IJCNLP 2021, January 2021
36. Xu, Y., Li, M., Cui, L., Huang, S., Wei, F., Zhou, M.: Layoutlm: pre-training of text and layout for document image understanding. In: Proceedings of the 26th ACM SIGKDD International Conference on Knowledge Discovery & Data Mining, pp. 1192–1200 (2020)
37. Xu, Y., et al.: Layoutxlm: multimodal pre-training for multilingual visually-rich document understanding. CoRR **abs/2104.08836** (2021). https://arxiv.org/abs/2104.08836
38. Yu, Y., et al.: An empirical study of pre-trained models on out-of-distribution generalization (2022). https://openreview.net/forum?id=2RYOwBOFesi
39. Zhang, P., Li, C., Qiao, L., Cheng, Z., Pu, S., Niu, Y., Wu, F.: VSR: a unified framework for document layout analysis combining vision, semantics and relations.

In: Lladós, J., Lopresti, D., Uchida, S. (eds.) ICDAR 2021. LNCS, vol. 12821, pp. 115–130. Springer, Cham (2021). https://doi.org/10.1007/978-3-030-86549-8_8

40. Zhang, P., et al.: Trie: end-to-end text reading and information extraction for document understanding. In: Proceedings of the 28th ACM International Conference on Multimedia, pp. 1413–1422 (2020)

GraphMLLM: A Graph-Based Multi-level Layout Language-Independent Model for Document Understanding

He-Sen Dai[1,2], Xiao-Hui Li[2(✉)], Fei Yin[2], Xudong Yan[3], Shuqi Mei[3], and Cheng-Lin Liu[1,2]

[1] School of Artificial Intelligence, University of Chinese Academy of Sciences, Beijing 100049, China
daihesen20@mails.ucas.ac.cn
[2] State Key Laboratory of Multimodal Artificial Intelligence Systems, Institute of Automation of Chinese Academy of Sciences, Beijing 100190, China
{xiaohui.li,fyin,liucl}@nlpr.ia.ac.cn
[3] T Lab, Tencent Map, Tencent Technology (Beijing) Co., Ltd., Beijing 100193, China
{owenyan,shawnmei}@tencent.com

Abstract. Self-supervised multi-modal document pre-training for document knowledge learning shows superiority in various downstream tasks. However, due to the diversity of document languages and structures, there is still room to better model various document layouts while efficiently utilizing the pre-trained language models. To this goal, this paper proposes a Graph-based Multi-level Layout Language-independent Model (GraphMLLM) which uses dual-stream structure to explore textual and layout information separately and cooperatively. Specifically, GraphMLLM consists of a text stream which uses off-the-shelf pre-trained language model to explore textual semantics and a layout stream which uses multi-level graph neural network (GNN) to model hierarchical page layouts. Through the cooperation of the text stream and layout stream, GraphMLLM can model multi-level page layouts more comprehensively and improve the performance of language-independent document pre-trained model. Experimental results show that compared with previous state-of-the-art methods, GraphMLLM yields higher performance on downstream visual information extraction (VIE) tasks after pre-training on less documents. Code and model will be available at https://github.com/HSDai/GraphMLLM.

Keywords: Visual information extraction · Self-supervised pre-training · Multi-level page layouts

1 Introduction

As an important task in Visual Document Understanding (VDU), Visual Information Extraction (VIE) focuses on automated information extraction through

E. H. Barney Smith et al. (Eds.): ICDAR 2024, LNCS 14804, pp. 227–243, 2024.
https://doi.org/10.1007/978-3-031-70533-5_14

Semantic Entity Recognition (SER) and Relationship Extraction (RE) from Visually-Rich Documents (VRD) including receipts, forms, reports, invoices, etc. It receives widespread attention from both industry and academia for its promising applications.

There have been numerous works of VIE reported in recent years. However, due to the heavy workload of manual annotation, the existing VIE datasets, such as FUNSD [9], XFUND [24], CORD [16] and EPHOIE [20], usually have small scales, severely limiting the performance of deep-learning based VIE methods trained from scratch. To overcome this limitation, many pre-training based methods, such as DocFormer [1], GraphDoc [26], SelfDoc [13], StructuralLM [11], and the LayoutLM [23] series, have been proposed. Different from BERT [4] designed for plain text, pre-training methods for VIE task usually take into consideration the natural multi-modal property of documents and utilize textual, visual and layout information when pre-training their models. Besides, many self-supervised pretext tasks are designed for self-supervised model learning, such as Masked Language Modeling (MLM) [4], Masked Image Modeling (MIM) [2], Masked Visual-Language Model (MVLM) [23], Text Image Alignment (TIA) [22], Text Image Matching (TIM) [22], etc.

The joint learning of multi-modal information can bring significant performance gain on downstream tasks, but it also brings some unexpected disadvantages including the huge data amount required by pre-training and the inflexibility when handling documents of languages not covered by the pre-trained model. Though some works, e.g. LayoutXLM [24], directly use multilingual documents for pre-training to achieve better performance on multilingual dataset XFUND [24], they require even more data for pre-training, yet the pre-trained model still lacks the ability to generalize to unseen languages. Considering that documents with different languages may share similar layouts, LiLT [19] proposes to use dual-stream transformer to decouple text and layout during pre-training, and then re-couples them for downstream task fine-tuning. Through this design, LiLT can be pre-trained on IIT-CDIP [10] which only contains English documents and then adapted to other languages during fine-tuning.

Document understanding can take advantage of the hierarchical nature of document layouts (see Fig. 1), which can provide important guidance for various document understanding tasks. To better excavate document information from multi-level layout structures, some methods including StrucTexT [14], FastStrucTexT [25], LayoutLMv2 [22] and ERNIE-mmLayout [21] have been proposed. These methods either model document hierarchy implicitly through token and segment 1D embedding [14,22] or explicitly exchange and integrate information from different levels and granularities [21,25].

To better exploit the hierarchical structure of layouts for visual document understanding, in this paper we propose a Graph-based Multi-level Layout Language-independent Model (GraphMLLM) that decouples text and layout through a dual-stream structure and explores multi-level layout information more efficiently. By decoupling text and layout, GraphMLLM can directly reuse existing pre-trained language models and greatly reduce the reliance on the amount of pre-training data. Meanwhile, graphs are used to model the multi-level

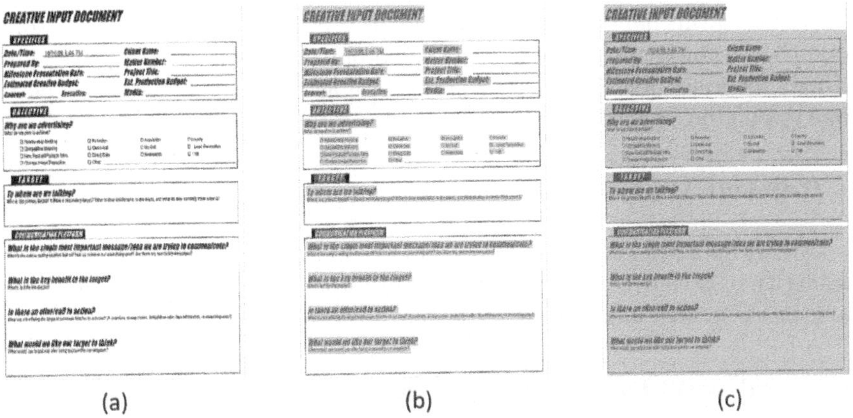

Fig. 1. The hierarchical layout structure of documents. Word-level and segment-level layouts in (a) and (b), are typically acquired via OCR engines. Region-level layout is derived through heuristic rules or layout analysis models.

layout information and interact with semantic modalities through a disentangled attention mechanism, enabling the language model to access different levels of layout information, so as to improve the performance of the entire model.

During pre-training, we use Masked Visual-Language Model, Key Point Location and Cross-modal Alignment Identification as pretext tasks and train our model on the monolingual IIT-CDIP dataset. While during fine-tuning, we conducted experiments on two monolingual datasets FUNSD, CORD and one multilingual dataset XFUND to demonstrate the effectiveness of our model. The experimental results show that despite using fewer data for pre-training, our approach can still outperform other multilingual pre-trained models and obtains competitive results on all tasks compared with state-of-the-art methods.

The contributions of this paper are summarized as follows:

(1) We propose a new multi-modal document pre-training model named GraphMLLM for document understanding, which contains a dual-stream structure to decouple the textual and layout information during pre-training to make it language independent.
(2) We use a multi-level graph neural network (GNN) to model hierarchical document layouts at different granularities, thus exploit the hierarchical structure of documents more efficiently.
(3) The combination of text-layout decoupling and hierarchical layout modeling can significantly reduce the required quantity of pre-training data while still remains high performance on downstream tasks.
(4) We evaluated on three benchmark datasets of different languages, and the experimental results show that GraphMLLM can obtain superior or competitive results compared with state-of-the-art methods.

The rest of this paper is organized as follows. Section 2 reviews related works. Section 3 introduces the architecture of the proposed model and the pre-training method. Sections 4 and 5 present experimental settings and results, and Sect. 6 draws concluding remarks.

2 Related Work

Here we briefly review existing methods closely related to our work based on the granularity of layout considered: word-level layout based model, segment-level layout based model, and multi-level layout based model.

2.1 Word-Level Layout Based Model

To perceive layout information, LayoutLM [23] adds token and word position embeddings as initial embeddings with 2D position awareness. In addition to layout information, LayoutLMv2 [22] integrates visual information by gridifing document images (e.g. 7 × 7) and achieves *soft alignment* between textual and visual information through pre-training tasks. By focusing on the relationship between texts, BROS [7] improves the attention mechanism by introducing relative position information. DocFormer [1] improves the attention mechanism in integrating textual, layout, and visual features, and proposes a pixel-level image reconstruction task. Furthermore, ERNIE-Layout [17] focuses on document layout information and enables the model to obtain correct reading order through serialization modules and pre-training task for reading order. LayoutXLM [24] emphasizes information extraction from multilingual documents and uses multilingual documents for pre-training. Such word-level layout based models can model fine-grained token-level information but are weak in macroscopic perspective when facing complex layouts.

2.2 Segment-Level Layout Based Model

Compared to words, segment-level layout (based on text lines, e.g.) is more informative for document understanding. To incorporate higher-level layout information, StructuralLM [11] uses segment-level layout information as the positional embedding of tokens. SelfDoc [13] uses segment-level semantic and layout features, and proposes a multimodal adaptation attention mechanism. Extended to image-centric Document task, LayoutLMv3 [8] adopts segment-level layout information and tokenizes images like the DiT [12] model, without relying on CNN-based visual encoders. Considering local dependencies between segments, GraphDoc [26] uses graph neural networks as the backbone of the pre-trained model. To achieve cross-language transfer capability, LiLT [19] uses a dual-stream Transformer structure with bi-directional attention complementation mechanism (BiACM) to decouple text and layout information. These methods focus on segment-level layout information but ignore the word-level layout information which is also crucial for the comprehension of certain documents such as forms.

2.3 Multi-level Layout Based Model

Multi-level layout based models aim to extract layout information of multi-level granularities for better document understanding. StrucTexT [14] adds the token embeddings and word-level layout embeddings as the initial token embeddings, and adds visual features and segment-level layout embeddings together as the initial visual token embeddings, and the word- and segment-level layout embeddings are integrated. As an extension of LayoutLMv2, ERINE-mmLayout [21] introduces additional segment and region information on the basis of LayoutLMv2 without pre-training again, to perceive multi-level layout information and effectively improves the model performance. These works show the effectiveness of integrating multi-level layout information. However, the multi-level layout information is still not utilized sufficiently in that the interactions between inter- and intra-level layouts are not considered in great detail.

In this paper, we try to model multi-level layout information of documents in pre-trained model with better integration and interaction between different levels, so that the model has better cross-language transfer capability at moderate model complexity and data reliance.

3 GraphMLLM

Inspired by the disentangled attention mechanism proposed in DeBERTa [6] and the decoupled modeling in LiLT, GraphMLLM (see Fig. 2) adopts a dual-stream structure to encode text features and layout features separately, and interacts information between two modalities through an attention based hierarchical interaction mechanism. In the following, the multi-level graph representation of documents is first introduced, followed by the text flow module, layout flow module and interaction mechanism between these two modalities.

3.1 Document Representation

First, all the texts in documents along with their coordinates are extracted using an OCR engine provided by ReSenseTech (other OCR software can be used alternatively)[1], and are transformed into text and layout representations using token embeddings and hierarchical layout embeddings, respectively.

Text Representation. Since the reading order obtained by OCR is noisy, we use XY Cut [5] to obtain a proper reading order of texts. Then, like many pre-trained language models, we serialize and tokenize the texts and add special tokens [CLS] and [SEP] at the beginning and end to obtain the text token sequence: $T = [t_1, ..., t_{N_t}]$, where N_t stands for the token sequence length. By adding token embeddings and 1D position embeddings, 1D-position aware token embeddings are obtained:

$$E_t = \text{LN}(E_{token} + E_{1D_{pos}}), \tag{1}$$

---

[1] http://www.resensetech.com

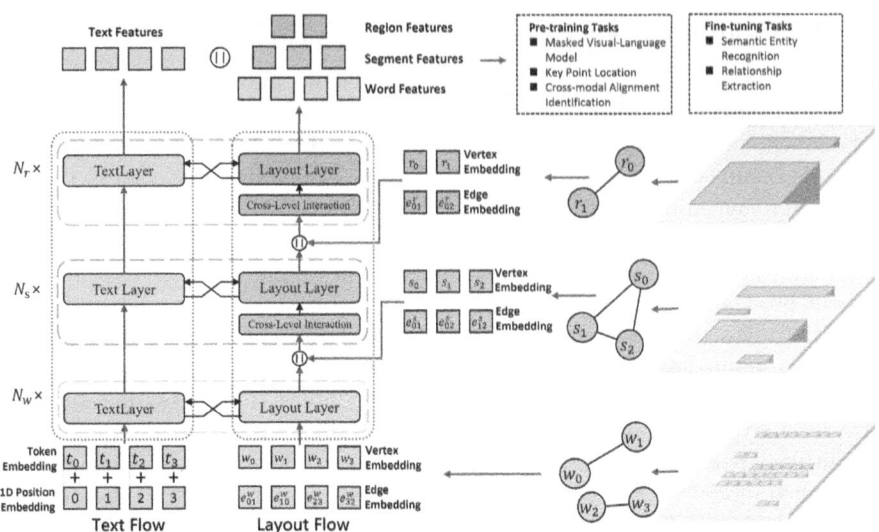

Fig. 2. The overall architecture of GraphMLLM, which consists of two streams: the *Text Flow* and the *Layout Flow*. It decouples text and layout information, and uses hierarchical graphs to model multi-level layouts. Higher level layouts are successively added to the model, N_w, N_s and N_r are layer numbers of each stage. In each layer, layout features are firstly interacted cross levels, then information from text and layout modalities are interacted through disentangled attention. Best viewed in zoomed-in.

where $\boldsymbol{E}_{token} \in \mathbb{R}^{N_t \times D_t}$ is token embedding matrix, $\boldsymbol{E}_{1D_{pos}} \in \mathbb{R}^{N_t \times D_t}$ is position embedding matrix, D_t is feature dimension and LN is Layer Normalization.

Multi-layer Layout Representation. Document layout information can be represented at three levels: word, segment, and region, represented by $W = \{w_1, ..., w_{N_w}\}$, $S = \{s_1, ..., s_{N_s}\}$ and $R = \{r_1, ..., r_{N_r}\}$ respectively. Here, N_w, N_s, N_r are the number of words, segments, and regions for the given document. The layout is represented as a graph structure $G = (V, E)$, where vertices $V = W \cup S \cup R$ denote layout elements (words, segments and regions), and edges $E = E_{ww} \cup E_{ss} \cup E_{rr}$ denote the connections between vertices of the same level.

Fellowing GraphDoc [26], the vertex and edge embeddings $\boldsymbol{E}_v \in \mathbb{R}^{N_v \times D_v}$, $\boldsymbol{E}_e \in \mathbb{R}^{N_e \times D_e}$ of graph are obtained as follows:

$$\boldsymbol{E}_v = Concat(\boldsymbol{E}_x(x_0, x_1, w), \boldsymbol{E}_y(y_0, y_1, h)), \tag{2}$$

where x_0, y_0, x_1, y_1 denote the left, top, right, bottom coordinates, and h, w denote height and width of bounding boxes, $\boldsymbol{E}_x, \boldsymbol{E}_y$ are learnable position embeddings. Similarly,

$$\boldsymbol{E}_e = \boldsymbol{E}_{tl}\boldsymbol{W}_{tl} + \boldsymbol{E}_{tr}\boldsymbol{W}_{tr} + \boldsymbol{E}_{bl}\boldsymbol{W}_{bl} + \boldsymbol{E}_{br}\boldsymbol{W}_{br}, \tag{3}$$

where $\boldsymbol{W}_{tl}, \boldsymbol{W}_{tr}, \boldsymbol{W}_{bl}, \boldsymbol{W}_{br}$ are learnable parameters, and $\boldsymbol{E}_{tl}, \boldsymbol{E}_{tr}, \boldsymbol{E}_{bl}, \boldsymbol{E}_{br}$ are sinusoidal position embeddings of distance of top-left, top-right, bottom-left and bottom-right coordinates of the vertex bounding boxes, which are calculated as follows:

$$\boldsymbol{E}_{dist} = Concat(\mathrm{PE}(x_{dist}), \mathrm{PE}(y_{dist})), \tag{4}$$

where $dist \in \{tl, tr, bl, br\}$, PE is a sinusoidal function [18], and x_{dist} and y_{dist} represent the horizontal and vertical distances of the corresponding coordinates.

For relationship between elements of different levels, we use matrixes $\boldsymbol{M}_{ws} \in \mathbb{R}^{N_w \times N_s}$ and $\boldsymbol{M}_{sr} \in \mathbb{R}^{N_s \times N_r}$ to represent the relationship between word-segment levels and between segment-region levels, respectively. Taking $\boldsymbol{M}_{ws}$ as an example, each m_{ij} in the matrix indicates whether the word w_i belongs (set as 1) to the segment s_j or not (set as 0).

Text-layout Alignment Relationship. The correspondence between text and layout is many-to-many. To facilitate the calculation, we use matrices $\boldsymbol{M}_{tw} \in \mathbb{R}^{N_t \times N_w}$, $\boldsymbol{M}_{ts} \in \mathbb{R}^{N_t \times N_s}$, $\boldsymbol{M}_{tr} \in \mathbb{R}^{N_t \times N_r}$ denote the correspondence of text-to-word-layout, text-to-segment-layout, text-to-region-layout, respectively. Taking $\boldsymbol{M}_{ts}$ as an example, each m_{ij} in the matrix indicates whether the text t_i belongs (set as 1) to the segment s_j or not (set as 0).

3.2 Text Flow

For extracting semantic features from text sequence, the backbone of text flow adopts a pre-trained language model consisting of several Transformer encoder layers [18]. Specifically, we input the token embeddings $\boldsymbol{E}_t$ into the text flow to derive semantic features with contextual information. As depicted in Fig. 3 and Fig. 4, for the k-th layer of the text flow:

$$\boldsymbol{H}_t^{k*} = \mathrm{LN}(\mathrm{MHA}_t(\boldsymbol{H}_t^k, \boldsymbol{H}_v^k) + \boldsymbol{H}_t^k), \tag{5}$$

$$\boldsymbol{H}_t^{k+1} = \mathrm{LN}(\mathrm{FFN}(\boldsymbol{H}_t^{k*}) + \boldsymbol{H}_t^{k*}), \tag{6}$$

where $\boldsymbol{H}_t^k \in \mathbb{R}^{N_t \times D_t}$ and $\boldsymbol{H}_v^k \in \mathbb{R}^{N_v \times D_v}$ are the token and layout features input to the k-th text layer, $\boldsymbol{H}_t^{k+1} \in \mathbb{R}^{N_t \times D_t}$ is the output of the k-th text layer, FFN is the Feed-Forward Network [18], MHA_t is the multi-head muti-modal self-attention mechanism, which will be described in detail in Sect. 3.4.

3.3 Layout Flow

In the layout flow, we use multi-level graph attention network to compute hidden representations of document layouts focusing on neighbouring features. We employ a bottom-up strategy to increasingly extract higher-level layout features. Specifically, the first k_w layers are designed to extract solely word-level layout features. Following this, the next k_s layers incorporate segment-level layout features into the model. Finally, the last k_r layers introduce region-level layout information to achieve a comprehensive representation of the document layout. The single-layer implementation of layout flow and inter-layer interaction are described separately in the following parts of this subsection.

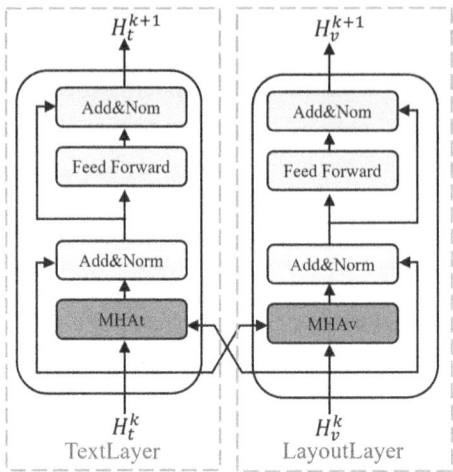

Fig. 3. Text-Layout Layer.

Layout Layer Following GraphDoc [26], we use graph attention layers to compute the hidden representation of layout tokens, by attending over its neighbors following a self-attention strategy. As depicted in Fig. 3 and Fig. 4, for the k-th layer of the layout flow:

$$H_v^{k*} = \text{LN}(\text{MHA}_v(H_v^k, H_t^k, E_e) + H_v^k), \tag{7}$$

$$H_v^{k+1} = \text{LN}(\text{FFN}(H_v^{k*}) + H_v^{k*}), \tag{8}$$

where $H_v^k \in \mathbb{R}^{N_v \times D_v}$ and $H_t^k \in \mathbb{R}^{N_t \times D_t}$ are the layout and text features input to the k-th layer, and $E_e \in \mathbb{R}^{N_e \times D_v}$ is the initialized edge embedding, and MHA_v is the graph attention mechanism, which will be described in Sect. 3.4.

Cross-level Interaction In order to enhance the multi-level layout representation ability of the layout flow, we consider interactions between different levels. Taking as an example the word-level layout feature H_w and segment-level feature H_s in one layer of the middle k_s layout layers:

$$H_s = AvePooling(H_w, M_{ws}) + H_s, \tag{9}$$

$$H_w = M_{ws}H_s + H_w, \tag{10}$$

where $AvePooling$ performs the average pooling operation among all words belonging to the same segments. Taking node s_i as an example, its corresponding feature is denoted as h_{s_i}:

$$h_{s_i} = \frac{1}{|\mathcal{E}(s_i)|} \sum_{w_j \in \mathcal{E}(s_i)} h_{w_j} + h_{s_i}, \tag{11}$$

where $h_{s_i} \in \mathbb{R}^{D_v}$, $h_{w_j} \in \mathbb{R}^{D_v}$, and $\mathcal{E}(s_i) = \{w_k | m_{ki} = 1, m_{ki} \in M_{wr}\}$.

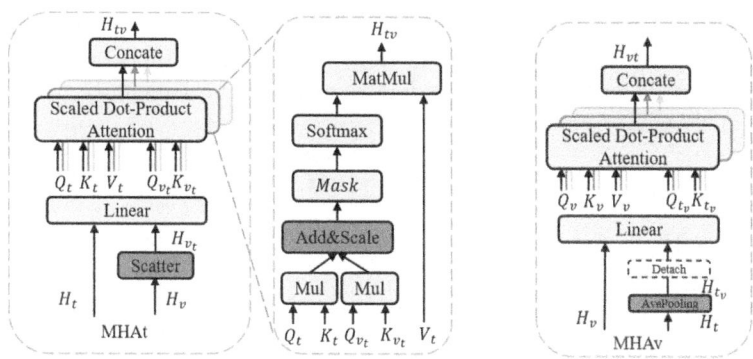

Fig. 4. Disentangled Interaction Mechanism.

3.4   Cross-modal Interaction Mechanism

As discussed earlier, it is possible to decouple layout and text features and interact only through attention scores. Here we give its details as follows.

Text Flow Attention Mechanism. MHA_t is a disentangled attention mechanism in which attention weights among tokens are computed using disentangled attention matrices on their contents and layouts.

First, the layout features of tokens $\boldsymbol{H}_{v_t} \in \mathbb{R}^{N_t \times D_t}$ are obtained through operation:

$$\boldsymbol{H}_{v_t} = Concate(\boldsymbol{M}_{tv}\boldsymbol{H}_v, \boldsymbol{E}_{1D_{lay}}),\tag{12}$$

where $\boldsymbol{M}_{tv} \in \{\boldsymbol{M}_{tw}, \boldsymbol{M}_{ts}, \boldsymbol{M}_{tr}\}$ according to the layout type, $\boldsymbol{E}_{1D_{lay}} \in \mathbb{R}^{N_t \times D_t}$ is 1D position embedding matrix similar to $\boldsymbol{E}_{1D_{pos}}$.

Then, we map token features $\boldsymbol{H}_t$ and layout features $\boldsymbol{H}_{v_t}$ to $\boldsymbol{Q}_t, \boldsymbol{K}_t, \boldsymbol{V}_t$, $\boldsymbol{Q}_{v_t}, \boldsymbol{K}_{v_t}, \boldsymbol{V}_{v_t}$:

$$\boldsymbol{Q}_t, \boldsymbol{K}_t, \boldsymbol{V}_t = \boldsymbol{H}_t\boldsymbol{W}_{Qt}, \boldsymbol{H}_t\boldsymbol{W}_{Kt}, \boldsymbol{H}_t\boldsymbol{W}_{Vt},\tag{13}$$

$$\boldsymbol{Q}_{v_t}, \boldsymbol{K}_{v_t}, \boldsymbol{V}_{v_t} = \boldsymbol{H}_{v_t}\boldsymbol{W}_{Q_{vt}}, \boldsymbol{H}_{v_t}\boldsymbol{W}_{K_{vt}}, \boldsymbol{H}_{v_t}\boldsymbol{W}_{V_{vt}},\tag{14}$$

where $\boldsymbol{W}_{*t} \in \mathbb{R}^{D_t \times D_t}$ and $\boldsymbol{W}_{*vt} \in \mathbb{R}^{D_{v_t} \times D_t}$ are learnable parameters.

Then, the contextualized representation output is obtained by taking a weighted sum of the values based on the attention weights:

$$\boldsymbol{A}_{tv} = \boldsymbol{Q}_t\boldsymbol{K}_t^T + \boldsymbol{Q}_{v_t}(\boldsymbol{K}_{v_t})^T,\tag{15}$$

$$\boldsymbol{H}_{tv} = Softmax(\frac{\boldsymbol{A}_{tv}}{\sqrt{D_t}})\boldsymbol{V}_t,\tag{16}$$

where $\boldsymbol{A}_{tv} \in \mathbb{R}^{N_t \times N_t}$ is the attention score matrix.

Layout Flow Attention Mechanism. MHA_v is also a disentangled attention mechanism that differs from MHA_t in that it relies only on local context.

First, the text features corresponding to each layout element $\boldsymbol{H}_{t_v} \in \mathbb{R}^{N_v \times D_t}$ are obtained by averaging pooling:

$$\boldsymbol{H}_{t_v} = AvePooling(\boldsymbol{H}_t, \boldsymbol{M}_{tv}), \tag{17}$$

Similarly to Eq. 13 and Eq. 14, we map text and layout features to get $\boldsymbol{Q}_{t_v}, \boldsymbol{K}_{t_v}, \boldsymbol{V}_{t_v}$ and $\boldsymbol{Q}_v, \boldsymbol{K}_v, \boldsymbol{V}_v$. Then, we use the following operations to obtain the attention matrix $\boldsymbol{A}_{vt} \in \mathbb{R}^{N_v \times N_v}$ and layout features $\boldsymbol{H}_{vt} \in \mathbb{R}^{N_v \times D_v}$ after fusing the neighbor information:

$$\boldsymbol{A}_{vt} = \boldsymbol{Q}_v \boldsymbol{K}_v^T + Reshape(\boldsymbol{Q}_v \boldsymbol{E}_e^T) + \boldsymbol{Q}_{t_v} \boldsymbol{K}_{t_v}^T, \tag{18}$$

$$\boldsymbol{A}_g = Mask(\boldsymbol{A}_{vt}, \boldsymbol{M}_v), \tag{19}$$

$$\boldsymbol{H}_{vt} = Softmax(\frac{\boldsymbol{A}_g}{\sqrt{D_v}})\boldsymbol{V}_v, \tag{20}$$

where $Mask$ is the masking operation [18] according to the matrix $\boldsymbol{M}_v$ which is the adjacency matrix of layout elements according to layout graph G, $Reshape$ is reshaping operation to match the shape of $\boldsymbol{A}_{vt}$.

3.5   Gradient Detach Operation

Following LiLT [19], to ensure the linguistic independence of the text flow, the gradient back propagation from the layout flow to the text flow needs to be terminated during pre-training. Specifically, during pre-training Eq. 17 is replaced as the following operation:

$$\boldsymbol{H}_{t_v} = Detach(AvePooling(\boldsymbol{H}_t, \boldsymbol{M}_{tv})) \tag{21}$$

where $Detach$ is the gradient detach operation, i.e., the gradient back propagation is not continued from here. The gradient detach operation can mitigate the gradient impact of the layout flow on the text flow, thus enhancing the cross-linguistic capability of model.

4   Experiment Setting

4.1   Pre-training Tasks

Masked Visual-Language Model. Masked Visual-Language Model (MVLM) [23] is a pre-training task for model to learn linguistic representations. During pre-training, 15% of the tokens are randomly masked, of which 80% are replaced by special tokens "[MASK]", 10% are replaced by random tokens sampled from the entire vocabulary, and the last 10% remain unchanged. The goal of this task is to predict the tokens masked in the text.

Key Point Location. Key Point Location (KPL) [19] is a pre-training task for model to learn the layout representation using surrounding layout information. During training, 15% of the text bounding boxes are randomly masked, of which 80% are replaced by [0,0,0,0], 10% are replaced by random boxes sampled from the same batch, and the last 10% remain unchanged. The target is to predict the key points (top left, center, bottom right) of each bounding boxes belonging to certain regions (the document is divided equally into 49 regions).

Cross-modal Alignment Identification. Cross-modal alignment identification (CAI) [19] is a pre-training task for aligning tokens and bounding boxes. In pre-training, the token-box pairs of encoded tokens are collected, and the training goal is to predict whether each pair has undergone a replacement operation. Since all three pre-training tasks are classification tasks, we use fully connected layers for classification and their losses are all calculated using cross-entropy loss.

4.2   Downstream Tasks

Semantic Entity Recognition. The goal of Semantic Entity Recognition (SER) [24] is to extract semantic entities from a set of tokens. Specifically, given a document D with a sequence of tokens $T = [t_0, t_1, ..., t_n]$ and target tags $C = \{c_0, c_1, ..., c_m\}$, it is required to predicted semantic entities:

$$\mathcal{E} = \{([x_0^0, ..., x_0^{n_0}], c_0), ..., ([x_k^0, ..., x_k^{n_k}], c_k)\} \tag{22}$$

where $x_k^{n_k} \in T$ and n_k is the length of the k-th extracted entity.

Relationship Extraction. Relationship Extraction (RE) [24] is to extract the relationships between entities. Specifically, given a set of entities of document D and the semantic relation labels $R = \{r_0, r_1, ..., r_m\}$, it is required to predict a set of semantic relations:

$$\mathcal{L} = \{([head_0, tail_0], r_0), ..., ([head_k, tail_k], r_k)\} \tag{23}$$

where $head_k$ and $tail_k$ are two semantic entities and r_k is the relation between them. In this work, we mainly focus on the key-value relation extraction..

5   Experiments

5.1   Datasets

IIT-CDIP: IIT-CDIP [10] is a large-scale dataset of scanned English document images, containing over 6 million documents and over 11 million scanned document images. This dataset is used for self-supervised pre-training.

FUNSD: FUNSD [9] is a scanned English form dataset for the form understanding task. It is divided into a training set containing 149 samples and a

Table 1. SER results on English datasets of FUNSD and CORD. "#Docs" represents the number of documents utilized for pre-training, measured in millions (M). "W", "S" and "R" denote word-level, segment-level and region-level layouts, respectively. **Bold** implicates the best results and <u>underline</u> the second best results.

Model	#Docs	Layout	FUNSD			CORD		
			Precision	Recall	F1	Precision	Recall	F1
BERT$_{BASE}$ [4]	–	–	0.5469	0.6710	0.6026	0.8833	0.9107	0.8968
RoBERTa$_{BASE}$ [15]	–	–	0.6349	0.6975	0.6648	–	–	–
LayoutLM$_{BASE}$ [23]	11M	W	0.7597	0.8155	0.7866	0.9437	0.9508	0.9472
LayoutLMv3$_{BASE}$ [8]	11M	S	–	–	**0.9029**	–	–	<u>0.9656</u>
GraphDoc [26]	0.32M	S	–	–	0.8795	–	–	**0.9693**
LayoutXLM$_{BASE}$ [24]	30M	W	–	–	0.794	–	–	–
LiLT [19]	11M	S	<u>0.8721</u>	**0.8965**	0.8841	<u>0.9598</u>	<u>0.9616</u>	0.9607
GraphMLLM	2M	W	0.7591	0.7955	0.7769	0.9313	0.9431	0.9372
GraphMLLM	2M	WS	0.8623	0.8830	0.8725	0.9515	0.9536	0.9525
GraphMLLM	2M	WSR	0.8616	0.8840	0.8727	0.9537	0.9558	0.9548
GraphMLLM	11M	WSR	**0.8835**	<u>0.8870</u>	<u>0.8852</u>	**0.9620**	**0.9656**	0.9638

test set containing 50 samples. Each document contains four types of entities: question, answer, heading, and other.

XFUND: XFUND [24] is a multilingual document understanding dataset extended from the FUNSD dataset. The languages of documents are extended from English (EN) to seven other languages, including Chinese (ZH), Japanese (JA), Spanish (ES), French (FR), Italian (IT), German (DE), and Portuguese (PT). Each language includes 199 forms, among which 149 forms are used for training and the other 50 forms are used for testing.

CORD: CORD [16] is a receipt dataset with a training set containing 800 samples, a validation set containing 100 samples, and a test set containing 100 samples. The dataset defines 30 fields under 4 categories and the task aims to label each word to the right field.

For pre-training on the IIT-CDIP dataset, the OCR engine is utilized to extract texts along with their bounding boxes. While for fine-tuning on FUNSD, CORD and XFUND, the official OCR annotations are used.

5.2   Evaluation Metrics

For the SER task and RE task, we use entity-level F1 score and pair-level F1 score as the evaluation metrics, respectively. Using the same settings as LayoutXLM [24] and LiLT [19], we evaluate the performance of the pre-trained model in three main settings: 1) fine-tuning and testing on a specific language; 2) fine-tuning on English data and then testing on multilingual data (zero-shot learning); 3) fine-tuning on all language data and testing on individual language data.

Table 2. Language-specific fine-tuning F1 accuracy on FUNSD and XFUND.

Task	Model	Pretrain Docs		FUNSD	XFUND							Avg.
		Language	Size	EN	ZH	JA	ES	FR	IT	DE	PT	
SER	LayoutXLM	Multilingual	30M	0.794	0.8924	0.7921	0.755	0.7902	0.8082	0.8222	0.7903	0.8056
	LiLT	English only	11M	0.8415	0.8938	0.7964	0.7911	0.7953	0.8376	0.8231	0.822	0.8251
	LiLT	English only	2M	–	–	–	–	–	–	–	–	0.7963
	GraphMLLM	English only	2M	0.8403	0.9080	0.8034	0.7954	0.8374	0.8458	0.8481	0.8301	0.8386
	GraphMLLM	English only	11M	0.8553	0.9041	0.7971	0.8222	0.8578	0.8666	0.8581	0.8444	0.8507
RE	LayoutXLM	Multilingual	30M	0.5483	0.7073	0.6963	0.6896	0.6353	0.6415	0.6551	0.5718	0.6432
	LiLT	English only	11M	0.6276	0.7297	0.7037	0.7195	0.6965	0.7043	0.6558	0.5874	0.6781
	GraphMLLM	English only	2M	0.6462	0.7734	0.7178	0.6832	0.6781	0.7172	0.6744	0.5888	0.6849
	GraphMLLM	English only	11M	0.7116	0.7657	0.7173	0.7327	0.7142	0.7123	0.6685	0.6100	0.7040

Table 3. Cross-lingual zero-shot transfer F1 accuracy on FUNSD and XFUND.

Task	Model	Pretrain Docs		FUNSD	XFUND							Avg.
		Language	Size	EN	ZH	JA	ES	FR	IT	DE	PT	
SER	LayoutXLM	Multilingual	30M	0.7940	0.6019	0.4715	0.4565	0.5757	0.4846	0.5252	0.5390	0.5561
	LiLT	English only	11M	0.8415	0.6152	0.5184	0.5101	0.5923	0.5371	0.6013	0.6325	0.6061
	GraphMLLM	English only	2M	0.8403	0.6102	0.5118	0.5104	0.6030	0.5446	0.5854	0.6387	0.6056
	GraphMLLM	English only	11M	0.8553	0.6404	0.5266	0.5374	0.6507	0.5953	0.6356	0.6353	0.6346
RE	LayoutXLM	Multilingual	30M	0.5483	0.4494	0.4408	0.4708	0.4416	0.4090	0.3820	0.3685	0.4388
	LiLT	English only	11M	0.6276	0.4764	0.5081	0.4968	0.5209	0.4697	0.4169	0.4272	0.4930
	GraphMLLM	English only	2M	0.6462	0.5667	0.5811	0.5453	0.5852	0.5022	0.4912	0.4330	0.5439
	GraphMLLM	English only	11M	0.7116	0.6395	0.6405	0.6169	0.6814	0.5919	0.5680	0.5255	0.6219

5.3   Implementation Details

For word-level and segment-level layout, we use the k-Nearest Neighbours algorithm to build the graph, with k values set to 100 and 50, respectively. While for regional-level layout, we initially employs a rule-based method which clustered regions based on the distance between segments. However, we found significant inconsistency between documents, thus we ultimately resorted to utilizing bounding boxes containing all text to provide the global layout feature of the whole document. We use a 12-layer, 12-heads text-layout model, where the hidden layer dimensions for text layers and layout layers are set as 768 and 192, respectively. The parameters k_w, k_s, and k_r of each layout level are all set to 4.

GraphMLLM is pre-trained using Adam with the learning rate $2e^{-5}$, weight decay $1e^{-2}$, and $(beta1, beta2) = (0.9, 0.999)$. The learning rate is linearly warmed up over the first 10% steps then linearly decayed. We set the batch size as 80 and train GraphMLLM for 2 epochs on the partial/full IIT-CDIP dataset using 4 NVIDIA A6000 48GB GPUs. Parameters of text flow are initialized with RoBERTa$_{BASE}$ [15], while parameters of layout flow are initialized from random. For multilingual downstream tasks, we load InfoXLM$_{BASE}$ [3] parameters to initialize the text flow, making GraphMLLM multi-language capable.

Table 4. Multitask fine-tuning F1 accuracy on FUNSD and XFUND.

Task	Model	Pretrain Docs		FUNSD	XFUND							Avg.
		Language	Size	EN	ZH	JA	ES	FR	IT	DE	PT	
SER	LayoutXLM	Multilingual	30M	0.7924	0.8755	0.7964	0.7798	0.8173	0.8210	0.8322	0.8241	0.8201
	LiLT	English only	11M	0.8574	0.9047	0.8088	0.8340	0.8577	0.8792	0.8769	0.8493	0.8585
	GraphMLLM	English only	2M	0.8737	0.9113	0.8079	0.8523	0.8854	0.8806	**0.8881**	0.8603	0.8700
	GraphMLLM	English only	11M	**0.8920**	**0.9178**	**0.8194**	**0.8573**	**0.9013**	**0.9033**	0.8830	**0.8699**	**0.8805**
RE	LayoutXLM	Multilingual	30M	0.6671	0.8241	0.8142	0.8104	0.8221	0.8310	0.7854	0.7044	0.7823
	LiLT	English only	11M	0.7407	0.8241	0.8345	0.8335	0.8466	0.8458	0.7878	0.7643	0.8125
	GraphMLLM	English only	2M	0.8298	**0.8946**	0.8456	**0.8533**	**0.8860**	**0.8641**	**0.8315**	**0.7836**	0.8486
	GraphMLLM	English only	11M	**0.8867**	0.8911	**0.8756**	0.8472	0.8791	0.8468	0.8301	0.7828	**0.8549**

5.4   Main Results

Language-Specific Fine-Tuning   First, we conducted experiments for GraphMLLM using different levels of layout information. The SER results of English dataset fine-tuning in Table 1 show that GraphMLLM using multi-level layouts performs better than using word-level layout alone. This justifies that using multi-level layout features can effectively improve the model's ability to understand documents. With only text and layout information as inputs, GraphMLLM can achieve competitive results using less training data than previous methods.

To validate the cross-language capability of GraphMLLM, we also evaluate it on XFUND. Table 2 shows that GraphMLLM with less pre-training data meets or exceeds the performance of previous multilingual models, which illustrates the superior cross-language capability and efficient data utilization of GraphMLLM. In comparison to LiLT, which solely relies on single-granularity layout information, GraphMLLM exhibits superior performance even with less pre-training documents. This underscores the significance of multi-level graph-based structure in modeling document layout.

Zero-Shot Transfer Learning.   Table 3 presents the results of cross-language zero-shot transfer learning. According to the results, the GraphMLLM model has a outstanding zero-shot transfer capability without applying multiple language documents for pre-training, and it outperforms previous counterpart models. Due to the multi-level layout flow, the GraphMLLM model is able to effectively model the layout structure of documents.

Multi-task Fine-Tuning.   Table 4 shows the experimental results of GraphMLLM fine-tuned using data from eight languages. GraphMLLM achieves optimal results, indicating that the model is capable of efficiently processing multiple language documents simultaneously and benefits from this ability. Compared with the results in Table 2, multi-task fine-tuning can further improve the performance on dataset of each language, implicitly showing that GraphMLLM can benefit from more fine-tuning data even though they have different languages.

Table 5. Ablation study on the effect of multi-level page layouts.

Task	Layout	FUNSD	XFUND							Avg.
		EN	ZH	JA	ES	FR	IT	DE	PT	
SER	W	0.7243	0.8995	0.7984	0.6879	0.7362	0.7283	0.7526	0.7377	0.7596
	WS	0.8367	0.9067	0.8019	0.7979	0.8333	0.8409	0.8496	0.8245	0.8364
	WSR	0.8403	0.9080	0.8034	0.7954	0.8374	0.8458	0.8481	0.8301	0.8386

Table 6. Ablation study on the effect of dual-stream inter-modal interaction.

Task	Inter-modal	FUNSD	XFUND							Avg.
		EN	ZH	JA	ES	FR	IT	DE	PT	
SER	w/o	0.6996	0.8918	0.7833	0.6424	0.7146	0.6885	0.7009	0.7007	0.7277
	w/	0.8403	0.9080	0.8034	0.7954	0.8374	0.8458	0.8481	0.8301	0.8386

5.5 Ablation Studies

We pre-trained GraphMLLM with 2M documents randomly selected from the IIT-CDIP dataset for ablation experiments. The experiments were conducted mainly on the multilingual datasets FUNSD and XFUND, using the language-specific task setting.

The results in Table 5 show that adding segment-level layout information significantly improves the performance. This is due to the fact that the granularity of word-level layout is too small to capture the high-level layout features. However, the improvement from adding region-level layout is somewhat marginal, likely because the page-level region layout is too coarse to effectively represent structural information about the document. More robust region extraction methods can be considered in the future to obtain more accurate and meaningful regions.

To verify the effect of inter-modal interaction in our dual-stream, we also conducted a simple comparison experiment, results showing in Table 6. We can see without inter-modal interaction, GraphMLLM's performance will be severely degraded, showing that the interaction between different modalities is essential for the performance.

6 Conclusion

In this paper, we present GraphMLLM, a multi-modal document pre-training model that can integrate multi-level layout information for document understanding tasks. Following the idea of decoupling text and layout information, GraphMLLM utilizes a dual-stream structure and models multi-level layout features through hierarchical graph attention networks. The text flow can reuse existing pre-trained language model, which can effectively reduce the quantity of dataset required for pre-training. After pre-training with monolingual documents, GraphMLLM can generalize to multilingual downstream tasks by leveraging off-the-shelf multilingual language models, as long as the OCR engine can provide multilingual text and hierarchical layout information. Experimental results on multiple datasets demonstrate the effectiveness and superiority of our proposed approach.

Despite its effectiveness, GraphMLLM still has two main limitations. The first one is the dependence on OCR results, especially on segment-level layouts. Different OCR engines may generate different results, and the OCR quality may affect the model's performance significantly. The second one is the absence of visual features. Document images have rich visual features, which can be integrated into the model in the future to enhance the model's capability to perform end-to-end visual document understanding tasks.

Acknowledgements. This work has been supported by the National Key Research and Development Program Grant 2020AAA0109700, and the National Natural Science Foundation of China (NSFC) Grant U23B2029.

References

1. Appalaraju, S., Jasani, B., Kota, B.U., Xie, Y., Manmatha, R.: Docformer: end-to-end transformer for document understanding. In: International Conference on Computer Vision, pp. 973–983 (2021)
2. Bao, H., Dong, L., Piao, S., Wei, F.: Beit: bert pre-training of image transformers. In: International Conference on Learning Representations (2021)
3. Chi, Z., et al.: InfoXLM: an information-theoretic framework for cross-lingual language model pre-training. In: 2021 Conference of the North American Chapter of the Association for Computational Linguistics: Human Language Technologies, pp. 3576–3588 (2021)
4. Devlin, J., Chang, M., Lee, K., Toutanova, K.: BERT: pre-training of deep bidirectional transformers for language understanding. In: 2019 Conference of the North American Chapter of the Association for Computational Linguistics: Human Language Technologies, pp. 4171–4186 (2019)
5. Gu, Z., et al.: Xylayoutlm: towards layout-aware multimodal networks for visually-rich document understanding. In: IEEE/CVF Conference on Computer Vision and Pattern Recognition, pp. 4583–4592 (2022)
6. He, P., Liu, X., Gao, J., Chen, W.: Deberta: decoding-enhanced Bert with disentangled attention. In: The 9th International Conference on Learning Representations (2021)
7. Hong, T., Kim, D., Ji, M., Hwang, W., Nam, D., Park, S.: BROS: a pre-trained language model focusing on text and layout for better key information extraction from documents. In: The 36th AAAI Conference on Artificial Intelligence, pp. 10767–10775 (2022)
8. Huang, Y., Lv, T., Cui, L., Lu, Y., Wei, F.: Layoutlmv3: pre-training for document AI with unified text and image masking. In: The 30th ACM International Conference on Multimedia, pp. 4083–4091 (2022)
9. Jaume, G., Ekenel, H.K., Thiran, J.: FUNSD: a dataset for form understanding in noisy scanned documents. In: The 2nd International Workshop on Open Services and Tools for Document Analysis, pp. 1–6 (2019)

10. Lewis, D.D., Agam, G., Argamon, S., Frieder, O., Grossman, D.A., Heard, J.: Building a test collection for complex document information processing. In: The 29th Annual International ACM SIGIR Conference on Research and Development in Information Retrieval, pp. 665–666 (2006)
11. Li, C., et al.: StructuralLM: structural pre-training for form understanding. In: The 59th Annual Meeting of the Association for Computational Linguistics and the 11th International Joint Conference on Natural Language Processing, vol. 1: Long Papers, pp. 6309–6318 (2021)
12. Li, J., Xu, Y., Lv, T., Cui, L., Zhang, C., Wei, F.: Dit: self-supervised pre-training for document image transformer. In: The 30th ACM International Conference on Multimedia, pp. 3530–3539 (2022)
13. Li, P., et al.: Selfdoc: self-supervised document representation learning. In: IEEE/CVF Conference on Computer Vision and Pattern Recognition, pp. 5652–5660 (2021)
14. Li, Y., et al.: Structext: structured text understanding with multi-modal transformers. In: The 21st ACM Multimedia Conference on Multimedia, pp. 1912–1920 (2021)
15. Liu, Y., et al.: Roberta: a robustly optimized Bert pretraining approach. arXiv preprint arXiv:1907.11692 (2019)
16. Park, S., et al.: Cord: a consolidated receipt dataset for post-ocr parsing. In: Workshop on Document Intelligence at NeurIPS 2019 (2019)
17. Peng, Q., et al.: ERNIE-layout: layout knowledge enhanced pre-training for visually-rich document understanding. In: Findings of the Association for Computational Linguistics: EMNLP 2022, pp. 3744–3756 (2022)
18. Vaswani, A., et al.: Attention is all you need. Adv. Neural Inf. Process. Syst. **30**, 5998–6008 (2017)
19. Wang, J., Jin, L., Ding, K.: LiLT: a simple yet effective language-independent layout transformer for structured document understanding. In: The 60th Annual Meeting of the Association for Computational Linguistics, vol. 1: Long Papers, pp. 7747–7757 (2022)
20. Wang, J., et al.: Towards robust visual information extraction in real world: new dataset and novel solution. In: The AAAI Conference on Artificial Intelligence, pp. 2738–2745 (2021)
21. Wang, W., et al.: Ernie-mmlayout: multi-grained multimodal transformer for document understanding. arXiv preprint arXiv:2209.08569 (2022)
22. Xu, Y., et al.: LayoutLMv2: multi-modal pre-training for visually-rich document understanding. In: The 59th Annual Meeting of the Association for Computational Linguistics and the 11th International Joint Conference on Natural Language Processing, vol. 1: Long Papers, pp. 2579–2591 (2021)
23. Xu, Y., Li, M., Cui, L., Huang, S., Wei, F., Zhou, M.: Layoutlm: pre-training of text and layout for document image understanding. In: The 26th ACM SIGKDD Conference on Knowledge Discovery and Data Mining, pp. 1192–1200 (2020)
24. Xu, Y., et al.: Layoutxlm: multimodal pre-training for multilingual visually-rich document understanding. arXiv preprint arXiv:2104.08836 (2021)
25. Zhai, M., et al.: Fast-structext: an efficient hourglass transformer with modality-guided dynamic token merge for document understanding. arXiv preprint arXiv:2305.11392 (2023)
26. Zhang, Z., Ma, J., Du, J., Wang, L., Zhang, J.: Multimodal pre-training based on graph attention network for document understanding. IEEE Trans. Multimedia **25**, 6743–6755 (2023)

One-Shot Transformer-Based Framework for Visually-Rich Document Understanding

Huynh Vu The$^{(\boxtimes)}$ [ID], Van Pham Hoai [ID], and Jeff Yang [ID]

Cinnamon AI, Ho Chi Minh, Vietnam
vuthe_huynh@yahoo.com
https://cinnamon.is

Abstract. There is a growing need for efficient entity extraction (EE) from business documents. While recent EE models have shown good accuracy for a variety of document templates, fine-tuning these models and acquiring additional training data can be expensive. To address this problem, we propose a novel template-based system for the EE task which does not require model fine-tuning for new entities and templates. The system includes two one-shot transformer-based models: one for template recognition and the other for entity recognition. The document recognition model (OTDC) achieves high accuracy (over 93%) on more than 200 templates of public and private datasets. The entity recognition model (OTER) outperforms recent zero-shot models with regards to the full set of labeled entities in the public SROIE datasets. We have also gathered and annotated the public RVL-CDIP and invoice datasets to showcase the generalization of our OTER models for the EE task across a wide range of document templates, containing both single and multiple-region fields.

Keywords: Entity extraction · Visually-Rich document understanding · One-shot approaches

1 Introduction

Businesses handle a large number of Visually-Rich Documents (VrDs) in physical formats on a daily basis. Working with the physical documents can be time-consuming, which is why digital transformation processes are implemented to improve the efficiency of document-related tasks. The EE task from digital documents has been gaining attention due to its potential applications in organizing, archiving, and indexing a variety of documents. The methods for the EE task can be divided into two main categories: template-free and template-based approaches.

Template-free approaches [15, 16, 23, 32, 35–37, 39] for structured information extraction (IE) remove the burden of template preparation in the training phase. Although good results have been achieved, these approaches still face challenges related to data availability and high training and labeling costs. In real-world scenarios, sufficient training data is not always available while some data is masked for security issues. Additionally, training data may be provided infrequently over time. Therefore, a model is required to be fine-tuned several times to maintain good accuracy for both old and new templates as well as new target entities. This leads to increased training costs and labeling effort. Without fine-tuning on new datasets, these template-free approaches may not be able to extract new fields or fields on the new templates of those datasets.

Recent template-based approaches in one-shot learning, such as those proposed by [4,38], have shown promise in significantly reducing the training and labeling effort by requiring only a few training samples. These methods classify data into templates, with template recognition being a requirement during inference. However, it is unclear how accurate template recognition is, and some level of fine-tuning is still required. Furthermore, the results are demonstrated on a limited set of templates from the public SROIE dataset [17].

To tackle the challenges previously mentioned, we propose a novel non-tuning system for the EE task, evaluating on a wide range of document templates on public datasets. Our system consists of an OTDC model for document recognition and an OTER model for entity recognition. The OTDC model is trained to recognize 254 fine-grained document templates, each of which has a unique arrangement of keys that can be described as a 'fingerprint' for the template. Our recognition experiments show that this approach has the potential to accurately recognize new templates without any further tuning. The current template-free approaches struggle to train general EE models due to the increasing complexity of document patterns and key-value pair relations. Fine-grained document recognition can significantly reduce the search space for the EE task and improve performance.

The QATM model uses a QA-based approach to extract information from a target document, given a reference document and a reference template configuration. The configuration specifies the keys and values to be extracted, which can be either paired together or stand alone (as shown in Fig. 1). The model learns to map the reference keys and values to corresponding entities in the target document. The mapping is inspired by an observation that if reference entities are provided, corresponding ones can be found in the target document regardless of the templates as well as the language of the documents. This is valuable for the QATM model, as it can be generalized and applied to various document formats and templates. The model combines the entities via textual, visual, and layout features, and can leverage these fused features to achieve high mapping accuracy. The contribution of this paper is summarized as follow

- A novel transformer-based system for the EE task with simple configuration.
- A one-shot fine-grained document recognition model. To the best of our knowledge, the paper is the first to train the model on a large number of fine-grained document categories, which achieved high top1 score on both public and private datasets.

- A novel one-shot transformer-based QA-based model for entity recognition, which requires no further model fine-tuning for new templates and new entities. The model outperforms non-tuned model-based approaches for the EE task on the public datasets.
- A public dataset for one-shot document recognition which contains 1400 images that are categorized into 97 fine-grained templates. The dataset will be publicly available.
- Three labelled public datasets for the one-shot EE task. The datasets are classified into 100 templates, covering a range of document types from receipts to invoices. The datasets have been annotated for single and multiple-region fields and will be made available to the public.

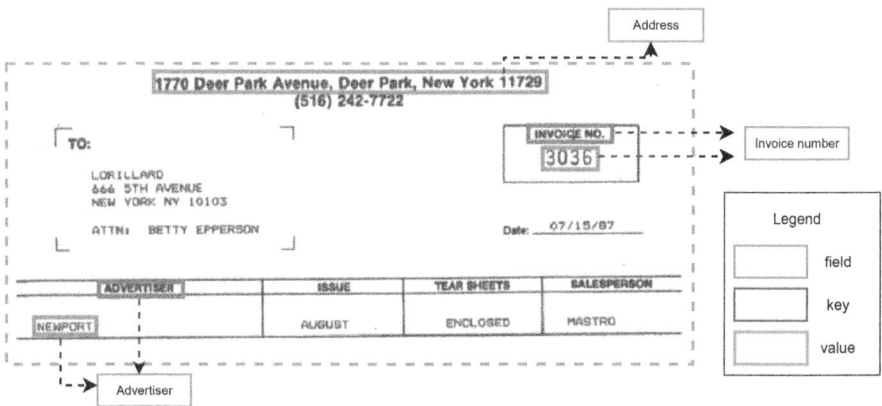

Fig. 1. The figure is an example of an invoice with an illustrated configuration. The configuration details the location of keys, values, and their corresponding fields. Some keys and values are paired together, such as the advertiser field, while others, such as the address field, stand alone.

2   Related Works

One-Shot Fine-Grained Document Recognition. Document recognition approaches, such as DocFormer [1] and LayoutLM [37], involve training a classification model of only 15 categories with considerable variation within each class. [30] attempted to perform one-shot document recognition with 74 classes, but they were limited to identifying only a specific type of document, the identity document. Creating a comprehensive set of fine-grained document categories is challenging and may result in class duplication. We trained a one-shot document recognition model using 254 fine-grained classes of Japanese documents, including invoices and receipts. Our OTDC model has the potential to create a generalized and robust one-shot document recognition model.

Template-Based Approach. The earliest approaches [6,7,27,29,31] to the EE task required a lot of knowledge engineering, such as creating rules based on regular expressions, numerical constraints, and lexical knowledge. While commercial products like smartFix [7] can handle a large amount of documents, they require experienced users and high levels of configuration. For smaller-scale EE tasks, INTELLIX [29] proposes a trainable system using indexers for common fields, but it is limited in its ability to handle different document layouts. One challenge of these initial rule-based approaches is extracting fields that are located in crowded regions or have multiple text regions.

Recent one-shot methods [4,38] for document analysis are model-based approaches. Cheng et al. [4] proposed an approach that uses an attention mechanism with conditional random fields [21] to map layout information and fields among document images. However, this approach does not work well for fields that are printed in unexpected positions, are slightly skewed, or do not match any fields in reference documents. To tackle this problem, Yao et al. [38] introduce a partial graph matching algorithm that leverages multi-modal features of position, text, and shape to compare the score of a field in a reference and input document. However, their approach cannot deal with multiple-region fields that have many text regions since it applies a one-to-one constraint. In a recent study, [12] proposed an approach that utilizes large language models to address the issue of multiple-fields regions. While the model requires fine-tuning with a few training samples, it has demonstrated high accuracy on public datasets. However, the large size of the model's weight compared to traditional EE models [16,36], can be a limiting factor when deploying it on commercial products.

In their proposal of the zero-shot approach, [2] suggest a two-stage architecture that can extract structural information from document forms. Their model was trained on Wikipedia Infobox, the largest online encyclopedia, in order to learn the relationship between key and value. The training dataset only contains easy and simple key-value alignments, with no difficult or complex relationships included. In addition, public evaluation of the approach is limited, only two over four fields of the public SROIE dataset [17] is evaluated.

Template-Free Approach Text-based frameworks such as those proposed by [25,28] have formulated the EE task as a sequential labeling problem. The text is segmented in optimal order before passing through a subsequent BiLSTM-CRF layer [22]. While the plain text framework provides a strong baseline for academic datasets, it has limitations in real-world applications. Text alone is not sufficient to provide comprehensive semantic information. For instance, numerical segments of text may represent varying amounts in invoices, and without spatial references, it is challenging to make accurate predictions across diverse layouts.

Various learning-based techniques that incorporate both text and spatial features have been proposed for the EE task. Graph neural networks have been used in these approaches [3,18,19,24]. In graph-based approaches, the document is represented as a graph, with text features as nodes and spatial relationships among the text segments (such as left, right, top, bottom, text width, and text height) as edges. This approach preserves the spatial relationship of document content. However, the shallow nature of the network constrains the propagation of relational information among edges, which limits the spatial modeling of distant nodes. To overcome this issue, graph-based attention has been applied to VrDU tasks [3, 9, 24] to model non-local entities in a document [33].

Recent research investigates the use of vision-based transformer architecture [20] to comprehensively understand the semantic structure of VrDs, integrating textual, visual, and layout information [15, 23, 34, 36, 37]. This approach is exemplified by LayoutLM [37] and LayoutLMV2 [36], which extract image features using a bottom-up approach through a detection model such as Faster-CNN [26], and then integrate these with textual and layout features.

3   System Overview

As shown in Fig. 2, the system is made up of three components: the OTDC feature extraction model, the document recognition module, and the OTER entity extraction model. To begin, an input document is processed by public text segmentation and OCR models to obtain its corresponding optical character recognition (OCR). The OCR contains a sequence of texts and corresponding coordinates that are then used to generate embeddings through the OTDC model, resulting in a representation of the input document. The document recognition module matches these embeddings with a database of embeddings to identify the template of the input document. Based on the detected template, the system looks up a corresponding template configuration that specifies the location of keys, values, and the name of the fields to be extracted (Fig. 1). The OTER models then extract the entities from the OCR information based on the reference configuration.

3.1   The OTDC Feature Extraction Model

The OTDC model is designed to generate a unique representation of a document, or 'fingerprint', through embeddings. Its architecture is similar to the OTER model, consisting of a text encoder, a location encoder, and the Transformer1, as described in following OTER model architecture. However, the OTDC model uses a fusion of multi-modal features, including layout, text, and visual features, instead of only visual features. The use of GRU in the text feature encoder and a small number of Transformer1 layers results in a compact model with only 21M parameters, compared to the 113M parameters in LayoutLM [37] or LayoutLMv2 [36].

3.2   The Document Recognition Module

The module matches input embeddings with all embeddings stored in a database to identify a matching template. Each template in the database has a number of corresponding embeddings. The accuracy of the recognition module is influenced by the feature extraction of the OTDC models and the number of embeddings associated with each template.

3.3   The OTER Entity Extraction Model

The overall model architecture can be seen in Fig. 3. The architecture includes a text encoder, a location encoder, and two transformers. The input to the model is a combination of the OCR information from a reference document and an input document, which is encoded by the text and location encoders to generate text embeddings (TE) and location embeddings (LE). The combined embeddings (TE + LE) are self-attended by Transformer1 to produce the output fused feature (F). In the final layers of the model, certain segments (e.g., the $Fr4$ in Fig. 3) of the fused features F are extracted as queries based on reference keys or values specified in the template configuration. These queries are then attended to the fused features (F) and passed through a classifier to identify entities in the input document.

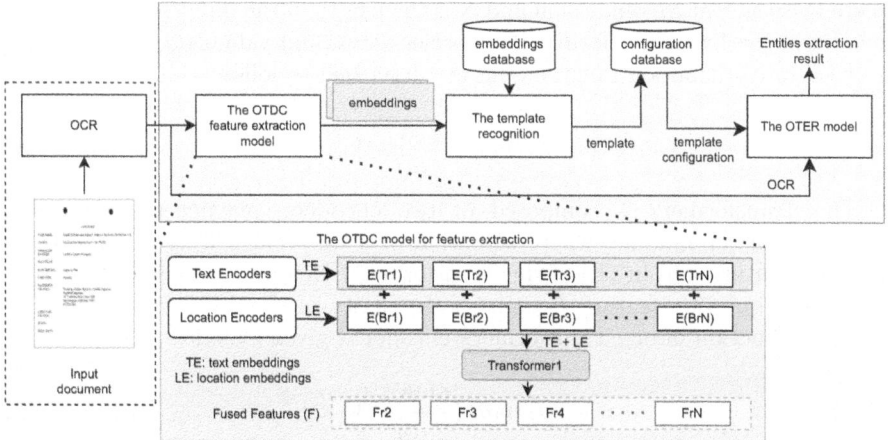

Fig. 2. The components of the proposed EE system, which are the OTDC module, the document recognition module, and the OTER module.

Text Feature Encoder. While previous work [23,36,37] has relied on pre-trained BERT-based language models [8] to encode text features, we propose a lightweight alternative. Our text feature encoder utilizes learnable embedding layers and two bidirectional gated recurrent units (GRUs) [5]. It takes as input a sequence of text segments, denoted as

$$S_{text} = \{Tr_1, Tr_2, .., Tr_{pad}, Ti_1, Ti_2, .., Ti_{pad}\} \tag{1}$$

where Tr_2 refers to the second segment of the reference sequence and Ti_2 refers to the second segment of the input sequence. To ensure equal lengths among different text sequences in batch implementation with several target images, padding (T_{pad}) is added. The text embeddings (TE) is obtained by the formula.

$$TE = Emb_{text}(S_{text}) \tag{2}$$

Location Feature Encoder. The location feature encoder of the model includes several embedding layers that can learn and adapt to the input sequence of location boxes. The input sequence S_{loc} consists of a set of boxes, each represented by four corner points. These corner points are denoted as Br_j, and each box has four corner points denoted as $v_j^{tl}, v_j^{tr}, v_j^{b}, v_j^{br}$. The coordinates of each point are normalized by the image height and width before the embedding layers are applied. The location embeddings (LE) is obtained by the following formula

$$LE = Linear(Concat(Norm((v_i^{tl}, v_i^{tr}, v_i^{bl}, v_i^{br})))) \tag{3}$$

Transformers. The system includes two Transformers, each with 8-head encoders and a hidden size of 256. The Transformer1 also has 8 encoders and takes a combination of text and location embeddings $LTE \in R^{N \times D}$ as input, where D is the feature dimension and N is the length of the input sequence. Each encoder in the first Transformer has queries, keys, and values (Q_1, K_1, V_1) that are used in its attention function and are described as follow

$$softmax(\frac{Q_1 K_1^T}{\sqrt{d}})V_1 \tag{4}$$

The Transformer2 is composed of two encoders, playing the role as a Question-Answer module, where attention of a question feature is applied on the fused feature output (F) of the Transformer1. Denote $K_2 and V_2$ as keys and values of a question feature F and Q_2 as queries of the fused feature F. The attention function of one Transformer2 encoder is given as follow

$$softmax(\frac{Q_2 K_2^T}{\sqrt{d}})V_2 \tag{5}$$

4   Experiments

4.1   Datasets for the Document Recognition (DC) Task

Private Dataset (DCPrivate). The dataset is made up of around 2850 Japanese business documents, including invoices, receipts, contracts, and more. These documents are divided into 525 specific categories, each containing at least two samples. During training, 254 of these categories are used, while the remaining 271 are held for testing purposes.

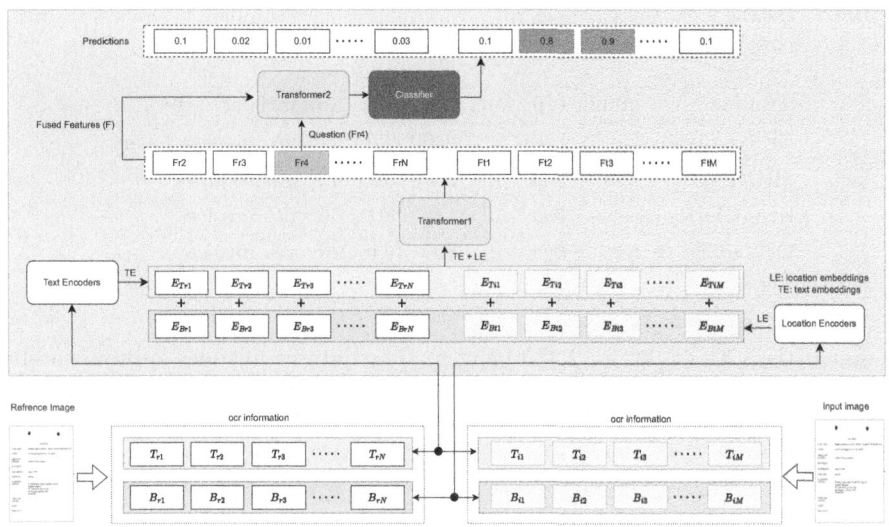

Fig. 3. The OTER model consists of a text encoder, a location encoder, and two transformers. The text and location encoders encode the input which is a combination of OCR information from a reference document and an input document. The encoders generates text embeddings (TE) and location embeddings (LE). Transformer1 self-attends to the combined embeddings (TE + LE) to produce the output fused feature (F).

Public Dataset (DCPublic). The dataset is gathered from three different public English datasets: SROIE [17], FUNSD [10], and the invoice test set of RVL-CDIP [11]. The dataset is comprised of 100 categories and approximately 670 images in total. Of these categories, 42 came from SROIE, 15 came from FUNSD, and 43 came from RVL-CDIP.

To ensure a good evaluation, the categories in the training and testing sets of DCPrivate should not overlap. Furthermore, the language used in the documents of DCPublic and DCPrivate is different, and the DCPrivate training set is used as a pre-training dataset for evaluating performance on the DCPublic dataset. We evaluate the document recognition on both public and private dataset.

4.2 Datasets for the EE Task

As shown in Table 1, the EE dataset contains two private datasets (EE_os_train, EE_os_test) and three public datasets that will be released to the public.

EE_os_train. This dataset comprises 2,850 Japanese documents, such as health checkups, invoices, insurance contracts, and medical receipts. It is not publicly available and is organized into 525 templates, with 454 templates allocated for

Table 1. Datasets for the EE task with #ns denotes as #samples, #ne as #entities, #nt as #templates, and mrf as multi-region fields

datasets	public	type	lang.	#ns	#ne	#nt	mrf	doc. type
EE_os_train	no	train	ja	2400	41700	454	yes	invoice, forms
SROIE_os	yes	test	en	208	1500	42	yes	receipts
RLV-CDIP_os	yes	test	en	142	480	39	no	invoice
INVOICE_os	yes	test	en	80	1640	20	yes	invoice

training (the EE_os_train, 2,400 images). The dataset includes approximately 45,000 annotated entities, encompassing both single-region and multiple-region fields.

SROIE_os. This dataset contains 208 English receipts from the test set of the public SROIE dataset [17]. Each receipt is annotated with value-only information about four types of entities: company, address, date, and total. The receipts are sorted into 42 templates, with a total of 1500 labelled entities across the dataset.

RLV-CDIP_os. We obtained our dataset by selecting 142 samples from the publicly available RLV-CDIP dataset [11]. These samples belong to the invoice group and are categorized into 39 templates. We annotated a total of 480 single-region entities in the samples.

INVOICE_os. We analyzed 20 publicly available excel templates and created 80 invoice samples, with each template containing 3 to 6 samples. These samples were annotated for both single-region and multiple-region fields.

4.3   Dataset Preparation and Experiment Setup

The experiments were performed on a single NVIDIA Tesla T4 with 16G of memory. During training, the learning rate was initially set to 10^{-4} and then gradually decreased using the CosineAnnealingLR scheduler. The KLDivLoss was used during training, and AdamW was used as the optimizer.

For the OTDC Model. Template labels are required in the training and evaluation of the OTDC model. A grouping tool is utilized to label templates for the DC task. By using the feature layers of the OTDC models, the tool extracts embeddings from input images. With a reference image, the tool sorts the remaining images based on the closest embedding distances. The sorting results highlight images that have small embedding distances with the reference image. Users can then visually examine the top-ranking results to determine if an image belongs to the same group as the reference image. This iterative process allows for the recursive acquisition of a set of detailed templates.

During each training step of the OTDC models, the data-loader selects three samples: two from the sample template and one from a different template, and the training process is supervised using triplet loss [14].

For the OTER Model. The OTER task involves different types of labels, including template labels, OCR labels, and entity labels. Template labels indicate the category of a specific sample, while OCR labels provide the corresponding text and its location. Entity labels specify field names such as address, invoice number, or advertiser. During training, the training data-loader selects a pair of samples from the same template, with one serving as the reference sample and the other as the target sample. During testing, to ensure consistent comparison of results across experiments on a test dataset, we choose a reference sample for each template, while the remaining samples of the template serve as the target samples.

4.4 Experiment Results for the DC Task

On the DCprivate Dataset. We compare the performance of our OTDC model, which takes in a sequence of location and text as input, with the baseline Resnet50 model, which takes in an image as input, on the private DCPrivate dataset. The results, shown in Table 2, indicate that the OTDC model outperforms Resnet50 by 3% (96.3% vs. 93.4%) while having only 14.2M parameters, which is half the number of parameters of Resnet50 with 25.6M parameters. The OTDC model also achieves a 0.5% higher top1 accuracy than Resnet50 when using image-based feature instead of the text and location features.

On the DCPublic Dataset. The OTDC model obtains a top1 score of 93.8% on the DCpublic dataset, which mainly consists of English documents. It is worth noting that the pre-training dataset contains Japanese documents. This result shows the robustness of the OTDC model across document templates and languages.

On the Number of Training Templates. The study investigates the effect of the number of training templates on the accuracy of the OTDC model on the DCprivate dataset. The results on Table 3 show that increasing the number of training templates resulted in higher accuracy. A 50-training-class model had a top1 accuracy of only 92.8%, while a 200-training-class model had a top1 accuracy of 96%.

Table 2. Results of the OTDC models on the DCPrivate and DCPublic dataset. Denote OTDC_im as the OTDC using image-based feature instead of the text and location features.

Methods	Dataset	top1	Params.
Resnet50 [13]	DCPrivate	93.4	25.6M
OTDC_im	DCPrivate	93.9	35.2M
Our OTDC	DCPrivate	**96.3**	**14.2M**
Our OTDC	DCPublic	93.8	14.2M

However, it is difficult to prepare a large number of fine-grained document templates without class duplication in real-world applications. Therefore, the other way to improve the model is by incorporating the layout of key into the model, since it can be represented as a document fingerprint.

Table 3. The effect of the number of categories on the accuracy of the OTDC model on the DCprivate dataset.

No. of categories	top1
50 training classes	92.8
100 training classes	94
150 training classes	95.8
200 training classes	96

Challenges in the DC Task. In Figs. 4 and 5, the challenges in the DC task for similar documents are illustrated. The green rectangles represent the regions with similar formats, while the red rectangle indicates the regions with different formats. In Fig. 5, the different-format region is small and poses difficulty for the models to distinguish. Consequently, the model treats the two images as a single template. In real-world scenarios, the definition of a template may lack clarity, leading to the categorization of the two images in Fig. 5 as the same format. To differentiate the two images as separate formats, it may be necessary to leverage the key layout from both images and utilize it for pattern classification. The key layout can be obtained through the EE task.

4.5   Experiment Results for the EE Task

Table 4 displays the experiment results on the public SROIE_os datasets. While recent SOTA models such as LayouLM_v3 [16] struggle to extract certain entities without fine-tuning, our OTER models outperform the zero-shot model

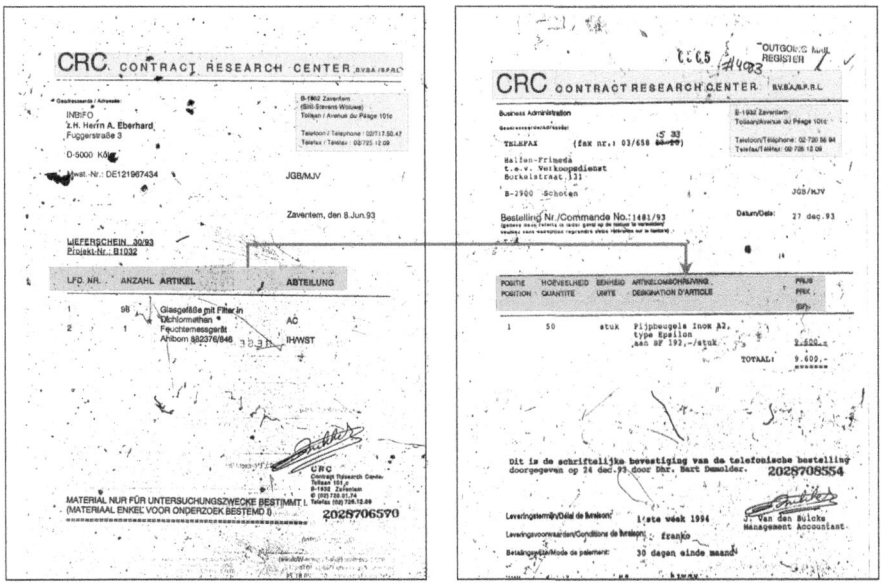

Fig. 4. The challenge of recognizing similar documents.

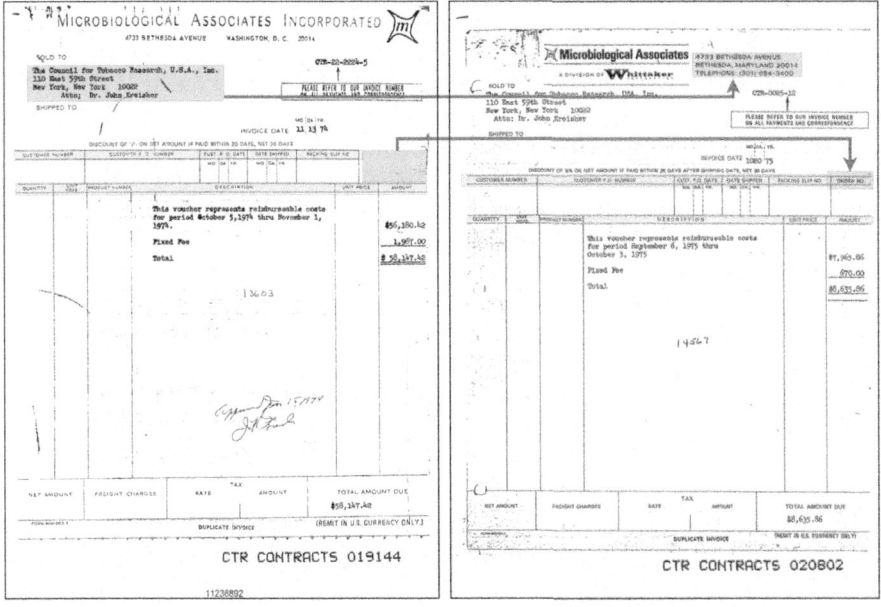

Fig. 5. The challenge of recognizing similar documents.

KATA-wiki [2] on the two fields of the SROIE dataset (telephone and total), achieving a large margin of success (79.7% vs 70.7%). In fact, KATA-wiki fails to obtain any result on the full set of four fields as originally labeled in the dataset. Our OTER models obtain F1-scores of 81.4% on the four fields using pre-trained models on the EE_os_train dataset.

In Table 5, we showcase the generalization ability of our OTER model by testing it on multiple public datasets, pre-training and evaluating it on different document types and languages. Our OTER models achieve good F1-scores (81.4%, 72.7%, 71.6%) on 100 new templates of public English datasets (SROIE_os, RLV-CDIP_os, INVOICE_os) even after being pre-trained on private Japanese documents of the EE_os_train dataset. To demonstrate that our models can generalize across different document types, we conduct a pre-training on INVOICE_os, which consists of business invoices, and evaluate the model on receipts documents of the SROIE_os dataset. Our model obtains high F1-scores of 68.15% on SROIE_os and 51.2% on RLV-CDIP. The lower results obtained from using pre-training on INVOICE_os compared to EE_os_train can be attributed to the limited amount of training data available in the INVOICE dataset. Our experiments indicate that our model can be applied to documents of any language.

Multiple-Region Fields. We have prepared the public INVOICE_os to evaluate the performance of OTER models on multiple-region fields, which constitute approximately 80% of the entities in the dataset. These fields are located within table areas, and there may be two or three tables within a single document. Table 5 displays that the OTER models achieve an F1 score of 71.6% on the multiple-region fields of the dataset. However, we have observed that the accuracy depends on the selection of reference document. Documents with a high number of table rows are more likely to generate higher accuracy on other documents of the same template. Additionally, the alignment of column values within a table may also impact the accuracy. In the future, we plan to train the models using multiple reference documents instead of a single one.

Key Fields. In analyzing fields, two types are considered: key fields and value fields. Table 5 shows that key fields have higher accuracy than value fields. The F1-score for both key and value fields on RLV-CDIP and INVOICE_os are only 72.7% and 71.6% respectively, while the number for key-only fields is 91% and 83.1%, respectively. Key fields have specific semantics, so it is easier for models to recognize them than value fields. The value of a field, such as a number, can be similar across different fields, which can confuse the models. The high accuracy of the key fields can be applied to detect the key layout of a document or can be used for document recognition, wherein each document template is represented by a specific layout of keys.

The key-only visualization of the EE task on the RLV-CDIP_os dataset is displayed in Fig. 6. The reference document is presented on the left, and the input document is shown on the right. The pink box in the input document signifies a True Positive (TP) of key extraction result. The results in the figure demonstrate that the key is extracted correctly despite the low image quality and the presence of duplicated keys. As depicted in the right document in the Figure, there are two "Amount" and two "ISSUE" texts in the input document, but only the "amount" and "ISSUE" that align with those in the reference document are extracted. Our EE models are capable of extracting the correct "amount" and "ISSUE". This ability may be attributed to the use of the fused feature of text and layout in the model architecture. One interesting aspect of key-only extraction is that the key layout of a document can be viewed as a document footprint, which can be used to differentiate among different document templates. Additionally, based on the detected keys and the document templates, we can infer the values.

Table 4. Results are available for the two fields (2F) (telephone and total) and the four fields (4F) (telephone, total, company, and address) of the public SROIE_os dataset. The KATA-wiki [2] approach provides results only for the two fields of the dataset.

Methods	Pre-train	F1
LayoutLM_v3 [16]	–	0
KATA-wiki (2F) [2]	–	70.7
The OTER (2F)	EE_os_train	**79.7**
The OTER (4F)	EE_os_train	**81.1**

Table 5. Results of the OTER on public and private datasets. *Note*: * indicates that the SROIE_os dataset contains value-only fields, ** indicates that the RLV-CDIP contains only single-region fields and the single output constraint with highest probability has been applied on this dataset.

Pre-train	Test	F1 (key&value)	F1 (key)
EE_os_train	SROIE_os *	81.4	–
EE_os_train	RLV-CDIP_os **	72.7	83.1
EE_os_train	INVOICE_os	71.6	91
INVOICE_os	SROIE_os	68.15	–
INVOICE_os	RLV-CDIP_os **	51.2	–

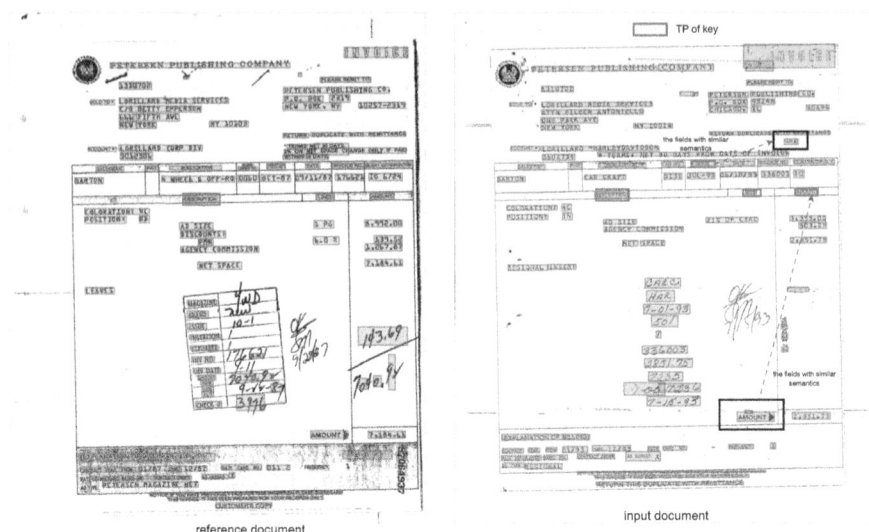

Fig. 6. The key-only EE results for images in the RLV-CDIP_os dataset.

5 Conclusion

We have developed a transformer-based system for the EE task that eliminates the need for model fine-tuning. Our system is capable of handling challenges associated with extracting multiple-region fields. It can perform one-shot document recognition on fine-grained document templates, achieving high accuracy (over 93%) on more than 200 templates of public and private datasets. Furthermore, our system can perform EE tasks for both new templates and new entities without the need for fine-tuning. Our EE models outperform non-tuned model-based approaches by a significant margin on various templates of public datasets.

Acknowledgments. We would like to express our deepest gratitude to Cinnamon AI for their immense support and invaluable assistance throughout this study. Their unwavering dedication and contributions have played a pivotal role in advancing our research endeavors and ensuring the successful completion of this project. We are truly thankful for their commitment to advancing knowledge in this field.

References

1. Appalaraju, S., Jasani, B., Kota, B.U., Xie, Y., Manmatha, R.: Docformer: end-to-end transformer for document understanding. In: Proceedings of the IEEE/CVF International Conference on Computer Vision, pp. 993–1003 (2021)
2. Cao, R., Luo, P.: Extracting zero-shot structured information from form-like documents: pretraining with keys and triggers. Proc. AAAI Conf. Artif. Intell. **35**, 12612–12620 (2021)

3. Carbonell, M., Riba, P., Villegas, M., Fornés, A., Lladós, J.: Named entity recognition and relation extraction with graph neural networks in semi structured documents. In: 2020 25th International Conference on Pattern Recognition (ICPR), pp. 9622–9627. IEEE (2021)
4. Cheng, M., Qiu, M., Shi, X., Huang, J., Lin, W.: One-shot text field labeling using attention and belief propagation for structure information extraction. In: Proceedings of the 28th ACM International Conference on Multimedia, pp. 340–348 (2020)
5. Chung, J., Gulcehre, C., Cho, K., Bengio, Y.: Empirical evaluation of gated recurrent neural networks on sequence modeling. arXiv preprint arXiv:1412.3555 (2014)
6. d'Andecy, V.P., Hartmann, E., Rusinol, M.: Field extraction by hybrid incremental and a-priori structural templates. In: 2018 13th IAPR International Workshop on Document Analysis Systems (DAS), pp. 251–256. IEEE (2018)
7. Deckert, F., Seidler, B., Ebbecke, M., Gillmann, M.: Table content understanding in smartfix. In: 2011 International Conference on Document Analysis and Recognition, pp. 488–492. IEEE (2011)
8. Devlin, J., Chang, M.W., Lee, K., Toutanova, K.: Bert: Pre-training of deep bidirectional transformers for language understanding. arXiv preprint arXiv:1810.04805 (2018)
9. Gui, T., et al.: A lexicon-based graph neural network for Chinese NER. In: Proceedings of the 2019 Conference on Empirical Methods in Natural Language Processing and the 9th International Joint Conference on Natural Language Processing (EMNLP-IJCNLP), pp. 1040–1050 (2019)
10. Jaume, G., Hazim Kemal Ekenel, J.P.T.: Funsd: a dataset for form understanding in noisy scanned documents. In: Accepted to ICDAR-OST (2019)
11. Harley, A.W., Ufkes, A., Derpanis, K.G.: Evaluation of deep convolutional nets for document image classification and retrieval. In: International Conference on Document Analysis and Recognition (ICDAR) (2015)
12. He, J., et al.: Icl-d3ie: in-context learning with diverse demonstrations updating for document information extraction. arXiv preprint arXiv:2303.05063 (2023)
13. He, K., Zhang, X., Ren, S., Sun, J.: Deep residual learning for image recognition. In: Proceedings of the IEEE Conference on Computer Vision and Pattern Recognition, pp. 770–778 (2016)
14. Hoffer, E., Ailon, N.: Deep metric learning using triplet network. In: Feragen, A., Pelillo, M., Loog, M. (eds.) SIMBAD 2015. LNCS, vol. 9370, pp. 84–92. Springer, Cham (2015). https://doi.org/10.1007/978-3-319-24261-3_7
15. Hong, T., Kim, D., Ji, M., Hwang, W., Nam, D., Park, S.: Bros: a pre-trained language model for understanding texts in document (2020)
16. Huang, Y., Lv, T., Cui, L., Lu, Y., Wei, F.: Layoutlmv3: pre-training for document AI with unified text and image masking. arXiv preprint arXiv:2204.08387 (2022)
17. Huang, Z., et al.: Icdar2019 competition on scanned receipt OCR and information extraction. In: 2019 International Conference on Document Analysis and Recognition (ICDAR), pp. 1516–1520. IEEE (2019)
18. Hwang, W., Yim, J., Park, S., Yang, S., Seo, M.: Spatial dependency parsing for semi-structured document information extraction. arXiv preprint arXiv:2005.00642 (2020)
19. Katti, A.R., et al.: Chargrid: towards understanding 2d documents. arXiv preprint arXiv:1809.08799 (2018)
20. Khan, S., Naseer, M., Hayat, M., Zamir, S.W., Khan, F.S., Shah, M.: Transformers in vision: a survey. ACM Comput. Surv. 54(10s), 1–41 (2022)

21. Lafferty, J., McCallum, A., Pereira, F.C.: Conditional random fields: probabilistic models for segmenting and labeling sequence data (2001)
22. Lample, G., Ballesteros, M., Subramanian, S., Kawakami, K., Dyer, C.: Neural architectures for named entity recognition. arXiv preprint arXiv:1603.01360 (2016)
23. Li, Y., et al.: Structext: structured text understanding with multi-modal transformers. In: Proceedings of the 29th ACM International Conference on Multimedia, pp. 1912–1920 (2021)
24. Liu, X., Gao, F., Zhang, Q., Zhao, H.: Graph convolution for multimodal information extraction from visually rich documents. arXiv preprint arXiv:1903.11279 (2019)
25. Palm, R.B., Winther, O., Laws, F.: Cloudscan-a configuration-free invoice analysis system using recurrent neural networks. In: 2017 14th IAPR International Conference on Document Analysis and Recognition (ICDAR), vol. 1, pp. 406–413. IEEE (2017)
26. Ren, S., He, K., Girshick, R., Sun, J.: Faster r-cnn: towards real-time object detection with region proposal networks. Adv. Neural Inf. Process. Syst. **28** (2015)
27. Rusinol, M., Benkhelfallah, T., Poulain dAndecy, V.: Field extraction from administrative documents by incremental structural templates. In: 2013 12th International Conference on Document Analysis and Recognition, pp. 1100–1104. IEEE (2013)
28. Sage, C., Aussem, A., Elghazel, H., Eglin, V., Espinas, J.: Recurrent neural network approach for table field extraction in business documents. In: 2019 International Conference on Document Analysis and Recognition (ICDAR), pp. 1308–1313. IEEE (2019)
29. Schuster, D., et al.: Intellix–end-user trained information extraction for document archiving. In: 2013 12th International Conference on Document Analysis and Recognition, pp. 101–105. IEEE (2013)
30. Simon, M., Rodner, E., Denzler, J.: Fine-grained classification of identity document types with only one example. In: 2015 14th IAPR International Conference on Machine Vision Applications (MVA), pp. 126–129. IEEE (2015)
31. Sunder, V., Srinivasan, A., Vig, L., Shroff, G., Rahul, R.: One-shot information extraction from document images using neuro-deductive program synthesis. arXiv preprint arXiv:1906.02427 (2019)
32. Tang, G., et al.: Matchvie: exploiting match relevancy between entities for visual information extraction. arXiv preprint arXiv:2106.12940 (2021)
33. Veličković, P., Cucurull, G., Casanova, A., Romero, A., Lio, P., Bengio, Y.: Graph attention networks. arXiv preprint arXiv:1710.10903 (2017)
34. Wang, J., et al.: Towards robust visual information extraction in real world: new dataset and novel solution. Proc. AAAI Conf. Artif. Intell. **35**, 2738–2745 (2021)
35. Wei, M., He, Y., Zhang, Q.: Robust layout-aware IE for visually rich documents with pre-trained language models. In: Proceedings of the 43rd International ACM SIGIR Conference on Research and Development in Information Retrieval, pp. 2367–2376 (2020)
36. Xu, Y., et al.: Layoutlmv2: multi-modal pre-training for visually-rich document understanding. arXiv preprint arXiv:2012.14740 (2020)
37. Xu, Y., Li, M., Cui, L., Huang, S., Wei, F., Zhou, M.: Layoutlm: pre-training of text and layout for document image understanding. In: Proceedings of the 26th ACM SIGKDD International Conference on Knowledge Discovery and Data Mining, pp. 1192–1200 (2020)

38. Yao, M., Liu, Z., Wang, L., Li, H., Zhuang, L.: One-shot key information extraction from document with deep partial graph matching. arXiv preprint arXiv:2109.13967 (2021)

39. Zhang, P., et al.: TRIE: end-to-end text reading and information extraction for document understanding. In: Proceedings of the 28th ACM International Conference on Multimedia, pp. 1413–1422 (2020)

EntityLayout: Entity-Level Pre-training Language Model for Semantic Entity Recognition and Relation Extraction

Chun-Bo Xu[1,2], Yi-Ming Chen[2], and Cheng-Lin Liu[1,3,4(✉)]

[1] School of Information Science and Technology, ShanghaiTech University, Shanghai 201203, People's Republic of China
xuchb2022@shanghaitech.edu.cn

[2] Hundsun Technologies Inc., Hundsun Center, 1888, Binxing Road, Binjiang District, Hangzhou, China
tony_chenheu@hotmail.com

[3] State Key Laboratory of Multimodal Artificial Intelligence Systems, Institute of Automation of Chinese Academy of Sciences, Beijing 100190, People's Republic of China

[4] School of Artificial Intelligence, University of Chinese Academy of Sciences, Beijing 100049, People's Republic of China
liucl@nlpr.ia.ac.cn

Abstract. Semantic entity recognition (SER) and relation extraction (RE) are the core tasks of information extraction from visually-rich documents (VrDs). Although self-supervised pre-training models have advanced the performance of these tasks, existing methods are insufficient in fusing the token features and modeling SER and RE jointly. In this paper, we propose an entity-level language model, named EntityLayout, and use a graph-based approach for efficient entity labeling and linking jointly on a continual pre-training model. In EntityLayout, a Token Fusion Module (TFM) is proposed to fuse the token feature and learn an entity-level representation. Then, we use the entity-level representation to build a graph and propose a Graph Pruning Module (GPM) to effectively prune the invalid links. Finally, SER and RE are accomplished simultaneously by a joint information extraction Task Module. Experimental results on public datasets FUNSD and CORD demonstrate that the proposed EntityLayout achieves competitive performance in SER and state-of-the-art performance in RE, i.e., SER F1 scores of 0.9108 and 0.9650, respectively, RE F1 scores of 0.8212 and 0.9898, respectively.

Keywords: Visual Information Extraction · Semantic Entity Recognition · Relation Extraction · Graph Pruning · Graph Neural Network

---

C.-B. Xu and Y.-M. Chen—Equal contributions. Work was done when Chun-Bo Xu was a Research Intern at Hundsun Technologies Inc.

E. H. Barney Smith et al. (Eds.): ICDAR 2024, LNCS 14804, pp. 262–279, 2024.
https://doi.org/10.1007/978-3-031-70533-5_16

1   Introduction

Semantic entity recognition (SER, also named Entity Labeling) [1–7] and relation extraction (RE, also named Entity Linking) [2,5,7–12] are playing an increasingly important role in document analysis in finance, education, and other fields. As shown in Fig. 1, the SER task determines the categories of entities in the text and the RE task is to judge the relationships between different semantic entities. The key to these tasks lies in their ability to precisely identify and associate text information from visually-rich documents (VrDs) [13–15], thereby significantly enhancing the efficiency and accuracy of data processing.

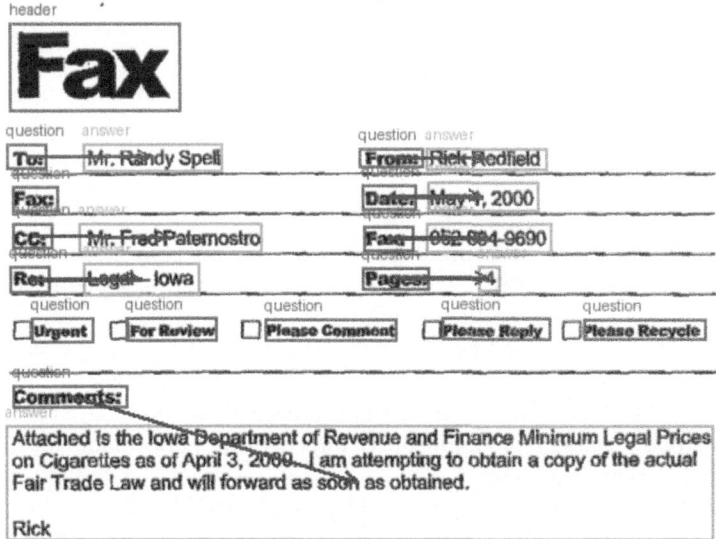

Fig. 1. Illustration of semantic entity recognition (SER) and relation extraction (RE). Semantic entities are classified into three labels: header, question, and answer, denoted by bounding boxes in red, green, and orange, respectively. The results of relation extraction are presented as key-value pairs, linked by blue arrows in the figure. Best viewed in color. (Color figure online)

SER and RE are at the core of visual information extraction (VIE). There are two main categories of approaches: graph neural networks (GNN) and Transformer Encoder based methods. GNNs [1–4,8–10,12] are widely applied in the modeling of structured documents. Unlike traditional deep neural networks, GNNs can operate directly on graph structures, effectively handling relationships between nodes (contextual information). The advantage of GNNs lies in aggregating the information of nodes with that of their neighboring nodes, updating the representation of each node through layerwise propagation. In SER and RE tasks, the entities and relationships are represented by nodes and edges, respectively. Using the Transformer Encoder, self-supervised pre-training [5,6,16–22] can be accomplished without relying on manual annotations, through, e.g., masked language modeling (MLM). These Transformer Encoder models have achieved remarkable success in mining deep semantic features. Models such as the LayoutLM series [5,6,23], StrucText series [18,19], BROS [20], LiLT [21], and GeoLayoutLM [7] significantly enhance their representation learning capabilities on visually rich documents through pre-training.

Despite the rapid development of self-supervised pre-training models in VIE tasks, existing models are lacking in some respects: 1) Existing pre-training models employ 1D position embedding to represent the text sequences. This is insufficient to integrate the semantic and layout features of visually rich documents, particularly those with complex, non-linear text arrangements. 2) The existing models tend to focus only on the first token of an entity, neglecting the information in the whole sequence of tokens. 3) Many existing models treat SER and RE tasks separately. This neglects the intrinsic correlation between SER and RE tasks, failing to harness the potential synergies that could improve the performance of two tasks.

In this paper, we propose an entity-level multimodal pre-training model named EntityLayout, aimed to model SER and RE tasks jointly. The model captures features at both the token level and entity level through a pre-training module, which is a continual pre-training model. EntityLayout introduces the Token Fusion Module (TFM), which fuses multiple token features within an entity into an entity-level feature, significantly enhancing the model's capability to process long text entities. For optimizing the graph structure, we also present the Graph Pruning Module (GPM), which prunes invalid links using the features generated by TFM. Via fusing the entity-level features and modeling SER and RE jointly, our method can significantly improve the performance of SER and RE in visually rich documents.

The contributions of this paper can be summarized as follows.

- A Token Fusion Module (TFM) is proposed to learn an entity-level representation by fusing token-level features after pre-training the Transformer Encoder.
- An efficient graph module is proposed for modeling SER and RE jointly, with a Graph Pruning Module (GPM) to effectively prune invalid links.
- We report competitive results on public datasets FUNSD and CORD, with SER F1 scores of 0.9108 and 0.9650, respectively, and RE F1 scores of 0.8212 and 0.9898, respectively.

2   Related Works

SER and RE tasks involve extracting key text content from documents that contain rich text, layout, and visual information, and further identifying the categories of entities and their inter-relations. Initially, researchers relied on graph neural networks (GNNs) to capture the relationships between text and layout features. In recent years, through the construction of a variety of pre-training tasks, self-supervised pre-training models significantly improve the learning efficiency of these features. Models based on the Transformer structure are drawing dominant attention.

2.1 GNN-Based Methods

Graph Neural Networks (GNNs) [24] are widely applied in modeling various structured data represented as graphs. GNNs can effectively aggregate contextual information through layerwise propagation to support inference by incorporating contexts. In SER and RE tasks, each entity corresponds to a node, and the relationships between entities are represented by the edges between nodes.

In the works of graph-based VIE, Liu et al. [1] represents the document as a fully connected graph, encoding the semantic and layout information of each text segment as the features of the graph's nodes and edges, with the prediction of entity categories obtained through a BiLSTM-CRF [25]. PICK [3] uses Transformer and CNN to embed text and visual features separately and realizes the interaction of these features with layout features through a graph network, ultimately predicting entity categories through BiLSTM-CRF as well. GraphDoc [4] encodes text and visual information with BERT [26] and Swin Transformer [27], respectively, followed by an attention-based graph network for feature interaction between entities. Methods [1,3,4] all utilize GNNs for the interaction of entity features, ultimately determining the entity categories. These methods have applied different modal types and explored various methods of modality encoding and alignment.

To enhance the GNN models' ability to learn layout information, several methods have studied how to prune the edges in the graph structure for improving the computation efficiency. Works [1,8] introduced edge features determined by the distance between bounding boxes of two entities. The methods in [1,3,4,8,28] connect all nodes with each other thus generating a massive graph containing a large number of redundant edges. During the execution of SER tasks [1,3,4,28], this full connectivity consumes a considerable burden of computation, and the noisy connections in the graph may deteriorate the performance of information extraction.

To address this issue, previous work [2,9–12] has employed various pruning methods. The method in [2] used the k-NN rule to create graph edges based on distances calculated from the top-left corner of the word bounding boxes. The method in [9] employed the beta-skeleton method, constructing the graph by allowing the propagation of document structure information. The methods in [10,11] use a line-of-sight heuristic strategy to establish relationships in information extraction tasks. The method in [12] considered layout information such as angles, distances, and IoU indices on top of k-NN rule to deal with entities that are in the same row or column in tabular documents. Although these heuristic-based pruning methods are effective, they cannot guarantee the completeness of useful connections in graph construction.

2.2 Transformer-Based Methods

LayoutLM [23] was the first pre-training model for VIE that interacts and jointly considers text and layout information. It uses axial embeddings to encode the absolute positions of text blocks and learns a token-level masked language model. Following [23], more pre-training models [5,6,16–19] have been proposed to integrate visual features with text features. The addition of visual information requires extra computation to process the raw document images but obtains only limited improvements compared to text and layout information. The method in [20] focuses on the combination of text and layout without relying on visual features, indicating that the effective integration of text and layout is still competitive. Current models [20] introduce the variable of input order directly through 1D position embedding, yet 1D position embedding is more suitable for continuous text input rather than discrete entity input scenarios in VrDs. The method in [22] has improved the VIE accuracy by changing the way the input entity order is altered. The existing models' pre-training transformer layers usually output text representations at the token dimension, while for entity-level SER and RE tasks, only the first token feature inside the entity is used, missing the information contained in subsequent tokens. This is unfavorable for entity-level information extraction. Our work aims to overcome this problem by fusing the token-level features into entity-level representation.

3 Methodology

3.1 Overall Arhitecture

The proposed EntityLayout model (as shown in Fig. 2) consists of three main parts: the Pre-training Module, the Graph Module, and the Task Module for information extraction tasks. The Pre-training Module, based on the Transformer Encoder architecture, is used to learn semantic and layout features. In the Graph Module, the token features extracted by the pre-training model are fused into entity-level features, so the entire document is modeled as a fully connected graph with entities as nodes. The fully connected graph is pruned using the Graph Pruning Module (GPM) so as to eliminate the invalid links. In the Task Module, SER is defined as a graph node classification task and RE as a graph edge classification task.

3.2 Pre-training Module

Pre-training tasks play a crucial role in learning contextual representations, where the model employs a Masked Language Model (MLM) as the pre-training mechanism. Since pre-training from scratch requires computing OCR results, which consumes substantial computational resources for result alignment, EntityLayout employs continual pre-training [29], which has the following benefits:

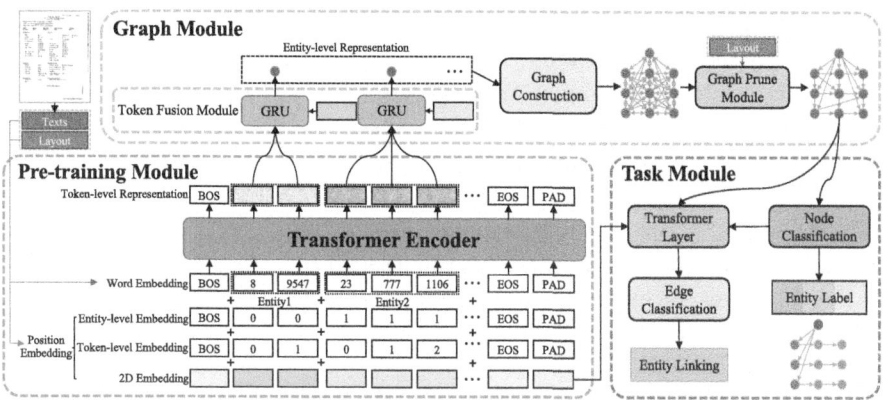

Fig. 2. An overview of EntityLayout, consisting of a Pre-training Module, a Graph Module, and a joint information extraction Task Module. The Pre-training Module additionally learns Token-level and Entity-level Position Embedding by Masked Language Modeling (MLM). The output token-level features are fused into entity-level through the Token Fusion Module (TFM). Then the entity-level graph is pruned by the Graph Pruning Module (GPM) to reduce invalid links. SER and RE are accomplished simultaneously by a joint information extraction Task Module.

- It saves a considerable amount of data processing costs.
- It retains most of the semantic understanding capabilities of the original pre-trained model.
- It concentrates on updating the token-level and entity-level position embeddings that we propose.

Inspired by two pre-trained models LayoutLMv3 [6] and BROS [20], we use the Masked Language Model (MLM) for pre-training. Based on the context information from the unmasked regions in the text, the Transformer architecture learns to predict the indices of text tokens in the masked text regions with cross-entropy loss $\ell_{\text{Pre-train}}^{\text{mask}}$ [30].

$$\ell_{\text{Pre-train}}^{\text{mask}} = -\sum_{c=1}^{V} y_{o,c}^{\text{mask}} \log(\hat{y}_{o,c}^{\text{mask}}), \tag{1}$$

where $y_{o,c}^{\text{mask}}$ is a one-hot vector representing the true class c for the masked observation o, $\hat{y}_{o,c}^{\text{mask}}$ is the predicted probability of class c for the masked token, and V is the vocabulary size.

To mitigate the forgetting of semantic information, we froze most of the parameters during the continual pre-training and only update the embedding parameters. The input of the pre-trained model mainly consists of token embedding, 2D position embedding, word-level position embedding, and entity-level position embedding.

The token embedding $\mathcal{E}_{\text{token}}$ is obtained as follows:

$$\mathcal{E}_{\text{token}} = \text{Embed}^{V_{\text{word}} \times d} (token_i), \tag{2}$$

where V_{word} is the word vocabulary size, d is the dimension of the embeddings, and $token_i$ is the index of the $i-$th token in the vocabulary after tokenization.

The 2D position embedding $\mathcal{E}_{2D\ position}$ is derived from the bounding box coordinates and their dimensions:

$$\mathcal{E}_{2D\ position} = \text{Embed}^{V_{coordinate} \times d} \left(f(x_{0,i}, y_{0,i}, x_{1,i}, y_{1,i}) \right), \tag{3}$$

where $V_{coordinate}$ is the coordinate size, d is the dimension of the layout embedding vector, and f is a function that normalizes all coordinates of $i-$th token by the size of images, including x-axis $[x_{0,i}, x_{1,i}]$, y-axis $[y_{0,i}, y_{1,i}]$, width and height $[x_{1,i} - x_{0,i}, y_{1,i} - y_{0,i}]$ features.

Unlike the LayoutLM [23], BROS [20], and similar series, in order to better learn the representations of both tokens and entities during the pre-training phase, EntityLayout's embedding uses a combination of word embedding and position embedding, as shown in the bottom-left part of Fig. 2. Position embedding includes token-level embedding, entity-level embedding, and 2D position embedding. Compared with LayoutLMv3, EntityLayout adopts a combination of token-level and entity-level position embeddings. The token-level and entity-level position embedding helps the model to differentiate the token order within an entity and to distinguish between different entities, respectively. The token-level and entity-level position embedding are obtained as follows:

$$\mathcal{E}_{token\text{-}level\ position} = \text{Embed}^{V_{token\text{-}level\ position} \times d} \left(\text{ID}_{token,i} \right), \tag{4}$$

where $V_{token\text{-}level\ position}$ denotes the size of the position vocabulary, actually the length of the entity. And d is the embedding dimension. The term $\text{ID}_{token,i}$ corresponds to the index of the $i-$th token's position within the entity.

$$\mathcal{E}_{entity\text{-}level\ position} = \text{Embed}^{V_{entity\text{-}level\ position} \times d} \left(\text{ID}_{entity,i} \right), \tag{5}$$

where $V_{entity\text{-}level\ position}$ denotes the size of the position vocabulary, actually the length of the sequence. And d is the embedding dimension. The term $\text{ID}_{entity,i}$ corresponds to the index of the $i-$th entity's position within the sequence.

3.3   Graph Module

This module uses a graph structure to represent the relationships between semantic entities within a document. Specifically, EntityLayout achieves efficient graph construction at the entity level through the Token Fusion Module (TFM) and the Graph Pruning Module (GPM).

Token Fusion Module. Here, we introduce the Token Fusion Module (TFM), as shown in Fig. 2. TFM can fuse token features within an entity into a single entity feature, thereby enhancing the model's capability to extract features from long text entities. For the output of the pre-training Transformer Encoder, we use a gated recurrent unit (GRU) [31] in the TFM to extract and capture entity dimension features.

Meanwhile, drawing on the skip connection technique from the residual networks (ResNet) [32], we skip-connect each entity's 2D position embedding to the corresponding TFM task, passing it as hidden states into the TFM, ultimately obtaining entity-level features. The TFM input contains a variable-length token sequence $[t_{1,i}, t_{2,i}, ..., t_{k,i}]$ and the skip-connect 2D position embedding $\mathcal{E}_{\text{2D position}}$ (Eq. 3) as the hidden state.

We use unsupervised SimCSE [33] in terms of data augmentation to help TFM robustly learn the entity-level representation. We get the representation of $i-$th entity:

$$\mathcal{F}_i^z = \text{TFM}([t_{1,i}, t_{2,i}, ..., t_{k,i}], \mathcal{E}_{\text{2D position}}, z), \tag{6}$$

where z is a random mask for dropout.

We simply feed the same input to the encoder twice and get two embeddings $\mathcal{F}_i$ and $\mathcal{F}_i^+$ with different dropout masks z, z'. The positive pair takes exactly the same entity, and their embeddings only differ in dropout masks. We increase the similarity between positive cases and decrease the similarity between negative cases and the training objective of SimCSE becomes:

$$\ell_i = -\log \frac{e^{\text{sim}(\mathcal{F}_i^{z_i}, \mathcal{F}_i^{z_i'})/\tau}}{\sum_{j=1}^{N} e^{\text{sim}(\mathcal{F}_i^{z_i}, \mathcal{F}_j^{z_j'})/\tau}}, \tag{7}$$

where N is the number of entities, actually the batch size. τ is a temperature hyperparameter and $\text{sim}(\mathcal{F}_1, \mathcal{F}_2)$ is the cosine similarity $\frac{\mathcal{F}_1^T \mathcal{F}_2}{\|\mathcal{F}_1\| \|\mathcal{F}_2\|}$.

After TFM, we model these semantic entities into a fully connected graph structure, which turns the SER task into a graph node classification problem and the RE task into a graph edge classification problem.

Graph Pruning Module. Through the TFM, EntityLayout fuses token-level features into entity-level features and models them into a fully connected graph structure. This approach introduces a serious issue for the RE task, where there could be a large number of invalid negative samples. To address the issue of an excess of irrelevant negative samples, traditional methods [2,9–12] usually use heuristic approaches to prune invalid edges, such as k-NN, line-sight, beta-skeleton, etc. Compared with heuristic methods, we believe that a network-trained pruning method is more robust and has better generalization capabilities. Therefore, we propose the Graph Pruning Module (GPM) for more effective pruning, its process shown in Fig. 3.

GPM represents the semantic entities as graph nodes and the potential directional relations between pairs of nodes as graph edges. The feature n_i of each node is determined by concatenating the following two features: (1) The feature vector v_{BERT} extracted by the BERT model with the input of the textual information of each semantic entity. (2) Normalized coordinates and bounding box features (width and height). Therefore, the $i-$th node feature can be expressed as follows:

$$n_i = \{v_{\text{BERT},i}, x_{center,i}, y_{center,i}, w_i, h_i, x_{0,i}, y_{0,i}, x_{1,i}, y_{1,i}\}, \tag{8}$$

where $v_{\text{BERT},i}$ is the vector extracted by BERT, $x_{center,i}, y_{center,i}$ are the central coordinates of the corresponding bounding box with width of w^i and the height of h^i, and $[x_{0,i}, y_{0,i}], [x_{1,i}, y_{1,i}]$ are the top-left and bottom-right coordinates of the bounding box.

We pass the features to a bilinear layer, thus generating a relationship matrix that represents the relationships between nodes. Then, we input all relationship pairs judged as positive in the relationship matrix into the local transformer block (LTB) [12] for further prediction.

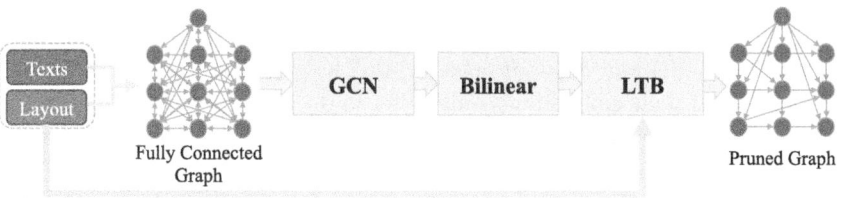

Fig. 3. An overview of the Graph Pruning Module (GPM). Based on text and layout information, a fully connected graph is constructed and a bilinear layer is used to generate the relationship matrix to be input to the LTB for positive edge prediction. Layout information is reused in LTB.

During the training process, for the Bilinear loss $\ell_{Bilinear}$, we adopt a separate-class loss to address the problem of imbalance between positive and negative samples. This means calculating loss separately for the positive samples and negative samples and then adding them together. We use the binary cross-entropy (BCE) loss [34] in Bilinear:

$$\ell_{\text{Bilinear}} = -\left(\frac{1}{n_{pos}} \sum_{i=1}^{n_{pos}} \log \hat{y}_i + \frac{1}{n_{neg}} \sum_{j=1}^{n_{neg}} \log(1 - \hat{y}_j) \right), \tag{9}$$

where n_{pos} is the number of positive entity linkings, n_{neg} is the number of negative entity linkings, $\hat{y}$ is the Bilinear prediction. The mean value of positive linkings and negative linkings is calculated separately to offset the gradient deviation.

To handle the severe imbalance problem between the positive and negative samples, the LTB loss ℓ_{LTB} uses a loss named random-under-sampling loss. During backpropagation, we randomly sample the same number of negative samples as positive samples and only regress the loss for this part of negative samples:

$$\ell_{\text{LTB}} = -\sum_{i=1}^{pos+neg_{selected}} y_i \log(\hat{y}_i), \tag{10}$$

where y_i is the target of entity linking, $\hat{y}_i$ is the LTB prediction, and the total number of linkings used for prediction is $pos + neg_{selected}$.

The total loss ℓ_{GPM} for GPM is the sum of ℓ_{Bilinear} and ℓ_{LTB}:

$$\ell_{\text{GPM}} = \ell_{\text{Bilinear}} + \ell_{\text{LTB}}, \tag{11}$$

With such a mechanism of Bilinear followed by LTB, GPM is able to preserve the true positive node relationships as much as possible while eliminating a large number of incorrect node relationships, significantly improving the efficiency and accuracy of the main network that constructs the graph with entities as nodes.

3.4   Information Extraction Task Module

EntityLayout considers the SER and RE tasks to be highly related and integrates these two downstream tasks into a joint training task, with both tasks outputting simultaneously in the classification head after sharing the same pre-training and graph neural network modules.

As shown in the bottom-right of Fig. 2, in handling the tasks of semantic entity recognition (SER) and relation extraction (RE), the SER task is first processed through the graph neural network, followed by a classifier to determine the categories of entities. The entity category logits are then input into the RE classifier to assist in relation determination.

In the RE task, the final embedding $\mathcal{E}_{\text{final}}$ for an entity linking $(entity_i, entity_j)$ combines four embeddings: $\mathcal{E}_{\text{order}}$ for the directional link between entities, $\mathcal{E}_{\text{logits}\oplus}$ for the concatenation of embeddings of SER logits, $\mathcal{F}_{\text{entity}\oplus}$ merging the individual representations of $entity_i$ and $entity_j$, and $\mathcal{E}_{\text{2D position}\oplus}$ used to learn layout information again. In this paper, the symbol '$\oplus$' denotes the operation of concatenation.

We use a sequence order embedding $\mathcal{E}_{\text{order}}$:

$$\mathcal{E}_{\text{order}} = \text{Embed}^{2 \times 2d} \left(order_{i,j}\right), \tag{12}$$

where $2d$ is the dimension of the order sequence embeddings and $order_{i,j} \in [0,1]$. A $order_{i,j}$ value of 0 indicates that $entity_i$ links to $entity_j$, whereas a value of 1 means the opposite linkage direction. By default, the linkage is assumed to be from $entity_i$ to $entity_j$.

We input the entity category logits from the SER classifier into the RE classifier for logit embedding:

$$\mathcal{E}_{\text{logits}\oplus} = \text{Embed}^{V_{\text{SER}} \times d} \left(Predict_i^{\text{SER}}\right) \oplus \text{Embed}^{V_{\text{SER}} \times d} \left(Predict_j^{\text{SER}}\right), \tag{13}$$

where we set V_{SER} as the SER label vocabulary size and d as the dimension of the embeddings. The term $Predict^{\text{SER}}$ refers to the entity category logit of SER.

We set the entity concatenated embedding:

$$\mathcal{F}_{\text{entity}\oplus} = \mathcal{F}_i \oplus \mathcal{F}_j, \tag{14}$$

where $\mathcal{F}$ is the entity-level representation modeled by TFM, as shown in Eq. 6.

Drawing on the skip connection technique from the residual networks (ResNet) [32], we implement skip connection operations for the 2D position embedding $\mathcal{E}_{\text{2D position}}$, as described in Eq. 3, to the RE classifier part. We adopt entity-level layout positions that tokens in an entity share the same 2D position since the tokens usually express the same semantic meaning. We set the 2D position concatenated embedding:

$$\mathcal{E}_{\text{2D position}\oplus} = \mathcal{E}_{\text{2D position}}\left(entity_i\right) \oplus \mathcal{E}_{\text{2D position}}\left(entity_j\right), \tag{15}$$

The final embedding $\mathcal{E}_{\text{final}}$ is as follows:

$$\mathcal{E}_{\text{final}} = \mathcal{E}_{\text{sequence}} + \mathcal{E}_{\text{logits}\oplus} + \mathcal{F}_{\text{entity}\oplus} + \mathcal{E}_{\text{2D position}\oplus}, \tag{16}$$

We pass $\mathcal{E}_{\text{final}}$ to a transformer layer and a linear layer to get the final relation extraction.

While training RE, we use the separate-class loss ℓ_{RE} by calculating the loss separately for positive and negative samples to address the issue of class imbalance. Positive samples use cross-entropy loss [30], while negative samples use Top-k loss [35]. The Top-k loss highlights a portion of the samples with the highest loss in the model, thus guiding the model to prioritize learning these ambiguous samples. While training SER, we use the cross-entropy loss ℓ_{SER}.

$$\ell_{pos}^{\text{RE}} = -\frac{1}{n_{pos}} \sum_{i=1}^{n_{pos}} (t_i^{pos} \times \ln \hat{t}_i^{pos} + (1 - t_i^{pos}) \times \ln(1 - \hat{t}_i^{pos})), \tag{17}$$

$$\ell_{neg}^{\text{RE}} = \frac{1}{n_{neg}} \sum_{i=1}^{n_{neg}} \left(-\log \left(\sum_{j \in \text{top-k}(t_i^{neg})} p_{ij} \right) \right), \tag{18}$$

$$\ell_{\text{SER}} = -\frac{1}{n_{\text{SER}}} \sum_{i=1}^{n_{\text{SER}}} (l_i \times \ln \hat{l}_i + (1 - l_i) \times \ln(1 - \hat{l}_i)), \tag{19}$$

where for Eq. 17, n_{pos} is the number of positive linkings, t_i^{pos} is the target of entity linking, and $\hat{t}_i^{pos}$ is the prediction of the RE task. For Eq. 18, n_{neg} is the number of negative linking samples, t_i^{neg} is the target of i−th linking, p_{ij} is the probability of i−th linking negative. For Eq. 19, n_{SER} is the number of entities, l_i is the target of the entity label and $\hat{l}_i$ is the prediction of the SER task.

The final loss ℓ is the sum of ℓ_{RE} and ℓ_{SER}:

$$\ell = \ell_{\text{RE}} + \alpha \cdot \ell_{\text{SER}}. \tag{20}$$

where α is determined through experimentation.

4 Experiments

4.1 Datasets

We conducted experiments using three datasets: IIT-CDIP [13] for continual pre-training, FUNSD [14], and CORD [15] for evaluating SER and RE tasks.

IIT-CDIP [13] is a large-scale public dataset of scanned images, containing about 11 million multi-page text images, totally around 42 million single-page images.

FUNSD [14] contains 199 fully annotated form documents sampled from the RVL-CDIP [36], with 149 for training and 50 for testing, totaling 31,485 words and 9,707 semantic entities. The form images are acquired by scanning, usually with low resolution and complex layouts with varied styles. Each form contains three key pieces of information: headers, questions, and answers, with multiple header-question and question-answer key-value pairs.

CORD [15] is an English receipt dataset, containing 1,000 samples, with 800 for training, 100 for validation, and 100 for testing. Receipt images are captured by photography, involving paper bending and background noise. The dataset has high-quality annotations, including text block and word-level OCR results, text block key information categories, entity relationships, and other detailed labels. The dataset contains four main key information categories such as payment information and menu, which can be further subdivided into a total of 30 key information categories.

4.2   Implementation Details

We select a subset comprising 10 thousand images from the IIT-CDIP dataset [13] for continual pre-training. To obtain bounding boxes from the text images, we apply the PaddleOCR [37]. Our transformer structure is fundamentally aligned with that of LayoutLMv3$_{BASE}$ [6], featuring a hidden size of 768, a total of 12 self-attention heads, and 12 Transformer layers. The model was initialized with the pre-trained weights from LayoutLMv3$_{BASE}$, while the weights for token-level and entity-level position embeddings were initialized randomly. We maintained the original word embedding vocabulary from LayoutLMv3$_{BASE}$ for consistency. We expand the input text length from 512 to 1024. During the continual pre-training phase, we randomly mask 30% of the tokens. Among these, 80% are substituted with the "[MASK]" token, 10% are replaced by random tokens from the text corpus, and the remaining 10% remain unchanged. We utilize the Adam optimizer, with a batch size of 64, across 4 NVIDIA TESLA A800 80GB GPUs for 300 steps. The training is regularized with a weight decay of 1e-2, and the optimizer's momentum parameters $(\beta_1, \beta_2) = (0.9, 0.99)$. We adopt a learning rate of 1e-5 for this stage of the training.

The number of GRU layers in Eq. 6 is 1. We experiment with several recurrent neural networks, such as GRU, LSTM [38], Transformer, etc. The result shows that a single-layer GRU performs best on fusing token representations. And the dropout ratio of Eq. 7 is set to 0.1.

We adopt a learning rate of 1e-3 and a batch size of 8 in the Graph Pruning Module. The training process includes a learning rate scheduler. Based on the plateau strategy, the learning rate reduces with patience of 5 epochs and a reduction factor of 0.95. For Eq. 10, we set the number of negative samples $neg_{selected}$ to be the minimum between the count of positive samples and that of negative samples due to the severe imbalance problem between the positive and negative samples.

We set the fine-tuning learning rate to be 5e-5 and a batch size of 16. The number of Transformer layers in Eq. 16 is 1. For Eq. 18, when calculating the Top-k loss, k is set to the top 70%. For Eq. 20, when calculating the sum of ℓ_{RE} and ℓ_{SER}, α is set to 1.

4.3   Graph Building by Graph Pruning Module

Table 1 presents a comparative analysis of the graph pruning effects between the Graph Pruning Module (GPM) and various other methods on the FUNSD training dataset [14]. For heuristic-based methods, there is a trade-off between efficiency and accuracy. A higher recall often leads to building more false edges. In contrast, GPM employs a model-driven approach for edge pruning. It outperforms heuristic-based counterparts by significantly enhancing the pruning ratio while maintaining an impressive recall rate of up to 100%. In practice, the goal of GPM is to achieve nearly 100% recall and to attain higher precision compared to other graph pruning methods.

Table 1. Results of different pruning strategies on FUNSD training set.

Method Category	Method	Positive/Total	#False Edges
Heuristic-based	Full Connection [28]	4230/4230	3.26×10^5
	β-Skeleton Method [9,39]	4050/4230	1.83×10^5
	k-NN (k = 60) [2]	4137/4230	1.98×10^5
	k-NN (k = 180) [2]	4230/4230	2.35×10^5
	Visibility-Based [10,11]	3913/4230	1.57×10^5
	Layout-based k-NN [12]	4230/4230	1.58×10^5
Model-driven	GPM (proposed)	**4230/4230**	$\mathbf{3.81 \times 10^4}$

4.4   Ablation Study

To rigorously evaluate the individual contributions of the Graph Pruning Module (GPM), Token Fusion Module (TFM), and Continual Pre-training within the EntityLayout framework, we conduct plentiful ablation studies. The outcomes are detailed in Table 2.

#0, which encompasses GPM and is trained from scratch with randomly initialized transformer parameters, establishes our baseline. #1 leverages pre-trained transformer weights from LayoutLMv3$_{BASE}$ [6], demonstrating a significant performance improvement relative to #0. This illustrates the importance of the transformer in learning semantic and layout information, and also highlights the effectiveness of GPM.

Compared with #1, Model #2 further refines performance by utilizing parameters optimized through continual pre-training. In scenarios where all other modules remain unchanged, the unique transformer parameters in our EntityLayout yield superior results due to the integration of token-level and entity-level embeddings, which effectively facilitate the learning and incorporation of semantic and layout features in visually rich documents.

In comparison, #2 leverages the refined parameters from continual pre-training, leading to an advancement in results. In scenarios where all other modules remain unchanged, the unique transformer parameters in our EntityLayout yield superior results. This benefit is attributable to the integration of token-level and entity-level position embeddings. Compared with common 1D position embeddings, token-level and entity-level position embeddings can better facilitate the learning and incorporation of semantic and layout features in visually rich documents.

Model #3 extends #2 by adding the TFM. Thanks to the unsupervised SimCSE [33], the TFM can robustly enhance entity-level representation. Consequently, the SER score increases by 0.09, and the RE score reaches a remarkable 82.12, as shown in the table.

Table 2. Ablation study on SER and RE task on FUNSD dataset. The first column labels the experiment settings.

#	GPM	Continual Pre-train	TFM	SER F1(%)	RE F1(%)
0	✓			65.33	27.81
1	✓	*		89.09	70.10
2	✓	✓		90.09	75.01
3	✓	✓	✓	**91.08**	**82.12**

* means reusing pre-trained transformer parameters from LayoutLMv3$_{BASE}$ [6].

4.5   Comparison with State-of-the-Art

In our paper, we assess the performance of our semantic entity recognition (SER) and relation extraction (RE) models using two public datasets, namely FUNSD [14] and CORD [15]. The comparative results are detailed in Table 3.

Typically, two-stage algorithms assume that the SER inputs to RE are flawless. However, in practical scenarios, the quality of RE is contingent on the accuracy of SER outcomes. Therefore, for such two-stage methodologies, we calculate the RE performance using the formula: RE F1 = SER F1 $\times$ F1$_{RE}$, reflecting the dependency of RE results on the SER detection accuracy.

The comparative performance presented in Table 3 clearly indicates that our model surpasses the performance of the existing models, particularly with its capacity to simultaneously generate SER and RE results. While

Table 3. Performance of EntityLayout and other two-stage methodologies on FUNSD and CORD dataset.

Model	SER F1(%)		RE F1(%)		# Params	Jointly
	FUNSD	CORD	FUNSD	CORD		
BERT$_{BASE}$ [26]	60.92	93.13	16.84	86.45	110M	No
LayoutLM$_{BASE}$ [23]	78.54	96.26	36.02	91.64	160M	No
LayoutLMv2$_{BASE}$ [5]	81.89	96.05	35.14	91.81	200M	No
BROS$_{BASE}$ [20]	83.05	96.50	59.35	92.35	110M	No
StrucTexT [18]	83.09	–	36.64	–	107M	No
LiLT [21]	85.74	96.07	63.51	–	–	No
LayoutLMv3$_{BASE}$ [6]	90.29	**96.56**	57.66	–	133M	No
EntityLayout	**91.08**	96.50	**82.12**	**98.98**	154M	Yes

LayoutLMv3$_{BASE}$ exhibits marginally superior performance in the SER task on the CORD dataset, EntityLayout excels due to its effective pruning mechanism and efficient token fusion. EntityLayout maintains robust performance in SER tasks without relying on visual modalities and achieves state-of-the-art results in the RE task.

5   Conclusion

The proposed EntityLayout is a multimodal pre-trained model designed for joint semantic entity recognition (SER) and relation extraction (RE) in visually rich documents (VrDs). Unlike existing models, EntityLayout introduces entity-level and word-level embeddings that are crucial for the recognition of entities and word sequences during continual pre-training. The fusion of token-level features into entity-level representation by the Token Fusion Module (TFM) is effective in improving the VIE performance, and the Graph Pruning Module (GPM) reduces invalid links largely and improves efficiency. On the FUNSD and CORD datasets, EntityLayout demonstrates competitive performance in both situations for both SER and RE tasks.

References

1. Liu, X., Gao, F., Zhang, Q., Zhao, H.: Graph convolution for multimodal information extraction from visually rich documents (2019). arXiv preprint arXiv:1903.11279
2. Carbonell, M., Riba, P., Villegas, M., Fornés, A., Lladós, J.: Named entity recognition and relation extraction with graph neural networks in semi structured documents. In: Proceedings of the 25th International Conference on Pattern Recognition (ICPR), pp. 9622–9627 (2021)

3. Yu, W., Lu, N., Qi, X., Gong, P., Xiao, R.: PICK: processing key information extraction from documents using improved graph learning-convolutional networks. In: Proceedings of the 25th International Conference on Pattern Recognition (ICPR), pp. 4363–4370 (2021)
4. Zhang, Z., Ma, J., Du, J., Wang, L., Zhang, J.: Multimodal pre-training based on graph attention network for document understanding. IEEE Trans. Multimedia **25**, 6743–6755 (2023)
5. Xu, Y., et al.: LayoutLMv2: Multi-modal pre-training for visually-rich document understanding (2020). arXiv preprint arXiv:2012.14740
6. Huang, Y., Lv, T., Cui, L., Lu, Y., Wei, F.: LayoutLMv3: pre-training for document AI with unified text and image masking. In: Proceedings of the 30th ACM International Conference on Multimedia, pp. 4083–4091 (2022)
7. Luo, C., Cheng, C., Zheng, Q., Yao, C.: GeoLayoutLM: geometric pre-training for visual information extraction. In: Proceedings of the IEEE/CVF Conference on Computer Vision and Pattern Recognition, pp. 7092–7101 (2023)
8. Zhang, Y., Zhang, B., Wang, R., Cao, J., Li, C., Bao, Z.: Entity relation extraction as dependency parsing in visually rich documents (2021). arXiv preprint arXiv:2110.09915
9. Lee, C.Y., et al.: FormNet: Structural encoding beyond sequential modeling in form document information extraction (2022). arXiv preprint arXiv:2203.08411
10. Davis, B., Morse, B., Cohen, S., Price, B., Tensmeyer, C.: Deep visual template-free form parsing. In: Proceedings of the International Conference on Document Analysis and Recognition (ICDAR), pp. 134–141 (2019)
11. Déjean, H., Clinchant, S., Meunier, J.L.: LayoutXLM vs. GNN: An empirical evaluation of relation extraction for documents (2022). arXiv preprint arXiv:2206.10304
12. Chen, Y.M., Hou, X.T., Lou, D.F., Liao, Z.L., Liu, C.L.: DAMGCN: entity linking in visually rich documents with dependency-aware multimodal graph convolutional network. In: Fink, G.A., Jain, R., Kise, K., Zanibbi, R. (eds.) Document Analysis and Recognition – ICDAR 2023. ICDAR 2023. LNCS, vol. 14189. Springer, Cham (2023). https://doi.org/10.1007/978-3-031-41682-8_3
13. Lewis, D., Agam, G., Argamon, S., Frieder, O., Grossman, D., Heard, J.: Building a test collection for complex document information processing. In: Proceedings of the 29th Annual International ACM SIGIR Conference on Research and Development in Information Retrieval, pp. 665–666 (2006)
14. Jaume, G., Ekenel, H.K., Thiran, J.P.: FUNSD: a dataset for form understanding in noisy scanned documents. In: Proceedings of International Conference on Document Analysis and Recognition Workshops (ICDARW). vol. 2, pp. 1–6 (2019)
15. Park, S., Shin, S., Lee, B., Lee, J., Surh, J., Seo, M., Lee, H.: CORD: a consolidated receipt dataset for post-OCR parsing. In: Workshop on Document Intelligence at NeurIPS 2019 (2019)
16. Appalaraju, S., Jasani, B., Kota, B.U., Xie, Y., Manmatha, R.: DocFormer: end-to-end transformer for document understanding. In: Proceedings of the IEEE/CVF International Conference on Computer Vision, pp. 993–1003 (2021)
17. Li, P., et al.: SelfDoc: self-supervised document representation learning. In: Proceedings of the IEEE/CVF Conference on Computer Vision and Pattern Recognition, pp. 5652–5660 (2021)
18. Li, Y., et al.: StrucText: structured text understanding with multi-modal transformers. In: Proceedings of the 29th ACM International Conference on Multimedia, pp. 1912–1920 (2021)

19. Yu, Y., et al.: StrucTexTv2: masked visual-textual prediction for document image pre-training. In: Proceedings of the 11th International Conference on Learning Representations (2023)
20. Hong, T., Kim, D., Ji, M., Hwang, W., Nam, D., Park, S.: BROS: a pre-trained language model focusing on text and layout for better key information extraction from documents. In: Proceedings of the AAAI Conference on Artificial Intelligence. vol. 36, pp. 10767–10775 (2022)
21. Wang, J., Jin, L., Ding, K.: LiLT: A simple yet effective language-independent layout transformer for structured document understanding (2022). arXiv preprint arXiv:2202.13669
22. Gu, Z., et al.: XYLayoutLM: towards layout-aware multimodal networks for visually-rich document understanding. In: Proceedings of the IEEE/CVF Conference on Computer Vision and Pattern Recognition, pp. 4583–4592 (2022)
23. Xu, Y., Li, M., Cui, L., Huang, S., Wei, F., Zhou, M.: LayoutLM: pre-training of text and layout for document image understanding. In: Proceedings of the 26th ACM SIGKDD International Conference on Knowledge Discovery & Data Mining, pp. 1192–1200 (2020)
24. Kipf, T.N., Welling, M.: Semi-supervised classification with graph convolutional networks. In: Proceedings of the 11th International Conference on Learning Representations (2017)
25. Huang, Z., Xu, W., Yu, K.: Bidirectional LSTM-CRF models for sequence tagging (2015). arXiv preprint arXiv:1508.01991
26. Devlin, J., Chang, M.W., Lee, K., Toutanova, K.: BERT: Pre-training of deep bidirectional transformers for language understanding (2018). arXiv preprint arXiv:1810.04805
27. Liu, Z., Lin, Y., Cao, Y., Hu, H., Wei, Y., Zhang, Z., Lin, S., Guo, B.: Swin transformer: hierarchical vision transformer using shifted windows. In: Proceedings of the IEEE/CVF International Conference on Computer Vision, pp. 10012–10022 (2021)
28. Gemelli, A., Biswas, S., Civitelli, E., Lladós, J., Marinai, S.: Doc2Graph: a task agnostic document understanding framework based on graph neural networks. In: Karlinsky, L., Michaeli, T., Nishino, K. (eds.) Computer Vision - ECCV 2022 Workshops. ECCV 2022. LNCS, vol. 13804. Springer, Cham (2023). https://doi.org/10.1007/978-3-031-25069-9_22
29. Sun, Y., et al.: ERNIE 2.0: a continual pre-training framework for language understanding. In: Proceedings of the AAAI Conference on Artificial Intelligence. vol. 34, pp. 8968–8975 (2020)
30. Zhang, Z., Sabuncu, M.: Generalized cross entropy loss for training deep neural networks with noisy labels. In: Advances in Neural Information Processing Systems. vol. 31, pp. 8778–8788 (2018)
31. Chung, J., Gulcehre, C., Cho, K., Bengio, Y.: Empirical evaluation of gated recurrent neural networks on sequence modeling (2014). arXiv preprint arXiv:1412.3555
32. He, K., Zhang, X., Ren, S., Sun, J.: Deep residual learning for image recognition. In: Proceedings of the IEEE/CVF Conference on Computer Vision and Pattern Recognition, pp. 770–778 (2016)
33. Gao, T., Yao, X., Chen, D.: SimCSE: simple contrastive learning of sentence embeddings. In: Proceedings of the 2021 Conference on Empirical Methods in Natural Language Processing, pp. 6894–6910 (2021)
34. Ruby, U., Yendapalli, V.: Binary cross entropy with deep learning technique for image classification. Int. J. Adv. Trends Comput. Sci. Eng. 9(10), 5393–5397 (2020)

35. Fan, Y., Lyu, S., Ying, Y., Hu, B.: Learning with average top-k loss. In: Advances in Neural Information Processing Systems. vol. 30, pp. 497–505 (2017)
36. Harley, A.W., Ufkes, A., Derpanis, K.G.: Evaluation of deep convolutional nets for document image classification and retrieval. In: Proceedings of the 13th International Conference on Document Analysis and Recognition (ICDAR), pp. 991–995. IEEE (2015)
37. Li, C., et al.: PP-OCRv3: More attempts for the improvement of ultra lightweight OCR system (2022). arXiv preprint arXiv:2206.03001
38. SHI, X., Chen, Z., Wang, H., Yeung, D.Y., Wong, W.K., WOO, W.C.: Convolutional LSTM network: a machine learning approach for precipitation nowcasting. In: Advances in Neural Information Processing Systems. vol. 28, pp. 802–810 (2015)
39. Wang, R., Fujii, Y., Popat, A.C.: Post-OCR paragraph recognition by graph convolutional networks. In: Proceedings of the IEEE/CVF Winter Conference on Applications of Computer Vision, pp. 493–502 (2022)

Embedding Layout in Text for Document Understanding Using Large Language Models

Mohammad Minouei[1,2]([✉])[iD], Mohammad Reza Soheili[2,3][iD],
and Didier Stricker[1,2]

[1] Department of Computer Science, RPTU Kaiserslautern-Landau,
Kaiserslautern, Germany
[2] German Research Institute for Artificial Intelligence (DFKI),
67663 Kaiserslautern, Germany
{mohammad.minouei,didier.stricker}@dfki.de
[3] Department of Electrical and Computer Engineering, Kharazmi University,
Tehran, Iran
Soheili@khu.ac.ir

Abstract. In this paper, we address the challenge of effectively utilizing Large Language Models (LLMs) for Visually Rich Document Understanding (VRDU), a key part of intelligent document processing systems. While LLMs excel in various Natural Language Processing (NLP) tasks, their application for extracting information from complex structured documents like invoices and forms is limited. This limitation arises from the difficulty in contextually understanding these documents, largely due to the lack of layout information. Our research is dedicated to unlocking the full potential of LLMs for VRDU by integrating OCR data into an HTML format, which preserves the essential spatial layout for accurate information extraction. The empirical results show a notable improvement, with a more than 20% increase over baseline performances. This research highlights the promising potential of LLMs in VRDU and sets the stage for further innovations in automated document processing.

Keywords: Document Understanding · Large Language Model · Information Extraction

1 Introduction

Document understanding aims at interpreting and extracting meaningful information from documents. Being an active area of research, a wide range of approaches have been studied in the literature that utilize document images, text, or a combination of both [25]. The field has evolved from initial heuristic methods [11], to the modern specialized deep neural network techniques [14]. The complexity of this task lies in the endless variations in document layouts, such as invoices, tax forms, and many more. These structured documents often

© The Author(s), under exclusive license to Springer Nature Switzerland AG 2024
E. H. Barney Smith et al. (Eds.): ICDAR 2024, LNCS 14804, pp. 280–293, 2024.
https://doi.org/10.1007/978-3-031-70533-5_17

feature elements such as tables and key-value pairs, needing advanced methods for information extraction.

The advent of LLMs has brought about a significant transformation in the field of natural language processing, surpassing previous state-of-the-art (SOTA) methods [21]. Models like Chat-GPT [3] and Llama [26] have gained significant recognition for their remarkable text understanding and generation capabilities. These models are trained on large datasets, allowing them to recognize complex patterns and subtleties in natural language. With their substantial size, often comprising hundreds of billions of parameters, LLMs can handle a diverse range of tasks without the need for task specific data, also known as zero-shot ability.

LLMs are not limited to just generating text; they have the potential to transform how we understand documents. Their applications are diverse, including tasks such as extracting information, conducting semantic searches, and summarizing documents [21]. This understanding can be further refined with a limited number of examples to adapt to specific document related tasks. The few-shot learning ability of LLMs is especially useful in processing documents from fields where labeled data is rare or costly to gather [5].

This has inspired researchers to explore new ways to effectively leverage LLMs for document understanding tasks. Despite their impressive capabilities, using LLMs to interpret visually dense, structured documents is still understudied. A major difficulty is the absence of layout information in the text, which is vital for effective information extraction. To make the most of LLMs, it is crucial to prepare the input data carefully. How well these models perform largely depends on how the input prompts are structured. If the preparation is not done properly, it can result in responses that are either irrelevant or incorrect [13].

Recently, researchers introduced the Visually-Rich Document Understanding (VRDU) benchmark dataset [27] to evaluate how well models perform in this area. VRDU includes two kinds of documents of *purchases* and *registration* forms. The VRDU benchmark provides challenging tasks to assess the capabilities of models in various scenarios, including test sets with mixed templates or templates that have not been seen before. It also evaluates the performance of models in situations with limited data (few-shot settings) and their ability to identify nested or repeated entities. This benchmark offers an excellent opportunity to showcase the potential of language models in understanding documents [27].

The significance of the VRDU benchmark is especially apparent when evaluating advanced document AI systems like LayoutLM [29] and FormNet [16]. These models demonstrate substantial progress on document understanding, yet struggle with VRDU's complex dataset. The LayoutLM series, represents a major step forward in understanding document images. They combine language models' capabilities with spatial and visual contexts. The LayoutLM model improved upon BERT by adding 2-D positional and image embedding for tokens. It showed skill in tasks like extracting information and classifying documents. LayoutLMv2 [28] went further, improving how it integrates visual data during pre-training and using a multi-modal transformer architecture. This development included a spatial-aware self-attention mechanism, enhanced through tasks like masked visual-language modeling and text-image matching, improving its understanding of visually complex documents. LayoutLMv3 [14], an improvement over LayoutLMv2, adopts patch embedding along with Vision Transformers [7] instead of using a CNN backbone. It simplifies the structure and pre-training process, focusing on masked language modeling (MLM), masked image modeling (MIM), and word-patch alignment (WPA), boosting its document understanding capabilities.

FormNet [16], another innovative system, combines sequence and convolutional approaches for impressive results. It introduces rich attention and supertokens. Rich attention calculates attention scores by considering the spatial relationships between tokens, capturing the document's structural details. Supertokens are created for each word in a form, incorporating embeddings from neighboring tokens using graph convolutions.

Incorporating layout with text in the network has been studied in various works [8,9,17–19]. However, Donut [15] proposed an end-to-end encoder-decoder model that leverages transformer architecture to directly map raw input images to desired outputs, bypassing the need for OCR.

In [23], the authors introduce a novel method called LMDX for extracting information from semi-structured documents using LLMs. The LMDX approach addresses the challenges of information extraction by incorporating text position encoding and a grounding mechanism along with their LLM. It uses a five-stage process: OCR, chunking, generating prompts, LLM inference, and decoding. This pipeline is designed to efficiently identify and locate entities in the documents.

Facing the challenge of using LLMs for understanding documents, we propose using a machine-friendly data representation. HTML stands out for its adaptability and proves to be highly effective in this context. In [10], authors extensively analyze the application of LLMs in tasks related to understanding HTML. The study highlights that, when appropriately fine-tuned, LLMs demonstrate outstanding performance on benchmark tests assessing HTML comprehension. In our specific use-case, the conversion of raw text into HTML allows us to retain both the textual content and spatial layout of the documents, which plays a key role in achieving accurate and precise analysis. This approach is particularly useful for documents with standard layouts, such as forms with key-value pairs.

In summary, our research aims to bridge the gap between the capabilities of LLMs and the practical requirements of visually-rich document understanding. We make the following contributions:

- We present a new method for transforming OCR document outputs into structured HTML representations that preserve spatial relationships and layout context.
- Through extensive experiments, we show that instruction-based prompting, which includes HTML representations, improves LLMs' capacity to comprehend complex visual layouts.

Overall, our research contributes to the ongoing efforts to leverage LLMs in real-world VRDU applications.

2   Approach

The field of large language models is rapidly evolving, with new models being introduced regularly. However, for our specific task, we require a LLM that can handle long contexts, understand HTML and JSON, and follow instructions precisely. We have identified a variant of Llama 2 [26], known as CodeLlama [24], which meets these requirements. CodeLlama is fine-tuned with massive datasets containing code, markup languages, and natural language text related to coding. It is specifically designed to handle long input contexts, allowing it to process longer sequences of up to 16,384 tokens. Moreover, CodeLlama excels at following detailed instructions, making it ideal for tasks related to programming and data manipulation.

In our approach, we use tailored instruction prompts to fine-tune the LLM for document understanding. We begin by preparing the data, converting the document's OCR output into an HTML representation. This HTML format serves as the input for the LLM, accompanied by an instruction prompt that describes the task and outlines the expected JSON output structure. This output JSON is derived from the ground truth, which includes key-value pairs for each page. By using these specific prompts, we fine-tune the LLM to better understand and process document content. Figure 1 presents an illustration of the overall process. The following sections will explain these steps in more detail.

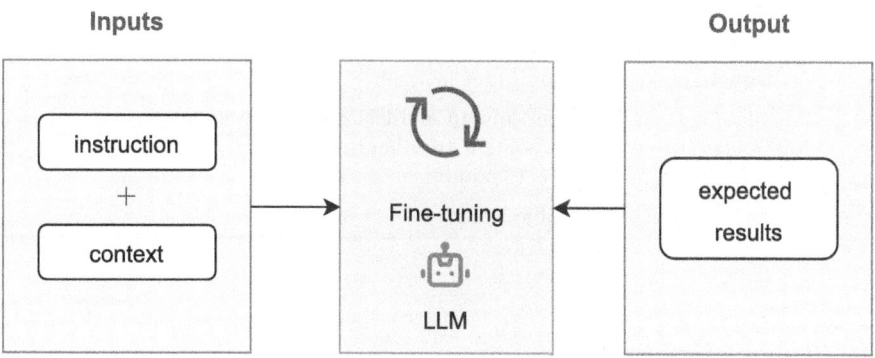

Fig. 1. An overall representation of our approach. Having prepared the inputs and expected outputs of the documents, the model is fine-tuned in a supervised manner.

2.1   HTML Representation

HTML is an ideal format for representing complex layout structures of documents. In the past, OCR engines such as Tesseract [2] have offered a specialized HTML representation in hOCR format [4]. Unlike plain text, HTML elements can capture how textual components spatially relate to one another within a rich formatting structure, which is particularly useful in scenarios that involve key-value inputs and require maintaining the relationships between words. As studied in [10], HTML serves as an interpretable structured medium for LLM.

In our application, we implemented a pre-processing step that transforms OCR output into HTML format. The details of this conversion is outlined in Algorithm 1. We use bounding box coordinates to arrange the text elements into a `<table>` layout, with `<tr>` rows and `<td>` cells, based on their relative positions. The algorithm then sorts and organizes these elements to create a coherent HTML structure that retains original spatial positioning relationships. Figure 2 shows a sample document and corresponding HTML encoding generated by this process.

Algorithm 1: Convert OCR results to HTML Table

Data: List of texts and bounding boxes
Result: HTML table
1 **foreach** *bounding box* **do**
2   Calculate row and column based on bounding box coordinates;
3   Append (row, column, text) to data list;

4 Sort data_list by row and then by column;
5 Initialize table_html;
6 **foreach** *(row, column, text) in sorted_data* **do**
7   **if** *row* $\neq$ *current_row* **then**
8   **if** *current_row* $\neq$ *0* **then**
9   Add "`</tr>`" to table_html;

10   Add "`<tr>`" to table_html;
11   Update current_row and reset last_col;

12   Calculate colspan based on column and last_col;
13   Add `<td></td>` with text content to table_html;
14   Update last_col with current column;

15 Add "`</tr> </table>`" to table_html;

2.2 Prompt Generation

LLMs can be guided to perform specific tasks by using instruction prompts that clearly define the expected behavior. The Llama LLM uses two types of prompts: a system prompt and an instruction prompt. The system prompt sets the general tone and expectations for the interaction and is placed at the beginning. Meanwhile, the instruction prompt specifies the particular task or type of response expected from the model for that exchange.

We have designed the system prompt as follows:

"Below is an instruction that describes a task, paired with an input that provides further context. Your response is a JSON object that appropriately completes the request. The JSON must be between [JSON] and [/JSON] tags."

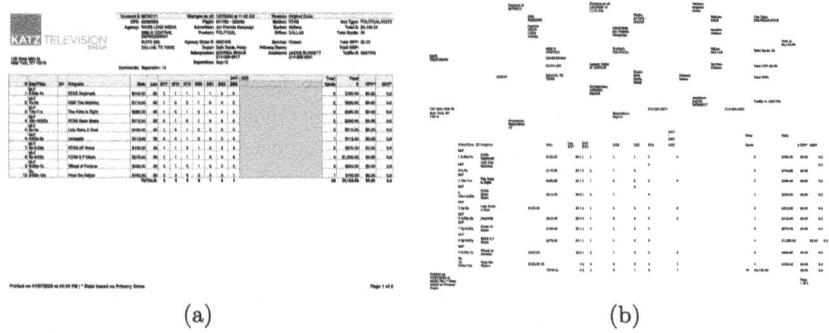

(a)   (b)

Fig. 2. Comparison between the original document (a) and its corresponding HTML representation (b). Sample from VRDU benchmark.

This prompt sets the format and expectations for the model's response.

Following this, the instruction prompt provides specific directions for the task, guiding the language model to accurately extract and organize the necessary information:

> *"Given the following HTML table, extract key details and organize them into a single JSON object. Please provide values for the fields including 'advertiser', 'property', 'agency', 'tv_address', 'contract_num', 'product', 'gross_amount', 'flight_from', 'flight_to', and 'line_item' is an array (with 'channel', 'program_desc', 'program_end_date', 'program_start_date', and 'sub_amount'). Ensure that the extracted information accurately reflects the content of the HTML. Output must be JSON."*

With these instructions and the aforementioned HTML representation of the document, the final prompt is formed.

2.3 Implementation Details

For fine-tuning we used parameter-efficient fine-tuning (PEFT) [20] library with Low-Rank Adaptation (LoRA) [12] configuration. The PEFT library offers multiple choices for fine-tuning just a few additional parameters of LLMs, significantly reducing computational costs. LoRA provides an efficient method for fine-tuning LLM by introducing smaller low-rank matrices to each layer instead of modifying the original weight matrices. Following best practices in [6], we applied LoRA to all linear layers, including the query, key, and value layers of each attention block. For our implementation, we chose an alpha value of 64, which scales the low-rank updates applied to the model weights. The rank of the low-rank matrices was set to 16, determining the size of the trainable matrices. Additionally, we integrated a dropout rate of 0.05 in our LoRA layers. Training was conducted for 1K iterations on an A100 GPU. We used a baseline learning rate of 1e-4 with a cosine learning rate scheduler. The scheduler decreases the rate following a cosine curve over the training span.

As illustrated in Fig. 3, during inference, documents with multiple pages are split and individually converted to HTML duplicates and then fed into the LLM with the appropriate instruction prompt. Afterwards, we combine the predicted outcomes from all pages into a single JSON object with the structure that the evaluation code demands. To ensure accurate evaluation, we sort the key values in both the JSON object and the ground truth. Our implementation is available at: https://github.com/minouei-kl/llm4vrdu.

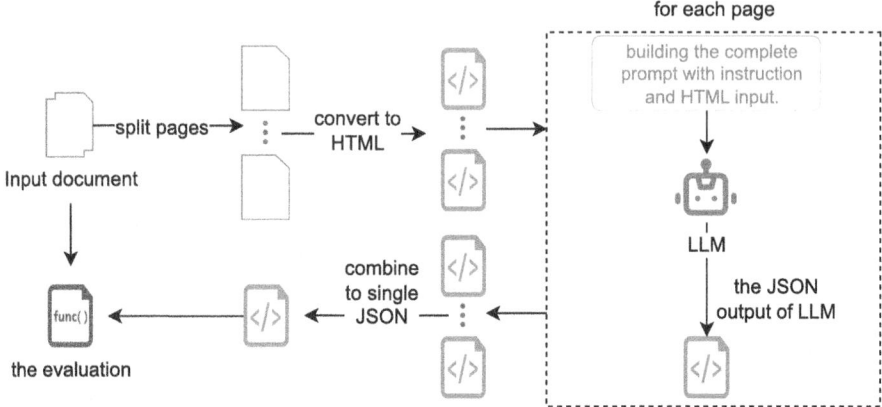

Fig. 3. Evaluation workflow: Documents are split into pages, converted to HTML, processed by the LLM with custom prompt, and outputs are compiled into a single JSON. The final output is compared with the ground truth for assessment.

3   Experiments and Results

In this section, we review the experiments and their results. First, we evaluate the Ad-buy dataset from VRDU under various settings defined by the benchmark. Next, we focus on a specific subset of the Ad-buy dataset, comprising only 100 training samples and a test set with unseen templates. This subset serves as challenging testing ground to conduct additional experiments. Using this subset, we assess the model's performance using a different input encoding, evaluate how the LLM performs without any training, and compare the results with another LLM. Lastly, we test the Consolidated Receipt Dataset (CORD) [22] to determine the versatility of our encoding approach.

3.1   Datasets

The VRDU benchmark includes two datasets: Ad-buy Forms and Registration Forms. The Registration Forms dataset has simpler layout with fewer details to extract. In contrast, the Ad-buy Forms dataset is more challenging, containing 641 documents primarily made up of invoices and receipts related to political advertisements. These documents have complex layouts with tables and detailed elements such as product names, flight dates, and total prices typical of invoices. The main challenge of this dataset is the accurate extraction and interpretation of structured data from these documents. This involves a variety of data types, including prices, dates, addresses, and nested entities, as well as complex features like tables, multi-column layouts, and key-value pairs.

The benchmark provides high-quality OCR extraction results for text and their corresponding positions within the documents. It includes two tasks: Mixed Template Learning (MTL) and Unseen Template Learning (UTL). MTL evaluate the models' ability to handle various templates by incorporating multiple templates across training and testing sets. UTL evaluates the models' capacity to adapt to templates not seen during training. Each task in the VRDU dataset consists of 300 documents in the testing set, with four different training sets of 10, 50, 100, and 200 samples, respectively. This structure allows for assessing models on their efficiency with data and their performance with limited training data. Additionally, the authors implement a type-aware matching algorithm to accurately assess performance. The algorithm uses specific matching functions tailored to each entity's data type. For example, it employs numeric comparisons for monetary values to ensure that differences in formatting do not affect the matching results.

Additionally, the CORD [22] contains a thousand of Indonesian receipt images receipts. It comes with rich annotations for OCR and multi-level semantic labels for each word. The dataset is divided into training (800 receipts), validation (100 receipts), and test sets (100 receipts).

3.2   Evaluation on VRDU Benchmark

Table 1 compares our proposed model with others, including LMDX, FormNet, and different versions of LayoutLM, evaluated on the Ad-buy dataset. It shows the performance of these models with varying data sizes and whether the templates in training/testing were mixed or unseen. The performance metrics include Micro-F1 and Line-Item F1 scores as defined in [27].

Our model shows significant improvement over the baseline methods such as the FormNet and the LayoutLM family in all settings. As the size of the training set increases, there is a consistent improvement in performance. The extraction of line items, which contain nested or itemized information, is particularly challenging because the evaluation process is strict; missing even a single item in a group results in it being marked completely incorrect.

Although the proposed method performs well, it has not reached the top performance achieved by LMDX due to several factors. First, LMDX has a larger architecture with greater processing capabilities. Additionally, LMDX benefits from pre-training on a private dataset, enhancing its performance. LMDX also utilizes multiple inferencing techniques, leading to higher accuracy at a higher computational cost. Lastly, LMDX undergoes more training with 4,000 iterations compared to our 1,000 iterations. These factors, considering the computational cost and limitations in our experiments, explain the superior performance of the LMDX model in this context.

Table 1. Performance on Ad-buy dataset across various train sizes and template setting in train/test (mixed, unseen). The reported numbers are sourced from [23].

Size	Model	Mixed Template		Unseen
		Micro-F1	Line Item F1	Micro-F1
10	FormNet [16]	20.47	5.72	20.28
	LayoutLM [29]	20.20	6.95	19.92
	LayoutLMv2 [28]	25.36	9.96	25.17
	LayoutLMv3 [14]	10.16	5.92	10.01
	LMDX $_{PaLM\ 2-S}$ [23]	54.35	39.35	54.82
	Proposed	38.06	19.66	37.76
50	FormNet [16]	40.68	19.06	39.52
	LayoutLM [29]	39.76	19.50	38.42
	LayoutLMv2 [28]	42.23	20.98	41.59
	LayoutLMv3 [14]	39.49	19.53	38.43
	LMDX $_{PaLM\ 2-S}$ [23]	75.08	65.42	75.70
	Proposed	58.16	42.72	56.87
100	FormNet [16]	40.38	18.80	39.88
	LayoutLM [29]	42.38	21.26	41.46
	LayoutLMv2 [28]	44.97	23.52	44.35
	LayoutLMv3 [14]	42.63	22.08	41.54
	LMDX $_{PaLM\ 2-S}$ [23]	78.05	69.77	75.99
	Proposed	65.9	52.51	63.71
200	FormNet [16]	43.23	21.86	42.87
	LayoutLM [29]	44.66	23.90	44.18
	LayoutLMv2 [28]	46.54	25.46	46.31
	LayoutLMv3 [14]	45.16	24.51	44.43
	LMDX $_{PaLM\ 2-S}$ [23]	79.82	72.09	78.42
	Proposed	74.74	64.24	71.82

Table 2 shows the detailed performance of our model on the Ad-buy dataset for different fields, under different template sample frequencies (10, 50, 100, 200). For both *mixed* and *unseen* templates, as the number of samples increases, there is an improvement in F1 scores across most fields. Some fields such as 'gross amount', 'product', 'agency', and 'advertiser' consistently show higher F1 scores across both template types and all data sizes, indicating the model's effectiveness in these areas. Conversely, fields like 'tv address', 'line item', have lower F1 scores, especially in template with fewer samples, which means the model struggles more with these fields. The address field often contain multi-line text, while line item refers to groups of information presented in tabular form within these documents.

Figure 4 shows a sample document, ground truth, and model predictions. While most details are accurately extracted, there are instances where parts of the program description are missed. Such mistakes lead to a decline in the performance of the line-item.

Table 2. F1-Scores per field on the Ad-Buy dataset across various train sizes and template setting in train/test (mixed, unseen).

Template	Size	Advertiser	Agency	Contract Num	Flight From	Flight To	Gross Amount	Product	TV Address	Property	Line Item	Macro	Unrepeated entities	Micro
Mixed	10	82.19	76.18	76.75	71.09	72.55	88.57	84.64	44.77	67.85	19.66	68.43	74.21	38.06
Mixed	50	94.32	85.01	93.18	88.81	86.76	94.92	90.15	75.96	82.84	42.72	83.47	88.46	58.16
Mixed	100	94.14	88.05	95.17	91.41	92.08	96.63	93.2	79.17	86.98	52.51	86.94	91.15	65.9
Mixed	200	97.58	93.16	96.69	94.43	94.66	97.21	95.13	84.77	93.18	64.24	91.11	94.32	74.74
Unseen	10	78.48	71.95	77.64	73.48	73.55	90.85	83	42.77	68.33	19.33	67.94	73.9	37.76
Unseen	50	93.86	87.53	93.73	88.53	87.19	94.15	92.37	76.6	83.69	40.66	83.83	88.94	56.87
Unseen	100	94.24	91.81	93.78	89.29	90.42	96.02	93.32	78.99	86.58	49.26	86.37	90.63	63.71
Unseen	200	96.21	95.16	96.56	90.49	91.15	95.94	94.62	85.53	93.44	60.14	89.92	93.32	71.82

3.3 Evaluation with Coordinate Embedding

As presented in [23], one encoding approach is to directly embed the normalized x—y coordinate pair of each word into the text input. As the authors state, this spatial context helps language models infer document layout relationships. For comparison, we train the LLM with this "coordinate-in-text" representation on a subset of dataset. As Table 3 shows, our model generally outperforms the "coordinate" model in most fields, as indicated by higher F1 scores.

Table 3. Evaluation coordinate in text on unseen template 100 subset (F1-Scores).

Model	Advertiser	Agency	Contract Num	Flight From	Flight To	Gross Amount	Product	TV Address	Property	Line Item	Macro	Unrepeated entities	Micro
Coordinate	91.05	85.65	**96.04**	87.95	**90.54**	**96.51**	89.53	75	85.28	45.4	84.29	89.05	60.52
Proposed	**94.24**	**91.81**	93.78	**89.29**	90.42	96.02	**93.32**	78.99	**86.58**	**49.26**	**86.37**	90.63	**63.71**

290837745009137290 M. Minouei et al.

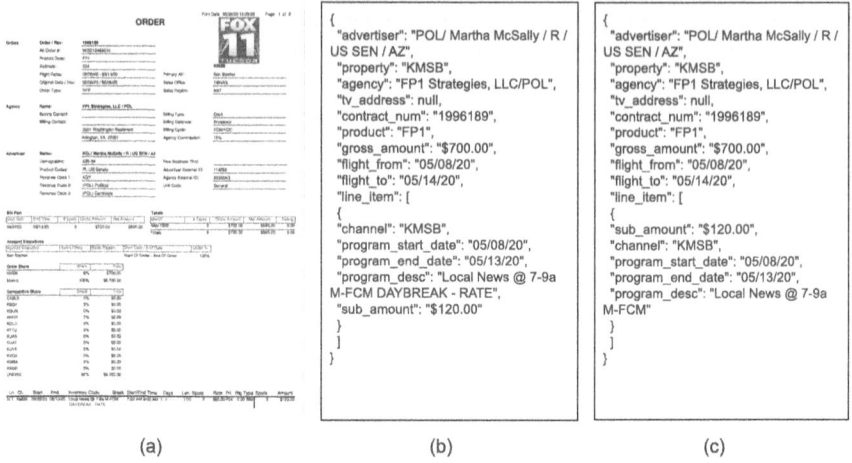

Fig. 4. (a) Sample image form VRDU. (b) Expected result. (c) Predicted result.

3.4 Zero-Shot Evaluation

In this experiment, we pass the input prompt to our LLM to evaluate its performance without tuning. Table 4 compares the proposed fine-tuned model against this zero-shot baseline on a subset of the VRDU dataset with unseen templates. We observe that certain information, such as advertiser and product names, can be extracted even without fine-tuning, due to their simple layout and straightforward availability for the LLM to detect. However, fine-tuning provides substantial gains, more than doubling scores across all categories by tailoring the model to the specific domain.

Table 4. Evaluation zero-shot on unseen template 100 subset (F1-Scores).

Model	Advertiser	Agency	Contract Num	Flight From	Flight To	Gross Amount	Product	TV Address	Property	Line Item	Macro	Unrepeated entities	Micro
CodeLlama [24]	72.5	44.03	46.61	49.2	57.75	50.39	84.32	13.23	42.81	2.48	46.33	51.66	21.99
Proposed	94.24	91.81	93.78	89.29	90.42	96.02	93.32	78.99	86.58	49.26	86.37	90.63	63.71

3.5 Evaluation of DeciLM-7B

To showcase the effectiveness of our encoding approach utilizing another LLM, we conducted a comparison with DeciLM-7B [1], a recently introduced instruction-following LLM that can handle long input contexts up to 8k. To ensure a fair comparison, we fine-tuned DeciLM-7B on the Ad-buy dataset using the same steps as our proposed model. Table 5 presents the results, showing that both models perform similarly. However, in our specific application, CodeLlama generally outperforms DeciLM-7B.

Table 5. Evaluation DeciLM-7B on unseen template 100 subset (F1-Scores).

Model	Advertiser	Agency	Contract Num	Flight From	Flight To	Gross Amount	Product	TV Address	Property	Line Item	Macro	Unrepeated entities	Micro
DeciLM-7B [1]	90.79	86.13	**95.61**	85.56	**90.8**	**96.04**	90.35	78.3	82.83	46.35	84.28	88.85	61.11
Proposed	**94.24**	**91.81**	93.78	**89.29**	90.42	96.02	**93.32**	**78.99**	**86.58**	**49.26**	**86.37**	**90.63**	**63.71**

3.6   Evaluation on CORD Dataset

We expanded our evaluation to include the CORD receipt dataset in two different settings: using only the first 50 samples to assess the model's few-shot learning capabilities, and using the complete dataset of 800 samples, in line with [23].

We followed the same procedure for prompt creation, training, and testing as applied to the VRDU dataset. However, this dataset presents an additional challenge as its lines in the OCR results are not aligned due to the presence of rotated and folded papers, making it difficult to construct the equivalent HTML.

Table 6 compares the n-TED accuracy [15] of various models on the CORD dataset, as reported in [23]. The results indicate that our model performs competitively in both training scenarios. With 50 samples, it achieves a higher n-TED accuracy compared to Donut and LayoutLMv3LARGE, but lower than LMDX-PaLM 2-S. With 800 samples, the model's accuracy increases and remains higher than Donut.

Table 6. Evaluation on CORD Dataset.

Size	Model	n-TED accuracy
50	Donut [15]	75.44
	LayoutLMv3LARGE [14]	87.29
	LMDX $_{PaLM\ 2\text{-}S}$ [23]	93.80
	proposed	89.9
800	Donut [15]	90.23
	LayoutLMv3LARGE [14]	96.21
	LMDX $_{PaLM\ 2\text{-}S}$ [23]	96.3
	proposed	91.4

4   Discussion

As we continue our research on using LLMs for various tasks, we encourage the community to adapt their inputs to be more easily understood by machines. LLMs that are familiar with HTML and JSON formats could be particularly advantageous. Our goal has been to demonstrate the practical utility of using HTML versions created from OCR data for information extraction with LLMs. For further accuracy, we recommend using any available proprietary software equipped with advanced features such as layout analysis, table detection, and superior OCR capabilities. Having additional metadata can greatly enhance the creation of precise HTML representation of a document.

The evaluation metric proposed by the dataset has a low tolerance for incomplete answers, as it does not accept partially correct responses. This might lead to an unfair comparison when a model's answer is marked incorrect due to missing a few characters.

5 Conclusion

In conclusion, we introduced a new approach for leveraging LLMs to extract information from documents with complex layouts. Our approach, which converts OCR outputs into HTML formats, effectively preserves the spatial layout and textual content, allowing LLMs to accurately extract information into a structured JSON format. Our experiments on the VRDU benchmark show significant improvement compared to baseline models and are comparable to SOTA results within computational limits. We have verified the effectiveness and flexibility of our method through testing on different inputs and models. Our findings highlight the importance of input formatting and the choice of LLM in the performance of information extraction. Future efforts can focus on improving text encoding and employing grounding techniques.

References

1. Deciai research team. 2024. DeciLM-7B-instruct. https://huggingface.co/Deci/DeciLM-7B-instruct
2. Tesseract open source OCR engine. https://github.com/tesseract-ocr/tesseract
3. Achiam, J., et al.: GPT-4 technical report (2023). arXiv preprint arXiv:2303.08774
4. Breuel, T.M.: The hOCR microformat for OCR workflow and results. In: Ninth International Conference on Document Analysis and Recognition (ICDAR 2007). vol. 2, pp. 1063–1067. IEEE (2007)
5. Brown, T., et al.: Language models are few-shot learners. Adv. Neural. Inf. Process. Syst. **33**, 1877–1901 (2020)
6. Dettmers, T., Pagnoni, A., Holtzman, A., Zettlemoyer, L.: QLORA: efficient fine-tuning of quantized LLMs. In: Advances in Neural Information Processing Systems, vol. 36 (2024)
7. Dosovitskiy, A., et al.: An image is worth 16×16 words: Transformers for image recognition at scale (2020). arXiv preprint arXiv:2010.11929
8. Garncarek, Ł, et al.: LAMBERT: layout-aware language modeling for information extraction. In: Lladós, J., Lopresti, D., Uchida, S. (eds.) ICDAR 2021. LNCS, vol. 12821, pp. 532–547. Springer, Cham (2021). https://doi.org/10.1007/978-3-030-86549-8_34
9. Gu, Z., et al.: XYLayoutLM: towards layout-aware multimodal networks for visually-rich document understanding. In: Proceedings of the IEEE/CVF Conference on Computer Vision and Pattern Recognition, pp. 4583–4592 (2022)
10. Gur, I., et al.: Understanding html with large language models (2022). arXiv preprint arXiv:2210.03945

11. Ha, J., Haralick, R., Phillips, I.: Recursive X-Y cut using bounding boxes of connected components. In: Proceedings of 3rd International Conference on Document Analysis and Recognition, vol. 2, pp. 952–955 (1995). https://doi.org/10.1109/ICDAR.1995.602059

12. Hu, E.J., et al.: LoRA: Low-rank adaptation of large language models (2021). arXiv preprint arXiv:2106.09685

13. Huang, L., et al.: A survey on hallucination in large language models: Principles, taxonomy, challenges, and open questions (2023). arXiv preprint arXiv:2311.05232

14. Huang, Y., Lv, T., Cui, L., Lu, Y., Wei, F.: LayoutLMv3: pre-training for document AI with unified text and image masking. In: Proceedings of the 30th ACM International Conference on Multimedia, pp. 4083–4091 (2022)

15. Kim, G., et al.: OCR-free document understanding transformer. In: Avidan, S., Brostow, G., Cissé, M., Farinella, G.M., Hassner, T. (eds.) Computer Vision – ECCV 2022. ECCV 2022. LNCS, vol. 13688. Springer, Cham (2022). https://doi.org/10.1007/978-3-031-19815-1_29

16. Lee, C.Y., et al.: FormNet: Structural encoding beyond sequential modeling in form document information extraction (2022). arXiv preprint arXiv:2203.08411

17. Li, Q., Li, Z., Cai, X., Du, B., Zhao, H.: Enhancing visually-rich document understanding via layout structure modeling. In: Proceedings of the 31st ACM International Conference on Multimedia, pp. 4513–4523 (2023)

18. Li, Y., et al.: StrucTexT: structured text understanding with multi-modal transformers. In: Proceedings of the 29th ACM International Conference on Multimedia, pp. 1912–1920 (2021)

19. Luo, C., Cheng, C., Zheng, Q., Yao, C.: GeoLayoutLM: geometric pre-training for visual information extraction. In: Proceedings of the IEEE/CVF Conference on Computer Vision and Pattern Recognition, pp. 7092–7101 (2023)

20. Mangrulkar, S., Gugger, S., Debut, L., Belkada, Y., Paul, S., Bossan, B.: PEFT: State-of-the-art parameter-efficient fine-tuning methods (2022). https://github.com/huggingface/peft

21. Naveed, H., et al.: A comprehensive overview of large language models (2023). arXiv preprint arXiv:2307.06435

22. Park, S., et al.: CORD: a consolidated receipt dataset for post-OCR parsing. In: Workshop on Document Intelligence at NeurIPS 2019 (2019)

23. Perot, V., et al.: LMDX: Language model-based document information extraction and localization (2023). arXiv preprint arXiv:2309.10952

24. Roziere, B., et al.: Code Llama: Open foundation models for code (2023). arXiv preprint arXiv:2308.12950

25. Sassioui, A., et al.: Visually-rich document understanding: concepts, taxonomy and challenges. In: 2023 10th International Conference on Wireless Networks and Mobile Communications (WINCOM), pp. 1–7. IEEE (2023)

26. Touvron, H., et al.: Llama 2: Open foundation and fine-tuned chat models (2023). arXiv preprint arXiv:2307.09288

27. Wang, Z., Zhou, Y., Wei, W., Lee, C.Y., Tata, S.: A benchmark for structured extractions from complex documents (2022). arXiv preprint arXiv:2211.15421

28. Xu, Y., et al.: LayoutLMv2: Multi-modal pre-training for visually-rich document understanding (2020). arXiv preprint arXiv:2012.14740

29. Xu, Y., Li, M., Cui, L., Huang, S., Wei, F., Zhou, M.: LayoutLM: pre-training of text and layout for document image understanding. In: Proceedings of the 26th ACM SIGKDD International Conference on Knowledge Discovery & Data Mining, pp. 1192–1200 (2020)

GeoContrastNet: Contrastive Key-Value Edge Learning for Language-Agnostic Document Understanding

Nil Biescas[1,2] , Carlos Boned[1,2] , Josep Lladós[1,2] ,
and Sanket Biswas[1,2(✉)]

[1] Computer Vision Center, Catalonia, Spain
{cboned,josep,sbiswas}@cvc.uab.es
[2] Computer Science Department, Universitat Autònoma de Barcelona,
Catalonia, Spain
Nil.Biescas@autonoma.cat

Abstract. This paper presents **GeoContrastNet**, a *language-agnostic* framework to structured document understanding (DU) by integrating a contrastive learning objective with graph attention networks (GATs), emphasizing the significant role of geometric features. We propose a novel methodology that combines geometric edge features with visual features within an overall two-staged GAT-based framework, demonstrating promising results in both link prediction and semantic entity recognition performance. Our findings reveal that combining both geometric and visual features could match the capabilities of large DU models that rely heavily on Optical Character Recognition (OCR) features in terms of performance accuracy and efficiency. This approach underscores the critical importance of relational layout information between the named text entities in a semi-structured layout of a page. Specifically, our results highlight the model's proficiency in identifying key-value relationships within the FUNSD dataset for forms and also discovering the spatial relationships in table-structured layouts for RVLCDIP business invoices. Our code is accessible on this GitHub (†https://github.com/NilBiescas/GeoContrastNet).

Keywords: Document Understanding · Graph Neural Networks · Contrastive Learning · Language-Agnostic Learning

1 Introduction

Visual information extraction (VIE) [7,8,10,17] has played a fundamental role in Document AI, drawing increasing attention from both industry and academia. The task mainly includes the recognition of semantic text entities (also known as *entity labeling* or *named entity recognition*) and the extraction of relationships between them (also referred to as *entity linking*) from Visually-rich Documents (VrDs) like administrative forms and invoices. Language-based DU approaches [13,15,21] have proven to be the current state-of-the-art for both

E. H. Barney Smith et al. (Eds.): ICDAR 2024, LNCS 14804, pp. 294–310, 2024.
https://doi.org/10.1007/978-3-031-70533-5_18

tasks. But these approaches have the following drawbacks: 1) They are impractical for deployment in real-world industrial scenarios, where computational
resources may be limited, and processing efficiency is vital. 2) They rely heavily
on large-scale pretraining to learn upon on a single language (mainly English)
making them constrained in terms of their applicability in multilingual scenarios 3) The presence of sensitive content within these business documents often
restricts access during the training phase, necessitating the development of DU
models capable of learning from only geometrical constraints and the perception
of the layout (e.g. coordinates of the word bounding boxes).

Given forms or invoices in an unfamiliar language, humans can generally
deduct the composed text entities and their relationships using layout cues and
some experience or prior which they try to approximate. Administrative documents frequently exhibit a semi-structured format, lacking a consistent layout
but containing a shared group of elements such as headers, footers, senders,
recipients and some entities. This spatial arrangement can often be perceived as
a table-like layout. Graphs can effectively capture such topological features of
documents with table-like layouts by representing the spatial and hierarchical
relationships between document elements in a structured manner.

Inspired by prior works on Graph Neural Networks (GNNs) [6,9,10,26,29]
to interpret administrative documents using mainly the layout information, we
propose a two-staged GNN model called *GeoContrastNet* that does not require
language information and could be potentially applied to visually similar languages without requiring fine-tuning. Also, existing GNN methods [9,10] have
mainly relied on learning a message passing between the different node components (e.g. classes of text segments like header, sender, recipient etc.) to predict
the pairwise relationship between the edges (entity linking). Contrary to this,
we propose a simple contrastive training strategy on the GNN [19] to learn some
robust edge features that include spatial proximity (how close elements are to
each other), hierarchical relationships (parent-child relationships between elements, such as a table and its cells), and the sequential order (the reading order
of text blocks). The key intuition is such representative grounded edge features
learnt during this contrastive training could essentially serve as a strong prior
in the second stage which uses a Graph Attention Network (GAT) [28] to solve
both node classification (for entity recognition) and edge classification (for entity
labeling) simultaneously.

Poor quality scans, variations in resolution, or inconsistencies in formatting
can often hinder the ability to accurately extract and utilize visual features as
observed in Doc2Graph [10]. To alleviate this issue, GeoContrastNet introduces
a new grounding mechanism that guides the graph attention to combine visual
and geometric features. With experimental evidence, we show the utility and
effectiveness of visual features when combined with rich representative geometric priors learnt during the contrastive stage. The novelties of this work can
be divided into four folds: 1) We propose a two-staged language-agnostic GNN
framework, GeoContrastNet, that introduces a simple contrastive learning strategy in the first stage to learn robust and generalized edge features over document

samples. 2) The framework also introduces graph attention in the second stage to ground the previously learnt edge features (representing key-value components) with the visual features. 3) A comprehensive analysis of the different sets of geometric features (both nodes and edges) has been studied and their utility for the entity recognition and labeling tasks. 4) To justify the effectiveness of the geometric features learnt during training, we also show a quantitative evaluation of our geometric-only model for invoice understanding task.

2   Related Work

Layout Representation Learning. The state-of-the-art DU foundation models [1,15,25,32,33] relies on large-scale pretraining focusing more on language rather than visual or geometrical elements for solving document intelligence tasks like classification [12], information extraction [16,17] and document visual question answering [23,24]. They introduced spatial layout information through 2D bounding box coordinates from an OCR as layout features to the language model. Other approaches [11,20] have focused on representing layout at the region level, identifying logical components such as paragraphs, figures, titles and tables which are essential for tasks like document layout analysis [3–5,22]. GeoLayoutLM [21] adds geometric constraints over LayoutLMv3 [15] using geometry-based pre-training objectives between the text segments in a self-supervised fashion. This helps them improve significantly on key-value entity linking tasks for forms [17]. Motivated by the effectiveness of layout features for several document understanding tasks, we propose a novel contrastive paradigm to learn geometrical layout representation for VrDs.

Language-Agnostic Document Understanding. Most of the DU foundation models are built upon heavy reliance on pretraining with language features to solve downstream tasks. Existing language-agnostic DU models like LayoutXLM [34] and LILT [31] also incorporate large-scaled multilingual pretrained models. Davis *et al.* [9] addressed this issue by achieving almost the same level of entity linking performance on the FUNSD [17] form understanding dataset by learning only visual features from the small provided training data using a Graph Convolutional Network (GCN) backbone. Voutharoja *et al.* [29] used a neuro-symbolic approach that uses an entity-relation graph for scanned forms. Although it achieves great results on the FUNSD benchmark, the features cannot be generalized in a more practical industrial scenario where data is online. Inspired by the graph language models [10,30], we build upon the promising geometric (both node and edge) representation learning which could help us build a language-agnostic DU model that does not utilize OCR but rather focuses on learning key-value relationships in document forms and invoices from a purely visual perspective.

3   Method

In this section, we will look closely into the proposed methodology of GeoContrastNet, the formulation of the problem, the two stages (task modules) incorporated in GeoContrastNet framework, and the learning objectives employed in the architecture shown in Fig. 1.

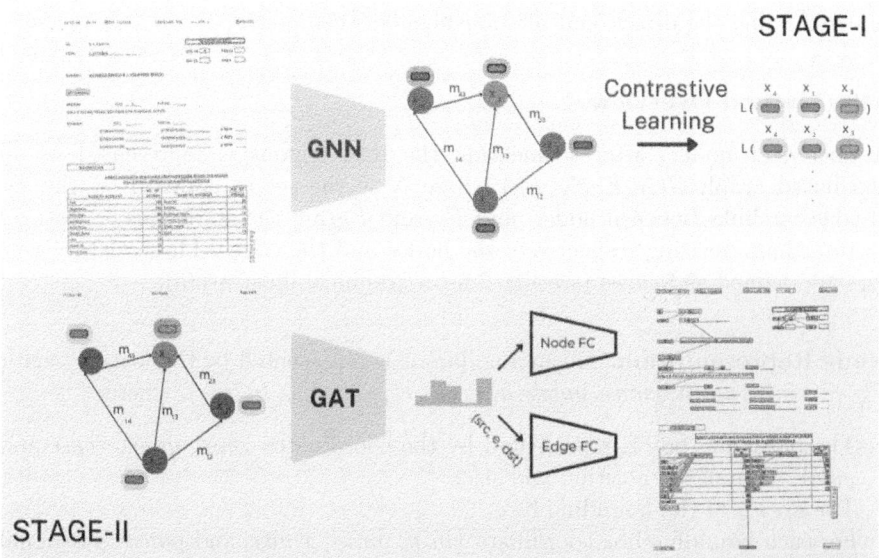

Fig. 1. Overall Architecture of the Proposed GeoContrastNet Framework. In Stage I, a GNN processes a document image represented as an attributed graph, learning key-value edge features in a contrastive setting using the triplet margin loss. Then, in Stage II, a GAT traverses these features to predict entity labels (nodes) and their relations (edges) as output.

3.1   Graph-Based Document Representation

Given a document image, it is represented as an attributed graph. The extraction of the document objects of the layout is performed by a layout segmentation process using an off-the-shelf Optical Character Recognition (OCR) provided in the ground truth or using the YOLOv5 [18] algorithm to get the bounding box regions of the page text regions. The image is segmented into text regions, represented by the corresponding bounding boxes. Graph nodes correspond to these text bounding boxes, attributed with the geometric attributes: its bounding box, the area of the bounding box, and a regional encoding that encapsulates the node's position within the document. Specifically, the regional encoding is designed to capture the spatial distribution of nodes, providing a comprehensive representation of each node's context. Graph edges represent the structural information of the document. The links between the nodes are constructed using the k-NN algorithm, where each node is linked with its k nearest neighbours in terms of spatial information. Each edge m_{ij} that connects node i to node j, is represented by a feature vector that includes several geometric attributes: the angle θ_{ji} between nodes i and j, measured in a standard reference frame; the Euclidean distance d_{ij} between the nodes, which reflects their spatial separation; discretized polar coordinates that offer a granular view of the relative positioning; and the relative positions of i and j, which provides information for understanding the directional relationships between nodes.

3.2 Method Overview

A document image, after segmenting the text regions is represented as an attributed graph $G(N, E, F_N, F_E)$ where N is the set of nodes, E is the set of edges or links between nodes, and F_N and F_E are the attributes or features vectors characterizing respectively the nodes and the edges. These feature vectors are defined as follows (see Fig. 2 for a graphical illustration):

Node Representation. Given a node i, it is represented by the feature vector $F_N^i = (xmin, ymin, xmax, ymax, a, r_{xmin}, r_{ymin}, r_{xmax}, r_{ymax})$, where:

- The bounding box is represented by the coordinates $xmin$, $ymin$, $xmax$ and $ymax$, defining its position and size.
- The area a of the bounding box.
- For each bounding box coordinate $xmin$, $ymin$, $xmax$, and $ymax$, a regional encoding is defined as r_{xmin}, r_{ymin}, r_{xmax}, and r_{ymax}. These encodings are derived from the normalize coordinates relative to the size of the image, with the largest dimension adjusted to a scale of 1. After this normalization, the document is conceptually divided into four equal sections along a single dimension. A code from the set $\{11, 12, 21, 22\}$ is assigned to each normalized coordinate based on which of these sections it falls into. This approach ensures that every coordinate is consistently and accurately represented within the normalized structure of the document.

Edge Representation. Given two nodes i, j, an edge (i, j) that links them is represented by the feature vector $F_E^{ij} = (\theta_{ji}, d_{ij}, pc_{ij}, rp_{ij})$, where:

- The angle θ_{ji} represents the orientation of the source node relative to the destination node, encapsulating the directional relationship between them.
- The Euclidean distance d_{ij} measures the direct spatial separation between two nodes, providing a quantitative assessment of their proximity.
- The relative positioning of nodes is determined using discretized polar coordinates pc_{ij}. Each source node is positioned at the center of a Cartesian plane, facilitating the encoding of its neighbors based on distance and angle relative to it. The space is divided into various bins (one-hot encoded), allowing for a selectable number of partitions.

– Relative positions rp_{ij} of nodes to provide more information of a global position of the nodes inside the document. The position is encoded based in a dictionary of language tokens that describe relative positional information: *left, right, top, bottom, vert-intersect, hor-intersect,sqr-intersect.*

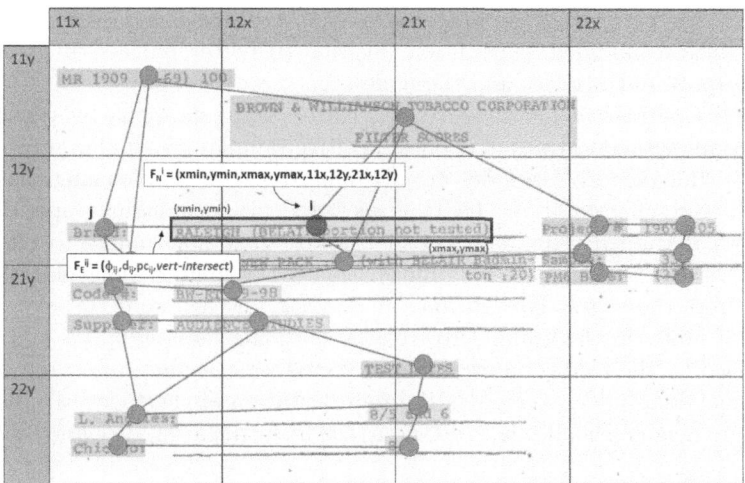

Fig. 2. Visual Illustration of Proposed Node and Edge Representation. Node features are defined by the bounding box coordinates, the area and regional encoding providing local and global structure position. Edge features provide information to geolocate each node relative to its neighbors.

Documents have a rich geometrical information that play a crucial role in understanding the overall structure of the entities. We propose two independent stages to address the geometric edge feature learning and the prediction phase.

– In *Stage-I*, during this phase, we explore the broad applications of the Graph Neural Network (GNN) model, focusing on its ability to process and learn from node and edge information. Key concepts in this phase include the implementation of triplet loss and a contrastive setting, which are instrumental in enhancing the model's understanding of the graph's topology and the relationships between its elements.
– In *Stage-II*, the model's learning objectives expand to encompass the joint learning of semantic entities and link prediction. We finetune a Unet to extract visually rich features, that then are combined with the features obtain in *Stage –I* to produce the input vector for the GAT layers.

In the following subsections we further describe the mentioned stages.

3.3   Stage I: Geometric Edge Feature Learning

GNNs are neural models designed to capture dependencies within the data that is represented in the graph nodes and edges. This is achieved by the concept of *message passing* consisting in propagating information between nodes and aggregating this information in node embeddings. In this way, at each layer n, each node is capturing the information of nodes that are n hops away, so the context is encoded in the node embedding. The Stage-1 architecture includes a GNN with a modified aggregation function to better process the geometric information from the edges and the nodes.

Graph Contrastive Learning (GCL) learns representations by contrasting positive samples against negative ones. Positive samples are similar graph nodes or edges, while negative samples are dissimilar ones. By maximizing the agreement between representations, GCL allows us to learn meaningful graph features. The GNN is trained using a triplet margin loss function, aiming to refine the representation of node geometric information by leveraging the geometric data derived from the edges. The *Message Formation* for each node u_i belonging to the set of nodes in the graph, involves the outgoing message m_{ij} $i \neq j$, being determined by the feature information of the edge, such that $m_{ij} = e_{ij}$.

The outgoing node aggregates the representation of the edge features of its immediate neighborhood, $\{h_v, \forall v \in \mathcal{N}(u)\}$. The aggregation is defined by:

$$h'_v = \frac{c}{|\Upsilon(i)|} \sum_{j \in \Upsilon(i)} m_{ij} \tag{1}$$

where $\Upsilon(i) = \{v \in \mathcal{N}(u) : \|u - v\| < \text{threshold}\}$, $\|u - j\|$ is the Euclidean distance of nodes u and v normalized between 0 and 1 and c is a constant scale factor. Following the aggregation process, the updated feature vector F_N^{i+1} is formed by concatenating the outgoing node's feature vector F_N^i with the newly aggregated value h'_v, this concatenated vector is then used as input to a linear transformation, normalized by a layer normalization and subsequently passed through a ReLU activation function to yield the updated node representation.

$$F_N^{i+1}(i) = \text{ReLU}\left(\text{LayerNorm}\left(\mathbf{W} \cdot [F_N^i \parallel h'_v]\right)\right) \tag{2}$$

where, $\mathbf{W}$ is the weight matrix associated with the linear transformation, $[F_N^i \parallel h'_v]$ denotes the concatenation of F_N^i and the aggregation result, and ReLU and LayerNorm represent the ReLU activation function and layer normalization, respectively. After the GNN, the new node representations based on geometric features are used for calculating the contrastive loss.

3.4   Stage II: Prediction Phase

The design for the second stage employs a two layer Graph Attention Network (GAT). GATs leverages attention layers to allow nodes to attend over their neighborhood features. By specifying different weights to different nodes in a neighborhood, GATs enhance expressiveness without prior knowledge of the graph structure. Our GAT receives enhanced geometric edge representations from the initial phase, integrating them with visual embeddings. The source of these visual embeddings is a U-net encoder, that is trained while performing both entity linking and semantic entity labeling tasks.

Semantic Entity Labeling. The output from the Graph Attention Network (GAT) layer, denoted as h, is subsequently processed through a sequence of five linear projection layers. These layers transform h to match the dimensionality corresponding to the desired number of prediction classes. The transformation is concluded with the application of a softmax function to obtain probability distributions over the classes, followed by the argmax operator to determine the predicted entity for each node.

Entity Linking. We adopt the edge representation as proposed in Doc2Graph [10]. Each edge is treated as a triplet structure (src, e, dst) where the edge representation h_e is defined by:

$$h_e = h_{src} \parallel h_{dst} \parallel cls_{src} \parallel cls_{dst} \parallel e_{polar} \tag{3}$$

Here, h_{src} and h_{dst} are sourced from the node embeddings produced by the final layer of our Graph Attention Network (GAT). The softmax probabilities, cls_{src} and cls_{dst}, come from the node prediction layer's output logits, and e_{polar} refers to the polar coordinates as outlined in the original paper's Sect. 3.2. We note as well as in Doc2Graph that using the class information is helpful for the model to predict if there exists a link between two types of nodes. Finally h_e is fed into five layers of the Fully-Connected (FC) classifier.

3.5   Learning Objectives

In this subsection, we discuss the key objective functions used in our model training for both Stage-I and Stage-II.

Stage – I Objectives. In this stage the learning objective is to construct a new geometric representation for each of the N entities present in the graph. To do so we use a *contrastive learning loss*, more specifically a triplet margin loss, defined as follows:

$$L(a, p, n) = \max\{d(a_i, p_i) - d(a_i, n_i) + \text{margin}, 0\} \tag{4}$$

where

$$d(x_i, y_i) = \|x_i - y_i\|_p \tag{5}$$

Stage – II Objectives. During this phase, the Graph Attention Network (GAT) and the Unet encoder were trained jointly on both objectives of link prediction and semantic entity recognition. The training utilized a cross-entropy loss function for both objectives, with the final loss being a summation of the individual losses from each task. The overall loss objective can be mathematically represented as:

$$\mathcal{L}_{\text{total}} = \mathcal{L}_{\text{entity}} + \mathcal{L}_{\text{link}} \tag{6}$$

where $\mathcal{L}_{\text{entity}}$ and $\mathcal{L}_{\text{link}}$, are described by the following:

$$\mathcal{L}_{\text{entity}} = -\sum_{i=1}^{N} y_{\text{entity},i} \log(\hat{y}_{\text{entity},i}) \tag{7}$$

$$\mathcal{L}_{\text{link}} = -\sum_{j=1}^{E} y_{\text{link},j} \log(\hat{y}_{\text{link},j}) \tag{8}$$

Here, N and E signify the total number of entities and links, respectively. The symbols $y_{\text{entity},i}$ and $y_{\text{link},j}$ correspond to the actual labels for entities and links, while $\hat{y}_{\text{entity},i}$ and $\hat{y}_{\text{link},j}$ denote the predicted probabilities for the entities and links, respectively.

4   Experimental Validation

In this section, we introduce the experimental validation of our proposed method. Finally, we show a series of ablation studies to justify the effectiveness of the different components in our model.

4.1   Dataset Description

We evaluate our method on two datasets, the FUNSD dataset and the RVL-CDIP Invoices. The FUNSD dataset consists of 199 authentic, fully annotated scanned forms. These documents are a curated selection from the broader RVL-CDIP [12] collection, which contains 400,000 grayscale images of diverse documents. The overall dataset is organized into a training set with 149 samples and a test set comprising 50 samples. For model validation, we employ a random partitioning strategy on the training set to create a validation subset. Our evaluation covers semantic entity recognition, categorizing entities into "question, "answer, "header," or "other, as well as link prediction tasks. In the work of Riba *et.al.* [26] another subset of RVL-CDIP has been released. The authors selected 518 documents from the invoices classes, annotating 6 different regions, namely: "invoice info", "other", "positions", "receiver, "supplier", "total". The task that can be performed are layout analysis, in terms of node classification, and table detection, in terms of bounding box IoU threshold greater than 0.5.

4.2   Evaluation Metrics

We present our findings across two main tasks: link prediction and semantic entity recognition for FUNSD. In the context of link prediction, we evaluate our model using three metrics: F1 score for non-entity links (F1 None), F1 score for key-value pairs (F1 Key-Value), and the Area Under the Curve (AUC).

For semantic entity recognition, we focus on the micro-averaged F1 score (F1 Micro) to assess overall performance. For RVL-CDIP Invoices we evaluate on table detection and on layout analysis.

4.3   Implementation Details

Stage-I. consists of a two-layer Graph Convolutional Network (GCN). The initial layer begins with a 9-dimensional node vector, which is merged with a 15-dimensional vector from message passing to form a 24-dimensional vector. This vector is processed through an MLP, followed by layer normalization and a ReLU activation, resulting in a 15-dimensional vector. The second layer adopts a similar approach, where it projects a 30-dimensional vector through an MLP, reducing it to 17 dimensions.

Stage-II. incorporates the learned representations from Stage-I along with visual features extracted from each bounding box using a MobileNet encoder [14] from a pretrained UNet. The first GAT layer projects dimensions from 1465 to 1500. The subsequent layer includes two heads for multi-head attention, expanding the dimensions from 1500 to 3000. To prevent overfitting, each GAT layer incorporates residual connections and applies a 20% dropout to both the features in the attention mechanism and the attention weights. For downstream tasks, two modules handle entity linking and semantic entity labeling, respectively. Each module takes the output features from the GAT and maps them to the respective number of classes required for each task. For semantic entity labeling, five MLPs project from 3000 to 4 classes. In the entity linking task, five MLPs are used to map from 6014 to 2 classes.

Hardware. We trained GeoContrastNet using a single NVIDIA GeForce RTX 3090. The entire training process lasts 1 h and 10 min overall, with the Stage-I phase taking only 10 min and Stage-II taking the rest 1 h.

4.4   Competitors

As shown in Table 1, we show a fair comparison of our proposed method with the existing SOTA. The results show that we achieve promising results on the semantic entity labeling task with 64.76% F1 score among language-agnostic approaches which is almost on par with FUDGE [9]. For the entity linking task, we achieve a decent score of 32.45%, although it lags a bit behind our competitors Doc2Graph [10] and FUDGE [9]. While Doc2Graph [10] uses the text supervisory signal massively to boost the performance of it's model. On the other hand, FUDGE [9] largely performs better mainly due to the effective interplay between the geometric and visual features in their GCN architecture pipeline. Although GeoContrastNet learns rich geometric representation in Stage-I, the simple feature concatenation adapted in Stage-II doesn't give it a huge boost when compared with FUDGE. On the other hand, Voutharaja *et al.* [29] is

Table 1. SOTA Comparison on FUNSD. The results have been shown for both semantic entity labeling (SEL) and entity linking (EL) tasks with their corresponding metrics where T:Text, G:Geometry, V:Visual

Method	Modalities	GNN	F_1 (↑) SER	EL	# Params ×10^6
BROS [13]	T + V	×	0.8121	0.6696	138
LayoutLM [33]	T + V	×	0.7895	0.4281	343
FUNSD [17]	T + G	✓	0.5700	0.0400	–
FUDGE [9]	V + G	✓	0.6507	0.5241	12
Doc2Graph [10]	T + G + V	✓	0.8225	0.5336	6.2
Voutharoja et al. [29]	G	✓	0.8225	0.8540	0.0000081
GeoContrastNet(Ours) + YOLO	G + V	✓	0.5260	0.2438	14
GeoContrastNet(Ours) + GT	G + V	✓	**0.6476**	**0.3245**	14

not a robust end-to-end differentiable approach as they construct a heuristical entity-relation graph with some heavily handcrafted priors to train their GNN for FUNSD. This gives them extremely high performance on the linking task but it's not designed for generalizability in a real-world practical setting.

Table 2. Ablation on Geometric Features. We report results for different combinations of geometric features used in Stage-I during message passing.

Edge Features				Node Features			F_1 Nodes (↑)	F_1 per classes (↑)		AUC-PR (↑)
Distance	Angle	Discretize	Polar	Bounding	Area	Regional		None	K-V	
✓	✓	✓	✓	✓	✓	✓	0.5049	0.8933	0.2017	0.5737
✗	✓	✓	✓	✓	✓	✓	0.4366	0.8825	0.1931	0.5767
✓	✗	✓	✓	✓	✓	✓	0.4051	0.8585	0.1808	0.5724
✓	✓	✗	✓	✓	✓	✓	0.5062	0.8805	0.1898	0.5761
✓	✓	✓	✗	✓	✓	✓	0.5244	0.8923	0.2059	0.5857
✓	✓	✓	✓	✗	✓	✓	0.5624	0.8779	0.1868	0.5850
✓	✓	✓	✓	✓	✗	✓	0.5412	0.9044	0.2168	0.5989
✓	✓	✓	✓	✓	✓	✗	**0.5750**	**0.9120**	**0.2290**	**0.6104**

4.5   Ablation Studies

We conducted extensive ablation studies to analyze the efficiency and generalizability of our method components. A series of tests were carried out using the GeoContrastNet framework.

Effectiveness of Geometric Features in Stage-I: To understand how various geometric features influence our results in the message passing in Stage-I, we examine the contribution of features derived from edges, nodes, or a combination of both, as detailed in Table 2. Our baseline incorporates all geometric features from both edges and nodes in the message passing, and we assess how each specific feature affects the performance in both entity linking and semantic entity recognition tasks. The best geometric features obtained in Stage-I correspond to the last line in Table 2. We also report results from using only the geometric features of the edges or the nodes in Table 3, where we observe that edge geometric information alone gives better results on both tasks compared to node geometric information. We hypothesize that edge information contains more spatial information of the surroundings, compared to node geometric information, enabling the model to perform better on both tasks.

Table 3. Node vs Edge Features. Results on both tasks when using either edge or node geometric information

Edge Features				Node Features			F_1 **Nodes** (↑)	F_1 per classes (↑)		**AUC-PR** (↑)
Distance	Angle	Discretize	Polar	Bounding	Area	Regional		None	K-V	
✓	✓	✓	✓	✗	✗	✗	0.5313	0.9056	0.2264	0.5871
✗	✗	✗	✗	✓	✓	✓	0.3532	0.8635	0.1798	0.5579

Table 4. Ablation Study for Modalities. Results of the different combination of modalities in GeoContrastNet.

Features		**AUC-PR** (↑)	F_1 per classes (↑)	
Stage-I Geometric	**Visual**		None	**Key-Value**
✗	✓	0.5483	0.8314	0.1581
✓	✗	0.5871	0.9056	0.2264
✓	✓	**0.6375**	**0.9825**	**0.3245**

Table 5. Table Detection in terms of F1 score. A table is considered correctly detected if its IoU is greater than 0.50. Threshold values refers to edges to not be cut: in our case is set to 0.50 by the softmax in use.

Method	Threshold	Metrics (↑)		
		Precision	**Recall**	F_1
Riba et al. [26]	0.5	0.1520	0.3650	0.2150
GeoContrastNet(Ours)	0.5	**0.2718**	**0.2669**	**0.2693**

Role of Different Modalities: In Table 4 we evaluate the effect of geometric features with combination with the visual features in the FUNSD dataset. We observe an incredible increment in F_1 scores thanks to the fusion of geometric and visual features.

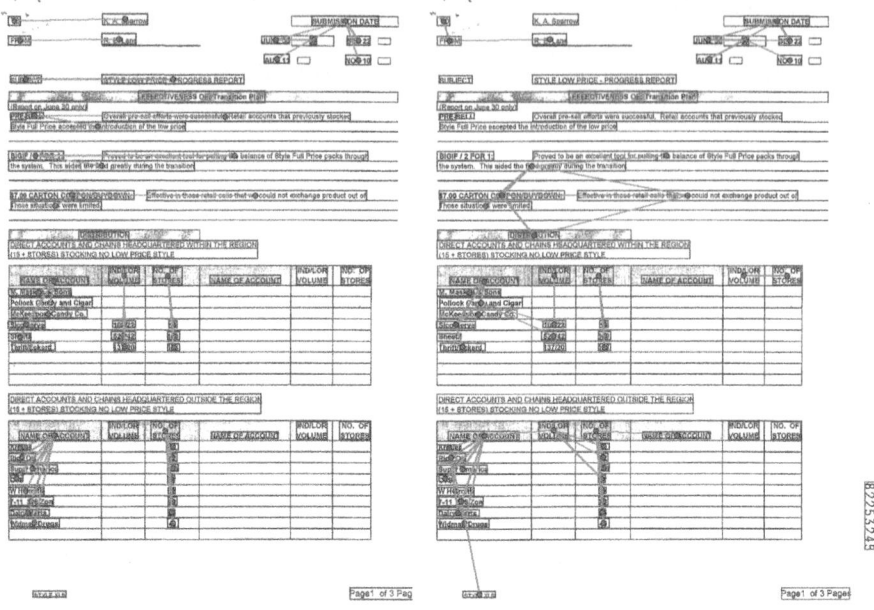

Fig. 3. Prediction on the link prediction task on **FUNSD dataset.** From (L to R) GT and predicted images respectively.

RVL-CDIP. We evaluate the proposed two stage model in the RVL-CDIP Invoices dataset. Our model outperforms in layout analysis and table detection as shown in Table 5. In particular, for table detection, we extracted the subgraph induced by the edge classified as 'table' (two nodes are linked if they are in the same table) to extract the target region. Riba et al. [26] formulated the problem as a binary classification: we report, for brevity, in Table 5 the threshold on confidence score they use to cut out edges, that in our multi-class setting ('none' or 'table') is implicitly set to 0.50 by the softmax.

4.6   Discussions and Analysis

Qualitative Discussion on FUNSD. As shown in Fig. 3 we see an example of a visually-rich form with ground-truth image compared with the predicted image from our model. The ground truth image (on the left) serves as a benchmark outlining the intended relationships and connections between semantic text entities within the document. It showcases the ideal mapping of links, providing a clear standard against which the predicted outcomes can be evaluated. On the other hand, the predicted image (on the right) our model's attempt to predict the links between the entities, to assess its capabilities. On closely examining, we see that there is a reason why our model gets a very high recall on the edges since it tries to predict a lot of relations in the page. There are some relations (which are not perfectly predicted) and hence it has a lower precision comparatively.

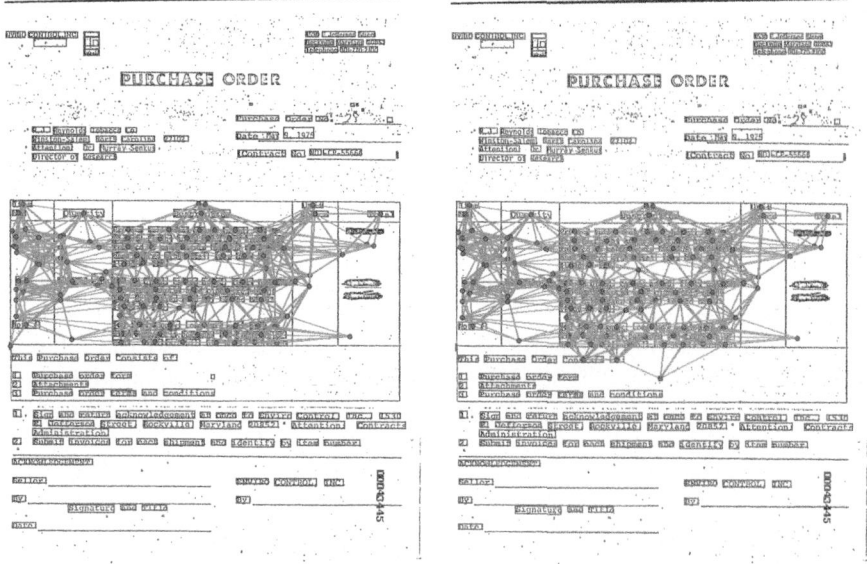

Fig. 4. Prediction on the link prediction task on **RVL-CDIP Invoices.** From (L to R) GT and predicted images respectively.

Qualitative Discussion on RVL-CDIP Invoices. As shown in Fig. 4, we see how the model is able to predict tables correctly. This also shows that the edge classifier performs really well and according to Table 5 attributes the very high accuracy numbers achieved for the layout analysis task. The much improved precision rate of the model compared to Riba *et al.* [26] is attributed to the geometric edge feature learning phase which results in more correctly predicted key-value links.

Information Extraction for Privacy-Preserving. In recent times, we have seen that Document AI is moving towards solving DU tasks on documents that contain sensitive or copyrighted content. A recent challenge has been launched for visual question answering [27] that deals in such scenario. Our geometric-only model (without integrating any visual or textual modality) contributes a step towards moving in this direction. We show results illustrated in Table 4 to actually show the potential of GeoContrastNet for the challenging table detection task without incoroporating any visual or textual information.

5 Conclusion and Future Work

In conclusion, GeoContrastNet offer a powerful and versatile framework for capturing the fine-grained topological features of documents with table-like layouts. By representing document text entities as nodes and explicitly learning their relationships (edge features) using a contrastive strategy, these models show highly promising results on both form and invoice understanding tasks. Also, we study the effectiveness of the visual features in this work by showing how the graph attention module help to align the layout structure with visual components of the page. This two-stage approach highlights the importance of geometric information inside structured documents, which can be beneficial to process complex document layouts in a privacy-preserving document understanding (e.g. visual question answering [27]) setting.

Future Scopes and Challenges. The graph construction, particularly through the use of the K-Nearest Neighbor (KNN) algorithm, presents certain limitations by conditioning both the structure of the graph and the information flow among nodes. The KNN method sets node connections using a fixed constant K, which conditions message transmission by determining the information available to each node at any given moment T. The choice of K can introduce biases that potentially affect the model's outcomes. GeoContrastNet, which employs the KNN algorithm to establish the graph's connections, may inadvertently incorporate these biases, impacting the learning process. Another limitation of GeoContrastNet is the fusion mechanism implemented which can be vastly improved using an attention mechanism [2] to align the vector spaces across the modalities. Also, we would like to explore the ability of our language-agnostic GNN model in multilingual settings [35] for future work.

Acknowledgment. Work co-supported by the Spanish projects PID2021-126808OB-I00 (GRAIL) and CNS2022-135947 (DOLORES), the Catalan project 2021 SGR 01559, the PhD Scholarship from AGAUR 2023 FI-3-00223 and the CVC Rosa Sensat Student Fellow. The Computer Vision Center is part of the CERCA Program/Generalitat de Catalunya.

References

1. Appalaraju, S., Jasani, B., Kota, B.U., Xie, Y., Manmatha, R.: DocFormer: end-to-end transformer for document understanding. In: Proceedings of the IEEE/CVF International Conference on Computer Vision, pp. 993–1003 (2021)
2. Bahdanau, D., Cho, K., Bengio, Y.: Neural machine translation by jointly learning to align and translate (2014). arXiv preprint arXiv:1409.0473
3. Banerjee, A., Biswas, S., Lladós, J., Pal, U.: SwinDocSegmenter: an end-to-end unified domain adaptive transformer for document instance segmentation. In: Fink, G.A., Jain, R., Kise, K., Zanibbi, R. (eds.) Document Analysis and Recognition - ICDAR 2023. ICDAR 2023. LNCS, vol. 14187. Springer, Cham (2023). https://doi.org/10.1007/978-3-031-41676-7_18
4. Biswas, S., Banerjee, A., Lladós, J., Pal, U.: DocSegTr: An instance-level end-to-end document image segmentation transformer (2022). arXiv preprint arXiv:2201.11438
5. Biswas, S., Riba, P., Lladós, J., Pal, U.: Beyond document object detection: instance-level segmentation of complex layouts. Int. J. Doc. Anal. Recogn. (IJDAR) **24**(3), 269–281 (2021)

6. Carbonell, M., Riba, P., Villegas, M., Fornés, A., Lladós, J.: Named entity recognition and relation extraction with graph neural networks in semi structured documents. In: 2020 25th International Conference on Pattern Recognition (ICPR), pp. 9622–9627. IEEE (2021)
7. Davis, B., Morse, B., Cohen, S., Price, B., Tensmeyer, C.: Deep visual template-free form parsing. In: 2019 International Conference on Document Analysis and Recognition (ICDAR), pp. 134–141. IEEE (2019)
8. Davis, B., Morse, B., Price, B., Tensmeyer, C., Wigington, C., Morariu, V.: End-to-end document recognition and understanding with dessurt. In: Karlinsky, L., Michaeli, T., Nishino, K. (eds.) Computer Vision – ECCV 2022 Workshops. ECCV 2022. LNCS, vol. 13804. Springer, Cham (2023). https://doi.org/10.1007/978-3-031-25069-9_19
9. Davis, B., Morse, B., Price, B., Tensmeyer, C., Wiginton, C.: Visual FUDGE: form understanding via dynamic graph editing. In: Lladós, J., Lopresti, D., Uchida, S. (eds.) ICDAR 2021. LNCS, vol. 12821, pp. 416–431. Springer, Cham (2021). https://doi.org/10.1007/978-3-030-86549-8_27
10. Gemelli, A., Biswas, S., Civitelli, E., Lladós, J., Marinai, S.: Doc2Graph: a task agnostic document understanding framework based on graph neural networks. In: Karlinsky, L., Michaeli, T., Nishino, K. (eds.) Computer Vision – ECCV 2022 Workshops. ECCV 2022. LNCS, vol. 13804. Springer, Cham (2023). https://doi.org/10.1007/978-3-031-25069-9_22
11. Gu, J., et al.: UniDoc: unified pretraining framework for document understanding. Adv. Neural. Inf. Process. Syst. **34**, 39–50 (2021)
12. Harley, A.W., Ufkes, A., Derpanis, K.G.: Evaluation of deep convolutional nets for document image classification and retrieval. In: International Conference on Document Analysis and Recognition (ICDAR)
13. Hong, T., Kim, D., Ji, M., Hwang, W., Nam, D., Park, S.: BROS: a pre-trained language model focusing on text and layout for better key information extraction from documents. In: Proceedings of the AAAI Conference on Artificial Intelligence. vol. 36, pp. 10767–10775 (2022)
14. Howard, A.G., et al.: MobileNets: Efficient convolutional neural networks for mobile vision applications (2017). arXiv preprint arXiv:1704.04861
15. Huang, Y., Lv, T., Cui, L., Lu, Y., Wei, F.: LayoutLMv3: pre-training for document AI with unified text and image masking. In: Proceedings of the 30th ACM International Conference on Multimedia, pp. 4083–4091 (2022)
16. Huang, Z., et al.: ICDAR2019 competition on scanned receipt OCR and information extraction. In: 2019 International Conference on Document Analysis and Recognition (ICDAR), pp. 1516–1520. IEEE (2019)
17. Jaume, G., Ekenel, H.K., Thiran, J.P.: FUNSD: a dataset for form understanding in noisy scanned documents. In: 2019 International Conference on Document Analysis and Recognition Workshops (ICDARW). vol. 2, pp. 1–6. IEEE (2019)
18. Jocher, G., et al.: ultralytics/yolov5: v6. 1-TensorRT, TensorFlow edge TPU and OpenVINO export and inference. Zenodo, Feb **22** (2022)
19. Kipf, T.N., Welling, M.: Semi-supervised classification with graph convolutional networks (2016). arXiv preprint arXiv:1609.02907
20. Li, P., et al.: SelfDoc: self-supervised document representation learning. In: Proceedings of the IEEE/CVF Conference on Computer Vision and Pattern Recognition, pp. 5652–5660 (2021)
21. Luo, C., Cheng, C., Zheng, Q., Yao, C.: GeoLayoutLM: geometric pre-training for visual information extraction. In: Proceedings of the IEEE/CVF Conference on Computer Vision and Pattern Recognition, pp. 7092–7101 (2023)

22. Maity, S.: et al.: SelfDocSeg: a self-supervised vision-based approach towards document segmentation. In: Fink, G.A., Jain, R., Kise, K., Zanibbi, R. (eds.) Document Analysis and Recognition - ICDAR 2023. ICDAR 2023. LNCS, vol. 14187. Springer, Cham (2023). https://doi.org/10.1007/978-3-031-41676-7_20

23. Mathew, M., Bagal, V., Tito, R., Karatzas, D., Valveny, E., Jawahar, C.: InfographicVQA. In: Proceedings of the IEEE/CVF Winter Conference on Applications of Computer Vision, pp. 1697–1706 (2022)

24. Mathew, M., Karatzas, D., Jawahar, C.: DocVQA: a dataset for VQA on document images. In: Proceedings of the IEEE/CVF Winter Conference on Applications of Computer Vision, pp. 2200–2209 (2021)

25. Powalski, R., Borchmann, Ł, Jurkiewicz, D., Dwojak, T., Pietruszka, M., Pałka, G.: Going full-TILT boogie on document understanding with text-image-layout transformer. In: Lladós, J., Lopresti, D., Uchida, S. (eds.) ICDAR 2021. LNCS, vol. 12822, pp. 732–747. Springer, Cham (2021). https://doi.org/10.1007/978-3-030-86331-9_47

26. Riba, P., Dutta, A., Goldmann, L., Fornés, A., Ramos, O., Lladós, J.: Table detection in invoice documents by graph neural networks. In: 2019 International Conference on Document Analysis and Recognition (ICDAR), pp. 122–127. IEEE (2019)

27. Tito, R., et al.: Privacy-aware document visual question answering (2023). arXiv preprint arXiv:2312.10108

28. Veličković, P., Cucurull, G., Casanova, A., Romero, A., Lio, P., Bengio, Y.: Graph attention networks (2017). arXiv preprint arXiv:1710.10903

29. Voutharoja, B.P., Qu, L., Shiri, F.: Language independent neuro-symbolic semantic parsing for form understanding (2023). arXiv preprint arXiv:2305.04460

30. Wang, D., Ma, Z., Nourbakhsh, A., Gu, K., Shah, S.: DocGraphLM: documental graph language model for information extraction. In: Proceedings of the 46th International ACM SIGIR Conference on Research and Development in Information Retrieval, pp. 1944–1948 (2023)

31. Wang, J., Jin, L., Ding, K.: LiLT: A simple yet effective language-independent layout transformer for structured document understanding (2022). arXiv preprint arXiv:2202.13669

32. Xu, Y., et al.: LayoutLMv2: Multi-modal pre-training for visually-rich document understanding (2020). arXiv preprint arXiv:2012.14740

33. Xu, Y., Li, M., Cui, L., Huang, S., Wei, F., Zhou, M.: LayoutLM: pre-training of text and layout for document image understanding. In: Proceedings of the 26th ACM SIGKDD International Conference on Knowledge Discovery & Data Mining, pp. 1192–1200 (2020)

34. Xu, Y., et al.: LayoutXLM: Multimodal pre-training for multilingual visually-rich document understanding (2021). arXiv preprint arXiv:2104.08836

35. Xu, Y., et al.: XFUND: a benchmark dataset for multilingual visually rich form understanding. In: Findings of the Association for Computational Linguistics: ACL 2022, pp. 3214–3224 (2022)

Transformers

Dynamic Relation Transformer for Contextual Text Block Detection

Jiawei Wang[1,3](✉), Shunchi Zhang[2,3], Kai Hu[1,3], Chixiang Ma[3], Zhuoyao Zhong[3], Lei Sun[3], and Qiang Huo[3]

[1] University of Science and Technology of China, Hefei, P.R. China
wangjiawei@mail.ustc.edu.cn
[2] Xi'an Jiaotong University, Xi'an, China
[3] Microsoft Research Asia, Beijing, China

Abstract. Contextual Text Block Detection (CTBD) is the task of identifying coherent text blocks in complex natural scenes. Previous methodologies have treated CTBD as either a visual relation extraction problem from the perspective of computer vision or as a sequence modeling problem from the perspective of natural language processing. We introduce a new framework that frames CTBD as a graph generation problem. This methodology consists of two essential procedures: identifying individual text units as graph nodes and discerning the sequential reading order relationships among these units as graph edges. Leveraging the cutting-edge capabilities of DQ-DETR for node detection, our framework innovates further by integrating a novel mechanism, a Dynamic Relation Transformer (DRFormer), dedicated to edge generation. DRFormer incorporates a dual interactive transformer decoder that manages a dynamic graph structure refinement process. Through this iterative process, the model gradually enhances the accuracy of the graph, ultimately resulting in improved precision in detecting contextual text blocks. Comprehensive experimental evaluations conducted on both SCUT-CTW-Context and ReCTS-Context datasets substantiate that our method achieves state-of-the-art results, underscoring the effectiveness and potential of our graph generation framework in advancing the field of CTBD.

Keywords: Graph Generation · Scene Text Detection · Text Region Detection

1 Introduction

Contextual Text Block Detection (CTBD) [39] aims to detect contextual text blocks (CTBs) within natural scenes, which are aggregates of one or more integral text units, such as characters, words, or text-lines, arranged in their natural

J. Wang, S. Zhang and K. Hu—Equal contribution
This work was done when J. Wang, S. Zhang, K. Hu were interns and C. Ma, Z. Zhong, L. Sun were full-time employees in Multi-Modal Interaction Group, Microsoft Research Asia, Beijing, China.

E. H. Barney Smith et al. (Eds.): ICDAR 2024, LNCS 14804, pp. 313–330, 2024.
https://doi.org/10.1007/978-3-031-70533-5_19

reading order. Unlike conventional scene text detectors that primarily focus on detecting individual words or text-lines, which results in capturing fragmented textual information devoid of its full context, CTBD endeavors to capture complete and coherent text messages by detecting contextual text blocks. This is essential for subsequent tasks including natural language processing and scene image understanding. An equivalent task in document layout analysis is text region detection, which concentrates on locating text blocks within document images. However, CTBD within natural scenes presents significantly greater challenges than text region detection within document images, attributable to three principal factors: 1) The extensive diversity in text font styles and sizes encountered in natural scenes; 2) The potential lack of clear alignment among text units that comprise a single CTB; 3) The prevalence of background noises within natural scenes that can obscure text.

The majority of existing methods in text region detection [6,11,15,33,47] have been designed with a focus on document images, frequently overlooking the complex text regions encountered in natural scenes. These methods are predominantly developed for document layout analysis, relying on OCR engines, or PDF parsers to determine the bounding boxes of text units. Such approaches often neglect the essential step of detecting text units themselves, a critical component of the analysis process. Moreover, there is a prevalent practice of recognizing only physical text blocks, disregarding the concept of logical text blocks. For example, a single logical block that spans across columns is typically divided into separate physical blocks, potentially resulting in a disjointed interpretation of the document's content.

Recently, HierText dataset [23] represents a significant advancement in acknowledging the hierarchical structure of text, offering a multi-level annotation schema that encompasses words, lines, and blocks in both natural scenes and document images. Despite this progress, HierText continues the trend of prior methodologies by not fully addressing the detection of logical text blocks. To address the challenge of capturing complete and meaningful text messages within the varying contexts of natural scenes, Xue et al. [39] introduced two comprehensive datasets, SCUT-CTW-Context and ReCTS-Context. These datasets are purposefully crafted for the detection of logical text blocks (aka contextual text blocks) within natural scenes. Leveraging these datasets, Xue et al. [39] developed Contextual Text Detector (CUTE), a method specifically designed to address contextual text block detection. CUTE adopts a natural language processing perspective to model the grouping and ordering of integral text units, extracting and transforming contextual visual features into feature embeddings to create integral text tokens, subsequently enabling the prediction of contextual text blocks. Despite these advancements, CUTE presents several limitations. Firstly, CUTE's performance is hampered in scenarios with densely packed text units, as its direct prediction of the subsequent text unit's index leads to an unwieldy and extensive index space. Secondly, the method exhibits shortcomings in precisely modeling the intricate relationships among text units that commonly arise in natural scenes. Thirdly, the utilization of an ROIAlign operator within CUTE for the extraction of local visual features proves to be insufficient for capturing a broader context required for accurate text block detection.

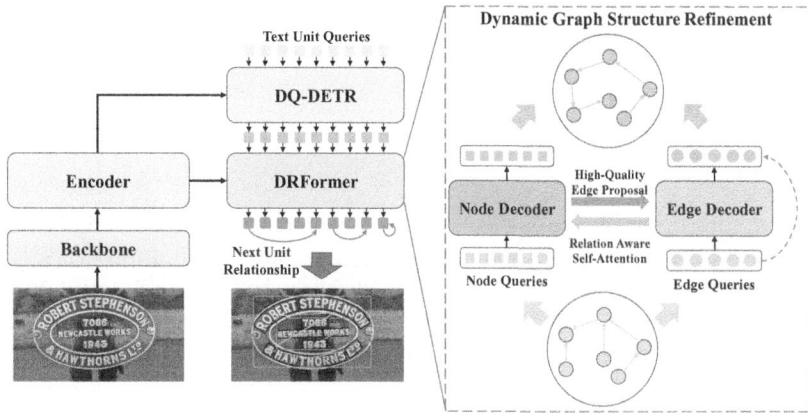

Fig. 1. Overview of our proposed framework for contextual text block detection.

In this paper, we introduce a novel approach to contextual text block detection by formulating it as a graph generation problem, where the graph comprises nodes and edges representing text units and the reading order relationships between them, respectively. As illustrated in Fig. 1, our approach comprises two primary components. Firstly, a DETR-like text detector (e.g., DQ-DETR [26]), which functions as an integral text detector, is tasked with the identification of text units that constitute the graph's nodes. Secondly, a novel *Dynamic Relation Transformer* based relation prediction module, called DRFormer, is dedicated to determining the reading order relationships among these detected text units, thereby facilitating the construction of the graph's edges. DRFormer employs a dual interactive transformer decoder to engage in a *dynamic graph structure refinement* process. The edge decoder enhances the node decoder's capabilities through *relation-aware self-attention*, which is informed by the graph's structure. Conversely, the node decoder contributes to the edge decoder's dynamic refinement process, leading to the generation of more accurate edge predictions. This iterative refinement procedure substantially improves contextual feature extraction from both node and edge queries. Moreover, we integrate a deformable attention mechanism [51] into our DRFormer, which broadens its capacity to incorporate diverse contextual information effectively. The experimental results demonstrate that our proposed method achieves state-of-the-art results on SCUT-CTW-Context and ReCTS-Context datasets.

The main contributions of this paper can be summarized as follows:

– We are the first to propose framing the task of contextual text block detection as a graph generation problem and to introduce an iterative refinement procedure that gradually improves the quality of the generated graph.

- We introduce a novel relation prediction module, DRFormer, which employs a dual interactive transformer decoder to engage in a dynamic graph structure refinement process.
- Our proposed approach has achieved state-of-the-art performance on SCUT-CTW-Context and ReCTS-Context benchmark datasets.

2   Related Work

2.1   Scene Text Detection

Scene text detection encompasses a spectrum of methodologies, broadly categorized into bottom-up and top-down strategies. Bottom-up approaches [1, 24, 27, 45] typically engage object detection frameworks to initially identify discrete text components, such as characters or text segments. These text components are subsequently aggregated to construct complete text instances. Conversely, top-down methods [21, 26, 28, 37, 46, 48, 50] consider entire words or text-lines as distinct object classes and employ a range of generic object detection or instance segmentation algorithms to identify them directly within a scene. For a comprehensive review of this topic, readers are referred to recent surveys in [22, 29, 52]. In this paper, we adopt a state-of-the-art text detector, DQ-DETR [26], as our text unit detection algorithm.

2.2   Text Region Detection

A text region is defined as a semantic unit of writing, which typically corresponds to a paragraph or a distinct block of text consisting of multiple text lines arranged according to a natural reading order. Recognizing these regions is a subtask within the broader domain of page object detection, which involves identifying and classifying various elements on a page, such as text regions, images, tables, and mathematical formulas. Text region detection can be approached through two primary methodological frameworks: bottom-up and top-down strategies.

Top-Down Methods. These methods leverage state-of-the-art top-down object detection or instance segmentation frameworks to tackle the challenge of text region detection. Early research by Yi et al. [42] and Oliveira et al. [31] adapted R-CNN [9] to identify and recognize text regions from document images. These pioneering attempts encountered limitations due to traditional strategies employed for region proposal generation. Progressing beyond these constraints, subsequent research has investigated the application of more advanced object detectors, including but not limited to Fast R-CNN [8], Faster R-CNN [34], Cascade R-CNN [3], SOLOv2 [38], YOLOv5 [13], and Deformable DETR [51] as investigated by Vo et al. [35], Zhong et al. [47], Li et al. [15], Biswas et al. [2], Pfitzmann et al. [33], and Yang et al. [41], respectively. Recently, it has been posited that the incorporation of textual modality information can substantially augment the efficacy of these detection frameworks. In this vein, Zhang et al. [44]

introduced a multi-modal Faster/Mask R-CNN model for text region detection that fused visual feature maps extracted by CNN with two 2D text embedding maps containing sentence and character embeddings. Furthermore, Li et al. [15], and Huang et al. [12] enhanced the capabilities of Faster R-CNN, Mask R-CNN, and Cascade R-CNN-based text region detectors by employing the pre-training of vision backbone networks on extensive corpora of document images via self-supervised learning algorithms. Although top-down methods have achieved significant success, they fall short in text region detection in natural scenes where text units within a same region can be widely dispersed and different regions may significantly overlap, presenting substantial challenges for accurate detection.

Bottom-Up Methods. These approaches formulate text region detection as a relation prediction problem, aiming to group text units (e.g., words, text-lines, connected components) into text regions by predicting relationships between them. Various approaches (e.g., [16,17,25,36]) represent each document page as a graph, where nodes correspond to basic text units, and edges denote inter-unit relationships. For instance, Li et al. [17] employed image processing techniques to initially generate line regions, subsequently employing two CRF models to predict whether pairs of line regions belong to a same text region, based on visual features extracted by CNNs. Advancing their research, Li et al. [16] substituted line regions with connected components as graph nodes and utilized a graph attention network (GAT) to refine the visual features of both nodes and edges. Wang et al. [36] focused on paragraph identification, developing a GCN-based approach to cluster text-lines into coherent paragraphs. Liu et al. [19] introduced a unified framework designed to concurrently perform text detection and paragraph (text-block) identification. Furthermore, Zhong et al. [49] advanced a multi-modal transformer-based model for the prediction of relations, thereby facilitating text region detection.

Despite achieving state-of-the-art results on several benchmark datasets, these methods predominantly target text region detection within document images, thereby neglecting the intricately structured text regions commonly found in natural scenes. Recent endeavors to bridge this gap have led to the introduction of datasets specifically designed for text region detection in natural scenes, such as HierText [23], SCUT-CTW-Context [39], and ReCTS-Context [39]. Long et al. [23] adopted an end-to-end instance segmentation model to detect arbitrarily shaped text and incorporated a multi-head self-attention layer to aggregate text regions. Xue et al. [39] approached text region detection from an NLP standpoint, conceptualizing text units as tokens that are subsequently aggregated into coherent token sequences corresponding to a same text region. Nevertheless, these approaches exhibit suboptimal performance in the presence of numerous text units. In this paper, we are the first to propose framing the task of text region detection within natural scenes as a graph generation problem and introducing an iterative refinement procedure that progressively improves the quality of the generated graphs.

3   Methodology

Our method conceptualizes the task of contextual text block detection as a graph generation problem, comprising two key components: (1) An integral text detector based on DQ-DETR [26], tasked with identifying individual text units and representing them as nodes within a graph; (2) A *dynamic relation transformer*, designed to discern the reading order relationships among these text units, thereby establishing the edges of the graph. These two components are jointly trained in an end-to-end manner. The overall architecture of our proposed approach is illustrated in Fig. 1. Subsequent subsections provide a detailed exposition of these components.

3.1   Integral Text Detector

Given an input image, we leverage a CNN backbone, such as ResNet [10], to generate a set of three multi-scale feature maps $\{C_3, C_4, C_5\}$. These feature maps are subsequently processed by a deformable transformer encoder [51], which enriches the pixel-level embeddings across each feature map. Leveraging the enhanced feature representations, we employ the DQ-DETR decoder to identify all text units of arbitrary shapes within the image, denoted by $\{t_1, t_2, t_3, ..., t_n\}$. For an in-depth understanding of the DQ-DETR framework, readers may consult the original publication [26]. The detected text units are then conceptualized as the nodes of a graph. These nodes are input into our dynamic relation transformer, which is tasked with predicting the reading order relationships among them, thus forming the graph's edges.

3.2   Dynamic Relation Transformer

As depicted in Fig. 2, our DRFormer introduces a *dynamic graph structure refinement* process, which iteratively enhances the quality of the generated graph. The process begins by generating initial node and edge queries through a *query initialization* module. These initial queries are then input to a *dual interactive decoder*, tasked with augmenting their embeddings. Following this, a relation prediction head and an edge classification head collaboratively work to refine the connections within the graph. Based on their output, incorrect edges are pruned and potential edge queries are formulated and added to the next decoder layer to assist in identifying further connections for the graph. Ultimately, the refined node and edge query embeddings produced by the terminal decoder layer are passed to the edge classification head, which plays a crucial role in filtering out false positives and in determining the correct reading order relationships among all the text units in the input image.

Query Initialization. For the node queries, we utilize the query embeddings and the bounding boxes of the detected text units produced by the DQ-DETR as the node embeddings $\{q_1^N, q_2^N, ..., q_n^N\}$ and corresponding node reference boxes

$\{B_1^N, B_2^N, ..., B_n^N\}$. Regarding the edge queries, we adopt the relation prediction head as proposed by Zhong et al. [49] to predict the subsequent text unit for each identified text unit. To bolster the recall rate of edge proposals, we select the top-K potential successors $\{t_{i_1}, t_{i_2}, ..., t_{i_k}\}$ for each text unit t_i, thereby establishing the initial edge proposals. The initialization of edge queries $\{q_{11}^E, q_{12}^E, ..., q_{nk}^E\}$ and their respective edge reference boxes $\{B_{11}^E, B_{12}^E, ..., B_{nk}^E\}$ is formalized as follows:

$$q_{ij}^E = FC_{init}([q_i^N, q_{i_j}^N]), \tag{1}$$

$$B_{ij}^E = Union\_Box([B_i^N, B_{i_j}^N]), \tag{2}$$

where FC_{init} denotes a dedicated fully-connected layer designed to merge the embeddings of q_i^N and $q_{i_j}^N$, and $Union\_Box$ is a function that computes the union bounding rectangle for the pair of bounding boxes B_i^N and $B_{i_j}^N$.

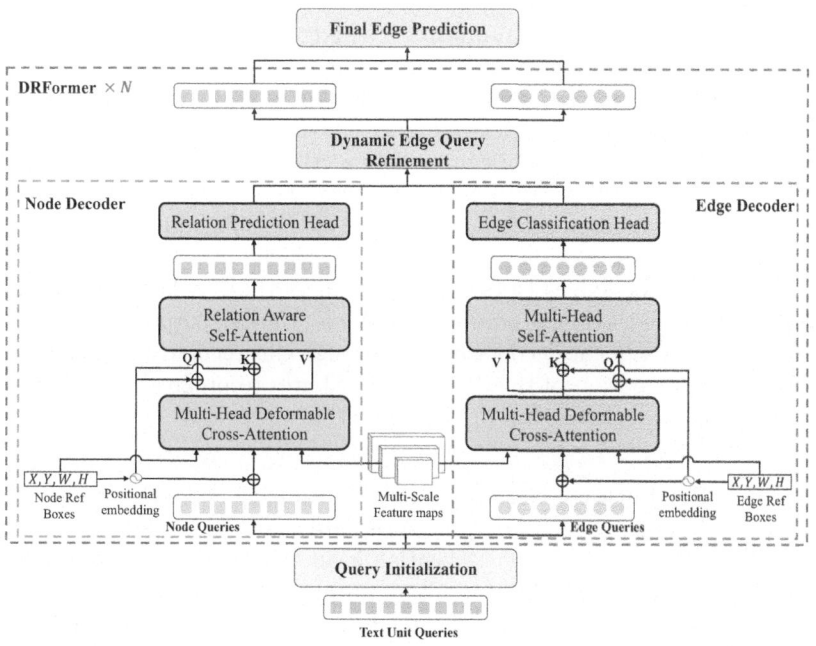

Fig. 2. The architecture of our proposed DRFormer, consisting of a dual interactive decoder. For clarity, we only show two attention layers of Transformer decoder and omit the FFN blocks.

Dual Interactive Decoder. As illustrated in Fig. 2, our DRFormer includes a dual interactive decoder, consisting of a node decoder and an edge decoder, each dedicated to refining the embeddings of their respective graph components. Inspired by the Mask2Former [5], our dual interactive decoder deviates from the vanilla deformable decoder architecture [51] by resequencing the cross-attention and self-attention operations within the decoder layers. The decoding process begins with the application of a deformable cross-attention module that dynamically samples relevant image features from the encoder's multi-scale feature maps. Following this, a self-attention mechanism is engaged to model the interactions among nodes and edges. To further concentrate the node decoder's attention on the complex relationships inherent to the graph structure, we incorporate a *relation-aware self-attention* module into the node decoding workflow. This novel module adheres to a guiding principle: it enables attention exchange exclusively among node pairs that are part of the same undirected connected sub-graph, while simultaneously restricting attention flow across nodes that belong to disparate connected sub-graphs. By employing this dual interactive attention approach, our system refines both node and edge embeddings with increased precision, thereby enhancing the overall accuracy of the graph generation.

Dynamic Edge Query Refinement. The edge query embeddings, updated at each iteration of the edge decoder, are input to a relation classification head designed to filter out incorrect edge proposals generated by the preceding layer. This classification head operates as a binary classifier, where proposals receiving low edge classification scores are pruned from the set of edge proposals. Post this filtration step, certain nodes may become isolated, devoid of any connecting edges. To address this, the enhanced node query embeddings from the node decoder are fed into a relation prediction head, which is tasked with generating the top-K most likely edges to potentially connect these isolated nodes. These candidate edges are then incorporated into the set of edge proposals as new edge queries for the subsequent decoder layer. Through this recursive refinement process, each successive layer of the decoder contributes to progressively improving the accuracy of the relationship prediction within the graph.

Final Edge Prediction. The node query embeddings and edge query embeddings, emitted by the terminal decoder layer, are channeled into both a relation prediction head and a relation classification head, which are responsible for generating the definitive edges of the resultant graph. In terms of architecture, both the relation prediction head and the relation classification head mirror their counterparts from the preceding layers.

3.3   Optimization

Loss for Integral Text Detector. The loss function for our detector, denoted as $L_{detector}$, aligns precisely with the loss function $L_{DQ-DETR}$ presented in [26]. This composite loss function consolidates a set of bounding box regression losses and classification losses stemming from the diverse prediction heads and denoising heads within the architecture. Specifically, the bounding box regression loss is formulated as a weighted sum of the L_1 loss and the GIoU loss [4], while the classification loss employs the focal loss [18] for its calculation. For a comprehensive understanding of these loss components, we direct the readers to the detailed descriptions provided in [26].

Loss for Relation Prediction Head. During the initial phase of edge query generation and at each stage of the decoder, we employ a relation prediction head to formulate edge proposals for the subsequent layer of the decoder. A consistent softmax cross-entropy loss is utilized across all relation prediction heads, which can be expressed as follows:

$$L_{relation}^{(l)} = \frac{1}{N} \sum_i L_{CE}\left(\boldsymbol{s}_i, \boldsymbol{s}_i^*\right) \tag{3}$$

where $\boldsymbol{s}_i$ denotes the predicted relation score vector that encapsulates the probability distribution for the text unit t_i linking to other text units, while $\boldsymbol{s}_i^*$ represents the corresponding ground truth label.

Loss for Edge Classification. For the task of edge classification, we employ a binary cross-entropy loss, which is articulated as follows:

$$L_{edge}^{(l)} = \frac{1}{N} \sum_i L_{BCE}\left(\boldsymbol{c}_i, \boldsymbol{c}_i^*\right) \tag{4}$$

where $\boldsymbol{c}_i$ signifies the logits corresponding to the i-th edge proposal as produced by the edge classification head, while $\boldsymbol{c}_i^*$ denotes the associated ground-truth label.

Overall Loss. All the components in our approach are jointly trained in an end-to-end manner. The overall loss is the sum of $L_{detector}$, $L_{relation}^{(l)}$ and $L_{edge}^{(l)}$:

$$L_{overall} = \lambda_1 L_{detector} + \lambda_2 \sum_{l=0}^{L} L_{relation}^{(l)} + \lambda_3 \sum_{l=1}^{L} L_{edge}^{(l)} \tag{5}$$

where L signifies the total number of decoder layers in the architecture, with $l = 0$ representing the initial phase of edge query generation, and $l = 1$ to L corresponding to the subsequent decoder layers. The coefficients $\lambda_1, \lambda_2, \lambda_3$ are uniformly set to 1 throughout the experiments reported in this study.

4 Experiments

4.1 Datasets

We conduct our experiments on SCUT-CTW-Context and ReCTS-Context datasets, both introduced in [39], to validate the effectiveness of our proposed framework.

SCUT-CTW-Context: The dataset, as annotated by Xue et al. [39], augments SCUT-CTW-1500 dataset [43] with contextual text blocks. It consists of a corpus of 940 training images and 498 test images. The majority of integral text units in this dataset are words, providing rich contextual information captured in various scenes.

ReCTS-Context: Similarly annotated by Xue et al. [39], this dataset enhances ICDAR2019-ReCTS [20] with reading order relationship annotations. It is partitioned into a training set of 15,000 images and a test set comprising 5,000 images. Characterized by Chinese script, the dataset presents characters as the fundamental textual elements, offering a unique challenge in the realm of reading order relationship prediction.

4.2   Evaluation Metrics

We adopt evaluation metrics proposed in [39] for the assessment of contextual text detection, which includes *Local Accuracy*, *Local Continuity*, and *Global Accuracy*. These metrics provide a comprehensive evaluation of the effectiveness of the proposed framework.

Local Accuracy (LA): This metric is employed to assess the accuracy of order prediction for neighboring integral text units, focusing on the local characteristics of integral text unit ordering.

Local Continuity (LC): LC evaluates the continuity of integral text units by computing a modified n-gram precision score, inspired by BLEU [32]. Similar to LA, LC also focuses on the local characteristics of integral text unit ordering.

Global Accuracy (GA): GA is introduced to evaluate the detection accuracy of contextual text blocks, representing a stringent constraint that provides a comprehensive measure of the overall performance of the proposed framework.

Moreover, a detected integral text unit is considered matched with a ground-truth text unit if the intersection-over-union (IoU) of their bounding boxes exceeds a certain threshold. We employ three commonly used IoU threshold standards in generic object detection tasks, including $IoU = 0.5$, $IoU = 0.75$, and $IoU = 0.5 : 0.05 : 0.95$ for comprehensive evaluation.

4.3   Implementation Details

Our methodology is implemented using PyTorch, and the experiments are conducted on a workstation equipped with 16 NVIDIA Tesla V100 GPUs (32 GB memory). To achieve the highest possible performance, we employ the latest state-of-the-art text detector, namely DQ-DETR [26], as the integral text detector in our framework. We use the same configuration as in DQ-DETR. The backbone in our framework is initialized with ResNet-50, which is pre-trained on ImageNet-1K dataset. The node decoder and edge decoder in DRFormer are both configured with 3 layers. Both are designed with the number of heads, the dimension of the hidden state, and the dimension of the feedforward network set as 8, 256, and 1024, respectively. The number of edge proposals for each node, denoted as K, is set to 3.

In accordance with the training strategy proposed in CUTE [39], our approach begins by training the integral text detector on the SCUT-CTW-Context and ReCTS-Context datasets separately. We adopt the same training configuration as DQ-DETR, with the only difference being that we do not pre-train our model on a mixture of SynthText150K [21], MLT2017 [30] and Total-Text [7]. Subsequently, we proceed to train the overall model while freezing the parameters in the backbone and DQ-DTER. To optimize the models, we employ Adam algorithm [14] with the following settings: a learning rate of 5e-4, betas set to (0.9, 0.999), epsilon set to 1e-8, and weight decay set to 1e-2. All models are trained for 5,000 iterations with a warmup strategy applied during the first 200 iterations. In each training iteration, we select 8 samples as a mini-batch for each GPU. We employ a similar multi-scale training strategy in DQ-DETR, randomly resizing shorter sides of an image to sizes ranging from 640 to 2,560 pixels without keeping aspect ratios. During inference, the longer side of each testing image is set to 1600.

4.4   Comparisons with Prior Arts

In this section, we conduct a comprehensive comparative analysis between DRFormer and several methods proposed in [39] across the SCUT-CTW-Context and ReCTS-Context datasets. To underscore the effectiveness of our proposed framework, we construct a solid baseline method, denoted as DRFormer-Node. The baseline configuration mirrors the structural pipeline of our framework, with the exception that it substitutes the DRFormer with a simpler 3-layer node decoder. Structurally, this baseline is akin to the text region detection architecture discussed in [49], with distinctive modifications. It diverges by replacing the multi-modal transformer encoder with a deformable decoder and by only leveraging visual modality information.

SCUT-CTW-Context. We benchmark DRFormer against existing baseline models on the SCUT-CTW-Context dataset, where integral text units are delineated at the granularity of individual words. Table 1 demonstrates that our robust baseline model, DRFormer-Node, which incorporates our proposed graph generation framework, exhibits marked superiority by outperforming the LINK-R101 [40] and CUTE-R101 [39] models. This showcases the effectiveness of our framework in the relevant domain. Building upon this robust foundation, our proposed DRFormer employs a *dynamic graph structure refinement* process to further enhance the accuracy of reading order relationships. This strategic refinement leads to DRFormer significantly outperforming the strong baseline, achieving noteworthy improvements across all metrics on the SCUT-CTW-Context dataset. Specifically, at an IoU threshold of 0.5, DRFormer achieves a 2% increase in Local Accuracy (LA), a 3.3% rise in Local Continuity (LC), and a 2% enhancement in Global Accuracy (GA). Notably, this superior performance is sustained even when faced with more stringent IoU thresholds. Within the challenging IoU evaluation spectrum of 0.5 to 0.95, incremented by 0.05, DRFormer continues to outpace the DRFormer-Node, showing increments of 1.2% in LA, 2.2% in LC, and 1.4% in GA, thereby confirming the model's robustness and consistent outperformance against more rigorous benchmarks.

Table 1. Quantitative performance comparison of DRFormer with state-of-the-art methods on SCUT-CTW-Context dataset. LA: Local Accuracy; LC: Local Continuity; GA: Global Accuracy.

Models	IoU = 0.5			IoU = 0.75			IoU = 0.5:0.05:0.95		
	LA	LC	GA	LA	LC	GA	LA	LC	GA
LINK-R50 [40]	25.5	3.3	18.9	20.3	3.2	14.7	19.3	2.9	14.3
CUTE-R50 [39]	54.0	39.2	30.7	41.6	31.2	23.7	39.4	29.0	22.1
LINK-R101 [40]	25.7	3.4	19.2	20.0	2.9	14.7	19.6	2.7	14.4
CUTE-R101 [39]	55.7	39.4	32.6	40.6	29.0	22.8	40.0	28.3	22.7
DRFormer-Node-R50	67.6	55.7	45.8	56.5	43.6	37.3	47.4	37.1	31.9
DRFormer-R50	**69.6**	**59.0**	**47.8**	**58.1**	**46.0**	**39.3**	**48.9**	**39.3**	**33.3**

Table 2. Quantitative performance comparison of DRFormer with state-of-the-art methods on ReCTS-Context dataset. LA: Local Accuracy; LC: Local Continuity; GA: Global Accuracy.

Models	IoU = 0.5			IoU = 0.75			IoU = 0.5:0.05:0.95		
	LA	LC	GA	LA	LC	GA	LA	LC	GA
LINK-R50 [40]	68.2	57.5	48.4	53.8	50.2	38.4	53.0	47.7	37.3
CUTE-R50 [39]	70.4	64.7	51.6	54.4	56.6	39.5	53.9	53.6	38.9
LINK-R101 [40]	70.8	59.1	49.9	54.5	51.0	39.0	53.4	48.3	37.9
CUTE-R101 [39]	72.4	67.3	53.8	55.1	**57.0**	40.2	54.6	**53.9**	39.4
DRFormer-Node-R50	82.2	71.4	69.6	63.2	50.0	52.8	56.4	46.0	47.6
DRFormer-R50	**83.3**	**74.6**	**71.8**	**67.6**	55.9	**56.9**	**59.4**	50.0	**50.6**

ReCTS-Context. Moreover, the effectiveness of our DRFormer is convincingly validated on the ReCTS-Context dataset, which is specifically designed to extract complex Chinese text displayed on signboards. The layout and arrangement of Chinese characters on signboards present unique challenges, differing significantly from other scenes. In the face of such intricacies, DRFormer demonstrates its superior performance capabilities, particularly within the stringent IoU range of 0.5 to 0.95. As summarized in Table 2, within this rigorous evaluative context, DRFormer delivers a noteworthy 3% improvement in Local Accuracy, a significant 4% enhancement in Local Continuity, and a remarkable 2.9%

augmentation in Global Accuracy, surpassing the strong baseline and solidifying its adeptness at handling complex text arrangements on signboards.

Table 3. Quantitative performance comparison of DRFormer with state-of-the-art methods on integral text grouping and ordering task on SCUT-CTW-Context and ReCTS-context datasets.

Models	SCUT-CTW-Context			ReCTS-Context		
	LA	LC	GA	LA	LC	GA
LINK-R50 [40]	30.2	4.5	22.8	83.8	68.4	61.1
CUTE-R50 [39]	71.5	58.5	49.7	92.1	82.8	76.0
LINK-R101 [40]	45.5	6.3	31.7	86.7	75.0	69.6
CUTE-R101 [39]	71.5	58.7	52.6	**93.1**	83.7	77.8
DRFormer-Node-R50	80.3	71.0	58.7	90.9	81.8	82.8
DRFormer-R50	**83.9**	**76.0**	**60.5**	92.8	**85.9**	**85.5**

Upper Bound Evaluation with GT Text Units. To validate the effectiveness of DRFormer solely in grouping and ordering integral text units, we introduce an additional evaluation criterion that assumes the accurate detection of all integral text units by leveraging the bounding boxes of ground-truth integral text units. As shown in Table 3, the proposed DRFormer demonstrates enhanced capability in the grouping and ordering of integral text units, outperforming all baseline models in this respect.

4.5 Ablation Studies

To thoroughly evaluate the effectiveness of various design aspects within DRFormer for the grouping and ordering of integral text units, we conducted a series of ablation experiments on the SCUT-CTW-Context dataset. These experiments assume the accurate detection of all integral text units by leveraging the bounding boxes of ground-truth integral text units.

Dynamic Graph Structure Refinement. Building upon a robust baseline model, our novel DRFormer integrates a dynamic graph structure refinement mechanism, which incrementally optimizes the quality of the constructed graph. This is achieved through the implementation of a dual interactive transformer decoder that facilitates a synergistic interaction between nodes and edges during the refinement process. As evidenced by the data presented in rows 1 and 2 of Table 4, the incorporation of the *Dynamic Graph Structure Refinement* leads to a measurable performance boost, with a 2% gain in Local Accuracy (LA) and a 1.6% increase in Local Continuity (LC). These statistics clearly demonstrate the critical contribution of our dynamic graph structure refinement approach in effectively predicting reading order relations within the generated graph.

Table 4. Ablation studies of various components within DRFormer on SCUT-CTW-Context dataset. (DGSR: Dynamic Graph Structure Refinement; CAF: Cross-Attention First; RASA: Relation-Aware Self-Attention)

#	Method	DGSR	CAF	RASA	LA	LC	GA
1	DRFormer-Node				80.3	71.0	58.7
2		✓			82.3	72.6	58.8
3		✓	✓		83.4	75.3	60.2
4	DRFormer	✓	✓	✓	**83.9**	**76.0**	**60.5**

Order of Self-attention and Cross-Attention. As outlined in Sect. 3.2, we propose reversing the conventional order of self-attention and cross-attention mechanisms within the transformer decoder architecture. In the context of our task, which involves predicting reading order relationships among text units, it is beneficial for the relation prediction model to first engage the cross-attention module to distill features from text units. Subsequently, the incorporation of the self-attention module empowers the model to skillfully decipher the interactions among these units, thereby augmenting its ability for accurate relationship prediction. The empirical results, as indicated in rows 2 and 3 of Table 4, reveal that the adoption of the *Cross-Attention First* strategy leads to a notable performance improvement, evidenced by a 1.1% rise in LA and a substantial 2.7% surge in LC.

Relation-Aware Self-attention. Within the structure of our proposed DRFormer, the edge decoder collaborates closely with the node decoder, imparting an understanding of the graph's structure. This collaboration empowers the node decoder to utilize *Relation-Aware Self-Attention*, enriching its comprehension of the intra-graph relationships. The outcomes of our ablation study, as shown in rows 3 and 4 of Table 4, indicate a performance enhancement when *Relation-Aware Self-Attention* is employed instead of the conventional self-attention mechanism. Specifically, we observe an increment of 0.5% in LA and 0.7% in LC, underscoring the benefits of our *Relation-Aware Self-Attention*.

4.6   Qualitative Results

To emphasize the effectiveness of our proposed methods, we offer a qualitative comparison between our baseline and DRFormer. As depicted in Fig. 3, DRFormer demonstrates superior accuracy in establishing edge relationships through dynamic graph structure refinement, significantly improving the overall quality of contextual text block detection compared to our baseline method.

4.7   Limitations of Our Approach

While our proposed DRFormer achieves outstanding performance in the task of contextual text block detection, as validated by previous experiments, it does

Fig. 3. Qualitative comparison between our proposed baseline (top) and DRFormer (bottom). Black bounding boxes indicate word-level or character-level integral texts, *i.e.*, graph's nodes, and brown arrows represent the reading order relationships between these integral texts, *i.e.*, graph's edges, which finally lead to contextual text blocks outlined in green bounding boxes. Best viewed in color.

come with certain limitations. For instance, our method necessitates explicit modeling of potential edges during the graph generation process, which may not be ideal in scenarios characterized by dense text units. Although we select the top-K candidate edges for each node, there are still potential efficiency concerns. Further experimentation is needed to fully assess the viability of our approach in text-heavy environments. Given that contextual text block detection is integral to logical layout analysis, where semantic information is vital, we plan to further enhance our approach by integrating language models in future developments.

5   Summary

In this paper, we introduce a novel framework for addressing the task of contextual text block detection by formulating it as a graph generation problem. To demonstrate the effectiveness of our framework, we propose *Dynamic Relation Transformer* (DRFormer), a novel relation prediction module that integrates the graph structure into the relation prediction process. Additionally, DRFormer introduces a *Dynamic Graph Structure Refinement* procedure through the incorporation of a dual interactive transformer decoder. This iterative process progressively augments the graph's fidelity, culminating in an enhanced precision of contextual text block demarcation. The promising results obtained in this study suggest that our proposed framework opens up promising avenues for future research in the field of scene text region detection.

In the future, we will explore the role of contextual text blocks in text spotting and introduce text embedding to assist in better relationship prediction. Additionally, we aim to explore the integration of the *Dynamic Graph Structure Refinement* concept into other tasks related to relationship prediction or graph generation.

References

1. Baek, Y., Lee, B., Han, D., Yun, S., Lee, H.: Character region awareness for text detection. In: CVPR, pp. 9365–9374 (2019)
2. Biswas, S., Banerjee, A., Lladós, J., Pal, U.: DocSegTr: an instance-level end-to-end document image segmentation transformer. arXiv preprint arXiv:2201.11438 (2022)
3. Cai, Z., Vasconcelos, N.: Cascade R-CNN: high quality object detection and instance segmentation. IEEE Trans. Pattern Anal. Mach. Intell. **43**(5), 1483–1498 (2019)
4. Carion, N., Massa, F., Synnaeve, G., Usunier, N., Kirillov, A., Zagoruyko, S.: End-to-end object detection with transformers. In: ECCV, pp. 213–229 (2020)
5. Cheng, B., Misra, I., Schwing, A.G., Kirillov, A., Girdhar, R.: Masked-attention mask transformer for universal image segmentation. In: CVPR, pp. 1290–1299 (2022)
6. Cheng, H., et al.: M6Doc: a large-scale multi-format, multi-type, multi-layout, multi-language, multi-annotation category dataset for modern document layout analysis. In: CVPR, pp. 15138–15147 (2023)
7. Ch'ng, C.K., Chan, C.S.: Total-text: a comprehensive dataset for scene text detection and recognition. In: ICDAR, pp. 935–942 (2017)
8. Girshick, R.: Fast R-CNN. In: ICCV, pp. 1440–1448 (2015)
9. Girshick, R., Donahue, J., Darrell, T., Malik, J.: Rich feature hierarchies for accurate object detection and semantic segmentation. In: CVPR, pp. 580–587 (2014)
10. He, K., Zhang, X., Ren, S., Sun, J.: Deep residual learning for image recognition. In: CVPR, pp. 770–778 (2016)
11. Hu, K., Zhong, Z., Sun, L., Huo, Q.: Mathematical formula detection in document images: a new dataset and a new approach. Pattern Recogn. **148**, 110212 (2024)
12. Huang, Y., Lv, T., Cui, L., Lu, Y., Wei, F.: LayoutLMV3: pre-training for document AI with unified text and image masking. In: ACM MM, pp. 4083–4091 (2022)
13. Jocher, G., et al.: ultralytics/yolov5: v5.0 - YOLOv5-P6 1280 models, AWS, Supervise.ly and YouTube integrations (2021)
14. Kingma, D.P., Ba, J.: Adam: A method for stochastic optimization. In: ICLR (2015)
15. Li, J., Xu, Y., Lv, T., Cui, L., Zhang, C., Wei, F.: DiT: self-supervised pre-training for document image transformer. In: ACM MM, pp. 3530–3539 (2022)
16. Li, X.H., Yin, F., Liu, C.L.: Page segmentation using convolutional neural network and graphical model. In: DAS Workshop, pp. 231–245 (2020)
17. Li, X., Yin, F., Liu, C.: Page object detection from pdf document images by deep structured prediction and supervised clustering. In: ICIP, pp. 3627–3632 (2018)
18. Lin, T.Y., Goyal, P., Girshick, R., He, K., Dollár, P.: Focal loss for dense object detection. In: ICCV, pp. 2980–2988 (2017)
19. Liu, S., Wang, R., Raptis, M., Fujii, Y.: Unified line and paragraph detection by graph convolutional networks. In: Uchida, S., Barney, E., Eglin, V. (eds.) DAS Workshop, vol. 13237, pp. 33–47. Springer, Cham (2022). https://doi.org/10.1007/978-3-031-06555-2_3
20. Liu, X., et al.: ICDAR 2019 robust reading challenge on reading Chinese text on signboard. arXiv preprint arXiv:1912.09641 (2019)
21. Liu, Y., Chen, H., Shen, C., He, T., Jin, L., Wang, L.: ABCNet: real-time scene text spotting with adaptive bezier-curve network. In: CVPR, pp. 9809–9818 (2020)

22. Long, S., He, X., Yao, C.: Scene text detection and recognition: the deep learning era. IJCV **129**, 161–184 (2021)
23. Long, S., Qin, S., Panteleev, D., Bissacco, A., Fujii, Y., Raptis, M.: Towards end-to-end unified scene text detection and layout analysis. In: CVPR, pp. 1049–1059 (2022)
24. Long, S., Ruan, J., Zhang, W., He, X., Wu, W., Yao, C.: TextSnake: a flexible representation for detecting text of arbitrary shapes. In: ECCV, pp. 20–36 (2018)
25. Luo, S., Ding, Y., Long, S., Poon, J., Han, S.C.: Doc-GCN: heterogeneous graph convolutional networks for document layout analysis. In: COLING, pp. 2906–2916 (2022)
26. Ma, C., Sun, L., Wang, J., Huo, Q.: DQ-DETR: dynamic queries enhanced detection transformer for arbitrary shape text detection. In: Fink, G.A., Jain, R., Kise, K., Zanibbi, R. (eds.) ICDAR, vol. 14188, pp. 243–260. Springer, Cham (2023). https://doi.org/10.1007/978-3-031-41679-8_14
27. Ma, C., Sun, L., Zhong, Z., Huo, Q.: ReLaText: exploiting visual relationships for arbitrary-shaped scene text detection with graph convolutional networks. Pattern Recognit. **111**, 107684 (2021)
28. Ma, J., et al.: Arbitrary-oriented scene text detection via rotation proposals. IEEE Trans. Multimedia **20**(11), 3111–3122 (2018)
29. Naiemi, F., Ghods, V., Khalesi, H.: Scene text detection and recognition: a survey. Multimed. Tools. Appl. **81**(14), 20255–20290 (2022)
30. Nayef, N., et al.: ICDAR2017 robust reading challenge on multi-lingual scene text detection and script identification-RRC-MLT. In: ICDAR, pp. 1454–1459 (2017)
31. Oliveira, D.A.B., Viana, M.P.: Fast CNN-based document layout analysis. In: ICCV Workshops, pp. 1173–1180 (2017)
32. Papineni, K., Roukos, S., Ward, T., Zhu, W.J.: BLEU: a method for automatic evaluation of machine translation. In: ACL, pp. 311–318 (2002)
33. Pfitzmann, B., Auer, C., Dolfi, M., Nassar, A.S., Staar, P.: DocLayNet: a large human-annotated dataset for document-layout segmentation. In: KDD, pp. 3743–3751 (2022)
34. Ren, S., He, K., Girshick, R., Sun, J.: Faster R-CNN: towards real-time object detection with region proposal networks. In: NeurIPS, pp. 91–99 (2015)
35. Vo, N.D., Nguyen, K., Nguyen, T.V., Nguyen, K.: Ensemble of deep object detectors for page object detection. In: IMCOM, pp. 1–6 (2018)
36. Wang, R., Fujii, Y., Popat, A.C.: Post-OCR paragraph recognition by graph convolutional networks. In: WACV, pp. 493–502 (2022)
37. Wang, X., Jiang, Y., Luo, Z., Liu, C., Choi, H., Kim, S.: Arbitrary shape scene text detection with adaptive text region representation. In: CVPR, pp. 6449–6458 (2019)
38. Wang, X., Zhang, R., Kong, T., Li, L., Shen, C.: SOLOv2: dynamic and fast instance segmentation. In: NeurIPS, vol. 33, pp. 17721–17732 (2020)
39. Xue, C., Huang, J., Zhang, W., Lu, S., Wang, C., Bai, S.: Contextual text block detection towards scene text understanding. In: Avidan, S., Brostow, G., Cissé, M., Farinella, G.M., Hassner, T. (eds.) ECCV, vol. 13688, pp. 374–391. Springer, Cham (2022). https://doi.org/10.1007/978-3-031-19815-1_22
40. Xue, C., Lu, S., Hoi, S.: Detection and rectification of arbitrary shaped scene texts by using text keypoints and links. Pattern Recognit. **124**, 108494 (2022)
41. Yang, H., Hsu, W.: Transformer-based approach for document layout understanding. In: ICIP, pp. 4043–4047 (2022)
42. Yi, X., Gao, L., Liao, Y., Zhang, X., Liu, R., Jiang, Z.: CNN based page object detection in document images. In: ICDAR, vol. 1, pp. 230–235 (2017)

43. Yuliang, L., Lianwen, J., Shuaitao, Z., Sheng, Z.: Detecting curve text in the wild: new dataset and new solution. arXiv preprint arXiv:1712.02170 (2017)
44. Zhang, P., Li, C., Qiao, L., Cheng, Z., Pu, S., Niu, Y., Wu, F.: VSR: a unified framework for document layout analysis combining vision, semantics and relations. In: Lladós, J., Lopresti, D., Uchida, S. (eds.) ICDAR 2021. LNCS, vol. 12821, pp. 115–130. Springer, Cham (2021). https://doi.org/10.1007/978-3-030-86549-8_8
45. Zhang, S.X., et al.: Deep relational reasoning graph network for arbitrary shape text detection. In: CVPR, pp. 9699–9708 (2020)
46. Zhang, X., Su, Y., Tripathi, S., Tu, Z.: Text spotting transformers. In: CVPR, pp. 9519–9528 (2022)
47. Zhong, X., Tang, J., Yepes, A.J.: PubLayNet: largest dataset ever for document layout analysis. In: ICDAR, pp. 1015–1022 (2019)
48. Zhong, Z., Jin, L., Huang, S.: DeepText: a new approach for text proposal generation and text detection in natural images. In: ICASSP, pp. 1208–1212 (2017)
49. Zhong, Z., et al.: A hybrid approach to document layout analysis for heterogeneous document images. In: Fink, G.A., Jain, R., Kise, K., Zanibbi, R. (eds.) ICDAR, vol. 14191, pp. 189–206. Springer, Cham (2023). https://doi.org/10.1007/978-3-031-41734-4_12
50. Zhou, X., et al.: EAST: an efficient and accurate scene text detector. In: CVPR, pp. 5551–5560 (2017)
51. Zhu, X., Su, W., Lu, L., Li, B., Wang, X., Dai, J.: Deformable DETR: deformable transformers for end-to-end object detection. In: ICLR (2021)
52. Zhu, Y., Yao, C., Bai, X.: Scene text detection and recognition: recent advances and future trends. Front. Comput. Sci. **10**, 19–36 (2016)

End to End Table Transformer

Yun Young Choi⬤, Taehoon Kim⬤, Namwook Kim⬤, Taehee Lee⬤,
and Seongho Joe$^{(\boxtimes)}$⬤

Samsung SDS, Seoul 05510, South Korea
drizzle.cho@samsung.com

Abstract. Table extraction (TE) task in document images is important
in deep learning for conveying structured information. TE was decomposed into three subtasks: table detection (TD), table structure recognition (TSR), and functional analysis (FA). Most of previous research
focused on developing models specifically tailored to each of these tasks,
leading to challenges in computational cost, model size, and performance
limitations. Transformer-based object detection models are being successfully applied to TE subtasks, yet they face inherent challenges due
to the one-to-one set matching approach for detecting objects. This
approach assigns only a few queries as positive samples, diminishing
the training efficacy of these samples and leading to a performance
bottleneck. Therefore, prior research in the object detection field has
introduced modifications to the Detection Transformer (DETR), adding
additional queries and training schemes that improve performance. In
this work, we introduce the End-to-end Table Transformer (ETT), a
specialized transformer-based object detection model designed for high-
performing TE from document images with single model. Our model
comprises three key components: a backbone, the Deformable DETR
(DDETR) model, and the novel layout analysis module with table layout
loss. This layout analysis module leverages explicit relationships between
table objects to enhance the table extraction task performance in multi
tables in images. We conduct rigorous experiments to assess the efficacy
of our proposed model against table extraction benchmark datasets, comparing it with other DETR variants, including vanilla DETR, DDETR,
and H-DETR. Empirical evaluations highlight that our model efficiently
secures state-of-the-art results in TE task.

Keywords: Document analysis systems · Document image
processing · Physical and logical layout analysis · Tables and charts

1 Introduction

Table extraction (TE) in document images is an important application of computer vision for analyzing and summarizing structured information. Previous
approaches divided the TE task into three subtasks: table detection (TD) for

---

Y. Y. Choi and T. Kim—Equal contribution.

© The Author(s), under exclusive license to Springer Nature Switzerland AG 2024
E. H. Barney Smith et al. (Eds.): ICDAR 2024, LNCS 14804, pp. 331–345, 2024.
https://doi.org/10.1007/978-3-031-70533-5_20

detecting table area in single document page; table structure recognition (TSR) for recognizing a table's components, such as rows, columns, and cells; and functional analysis (FA) for extracting a table's keys and values. Traditionally, these subtasks were studied using rule-based methods [5,17,25]; however, with the advancement of the computer vision field, Transformer-based [9,15,22,24] and other deep learning-based techniques [6,7,16,18,21,23,26,29] have been increasingly adopted. Recently, the transformer-based object detection model, namely Detection Transformer (DETR), has been successfully applied to TD, TSR, and FA tasks within the PubTables-1M dataset [24].

PubTables-1M, containing nearly one million tables sourced from scientific articles, provides a comprehensive testbed for evaluating models for each TE subtask due to its table structure and complexity diversity compared to other dataset [12,13,28]. Although DETR has marked considerable advancements in the domain of TE, several limitations remain. One primary concern is the efficiency of computational and memory resources, as it necessitates separate models for executing the TSR and TD tasks, leading to increased overhead. Additionally, the TSR task's reliance on the TD task's results introduces a potential vulnerability, as errors in detection can cascade, affecting the overall robustness of the system. Moreover, there is a noticeable drop in performance when applying TSR to comprehensive document images, which presents a significant challenge for practical applications. There are previous research in other object detection domain to increase the performance of DETR. Recent studies have proposed techniques that leverage additional queries or training process to enhance accuracy and robustness [3,10,11,14,31]. However, these approaches have the downside of increasing the memory and computational costs of training the model.

To address these issues, we introduce a unified architecture of End-to-End Table Transformer (ETT) architecture, combining TD, TSR, and FA tasks. We propose ETT based on the Deformable DETR (DDETR) framework due to its effectiveness in managing the size variations of objects in documents, where table components like columns and rows are significantly smaller compared to the table. Alongside the DDETR models, we integrate explicit encoding of layout information for candidate TSR objects through a layout analysis module with table layout loss. During the training process, table layout loss is calculated using the layout analysis module outputs based on queries assigned by one-to-one set matching and table components' ground truth adjacency matrix. This matrix is constructed using the principle that table components belonging to the same table are connected while those from different tables are not. The proposed architecture and layout loss enable TE with high performance, achieved without the need for additional matching processes or an increase in the number of queries. To justify the proposed model's performance on TE task, we also present a new dataset, E2E-PubTables, modified version of the PubTables-1M dataset specifically for the TE task. Then we perform experiments on the E2E-PubTables dataset aim to demonstrating the superior predictive performance and efficiency of our proposed model, compared to variant of DETRs Fig. 1.

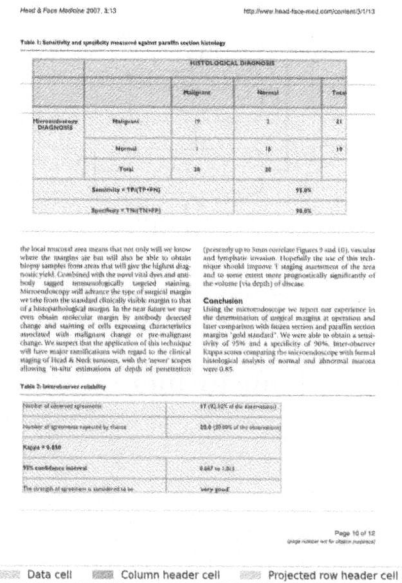

Fig. 1. An example of a TE task in single document page.

2   Related Works

Table extraction Previous approaches have explored various methods for the table structure recognition (TSR) task, including image-to-text approaches [12], graph-based methods [4,19], and object detection models [2,18,21,27]. One of the most successful approaches for the TE subtasks is utilizing DETR models for TD and TSR. The DETR models employed to perform TD, TSR, and FA tasks on PubTables-1M dataset, achieving state-of-the-art performance compared to Faster R-CNN [20], without needing any special customization. Recent studies also found that introducing DDETR can enhance the performance of the TD task in document images compared to CNN-based models [22].

Transformer-based object detection In object detection tasks, DETR [2] have achieved groud-braking success without hand-crafted components [8,20] using from previous researches. DETR consists of CNN-backbone, Transformer, and prediction heads with bipartite matching for direct set predictions. After the pioneering work, there are many variants studied to enhance the performance of DETR and increase the efficiency of the training procedure. DDETR proposes a deformable attention scheme for efficiently capturing multi-scale features to achieve better performance than DETR. Some previous works [3,10,11,31,31] focus on improving the one-to-one set matching part, which is a major cause of slow convergence and instability in the training process [11]. H-DETR [10] proposes a hybrid matching scheme consisting of one-to-one and auxiliary one-to-many matching assignments for increasing training efficiency.

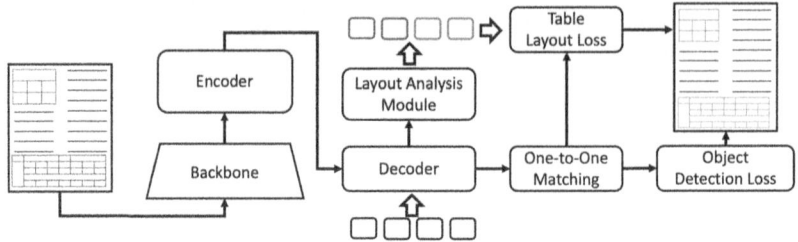

Fig. 2. Overall Architecture of ETT.

3 End to End Table Transformer

3.1 Preliminary

DETR Pipeline The DETR is a pioneering transformer-based object detection model for detecting objects in images. DETR consists of a CNN backbone, transformer encoder, transformer decoder, and prediction heads. Suppose we are given the image I, the backbone extracts image features X_0 as follows:

$$X_0 = \mathbf{Backbone}(I), \quad X_0 \in R^{C \times H \times W} \tag{1}$$

The transformer encoder generates hidden representations from X_0 and positional encodings through a multi-head self-attention module and a feed-forward network.

$$X = \mathbf{Encoder}(X_0), \quad X \in R^{d \times H \times W} \tag{2}$$

The DETR uses fixed-size object queries $Q_0 \in R^{n \times D}$ for learning the bounding box and class of objects. The candidates are processed through a transformer decoder, including both self-attention and cross-attention mechanisms as follows:

$$Q_N = \mathbf{Decoder}(X, Q_0), \quad X \in Q_N^{n \times D} \tag{3}$$

where N is the number of decoder layers. The outputs of decoder layers Q_N are passed through predictors for obtaining predicting bounding boxes and object classes Y.

DETR Training For end-to-end object detection, the training procedure of DETR uses bipartite matching between ground truth objects and predictions. The candidates need to be matched with ground truth objects to compute the loss. Given a set of predictions $P = \{p_1, p_2, ..., p_N\}$ and ground truth objects $G = \{g_1, g_2, ..., g_M\}$, we first compute a cost matrix C where each entry C_{ij} represents a predefined distance (often L1 loss) between prediction p_i and ground truth g_j:

$$C_{ij} = L1(p_i, g_j) \tag{4}$$

Using the Hungarian algorithm, we find the optimal assignment that minimizes the total cost. This gives us the matched pairs of predictions and ground truths. The overall loss, L_{DETR}, for DETR, is then computed using the matched pairs:

$$L_{\text{DETR}} = \sum_{(p,g)\in\mathcal{H}(P,G)} \text{loss}(p,g) \tag{5}$$

where $\mathcal{H}(P,G)$ be the set of matched pairs (prediction, ground truth) resulting from the Hungarian algorithm. Here, $\text{loss}(p,g)$ can be a combination of classification loss for the object class and regression loss for the bounding box coordinates based on the exact configuration of the DETR model.

Deformable DETR DDETR, representing an advanced model of DETR, incorporates several enhancements to mitigate limitations such as slow convergence and limited feature spatial resolution. Therefore, DDETR substitutes the standard multi-head self-attention and cross-attention with a deformable attention-based approach for handling many image features. To increase the performance of models, it employs an iterative refinement method and a dynamic query from the transformer encoder.

3.2   Model Architecture

We introduce the ETT, a unified approach for TE task on document images. The overall architecture of our model is illustrated in Fig. 2. ETT consists of a backbone, encoder, decoder, layout analysis module, and prediction heads based on DDETR framework [30]. In the architecture, various models can be employed for the backbone, which plays a role in feature extraction from the input images, such as ResNet-50 (R50) or Swin-T.

Layout analysis module To increase the performance of model for multiple tables in single images and improve training efficiency with positive samples, we have designed an additional module that explicitly utilizes the relationships between queries. The layout analysis module takes the output of the end of decoder layers $Q_N \in R^{n \times D}$. The predicted adjacency matrix A for all candidates is computed as follows:

$$A = \textbf{Sigmoid}(Q_N Q_N^T), \quad A \in R^{n \times n} \tag{6}$$

where n is the number of candidates and Q^T is the transpose of the matrix Q. This matrix A represents relationships between different queries, enabling us to discern which particular table each candidate query belongs to. Through the predicted adjacency matrix, ETT goes beyond simple object detection and can effectively determine the grouping of candidate queries.

3.3   Loss Function

Table layout loss Before calculating the table layout loss, we construct the ground truth adjacency matrix, A^{GT}, for documents containing multiple tables.

336 Y. Y. Choi et al.

We use two assumptions: elements within the same table are fully connected; there are no connections between elements of different tables. For any two objects e_i and e_j in pages, the adjacency matrix is defined as :

$$A_{i,j}^{\mathrm{GT}} = \begin{cases} 1 & \text{if } e_i \text{ and } e_j \text{ belong to the same table} \\ 0 & \text{otherwise} \end{cases} \tag{7}$$

Note that the size of A^{GT} is $M \times M$, and M is the number of ground truth table objects. To calculate table layout loss, we also use an assigned index set $\mathbf{I}$ from the one-to-one assignment process from calculating object detection loss. Given an predicted adjacency matrix A of size $n \times n$ and an assigned index set $\mathbf{I}$, the submatrix A' extracted from A using the indices in I is defined as:

$$A'_{i,j} = A_{k,l} \quad \text{for all } (k,l) \in \mathbf{I} \times \mathbf{I} \tag{8}$$

where (i,j) are indices of A' and (k,l) are the corresponding indices in A. The table layout loss for table structure recognition in documents is formulated as below :

$$L_{TL} = -\frac{1}{M^2} \sum_{i=1}^{M} \sum_{i=j}^{M} (y_{i,j} \log(p_{i,j}) + (1 - y_{i,j}) \log(1 - p_{i,j})) \tag{9}$$

where M is the total number of elements in the matrices, $y_{i,j}$ is the i-th element of A^{GT}, and $p_{i,j}$ is the i, j-th element of A'.

Total loss In the end-to-end training of the ETT, we compute the total loss as a weighted combination of two primary components: the object detection loss, which ensures accurate recognition of individual elements, and the table layout loss, which preserves the structural integrity and relationships within the table. As a result, the model's training target is defined as :

$$Loss = L_{DETR} + \lambda L_{TL} \tag{10}$$

where λ is weighting factors.

4 Experiments

Datasets To assess the effectiveness of our proposed model, particularly in terms of TE from document images, we have preprocessed the PubTables-1M dataset. This choice stems from the fact that while PubTables-1M provides separate datasets for TD and TSR+FA, it lacks an end-to-end formatted dataset specifically for TE tasks. Furthermore, no other public large dataset exists for this purpose. The pseudocode used for preprocessing is as follows Algorithm 1. Upon completing the preprocessing, we have termed the resulting dataset E2E-PubTables. A summary of the statistics for the E2E-PubTables dataset is presented in Table 1. On average, each page contains 1.19 tables, with a maximum of 15 tables. The number of objects per page ranges from a minimum of 4 to a maximum of 202. The reason for the reduced number of tables in the dataset, compared to the original PubTables-1M, is due to the exclusion of rotated tables and the removal of pages where annotations did not properly match.

Table 1. A summary of statistics of PubTables-1M and E2E-PubTables dataset for comparison of TD, TSR and TE tasks. Note that # Tables # and Objects per single document images.

Dataset	PubTables-1M	PubTables-1M	E2E-PubTables
Task	TD	TSR	TE
# document image	575305	947642	473109
# Tables in dataset	683056	947642	563175
Avg./Std. # Tables	1.1873 / 0.4463	1.0000 / 0.0000	1.1904 / 0.4490
Max./Min. # Tables	15 / 1	1 / 1	15 / 1
Avg./Std. # Objects	1.1873 / 0.4463	23.6920 / 12.7840	28.4704 / 15.7928
Max./Min. # Objects	15 / 1	100 / 4	202 / 4

Models To demonstrate the effectiveness of proposed model, we compare our approach on benchmark datasets with the object detection techniques [2,10,24, 30,31]. We impose that the most of architecture components in the DETR and DETR variants are identical to those of ETT. Difference in the number of model parameters among theses models are derived from the attention scheme, and additional queries for one-to-many matching in H-DETR. The total number of queries is 210, which is slightly larger than the maximum number of objects in single image. For the training of H-DETR, the total number of queries used was set to 630. Other hyperparameters are same as previous research [10] with two-stage option and iterative refinement prediction scheme. Further, we combine ETT with hybrid training process same with H-DETR to analyze effect of proposed methods with DETR variants. Note that we apply mixed query selection, and looking forward twice option [1].

Detail of implementations For evaluating table extraction performance, we use metrics in terms of object detection, Mean Average Precision (mAP), and TSR-specific metrics, **GriTS**, proposed by [25]. **GriTS** consists of three parts; cell topology recognition $\mathbf{GriTS_{Top}}$, cell content recognition ($\mathbf{GriTS_{Con}}$), and cell location recognition ($\mathbf{GriTS_{Loc}}$). Detail of the metric shown in reference [25]. There metrics are defined as :

$$\mathbf{GriTS}_f(\mathbf{A}, \mathbf{B}) = \frac{2 \cdot \sum_{i,j} f(\tilde{\mathbf{A}}_{i,j}, \tilde{\mathbf{B}}_{i,j})}{|\mathbf{A}| + |\mathbf{B}|} \tag{11}$$

where A and B are ground truth and predicted matrices of cells in table, respectively. The **GriTS** metric was originally designed for TSR task, operating

---

Algorithm 1. E2EPubTables Conversion

Require: Table Detection & Structure Recognition Dataset
Ensure: Converted E2E Dataset

1: **function** CONVERT(*td_paths*, *tsr_paths*)
2:    **for** each *td_path* in *td_paths* **do**
3:        Parse *td_path* to get *TdData*
4:        Calculate *rotated_table_object* in *Td_Data*
5:        **if** any *rotated_table_object* **then**
6:            Skip to next *td_path*
7:        **end if**
8:        GENE2EDATA(*td_path*, *tsr_paths*)
9:    **end for**
10: **end function**

11: **function** GENE2EDATA(*td_path*, *tsr_paths*)
12:    Parse *td_path* to get *TdData*
13:    **for** each *table_object* in *TdData* **do**
14:        Parse *tsr_paths* to get *TsrData*
15:        Get *TsrData* that matches the *table_object*
16:        Calculate translation from *table_object*
17:        **for** each *object* in *TsrData* **do**
18:            Translate *object* coordinates
19:            Append translated *object* to *TdData*
20:        **end for**
21:        Append *TdData* to *E2EData*
22:    **end for**
23:    **return** *E2EData*
24: **end function**

Descriptions:

- **td_path**: The path where the TD Annotation file (XML file) is located.
- **tsr_paths**: TSR annotation file path list that matches the file title of the TD.
- **TdData, TSRData, E2EData**: Annotation data.
- **table_object, rotated_table_object**: Table / Rotated Table Object in TD data.
- **object**: Object in TSR data.

Function Descriptions:

- **Convert**: Reads TD data and filters data that has at least one rotated table.
- **GenE2EData**: Matches and merges TD data and TSR data to generate new data.
  1. Read XML information from td_path to create TdData.
  2. Get the table object of TdData and find the matching TSR data.
  3. Apply coordinate shift operation to matched TSR data to match the coordinates of the TD image.

under the assumption that there is only one table present in an image. We have modified this metric to function effectively in document images containing multiple tables, $\mathbf{GriTS}_f^{TE}$. For the confidence threshold, DETR was set at 0.5, consistent with prior study [24], while the rest of the models were all fixed at 0.9. We use only scale augmentation for proposed model, which is same as the DETR model for the TD task [24]. The experiments were performed on 64 physical cores (Intel(R) Xeon(R) Gold 6240 CPU @ 2.60GHz) with 3 NVIDIA Tesla V100 SXM2 32GB GPUs.

Table 2. TD+TSR results obtained from benchmark using object detection metrics from E2E-PubTables. Results with grey color text are cited from available results obtained from pre-existing publications. Note that both TD and TSR are reported for comparison with effectivness of two-stage model from PubTables-1M dataset [24] .

Task	Model	Backbone	AP	AP_{50}	AP_{75}	AR
TD	Faster R-CNN	-	0.825	0.985	0.927	0.866
	DETR	R18	0.970	0.995	0.989	0.985
TSR	Faster R-CNN	-	0.722	0.815	0.785	0.762
	DETR	R18	0.912	0.971	0.948	0.942
TD+TSR (Single-stage)	DETR	R50	0.462	0.713	0.477	0.614
TD+TSR	DDETR	R50	0.831	0.916	0.894	0.889
	DDETR	Swin-T	0.851	0.944	0.926	0.907
	H-DETR	R50	0.911	0.975	0.967	0.949
	H-DETR	Swin-T	0.890	0.957	0.949	0.943
TD+TSR	ETT	R50	0.885	0.984	0.970	0.926
	ETT	Swin-T	**0.917**	**0.985**	**0.980**	**0.949**
	H-ETT	R50	0.913	0.984	0.979	0.913
	H-ETT	Swin-T	0.914	0.984	0.980	0.948

Results Table 2 shows comparisons between DETR, its variants, and proposed model after 20 epochs on the E2E-PubTables. Applying the single-stage DETR model to the TD+TSR task resulted in significantly lower mAP and AR values compared to the two-stage DETR model, which separates the TD and TSR tasks into individual models for each task. Our ETT model outperformed DDETR across all instances, with a notable increase in mAP by approximately 0.05, using identical backbones, encoders, and decoders. Compared to H-DETR, ETT exhibited a higher mAP, indicating that high AP can be achieved through one-to-one matching without the need for additional queries and a hybrid training process. Even when compared to the DETR model used for TSR, it was observed that the AP values were higher. However, when the proposed layout analysis module was integrated with H-DETR to create H-ETT, the performance of H-ETT was observed to be similar to that of H-DETR.

Table 3. TE task results are obtained from E2E-PubTables using TSR-specific metrics modified for TE. Results with grey color text are cited from available results obtained from pre-existing publications. TSR+FA task results are obtained from PubTables-1M dataset [24].

Task	Model	Backbone	Table Category	Acc_{Con}	$\text{GriTS}_{\text{Top}}$	$\text{GriTS}_{\text{Con}}$	$\text{GriTS}_{\text{Loc}}$	DAR_{Con}
TSR+FA	Faster R-CNN	-	Simple	0.0867	0.8682	0.8571	0.6869	0.8024
			Complex	0.1193	0.8556	0.8507	0.7518	0.7734
			All	0.1039	0.8616	0.8538	0.7211	0.7871
	DETR	R18	Simple	0.9468	0.9949	0.9938	0.9922	0.9893
			Complex	0.6944	0.9752	0.9763	0.9654	0.9667
			All	0.8138	0.9845	0.9846	0.9781	0.9774

Task	Model	Backbone	Table Category	$\text{Acc}_{\text{Con}}^{\text{TE}}$	$\text{GriTS}_{\text{Top}}^{\text{TE}}$	$\text{GriTS}_{\text{Con}}^{\text{TE}}$	$\text{GriTS}_{\text{Loc}}^{\text{TE}}$	$\text{DAR}_{\text{Con}}^{\text{TE}}$
TE (Single-stage)	DETR	R50	Simple	0.4245	0.9160	0.8896	0.8517	0.7845
			Complex	0.0909	0.8083	0.7862	0.7256	0.6466
			All	0.2375	0.8557	0.8316	0.7810	0.7072
TE	DDETR	R50	Simple	0.1289	0.7494	0.7506	0.6848	0.7140
			Complex	0.0752	0.7120	0.7381	0.6430	0.7241
			All	0.0988	0.7284	0.7436	0.6613	0.7197
TE	DDETR	Swin-T	Simple	0.2943	0.8906	0.8923	0.8082	0.8759
			Complex	0.1843	0.8570	0.8934	0.7730	0.8860
			All	0.2326	0.8717	0.8929	0.7885	0.8816
TE	H-DETR	R50	Simple	0.9572	0.9919	0.9939	0.9165	0.9929
			Complex	0.9216	0.9690	0.9938	0.8979	0.9853
			All	0.9373	0.9790	0.9938	0.9061	0.9886
TE	H-DETR	Swin-T	Simple	0.9707	0.9948	0.9958	0.9209	0.9944
			Complex	0.9523	0.9767	0.9948	0.9023	0.9882
			All	0.9631	0.9847	0.9952	0.9105	0.9909
TE	ETT	R50	Simple	0.9886	0.9949	0.9982	0.8928	0.9969
			Complex	0.9668	0.9795	0.9964	0.8795	0.9903
			All	0.9764	0.9862	0.9972	0.8853	0.9932
TE	ETT	Swin-T	Simple	0.9901	0.9965	0.9982	0.9212	0.9976
			Complex	0.9752	0.9844	0.9971	0.9111	0.9924
			All	**0.9817**	**0.9897**	**0.9976**	0.9156	**0.9947**
TE	H-ETT	R50	Simple	0.9857	0.9969	0.9978	0.9205	0.9973
			Complex	0.9658	0.9806	0.9961	0.9058	0.9904
			All	0.9746	0.9877	0.9968	0.9123	0.9934
TE	H-ETT	Swin-T	Simple	0.9865	0.9967	0.9976	0.9262	0.9968
			Complex	0.9666	0.9819	0.9958	0.9123	0.9908
			All	0.9753	0.9884	0.9966	**0.9184**	0.9935

In Table 3, we compared our proposed approach and variants of DETR based on a TSR-specific metric modified for TE. In these results, ETT also showed better performance than all the variants of DETR and single-stage DETR exhibited very poor performance. This superior performance was even evident when compared to DETR model for TSR+FA task. However, a lower cell location recognition($\mathbf{GriTS_{Loc}}$) value was noted. This is because, when performing TSR+FA, there is an additional assumption that each image contains only one table and that the images are filled with tables.

Comparison of training process We compare the training process in terms of convergence curves of the variants of DETR, and ETTs as shown in Fig. 3. In comparison to the variants of DETR, it is observed that convergence is achieved in fewer epochs, and higher AP values are attained. This improvement can be attributed to the integration of a layout analysis module with the assigned queries, enhancing discriminative power. This enhancement is particularly effective in reducing the instability of one-to-one matching process for images containing multiple tables.

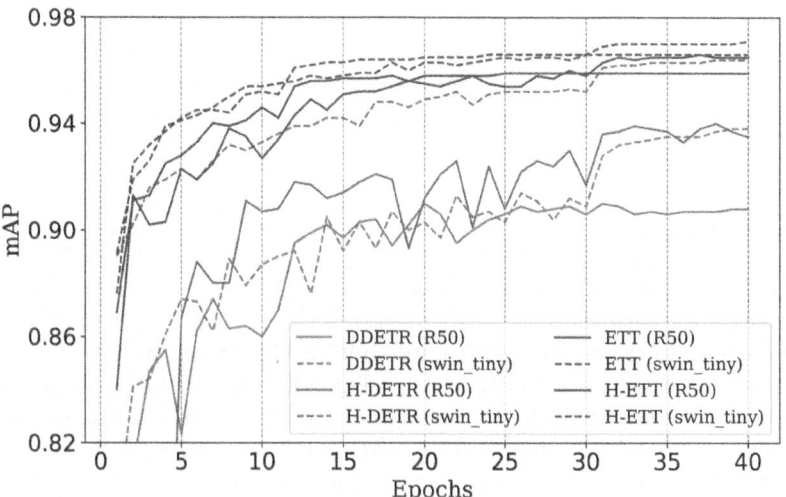

Fig. 3. Comparison of training process of ETT and DETR variants.

Analysis of GriTS across different confidence thresholds Unlike previous studies that set the confidence threshold at 0.5, we conduct a comparative analysis on the impact of different confidence thresholds on the ETT and H-ETT, as shown in Table 4. The results indicate that, at higher threshold values, there is superior performance in terms of the **GriTS** metric. Consequently, diverging from prior research, we have determined the optimal threshold to be 0.9 based on these results.

Table 4. GriTS comparison using different confidence thresholds values. Note that the best results for each model are marked in bold.

Model	Backbone	Threshold	$\text{Acc}_{\text{Con}}^{\text{TE}}$	$\text{GriTS}_{\text{Top}}^{\text{TE}}$	$\text{GriTS}_{\text{Con}}^{\text{TE}}$	$\text{GriTS}_{\text{Loc}}^{\text{TE}}$	$\text{DAR}_{\text{Con}}^{\text{TE}}$
ETT	Swin-T	0.5	0.9790	0.9518	0.9974	0.8819	0.9783
		0.6	0.9822	0.9675	0.9977	0.8960	0.9851
		0.7	0.9825	0.9757	0.9978	0.9033	0.9888
		0.8	**0.9830**	0.9820	0.9978	0.9088	0.9916
		0.9	0.9817	**0.9897**	**0.9976**	**0.9156**	**0.9947**
H-ETT	Swin-T	0.5	0.9720	0.9466	0.9970	0.8814	0.9769
		0.6	0.9762	0.9701	0.9973	0.9026	0.9866
		0.7	0.9768	0.9790	**0.9974**	0.9104	0.9904
		0.8	**0.9769**	0.9852	0.9973	0.9158	0.9930
		0.9	0.9753	**0.9884**	0.9966	**0.9184**	**0.9935**

Analyzing error and limitations We perform an error analysis on test images to identify error patterns of ETT. Figure 4 shows the ground truth and prediction results across different cell topology recognition($\text{GriTS}_{\text{Top}}$) values. In the first image, we observe that while there is a slight discrepancy in the width of the column predictions from the ground truth, the cell prediction results are very accurate. In the second image, it is evident that errors occur in parts with spanning cells due to their complex structure. In the third image, differences in patterns between the top and bottom parts of the table lead to a decrease in overall prediction accuracy. This analysis shows that the proposed model have limitations when dealing with tables that have complex patterns.

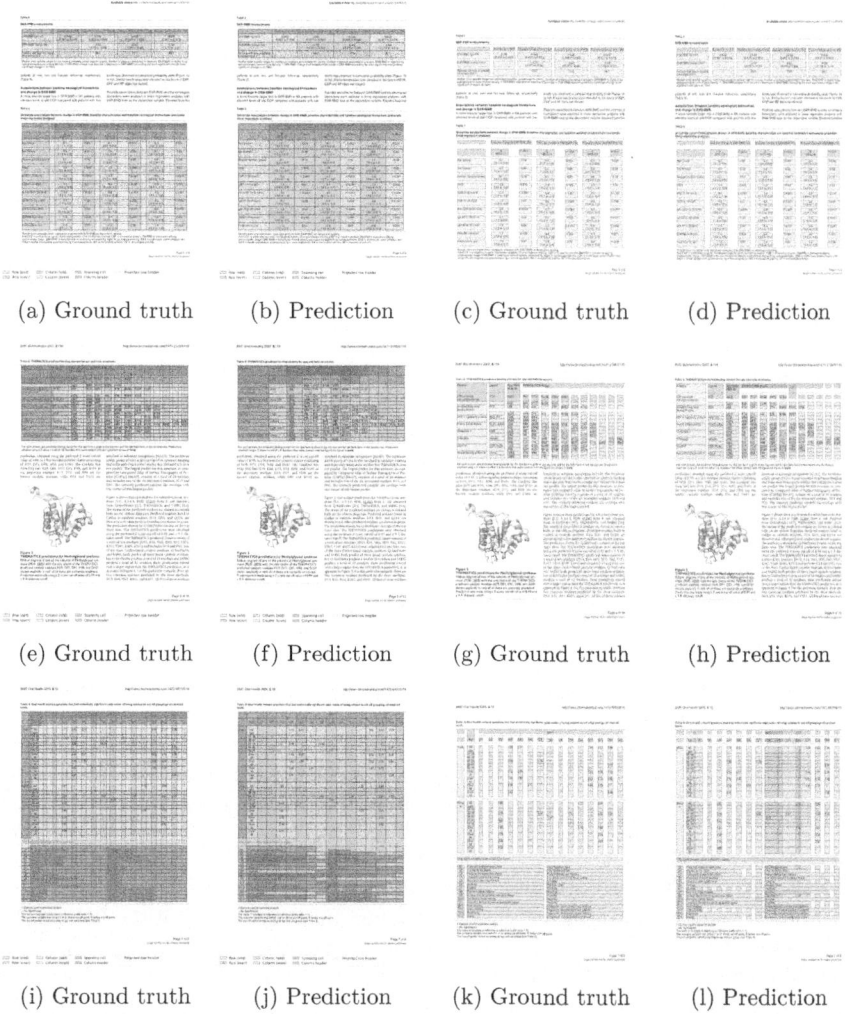

(a) Ground truth   (b) Prediction   (c) Ground truth   (d) Prediction

(e) Ground truth   (f) Prediction   (g) Ground truth   (h) Prediction

(i) Ground truth   (j) Prediction   (k) Ground truth   (l) Prediction

Fig. 4. Visualization of ETT(Swin-T) predictions and ground truth images across different cell topology recognition($\mathbf{GriTS_{Top}}$) values: (a)~(d) for 1.000, (e)~(h) for 0.977, and (i)~(l) for 0.842.

5   Conclusion

This paper presents a method that improves the performance of transformer-based object detection models in the TE task by adding a simple yet novel layout analysis module. Our approach explicitly leverages objects' relation between table and TSR objects in the TE task, thus eliminating the need for additional matching methods or queries, unlike other variants of DETR. As future work, we aim to apply this methodology across various document types and accelerate the advancement of transformer-based object detection methods for a wide range of vision tasks.

References

1. Bai, X., et al.: Transfusion: robust lidar-camera fusion for 3d object detection with transformers. In: Proceedings of the IEEE/CVF Conference on Computer Vision and Pattern Recognition, pp. 1090–1099 (2022)
2. Carion, N., Massa, F., Synnaeve, G., Usunier, N., Kirillov, A., Zagoruyko, S.: End-to-end object detection with transformers. In: Vedaldi, A., Bischof, H., Brox, T., Frahm, J.-M. (eds.) ECCV 2020. LNCS, vol. 12346, pp. 213–229. Springer, Cham (2020). https://doi.org/10.1007/978-3-030-58452-8_13
3. Chen, Q., Chen, X., Zeng, G., Wang, J.: Group DETR: fast training convergence with decoupled one-to-many label assignment. arXiv preprint arXiv:2207.13085 (2022)
4. Chi, Z., Huang, H., Xu, H.D., Yu, H., Yin, W., Mao, X.L.: Complicated table structure recognition. arXiv preprint arXiv:1908.04729 (2019)
5. Göbel, M., Hassan, T., Oro, E., Orsi, G.: A methodology for evaluating algorithms for table understanding in pdf documents. In: Proceedings of the 2012 ACM Symposium on Document Engineering, pp. 45–48 (2012)
6. Hashmi, K.A., Pagani, A., Liwicki, M., Stricker, D., Afzal, M.Z.: Cascade network with deformable composite backbone for formula detection in scanned document images. Appl. Sci. 11(16), 7610 (2021)
7. Hashmi, K.A., Stricker, D., Liwicki, M., Afzal, M.N., Afzal, M.Z.: Guided table structure recognition through anchor optimization. IEEE Access 9, 113521–113534 (2021)
8. Hosang, J., Benenson, R., Schiele, B.: Learning non-maximum suppression. In: Proceedings of the IEEE Conference on Computer Vision and Pattern Recognition, pp. 4507–4515 (2017)
9. Huang, Y., et al.: Improving table structure recognition with visual-alignment sequential coordinate modeling. In: Proceedings of the IEEE/CVF Conference on Computer Vision and Pattern Recognition, pp. 11134–11143 (2023)
10. Jia, D., et al.: DETRS with hybrid matching. In: Proceedings of the IEEE/CVF Conference on Computer Vision and Pattern Recognition, pp. 19702–19712 (2023)
11. Li, F., Zhang, H., Liu, S., Guo, J., Ni, L.M., Zhang, L.: DN-DETR: accelerate DETR training by introducing query denoising. In: Proceedings of the IEEE/CVF Conference on Computer Vision and Pattern Recognition, pp. 13619–13627 (2022)
12. Li, M., Cui, L., Huang, S., Wei, F., Zhou, M., Li, Z.: Tablebank: table benchmark for image-based table detection and recognition. In: Proceedings of the Twelfth Language Resources and Evaluation Conference, pp. 1918–1925 (2020)
13. Li, M., et al.: Docbank: A benchmark dataset for document layout analysis. arXiv preprint arXiv:2006.01038 (2020)
14. Liu, S., et al.: Dab-DETR: Dynamic anchor boxes are better queries for detr. arXiv preprint arXiv:2201.12329 (2022)
15. Long, S., Qin, S., Panteleev, D., Bissacco, A., Fujii, Y., Raptis, M.: Towards end-to-end unified scene text detection and layout analysis. In: Proceedings of the IEEE/CVF Conference on Computer Vision and Pattern Recognition, pp. 1049–1059 (2022)
16. Minouei, M., Hashmi, K.A., Soheili, M.R., Afzal, M.Z., Stricker, D.: Continual learning for table detection in document images. Appl. Sci. 12(18), 8969 (2022)
17. Paliwal, S.S., Vishwanath, D., Rahul, R., Sharma, M., Vig, L.: Tablenet: deep learning model for end-to-end table detection and tabular data extraction from scanned document images. In: 2019 International Conference on Document Analysis and Recognition (ICDAR), pp. 128–133. IEEE (2019)

18. Prasad, D., Gadpal, A., Kapadni, K., Visave, M., Sultanpure, K.: Cascadetabnet: an approach for end to end table detection and structure recognition from image-based documents. In: Proceedings of the IEEE/CVF Conference on Computer Vision and Pattern Recognition Workshops, pp. 572–573 (2020)
19. Qasim, S.R., Mahmood, H., Shafait, F.: Rethinking table recognition using graph neural networks. In: 2019 International Conference on Document Analysis and Recognition (ICDAR), pp. 142–147. IEEE (2019)
20. Ren, S., He, K., Girshick, R., Sun, J.: Faster r-CNN: towards real-time object detection with region proposal networks. Adv. Neural Inf. Process. Syst. **28** (2015)
21. Schreiber, S., Agne, S., Wolf, I., Dengel, A., Ahmed, S.: Deepdesrt: deep learning for detection and structure recognition of tables in document images. In: 2017 14th IAPR International Conference on Document Analysis and Recognition (ICDAR). vol. 1, pp. 1162–1167. IEEE (2017)
22. Shehzadi, T., Azeem Hashmi, K., Stricker, D., Liwicki, M., Zeshan Afzal, M.: Towards end-to-end semi-supervised table detection with deformable transformer. In: International Conference on Document Analysis and Recognition, pp. 51–76. Springer (2023). https://doi.org/10.1007/978-3-031-41679-8_4
23. Sinha, S., Hashmi, K.A., Pagani, A., Liwicki, M., Stricker, D., Afzal, M.Z.: Rethinking learnable proposals for graphical object detection in scanned document images. Appl. Sci. **12**(20), 10578 (2022)
24. Smock, B., Pesala, R., Abraham, R.: Pubtables-1m: towards comprehensive table extraction from unstructured documents. In: Proceedings of the IEEE/CVF Conference on Computer Vision and Pattern Recognition, pp. 4634–4642 (2022)
25. Smock, B., Pesala, R., Abraham, R.: Grits: grid table similarity metric for table structure recognition. In: International Conference on Document Analysis and Recognition, pp. 535–549. Springer (2023). https://doi.org/10.1007/978-3-031-41734-4_33
26. Wang, J., Hu, K., Zhong, Z., Sun, L., Huo, Q.: Detect-order-construct: A tree construction based approach for hierarchical document structure analysis. arXiv preprint arXiv:2401.11874 (2024)
27. Zheng, X., Burdick, D., Popa, L., Zhong, X., Wang, N.X.R.: Global table extractor (GTE): a framework for joint table identification and cell structure recognition using visual context. In: Proceedings of the IEEE/CVF Winter Conference on Applications of Computer Vision, pp. 697–706 (2021)
28. Zhong, X., Tang, J., Yepes, A.J.: Publaynet: largest dataset ever for document layout analysis. In: 2019 International Conference on Document Analysis and Recognition (ICDAR), pp. 1015–1022. IEEE (2019)
29. Zhong, Z., et al.: A hybrid approach to document layout analysis for heterogeneous document images. In: International Conference on Document Analysis and Recognition, pp. 189–206. Springer (2023). https://doi.org/10.1007/978-3-031-41734-4_12
30. Zhu, X., Su, W., Lu, L., Li, B., Wang, X., Dai, J.: Deformable detr: Deformable transformers for end-to-end object detection. arXiv preprint arXiv:2010.04159 (2020)
31. Zong, Z., Song, G., Liu, Y.: DETRs with collaborative hybrid assignments training. In: Proceedings of the IEEE/CVF International Conference on Computer Vision, pp. 6748–6758 (2023)

Charts and Tables

AltChart: Enhancing VLM-Based Chart Summarization Through Multi-pretext Tasks

Omar Moured[1,2], Jiaming Zhang[1,2(✉)], M. Saquib Sarfraz[1],
and Rainer Stiefelhagen[1,2]

[1] CV:HCI lab, Karlsruhe Institute of Technology, Karlsruhe, Germany
{omar.moured,jiaming.zhang,m.sarfraz,rainer.stiefelhagen}@kit.edu
[2] ACCESS@KIT, Karlsruhe Institute of Technology, Karlsruhe, Germany
https://cvhci.anthropomatik.kit.edu/

Abstract. Chart summarization is a crucial task for blind and visually impaired individuals as it is their primary means of accessing and interpreting graphical data. Crafting high-quality descriptions is challenging because it requires precise communication of essential details within the chart without vision perception. Many chart analysis methods, however, produce brief, unstructured responses that may contain significant hallucinations, affecting their reliability for blind people. To address these challenges, this work presents three key contributions: (1) We introduce the AltChart dataset, comprising 10,000 real chart images, each paired with a comprehensive summary that features long-context, and semantically rich annotations. (2) We propose a new method for pretraining Vision-Language Models (VLMs) to learn fine-grained chart representations through training with multiple pretext tasks, yielding a performance gain with ∼2.5%. (3) We conduct extensive evaluations of four leading chart summarization models, analyzing how accessible their descriptions are. Our dataset and codes are publicly available on our project page: https://github.com/moured/AltChart.

Keywords: Alternative Text · Text Semantics · Pretext Tasks

1 Introduction

We can find charts everywhere in newspapers, in scientific documents, and in many applications that track personal data. Charts provide compact and coherent information visualization and are often used to obtain significant insights from complex data. Given the importance and diverse use cases for charts, it's essential to guarantee equal access for a broad audience. This includes the 253 million Blind and Visually Impaired (BVI) individuals [1], who are often given no choice in understanding charts [22].

Visualizations are accessible to individuals with blindness through two methods: tactile modality, where data is rendered on printed materials with varied

© The Author(s), under exclusive license to Springer Nature Switzerland AG 2024
E. H. Barney Smith et al. (Eds.): ICDAR 2024, LNCS 14804, pp. 349–366, 2024.
https://doi.org/10.1007/978-3-031-70533-5_21

elevations [37–39]. However, the primary means by which they use is via textual summaries, commonly known as **alternative (alt) text** [55]. These summaries are embedded as hidden tags within documents' web pages. A study [10] revealed that over 80% of these do not adhere to accessibility best practices, often neglecting to include alt text. Furthermore, even when they are available, they often don't comply with recommended guidelines (e.g. WCAG [54], Diagram Center's [13]) and standards (e.g. W3C [52]). Recent studies have highlighted significant deficiencies in summaries found within publications [2,9].

Chart summaries are typically derived from two sources: they are either manually created by humans or generated (semi-)automatically by AI systems. While human-generated descriptions tend to be accurate, they are often overly simplified or too complex, lacking the necessary balance for effective communication [31]. **A high-quality chart summary should clearly convey the essential details of the visualization, meeting diverse user preferences. This includes providing not only the necessary contextual information but also visual aspects** [31] (e.g., chart type, data encodings, color schemes, etc.) to accommodate different interaction styles. Authoring a high-quality chart summary is a nontrivial task [32] in both document AI and document accessibility domains.

Generated descriptions through Vision-Language (VL) models, unlike deterministic systems, are faster to obtain and require no expertise, but are highly susceptible to hallucinations [28]. Recent advancements in large VL models have significantly improved chart analysis field, enabling tasks like chart2text [23], chart2table [8], and chart2code [35], among others. However, **these models are often trained on synthetically generated datasets or existing real chart corpus, which are either limited in size or do not meet accessibility guidelines**. This limitation can result in semantically weak summaries that are brief and potentially inaccurate. To address this, our work introduces AltChart, a dataset particularly suitable for VL models with 10,000 real chart images with human-authored chart summarize, adhering to accessibility guidelines and semantically rich.

VL models have shown improvement in overall performance when scaling the pretraining corpus [18,53]. The increased sample size enables these models to learn fine-grained representations. However, the use of synthetically generated data raises concerns about model robustness and biases towards certain visualization styles [3,5]. To address these limitations, we investigate whether vision encoders could develop better representations through different training means.

Pretext tasks are among the methods that have shown promising performance in preparing models for complex tasks [45]. They challenge models to resolve smaller image-level tasks as a preliminary step before the mainstream one. Consequently, we have conducted experiments with multi-pretext tasks and found that it can achieve state-of-the-art performance on widely recognised chart summarization benchmarks. We also find that other available annotation types in datasets (e.g., bounding boxes, segmentation masks, key points, etc.) could be made useful for vision language models with pretext tasks as they can be defined accordingly.

To foster further research on chart summarization, we will make the dataset, and codes publicly available to the community. The key findings and contributions of this paper can be summarized as:

- We introduce pretext tasks to train VL models, aiming to reduce hallucinations and develop fine-grained chart representation. Our model *AltChart* achieves state-of-the-art performance in popular chart summarization benchmarks.
- We collect and publicly release the AltChart dataset serving as a benchmark for semantically rich and long-context chart descriptions. The dataset features 10,000 real chart images sourced from HCI publications, spanning 8 categories and 10 key semantics.
- We conducted extensive evaluations of 4 leading chart summarization models, analyzing the impact of structured and lengthy summaries on their performance. Our findings highlight the effectiveness of our approach in learning better visual chart representations. However, further improvements are needed in specific areas, such as panel and multivariate charts, as well as for low-resolution data.

2   Background

High-quality alternative text is characterized by its semantic richness. Semantics could be words, phrases, or sentences grouped to define the theme of the overall text. Each semantic element helps further streamline the interpretation of the described content. For example, see Fig. 1. Defining and extracting these textual semantics guides the model to form a representation of the patterns that an accessible description could embody. In this work, we use the term 'semantic' to refer to the textual keywords as shown in the aforementioned figure.

We build on the work of Lundgard, A., et al. [23], who conducted a thorough examination of visualizations, focusing specifically on the semantic depth of effective chart descriptions. They expanded the summarization guidelines framework with a more general conceptual model covering four levels of semantic content:

- L1: Chart construction properties (e.g., axes, encodings, title).
- L2: Statistical concepts and relations (e.g., outliers, correlations, statistics).
- L3: Perceptual and cognitive phenomena (e.g., trends, patterns).
- L4: Domain-specific insights (e.g., context relevant to the data).

The study suggests two key points: Firstly, it indicates that captions should communicate key trends and statistics, while also considering the preferences of the reader. Secondly, it highlights the importance of using existing accessibility guidelines as a foundation to enrich chart summarization research.

3   Related Work

Recent attention on VL models has accelerated the development of chart summarization tasks [9,50]. In this section, we start by discussing the top related benchmarks for this task and then explore how VL models are trained to tackle this challenge.

3.1   Existing Datasets

Table 1 lists the top five related datasets for chart summarization task. The Chart-to-Text dataset [23] compiles descriptions for charts from Pew[1] and Statista[2], covering line, bar, pie, and area charts. ChartSumm [44] expands on this by nearly doubling the dataset size and including longer summaries. However, our analysis shows that both datasets focus mainly on Statistical and Perceptual (L2L3) sentences, with 91% and 94% of their content, respectively, missing foundational visual sentences (L1). AutoChart [58], while offering a balanced mix of sentence levels through synthesized charts and template-based captions, suffers from limited variation. These datasets fall short on accessibility standards. In contrast, our dataset uses real charts and summaries collected from accessible venues, doubling the chart categories to include new challenging types like Compose and Panel charts

More recently, datasets like HCI Alt Text [9] and VisText [50] have been developed to specifically address chart summarization for BVI individuals. Both datasets are rich with L1 and L2L3 semantics. VisText creates synthetic chart images using the Vega-Lite visualization tool, then used crowdsourcing for L2L3 summarize, while machine learning models are employed for L1 captions. In contrast, HCI Alt Text compiles figures from accessibility venues, filtering for those with alternative text. However, this dataset, intended primarily for analysis, comprises only 511 chart images. This limited size makes it challenging to train effective data-driven methods. To overcome these constraints, we adopted a similar methodology to HCI Alt Text, but expanded our collection to 10,000 chart images and manually annotated them with 10 text semantics.

Table 1. Overview of the five most related datasets. Our AltChart dataset includes real-charts and real-summarize, with a broader range of categories and semantics.

Name	Data Type		Categories	Semantics	Image Count
	Images	Descriptions			
Chart-to-Text [23]	real	real	4	✗	44,085
HCI Alt Text [9]	real	real	2	2	511
ChartSumm [44]	real	mixed	3	✗	84,363
AutoChart [58]	synthetic	synthetic	3	✗	23,543
Vistext [50]	synthetic	mixed	3	2	8,822
AltChart (Ours)	**real**	**real**	**8**	**10**	**10,000**

---

[1] https://www.pewresearch.org/.

[2] https://www.statista.com/.

3.2   Multimodal Foundation Models

Building a pretrained multimodal foundation model typically involves two steps. First, textual inputs are encoded using a language model such as T5 [43], BERT [12], or recent architectures like Llama 2.0 [51]. Second, a vision encoder processes the input image, which may focus on parts of images [7], by using FastRCNN [46], or use recent, larger transformers (e.g., VIT [14]) that encode the entire image [26]. These models are initially trained on extensive web content for comprehension tasks such as text-to-text [16] and image-to-text [42] are mainly trained with natural images. To adapt these models for specific domain tasks, a second fine-tuning iteration is often necessary to ensure that the model develops a meaningful latent space for the targeted task. For instance, Donut [24] introduced an OCR-free Transformer, trained end-to-end for document understanding. Subsequently, Nougat [4] fine-tuned Donut, making it effective for converting academic documents into markdown language. Charts, however, present a unique challenge compared to natural images or textual documents. The complexity of user questions often involves sophisticated mathematical calculations. As a result, multimodal foundation models often struggle when addressing tasks related to charts [15].

3.3   Chart-Specific Vision-Language Pretraining

Recent works have addressed chart-related tasks using various techniques. Some approaches involve modifying the architecture by developing adapters to interpret charts, while others introduce more comprehensive benchmarks for fine-tuning. Matcha [27] builds upon Pix2Struct [25], incorporating numerical reasoning knowledge into the image-to-text model by learning from textual math datasets. UniChart [34] employs a visual instruction tuning approach [30] and fine-tunes the Donut base model with real charts for multiple low-level tasks (e.g., extracting table data) and high-level tasks (e.g., generating summaries). Unlike UniChart, a recent model named ChartLlama [18], based on LLaVA-1.5 [29], proposes an extensive chart-related benchmark leveraging GPT-4. This benchmark is synthetically created with multiple steps to ensure high quality.

Although these models generate generally appealing outputs for sighted individuals, they raise significant concerns regarding accessibility. It is important to remember that **while a summary accessible to blind individuals is also accessible to sighted individuals, the reverse is not necessarily true.** J. Tang and Bogust et al. [50] were the first to experiment with the abilities of VLMs to generate accessible summaries, but only with synthetically generated charts. In contrast, our proposal, AltChart, ensures that our pretraining corpus comprises real charts from accessible resources, which are semantically rich for everyone.

The aforementioned state-of-the-art approaches mainly follow a similar strategy, extending the size of the pretraining corpus, which leads to higher performance on specific benchmarks but also tends to suffer from catastrophic forgetting [6,56] and lacks consistent summary structure among similar visual inputs. We instead question whether "less can be more." We demonstrate in our work how pretraining vision encoders with multiple pretext tasks such as, classification and colorization, can achieve state-of-the-art performance. Pretext tasks have already shown promising performance with vision models in previous studies [21,49]. Furthermore, Pretext tasks could enable the use of other annotations format that were previously not possible to train with VLMs, such as segmentation masks. We also believe that pretext tasks could help us address challenging samples (e.g., those with high loss values) with simpler substream tasks.

4   AltChart Dataset

We analyzed five current benchmarks and developed our dataset, designed to bridge some of the identified gaps in current research. Our dataset specifically targets L1 and combines L2 and L3 to simplify the annotation process. We decided to exclude L4 from our current dataset due to the domain knowledge required beyond input chart images, such as document-level topics. With the interest to explore this level in future research.

4.1   Dataset Construction

Considering the existing limitations in available data, such as the lack of semantically rich descriptions, short descriptions, or adherence to accessibility guidelines, we dedicated efforts to creating the AltChart dataset. We began by crawling HCI publications from five ACM[3] conferences (CHI, ASSETS, DIS, UIST, W4A) spanning 2015 to 2023. Our focus was on papers containing alt-text tags. This process yielded 8,000 PDFs and 43,510 images.

To capture high-quality images with alt-text in our corpus, we undertook three steps: (1) We fine-tuned a BERT-based classifier [12] on the HCI Alt Text dataset to determine whether the alt text corresponded to a chart and for sentence-level classification (L1/L2L3). The model achieved an F1 score of 93% on the test set. (2) We reviewed the predictions and filtered out the false positives. (3) Throughout the annotating phase, we further eliminated images lacking L1/L2L3 descriptions, those shorter than three sentences, or not adhering to alt-text guidelines (e.g., presenting incorrect information), ultimately retaining 10,000 images. These steps ensured that our corpus was semantically rich and included longer author-written descriptions. A comparison analysis was conducted to verify this, as illustrated in Table 2. We randomly split our dataset into training, validation, and test sets using chart IDs to prevent data leakage across sets, resulting in an approximate 80:10:10 ratio. Next, we discuss our dataset annotation process.

4.2   Data Annotations and Properties

For each image, we recorded the paper's DOI, the figure number, and both the image caption and its alt-text. To annotate descriptions, we followed the

---

[3] https://dl.acm.org/conferences/.

Table 2. Comparison of three leading datasets in terms of comprehensive summarization. *AltChart* stands out with significantly higher average sentence and word counts-nearly double those of the others-and showcases the most balanced L1 to L2/L3 sentence ratio.

Dataset	Avg. Summ.		L1:L2L3 Ratio
	Sentence Count	Word Count	
ChartSumm	2.0	45.44	1.17 : 98.83
Vistext	2.26	42.6	56.2 : 43.8
HCI Alt Text	3.66	77.0	74.2 : 25.8
AltChart	**5.67**	**136.35**	**44.9 : 55.1**

protocol outlined by Lundgard, A., et al. [31]. In a given batch of 300 images, each description was semantically tagged using 10 attributes keys as seen in Fig. 1. GPT-4.0[4] was then employed to tag the remaining descriptions. Each tagged result underwent verification by our annotators. While our primary aim in using these semantic tags was to facilitate our pretext tasks, the annotations can also be useful for analyzing the data structure and enhancing accessibility by identifying missing attributes. The AltChart dataset encompasses a range of eight chart types: line, bar, area, scatter, multivariate, panel, pie, and box charts. For clarification, multivariate and panel charts represent two new categories not previously addressed in earlier benchmarks. Multivariate charts refer to those displaying more than one data type (e.g., combining lines and bars), while panel charts (Fig. 1-b) are a collection of multiple charts within a single figure sharing common elements, such as a unified legend or axis.

5   Method

Although larger datasets may improve performance, as demonstrated by several SOTA works, the critical factor is how effectively the model learns from this synthetic data. The capability of these models to handle the variations and complexities of real-world charts remains a challenging issue. With this in mind, we pose a question: **Can we improve the vision encoder hidden representations to a degree that minimizes our reliance on synthetic data?** To address this, our approach leverages pretext tasks to guide the vision encoder in capturing essential covariant and invariant chart features, thereby reducing hallucinations in descriptions. Next, we discuss the details of our pretext task implementations, as outlined in Fig. 2.

5.1   Chart Pretext Tasks

An effective feature extraction process should include both covariant and invariant features [36]. Covariant features, which adapt to transformations such as

---

[4] https://openai.com/gpt-4.

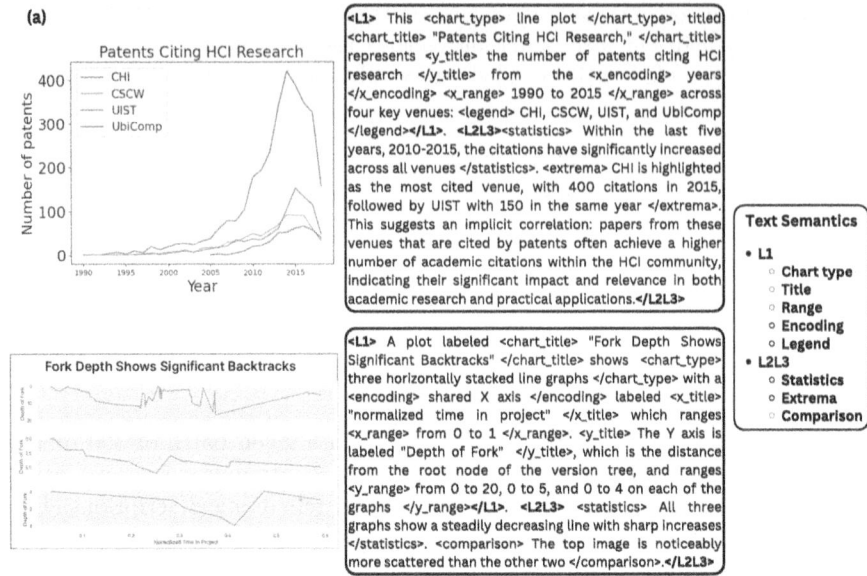

Fig. 1. Two chart samples from *AltChart* with their annotated summaries. Semantics are indicated by a color code, where <semantic-name> marks the beginning and </semantic-name> marks the end of the semantic segment.

scaling or rotation (e.g., vertical axis labels, font sizes), enhance vision encoders' ability to recognize objects despite spatial changes. Meanwhile, invariant features maintain consistency by capturing key characteristics that remain constant across various scenarios (e.g., multi-line charts often include legends). This duality is crucial to ensure a comprehensive and reliable interpretation of chart images. These features, covariant and invariant, are developed through pretext tasks using both self-supervised and supervised training methods, respectively.

Self-supervised Tasks. is to learn image representations directly from pixels, without relying on predefined semantic annotations. This process typically involves applying transformations to input images and training sub-models to predict the properties of the transformation. In this work, we have chosen three traditional tasks, namely:

1. **Rotation Prediction** [17]: rotating chart images by various degrees and having a sub-model to predict the angle of rotation. This helps in learning the orientation and geometry of chart components.
2. **Jigsaw Puzzle Solving** [40]: scrambling charts to multiple segments and training a sub-model to reorder them correctly. This teaches the model about the spatial relationships within chart elements.
3. **Colorization** [57]: feeding grayscale chart images into a sub-model for colorization. This helps in developing features that distinguish between chart components (e.g., different pies in a pie chart).

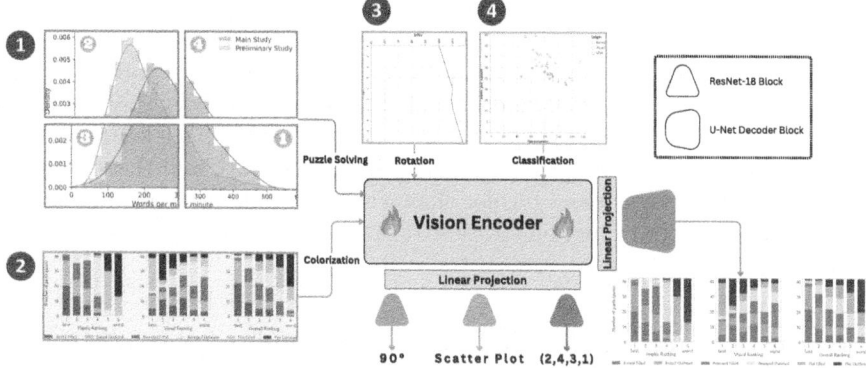

Fig. 2. Overview of our vision encoder's training approach, starting from the top-left with tasks including puzzle solving, colorization, rotation, and classification. Sample outputs for each corresponding task are shown on the bottom-right of the figure.

Supervised Tasks. defined as the utilization of a small amount of labelled data to capture consistent representations. Given the chart analysis topic, different datasets provide varying types of annotations. For example, AltChart and Vistext lack pixel-level semantic annotations, while the Chart-to-Text benchmark offers bounding box annotations for text. To ensure applicability across all benchmarks, we leverage the classification of chart categories task. However, one may explore additional approaches, such as masked element identification, when segmentations or bounding boxes are available.

5.2   Implementation Details

All pretext tasks need to be evaluated against quantitative metrics to ensure their effectiveness. To facilitate this evaluation, we employ a shallow backbone network complemented by a task-specific head. In the following sections, we will detail the transformation functions, sub-models, and our formation of the loss function.

Transformation Functions. Given an image I, we apply the transformation functions $g(.)$, $f(.)$ and $h(.)$ to generate a transformed image and corresponding ground-truth labels for rotation, puzzle solving, and colorization tasks, respectively.

For rotation, the image undergoes one of four rotational levels: $0°$, $90°$, $180°$, and $270°$. This process follows the methodology described in [17], where the network is tasked with identifying the correct rotation angle. In the puzzle-solving task, the image is divided into a 3×3 grid, resulting in nine 64×64 pixel patches. To avoid overfitting, each patch's location is randomly jittered by up to seven pixels, in line with the approach of N. Mehdi et al. [40], thus creating nine

distinct 64×64 pixel tiles. We define 100 possible permutations (puzzle configurations), each associated with a unique index. The function $f(.)$ outputs the
nine image tiles along with the index of the corresponding permutation, serving
as the ground-truth label. The model's objective is to accurately classify the correct permutation index. For the colorization operation, the image is transformed
into a grayscale image using the formula $\frac{R+G+B}{3}$ as in [11], and the sub-model
is trained to predict the a and b color channels in the CIE Lab color space. The
development of the loss function for these operations is discussed in the following
section.

Loss Function. three pretexts and one supervised task are present in our formulations. three of which; rotation, puzzle solving, and the supervised tasks
end with a softmax activation layer hence we utilized the traditional Cross-
Entropy Loss $\mathcal{L}_{rotation}$, $\mathcal{L}_{puzzle}$ and $\mathcal{L}_{categ}$ respectively to output probability
values between 0 and 1. For the colorization we utilized the conditional GAN
loss $\mathcal{L}_{cGAN}$ with regression mean absolute error loss $\mathcal{L}_{L1}$ as proposed in Pix2Pix
[20]. Given N images in a batch, The colorization loss $\mathcal{L}_{color}$ is then computed
as follows:

$$\mathcal{L}_{cGAN}(G, D) = \frac{1}{N} \sum_{i=1}^{N} \log D(I_g^i, I_{ab}^i) + \log\left(1 - D(I_g^i, G(I_g^i, z))\right) \quad (1)$$

$$\mathcal{L}_{L1}(G) = \frac{1}{N} \sum_{i=1}^{N} |I_{ab}^i - G(I_g^i, z)| \quad (2)$$

$$\mathcal{L}_{color} = \mathcal{L}_{cGAN}(G, D) + \alpha \mathcal{L}_{L1}(G) \quad (3)$$

In Eq. 1, the generator, G, takes a grayscale image I_g^i and produces a 2-
channel image I_{ab}. The discriminator, D, concatenates both images to decide
whether they are fake or real. It's important to note that both models are conditioned on the grayscale image, meaning the noise vector is omitted [20]. The
mean absolute error $\mathcal{L}_{L1}(G)$ in Eq. 2 aims for pixel-wise comparison between the
generated image and the ground truth to introduce a form of self-supervision.
Since $L1$ has been shown to produce contrastive results , the parameter α is
introduced to balance its overall impact. Our final loss $\mathcal{L}_{total}$ is formed as follows:

$$\mathcal{L}_{total} = \gamma_1 \mathcal{L}_{color} + \gamma_2 \mathcal{L}_{rotation} + \gamma_3 \mathcal{L}_{puzzle} + \gamma_4 \mathcal{L}_{categ} \quad (4)$$

In calculating the total loss, we sum each one with its respective gamma
parameter γ_{1-4}, allowing us to fine-tune their contributions.

Convolutional Network. In our experiments, we utilize a ResNet-18 network
[19]. Two modifications are applied: (1) The standard 3-channel convolutional
layer input is adjusted to align with the vision encoder output's shape (2) After
the final FC layer, following the final FC (Fully Connected) layer, we implement

average pooling and linearly project the output to match the number of classes for each specific pretext task. For the colorization task, a U-Net decoder [47] is concatenated to the vision encoder to function as the generator. The discriminator is again a ResNet-18.

Hyperparameters. For our pretext tasks, we use a default image resolution of 224×224, unless specified otherwise. We set the parameter $\alpha = 100$ in Eq. 3 and an equal impact for all losses, $\gamma_{1-4} = 0.25$ in Eq. 4. Both the ResNet and the U-Net decoder are initialized with their pre-trained weights, obtained from MMpretrain[5] and MMsegmentation[6] respectively.

6   Experiment

In this section, we conduct both quantitative and qualitative comparisons of four SOTA methods against AltChart.

6.1   Experimental Setups

In this study, we conducted all training processes on a cluster equipped with four NVIDIA-40 GPUs. We utilized publicly available source code from GitHub for each model. Initially, all models were trained on the Chart-2-Text dataset. We then fine-tuned these pre-trained models on the other datasets listed in Table 3 (Vistext and AltChart), with the exception of our baseline model, which is described subsequently. For fine-tuning, the training epochs were set to 5, and the LoRa adapter was employed. We did not alter the input resolution for any of the approaches from their initial configurations. Each experimental run took approximately 8–10 hours to complete.

6.2   Baselines and Evaluation Metrics

We compare our model against four baselines: (1) Vistext [50], a VL-T5-based model that achieves SOTA results in generating accessible chart summaries. (2) Matcha [27], an adaptation of Pix2Struct for charts, pre-trained on mathematical reasoning and chart data extraction tasks. (3) UniChart [34], a model based on Donut [24], further pre-trained on multiple chart analysis tasks, achieving SOTA on Chart-2-Text [23] and ChartQA [33]. (4) ChartLLama [18], a fine-tuned LLaVA 1.5 [29] model trained on a large chart corpus synthetically generated with GPT-4.

To conduct the comparison on our benchmarks, we first reproduced and ran inference with each of the baseline models, evaluating their summarization performance using the BLEU score [41]. Furthermore, given that the BLEU score primarily focuses on n-gram matching between the generated and reference texts,

---

[5] https://github.com/open-mmlab/mmpretrain.

[6] https://github.com/open-mmlab/mmsegmentation.

it may overlook essential aspects such as semantic similarity, informativeness, and factual correctness [48]. Hence, we also performed a qualitative evaluation and error analysis of the outputs.

6.3   Training Details

As a baseline, we utilized the Donut model, primarily chosen for its relatively low number of parameters, scaling to millions, in contrast to the LLaVA model, which scales to billions. Initially, we employed the base Donut weights, pretrained for text reading tasks as an alternative to OCR engines. These base weights cannot comprehend chart images. We initially train the transformer encoder for 3 epochs on pretext tasks as previously described, followed by training the entire model (vision+language components) for an additional 2 epochs on summarization tasks.

Table 3. Results of state-of-the-art methods on three datasets of chart summarization. The number of training parameters reported.

Model	#Params	VisText			Chart-2-Text			AltChart		
		L1	L2/L3	avg.	Pew	Statista	avg.	L1	L2/L3	avg.
Vistext - image guided [50]	224M	9.0	2.0	5.5	14.2	44.2	29.2	-	-	-
Matcha	282M	6.0	4.0	5.0	12.2	39.4	25.8	16.5	8.0	12.2
Unichart	201M	6.3	5.2	5.75	12.4	38.2	25.3	22.7	13.9	18.3
ChartLLama - 13B	500M	35.0	6.0	20.5	14.2	40.7	27.45	35.0	14.2	24.6
Ours (AltChart)	**180M**	**37.6**	5.6	**21.6**	**15.1**	**46.0**	**30.55**	**44.1**	**14.6**	**29.3**

6.4   Comparison of State-of-the-Art Models

Our novel chart summarization model achieves state-of-the-art performance on diverse datasets, pushing the boundaries of both efficiency and quality. As shown in Table 3, we evaluate our method against different chart models, such as MatCha, UniChart, and ChartLlama on three datasets: VisText, Chart-2-text, and our proposed AltChart. Each dataset offers unique challenges, with VisText and AltChart testing different summarization levels and Chart-2-text utilizing data from two distinct sources: Pew and Statista. Despite having fewer trainable parameters (180M) than other text-based models, our approach boasts superior efficiency and adaptability. It readily applies to different benchmarks and swiftly trains for downstream tasks.

Apart from the significant efficiency, our method has outstanding performance, yielding state-of-the-art scores on all three datasets. For instance, on VisText, it delivers a remarkable 37.6% score on L1 test, outperforming ChartLlama by 2.6%, and achieves exceptional results on L2/L3 test. Similar success manifests in the Chart-2-text dataset, where it scores 15.1% and 46.0% across the two data sources. It shows the genearlizability of the proposed method. Our proposed AltChart dataset unveils the potential of our method. In the structured summarization test, it surpasses the previous best score by 9.06% (reach 44.1% on L1) and obtains a best 14.6% on the more challenging L2/L3 test. These remarkable improvements across diverse datasets showcase the effectiveness and generalizability of our pre-training method. Our model sets a new standard for both efficiency and quality in chart summarization.

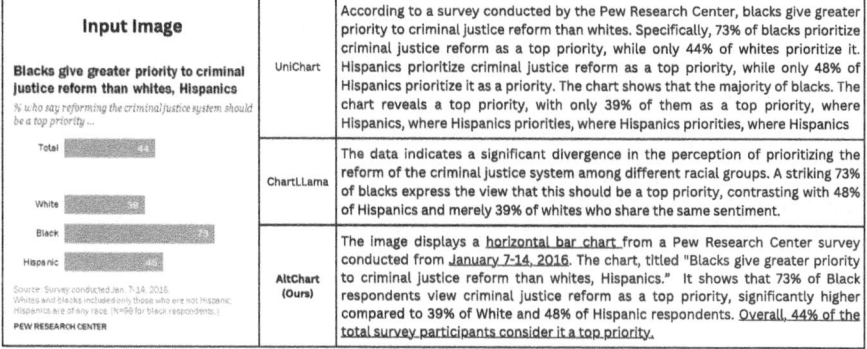

Fig. 3. Qualitative analysis of chart summarization.

6.5 Qualitative Analysis

To present the qualitative results of the proposed AltChart benchmark, we conduct a visualization comparison in Fig. 3. The visualization showcases a chart summarization example from the proposed AltChart dataset, comparing our method with other models like UniChart and ChartLlama. Each case displays the input chart image followed by summaries generated by all three models. On the left side of the visualization is the same input chart image that is a horizontal bar chart, and on the right side from top to bottom are the prediction from UniChart, ChartLLama, and ours. Among them, UniChart made some wrong description about this chart, such as "44% of whites" is incorrect. UniChart struggles in this case, producing repetitive and unclear summaries. ChartLlama also falters, offering basic and incomplete descriptions that miss key details like the total value. In contrast, our method delivers accurate and detailed summaries for the input bar chart. For instance, it can correctly identify the chart type as "horizontal bar chart" and timeframe "January 7–14, 2016". Apart from that, our method can provide insightful descriptions such as "Overall, 44% of the total survey participants consider it a top priority". This comparative analysis demonstrates our model's ability to generate precise and comprehensive chart summaries, surpassing the capabilities of existing methods.

6.6 Ablation Study

To isolate the impact of our proposed pre-training tasks on chart summarization performance, we conduct an ablation study on the proposed AltChart dataset.

As shown in Table 4, we divide the experiments into two groups: self-supervised and supervised training, allowing for clear comparisons between different pre-training paradigms. The reported score is the average from the L1 and L2/L3 summarizations. Self-supervised training can achieve a score of 27.9%, while supervised training alone can reach 25.3%. This 2.6% gain highlights the effectiveness of self-supervised learning in enriching the model's understanding of charts. Pushing the boundaries further, our combined approach that leverages both self-supervised and supervised training delivered the best score of 29.35% in structured chart summarization. This significant improvement proves the effectiveness of our carefully designed prefix tasks in propelling the model towards superior performance.

Table 4. Ablation study of the prefix tasks. Accuracy measured as the ratio of correct predictions to total predictions.

Self-supervised	Supervised	Result (avg L123)
✓		27.90
	✓	25.30
✓	✓	**29.35**

7 Conclusion

In this paper, we introduce AltChart, a SOTA chart summarization model that generates rich, accessible summaries. We employ pretext tasks as a pretraining technique for our model, without significant reliance on synthetic data. Our method facilitates the use of various annotation formats during the training process to acquire robust feature representations, which were absent in previous in earlier models. Additionally, we present the AltChart dataset, which is so far the largest accessibility-compliant, real chart summarization dataset with rich semantics. This dataset can be further utilized for other tasks, such as recommendation and classification systems. While our model sets a new record on the chart summarization tasks, the evaluation also suggests opportunities for further improvement.

Limitations. Nonetheless, our model faces several limitations. Despite being trained with twice the number of chart categories, the model still tends to hallucinate and produce factually incorrect statements, especially with visually complex charts, such as those with many lines, bars, or 3×3 panel charts. Therefore, there is a need for more generalizable chart models that can accurately interpret the diverse range of charts found in documents. Future directions may focus on enhancing the vision encoders, as they are the first component to digest the image.

Due to the limited, time-constrained computing resources, we didn't investigate how each pretext task affects the overall model performance. Furthermore, our experimentation was limited to training a single-vision encoder using a two-step training strategy. However, future work could include pretext tasks as part of an end-to-end training cycle and explore different vision encoders, such as CLIP. Finally, we believe that undertaking a robustness benchmark is a necessary measure, particularly considering the use of these models by blind and visually impaired individuals. We aim to carry out this investigation in our forthcoming research.

Acknowledgments. The authors would like to thank the HoreKa computing cluster at KIT for the computing resources used to conduct this research. We also thank everyone from CVHCI lab who contributed to the annotation phase.

Disclosure of Interests. The first author has received a fund grant from the European Union's Horizon 2020 research and innovation program under the Marie Sklodowska-Curie Grant No.861166.

References

1. Ackland, P., Resnikoff, S., Bourne, R.: World blindness and visual impairment: despite many successes, the problem is growing. Commun. Eye Health **30**(100), 71 (2017)
2. Alam, M.Z.I., Islam, S., Hoque, E.: SeeChart: enabling accessible visualizations through interactive natural language interface for people with visual impairments. In: Proceedings of the 28th International Conference on Intelligent User Interfaces, pp. 46–64 (2023)
3. Bansal, H., Grover, A.: Leaving reality to imagination: Robust classification via generated datasets. In: ICLR 2023 Workshop on Trustworthy and Reliable Large-Scale Machine Learning Models (2023)
4. Blecher, L., Cucurull, G., Scialom, T., Stojnic, R.: Nougat: neural optical understanding for academic documents. In: The Twelfth International Conference on Learning Representations (2023)
5. Cascante-Bonilla, P., et al.: Going beyond nouns with vision & language models using synthetic data. In: Proceedings of the IEEE/CVF International Conference on Computer Vision, pp. 20155–20165 (2023)
6. Chen, S., Hou, Y., Cui, Y., Che, W., Liu, T., Yu, X.: Recall and learn: Fine-tuning deep pretrained language models with less forgetting. In: Proceedings of the 2020 Conference on Empirical Methods in Natural Language Processing (EMNLP), pp. 7870–7881 (2020)
7. Chen, Y.-C., et al.: UNITER: UNiversal image-TExt representation learning. In: Vedaldi, A., Bischof, H., Brox, T., Frahm, J.-M. (eds.) ECCV 2020. LNCS, vol. 12375, pp. 104–120. Springer, Cham (2020). https://doi.org/10.1007/978-3-030-58577-8_7
8. Cheng, Z.Q., Dai, Q., Hauptmann, A.G.: ChartReader: a unified framework for chart derendering and comprehension without heuristic rules. In: Proceedings of the IEEE/CVF International Conference on Computer Vision, pp. 22202–22213 (2023)

9. Chintalapati, S.S., Bragg, J., Wang, L.L.: A dataset of alt texts from HCI publications: Analyses and uses towards producing more descriptive alt texts of data visualizations in scientific papers. In: Proceedings of the 24th International ACM SIGACCESS Conference on Computers and Accessibility, pp. 1–12 (2022)

10. Commission, D.R.: The Web: Access and Inclusion for Disabled People; A Formal Investigation. The Stationery Office (2004)

11. Deshpande, A., Rock, J., Forsyth, D.: Learning large-scale automatic image colorization. In: Proceedings of the IEEE International Conference on Computer Vision, pp. 567–575 (2015)

12. Devlin, J., Chang, M.W., Lee, K., Toutanova, K.: BERT: pre-training of deep bidirectional transformers for language understanding. In: Burstein, J., Doran, C., Solorio, T. (eds.) Proceedings of the 2019 Conference of the North American Chapter of the Association for Computational Linguistics: Human Language Technologies, Volume 1 (Long and Short Papers), pp. 4171–4186. Association for Computational Linguistics, Minneapolis, Minnesota (2019). https://doi.org/10.18653/v1/N19-1423, https://aclanthology.org/N19-1423

13. Diagram center: specific guidelines - graphs. http://diagramcenter.org/specific-guidelines-e.html (2022)

14. Dosovitskiy, A., et al.: An image is worth 16×16 words: transformers for image recognition at scale. In: International Conference on Learning Representations (2020)

15. Farahani, A.M., Adibi, P., Darvishy, A., Ehsani, M.S., Hutter, H.P.: Automatic chart understanding: a review. IEEE Access (2023)

16. Floridi, L., Chiriatti, M.: GPT-3: Its nature, scope, limits, and consequences. Minds Mach. **30**(4), 681–694 (2020). https://doi.org/10.1007/s11023-020-09548-1

17. Gidaris, S., Singh, P., Komodakis, N.: Unsupervised representation learning by predicting image rotations (2018)

18. Han, Y., et al.: ChartLlama: a multimodal LLM for chart understanding and generation. arXiv preprint arXiv:2311.16483 (2023)

19. He, K., Zhang, X., Ren, S., Sun, J.: Deep residual learning for image recognition. In: Proceedings of the IEEE Conference on Computer Vision and Pattern Recognition, pp. 770–778 (2016)

20. Isola, P., Zhu, J.Y., Zhou, T., Efros, A.A.: Image-to-image translation with conditional adversarial networks (2018)

21. Jaiswal, A., Babu, A.R., Zadeh, M.Z., Banerjee, D., Makedon, F.: A survey on contrastive self-supervised learning. Technologies **9**(1), 2 (2020)

22. Jung, C., Mehta, S., Kulkarni, A., Zhao, Y., Kim, Y.S.: Communicating visualizations without visuals: investigation of visualization alternative text for people with visual impairments (2021)

23. Kantharaj, S., et al.: Chart-to-text: a large-scale benchmark for chart summarization. In: Proceedings of the 60th Annual Meeting of the Association for Computational Linguistics (Volume 1: Long Papers), pp. 4005–4023 (2022)

24. Kim, G., et al.: OCR-free document understanding transformer. In: European Conference on Computer Vision, pp. 498–517 (2022)

25. Lee, K., et al.: Pix2Struct: screenshot parsing as pretraining for visual language understanding. In: International Conference on Machine Learning, pp. 18893–18912. PMLR (2023)

26. Li, J., Li, D., Xiong, C., Hoi, S.: BLIP: bootstrapping language-image pre-training for unified vision-language understanding and generation. In: International Conference on Machine Learning, pp. 12888–12900. PMLR (2022)

27. Liu, F., et al.: MatCha: enhancing visual language pretraining with math reasoning and chart derendering (2023)
28. Liu, H., et al.: A survey on hallucination in large vision-language models. arXiv preprint arXiv:2402.00253 (2024)
29. Liu, H., Li, C., Li, Y., Lee, Y.J.: Improved baselines with visual instruction tuning (2023)
30. Liu, H., Li, C., Wu, Q., Lee, Y.J.: Visual instruction tuning. In: Advances in Neural Information Processing Systems, vol. 36 (2024)
31. Lundgard, A., Satyanarayan, A.: Accessible visualization via natural language descriptions: a four-level model of semantic content. IEEE Trans. Visual Comput. Graphics 28(1), 1073–1083 (2021)
32. Mack, K., Cutrell, E., Lee, B., Morris, M.R.: Designing tools for high-quality alt text authoring. In: Proceedings of the 23rd International ACM SIGACCESS Conference on Computers and Accessibility, pp. 1–14 (2021)
33. Masry, A., Do, X.L., Tan, J.Q., Joty, S., Hoque, E.: ChartQA: a benchmark for question answering about charts with visual and logical reasoning. In: Findings of the Association for Computational Linguistics: ACL 2022, pp. 2263–2279 (2022)
34. Masry, A., Kavehzadeh, P., Do, X.L., Hoque, E., Joty, S.: UniChart: a universal vision-language pretrained model for chart comprehension and reasoning. In: Proceedings of the 2023 Conference on Empirical Methods in Natural Language Processing, pp. 14662–14684 (2023)
35. Meng, F., et al.: ChartAssisstant: a universal chart multimodal language model via chart-to-table pre-training and multitask instruction tuning. arXiv preprint arXiv:2401.02384 (2024)
36. Misra, I., Maaten, L.v.d.: Self-supervised learning of pretext-invariant representations. In: Proceedings of the IEEE/CVF Conference on Computer Vision and Pattern Recognition, pp. 6707–6717 (2020)
37. Moured, O., Alzalabny, S., Schwarz, T., Rapp, B., Stiefelhagen, R.: Accessible document layout: an interface for 2D tactile displays. In: Proceedings of the 16th International Conference on PErvasive Technologies Related to Assistive Environments, pp. 265–271 (2023)
38. Moured, O., Baumgarten-Egemole, M., Müller, K., Roitberg, A., Schwarz, T., Stiefelhagen, R.: Chart4Blind: an intelligent interface for chart accessibility conversion. In: Proceedings of the 29th International Conference on Intelligent User Interfaces, pp. 504–514 (2024)
39. Moured, O., Zhang, J., Roitberg, A., Schwarz, T., Stiefelhagen, R.: Line graphics digitization: a step towards full automation. In: Fink, G.A., Jain, R., Kise, K., Zanibbi, R. (eds.) International Conference on Document Analysis and Recognition, vol. 14191, pp. 438–453. Springer, Cham (2023). https://doi.org/10.1007/978-3-031-41734-4_27
40. Noroozi, M., Favaro, P.: Unsupervised learning of visual representations by solving jigsaw puzzles (2017)
41. Post, M.: A call for clarity in reporting bleu scores. In: Proceedings of the Third Conference on Machine Translation: Research Papers, pp. 186–191 (2018)
42. Radford, A., et al.: Learning transferable visual models from natural language supervision. In: International Conference on Machine Learning, pp. 8748–8763. PMLR (2021)
43. Raffel, C., et al.: Exploring the limits of transfer learning with a unified text-to-text transformer. J. Mach. Learn. Res. 21(1), 5485–5551 (2020)

44. Rahman, R., Hasan, R., Farhad, A.A.: ChartSumm: a large scale benchmark for Chart to Text Summarization. Ph.D. thesis, Department of Computer Science and Engineering (CSE), Islamic University (2022)
45. Rani, V., Nabi, S.T., Kumar, M., Mittal, A., Kumar, K.: Self-supervised learning: a succinct review. Arch. Comput. Methods Eng. **30**(4), 2761–2775 (2023)
46. Ren, S., He, K., Girshick, R., Sun, J.: Faster R-CNN: towards real-time object detection with region proposal networks. In: Advances in Neural Information Processing Systems, vol. 28 (2015)
47. Ronneberger, O., Fischer, P., Brox, T.: U-Net: convolutional networks for biomedical image segmentation. In: Navab, N., Hornegger, J., Wells, W.M., Frangi, A.F. (eds.) MICCAI 2015. LNCS, vol. 9351, pp. 234–241. Springer, Cham (2015). https://doi.org/10.1007/978-3-319-24574-4_28
48. Sai, A.B., Mohankumar, A.K., Khapra, M.M.: A survey of evaluation metrics used for NLG systems. ACM Comput. Surv. (CSUR) **55**(2), 1–39 (2022)
49. Schiappa, M.C., Rawat, Y.S., Shah, M.: Self-supervised learning for videos: a survey. ACM Comput. Surv. **55**(13s), 1–37 (2023)
50. Tang, B., Boggust, A., Satyanarayan, A.: VisText: a benchmark for semantically rich chart captioning. In: Proceedings of the 61st Annual Meeting of the Association for Computational Linguistics (Volume 1: Long Papers), pp. 7268–7298 (2023)
51. Touvron, H., et al.: Llama 2: open foundation and fine-tuned chat models. arXiv preprint arXiv:2307.09288 (2023)
52. W3C: Standards (2022). https://www.w3.org/standards/
53. Wang, Z., et al.: Scaling data generation in vision-and-language navigation. In: Proceedings of the IEEE/CVF International Conference on Computer Vision, pp. 12009–12020 (2023)
54. Web Content Accessibility Guidelines (WCAG): complex images (2022). https://www.w3.org/WAI/tutorials/images/complex/
55. WebAIM: Screen reader user survey 9 results (2021). https://webaim.org/projects/screenreadersurvey9/
56. Xu, Y., Zhong, X., Yepes, A.J.J., Lau, J.H.: Forget me not: reducing catastrophic forgetting for domain adaptation in reading comprehension. In: 2020 International Joint Conference on Neural Networks (IJCNN), pp. 1–8. IEEE (2020)
57. Zhang, R., Isola, P., Efros, A.A.: Colorful image colorization (2016)
58. Zhu, J., Ran, J., Lee, R.K.W., Li, Z., Choo, K.: AutoChart: a dataset for chart-to-text generation task. In: Proceedings of the International Conference on Recent Advances in Natural Language Processing (RANLP 2021), pp. 1636–1644 (2021)

Synthesizing Realistic Data for Table Recognition

Qiyu Hou[1] (ORCID), Jun Wang[1(✉)] (ORCID), Meixuan Qiao[2], and Lujun Tian[1]

[1] iWudao Tech, Nanjing, China
{houqy,jwang,tianlj}@iwudao.tech
[2] Huazhong University of Science and Technology, Wuhan, China
qiaomeixuan@hust.edu.cn

Abstract. To overcome the limitations and challenges of current automatic table data annotation methods and random table data synthesis approaches, we propose a novel method for synthesizing annotation data specifically designed for table recognition. This method utilizes the structure and content of existing complex tables, facilitating the efficient creation of tables that closely replicate the authentic styles found in the target domain. By leveraging the actual structure and content of tables from Chinese financial announcements, we have developed the first extensive table annotation dataset in this domain. We used this dataset to train several recent deep learning-based end-to-end table recognition models. Additionally, we have established the inaugural benchmark for real-world complex tables in the Chinese financial announcement domain, using it to assess the performance of models trained on our synthetic data, thereby effectively validating our method's practicality and effectiveness. Furthermore, we applied our synthesis method to augment the FinTabNet dataset, extracted from English financial announcements, by increasing the proportion of tables with multiple spanning cells to introduce greater complexity. Our experiments show that models trained on this augmented dataset achieve comprehensive improvements in performance, especially in the recognition of tables with multiple spanning cells.

Keywords: Table Data Synthesis · Data Augmentation · Table Recognition

1 Introduction

Tables, serving as a vital carrier of data, are prevalent across a wide range of digital documents. They efficiently store and display data in a compact and lucid format, encapsulating an immense volume of valuable information. However, recognizing the structures of tables within digital documents, such as PDF and images, and subsequently extracting structured data, present significant challenges due to the complexity and diversity of their structure and style [1]. In recent years, with the advancement of deep learning, new methodologies have surfaced, leading to significant progress in table recognition [16,18,19,21,24].

© The Author(s), under exclusive license to Springer Nature Switzerland AG 2024
E. H. Barney Smith et al. (Eds.): ICDAR 2024, LNCS 14804, pp. 367–388, 2024.
https://doi.org/10.1007/978-3-031-70533-5_22

Deep learning-based methods are adept at managing complex table structures and diverse styles more effectively. However, their reliance on large-scale, high-quality annotated table datasets for model training is pronounced [1,16]. The scarcity of comprehensive and intricately detailed, publicly accessible datasets emerges as a substantial barrier, impeding the further advancement of table structure recognition.

To create large-scale datasets for table recognition, some researchers have initially started by utilizing specific repositories of scientific papers or financial reports. Each document in these repositories contains tables that correspond to some form of structured source codes (such as LaTeX, XML, HTML). They facilitate large-scale annotation of table recognition data by automatically mapping the displayed tables to their corresponding structured source codes. The existing large-scale, real-world table annotation datasets [2,3,12,20,26,27] are few but have all been constructed using similar methodologies. However, the applicability of these methods for creating table annotation data is notably limited, as only a select number of fields have access to document repositories where structured source codes correspond to rendered tables. Given the substantial variations in table structures and styles across different domains and languages, the table styles featured in these datasets tend to exhibit similarities. This similarity poses challenges when attempting to apply these datasets to a broader range of domains [1,16]. These automatically annotated datasets frequently contain a significant number of annotation errors. For instance, in our sampling of over 10,000 tables from the FinTabNet dataset, we found that approximately 9% had obvious annotation errors. Furthermore, the annotation information required varies among different table recognition methods, and some of these table datasets do not completely fulfill the annotation requirements of several prevalent table recognition models [20].

To adapt more efficiently to a wider range of application domains, some researchers are exploring methods for synthesizing annotated data for table recognition. These methods largely depend on predefined structural templates and randomly selected text to generate table structures and fill in content. Moreover, they employ a web browser engine to render the synthesized tables and add annotations, following predefined style templates. However, tables rendered with HTML and CSS offer limited visual fidelity and often fall short of replicating the complex or unique appearances of tables in documents like PDFs. Consequently, they lack the richness and complexity required to accurately simulate the intricate table structures encountered in real-world scenarios.

After careful analysis of real-world tables, we discovered that generating appropriate templates for tables with complex structures, such as those found in the financial sector, is not easily achieved through random methods. The structures produced randomly often differ significantly from real complex tables, and the text randomly inserted into these tables usually deviates greatly from the context and semantics of actual tables. We have also observed that in specific application domains, such as finance, the primary challenge in table recognition stems from the fact that many tables, despite having similar structures and content, display significantly varied presentation styles. To address this issue, we propose a method that utilizes the structure and content of existing complex tables to generate high-quality, realistic synthetic datasets tailored to the target application domain. Firstly, in many scenarios, we can relatively easily access the structure and content of numerous existing tables. Taking the financial domain as an example, tables in U.S. financial disclosures are available in HTML format, allowing direct access to the table structure and the text content within the cells. Although Chinese financial announcements in PDF format do not have corresponding HTML files, they contain a large number of bordered tables. Using PDF parsing tools to extract and analyze the lines within these bordered tables, it is relatively easy to deduce the table structure and the text content in each cell. Secondly, reflecting the actual distribution of tables in the target application domain, tables can be categorized and summarized. This involves extracting and documenting the various attributes associated with the presentation styles of each table within every category. These attributes are then stored in a profile and incorporated into the candidate style set for that category. For a source table whose structure and content have been obtained, its corresponding table category can first be identified based on its content. From this category's candidate set, a profile is selected, and a small degree of randomness is introduced to the attributes of the selected profile to set the profile for the target table. This results in the transformation into a new table that is completely different in style from the original, yet still possesses a very realistic appearance. For example, Fig. 1 shows the transformation of an original bordered table into two borderless tables and one bordered table with different style, all of which have the same structure and content. Additionally, our method, based on image rendering techniques for synthesizing tables, is not constrained by the limitations of browser rendering. It can draw or paste real table appearance components from various documents, enabling the creation of more realistic complex table styles.

This paper makes the following contributions:

- A novel method for synthesizing annotation data specifically designed for table recognition has been proposed, which utilizes the structure and content of existing complex tables. This approach enables the straightforward synthesis of tables that closely resemble the actual table styles prevalent in the target domain, accompanied by comprehensive and complete annotation data.
- Utilizing the actual structure and content of tables from Chinese financial announcements, we synthesized the first large-scale table annotation dataset in the domain of Chinese financial announcements. On this basis, we trained several recent deep learning-based end-to-end table recognition models. Furthermore, we created the first benchmark for real-world complex tables in the Chinese financial announcement domain, which was used to evaluate the performance of models trained on synthetic data, thereby validating the practicality and effectiveness of the method proposed in this paper.
- Additionally, the method proposed in this paper was used to augment the table dataset extracted from English financial announcements, known as

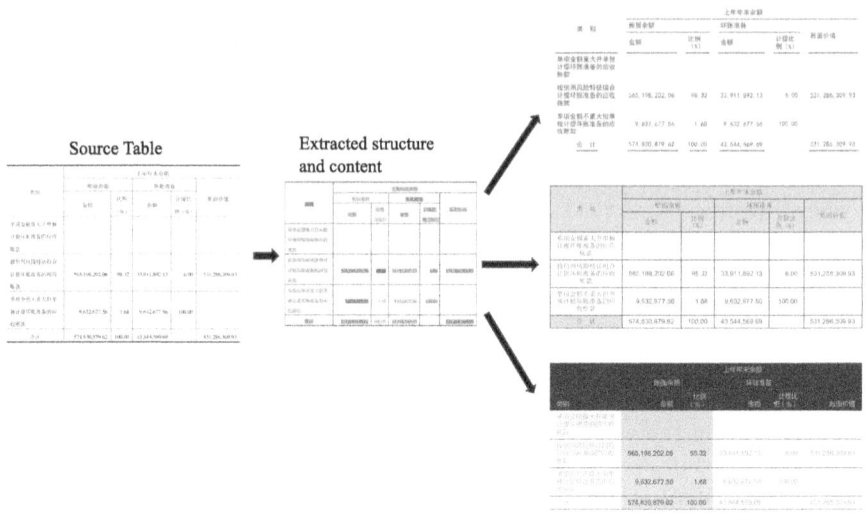

Fig. 1. The transformation of an original bordered table into two borderless tables and one bordered table with different style

FinTabNet, by synthesizing and increasing the proportion of more complex tables with multiple spanning cells. Experiments demonstrate that table recognition models trained on this enhanced dataset exhibit comprehensive improvements in performance, particularly in recognizing more complex tables that contain multiple spanning cells.

2   Related Work

2.1   Large-Scale Table Recognition Datasets

The current large-scale table recognition datasets are automatically annotated by utilizing a few document repositories, where the tables displayed in the documents can be associated with corresponding structured source codes.

TABLE2LATEX-450K [2], TableBank [12], and TabLeX [3] are datasets derived from articles in the Arxiv repository, where tables in PDF documents correspond to LaTeX source codes. Additionally, TableBank [12] has crawled some Word documents from the internet, linking tables in these documents to Office XML codes. PubTabNet [27] and PubTables-1M [20] were sourced from scientific papers in the PubMed Central Open Access (PMCOA) database, with tables in the PDF files paired with corresponding XML codes. Meanwhile, FinTabNet [26] compiled annual reports from S&P 500 companies, featuring tables in the PDF documents that can be associated with corresponding HTML codes. The datasets mentioned above are confined to scientific papers or financial reports, and models trained on them may not perform well in other domains.

As mentioned before, these automatically annotated datasets often contain a considerable number of annotation errors. These issues encompass a variety of problems, including inaccuracies in localizing table regions, omissions of table content in annotations, structural errors in annotating headers or table content, and inconsistencies in annotations for identical structures or content within tables, among others. Except for PubTables-1M, the annotations in the other datasets do not fully cover the annotation requirements needed by several common table recognition models [20]. Specifically, both PubTabNet and FinTabNet lack annotations for the coordinates of cell bounding boxes [26,27]. Additionally, TableBank and TabLeX provide only the overarching structure of the table, without annotations for the content and coordinates of text blocks within each cell [3,12].

2.2   Randomly Synthesized Data for Table Recognition

Another series of efforts to address the scarcity of large-scale table recognition datasets involves constructing table annotation datasets through the synthesis of HTML tables.

Qasim et al. [17] initially employed four types of table templates to create a synthetic table dataset based on HTML. TableGeneration [28] further expanded the method, maintaining support for four table templates while introducing a broader range of configurable parameters, which include cell type, the number of rows and columns in a table, the quantity of merged cells, and the provision for colored cells. WikiTableSet [14] constructed a Wikipedia table extractor to harvest tables (in HTML code format) from the Wikipedia dump, and then normalized these HTML tables to align with the PubTabNet format, which involved separating table headers from data and stripping CSS and style tags. SynthTabNet [16] took a parameter-driven approach to initially generate the table structure, detailing the total number of rows and columns, header row count, types of spans (including header-only, row-only, column-only, and both row and column spans), maximum span size, and the proportion of table area covered by spans. Appropriate content templates are then selected and combined with purely random text to produce synthetic content. A collection of styling templates is manually curated, and a style is randomly selected to determine the appearance of the synthesized table. In a similar vein, the ComplexTable dataset [1] is synthetically generated using an auto HTML table creator, which produces table images alongside corresponding structured HTML code. Notably, ComplexTable features a significantly higher frequency of complex tables compared to SynthTabNet, and it showcases a more diverse range of table styles within the dataset.

The aforementioned methods employ HTML for table synthesis and CSS to define table styles, followed by the use of a Web Browser engine to render the table images. However, this approach has its limitations. Often, it falls short in replicating the complex or distinctive visual effects that tables in documents, such as PDFs, typically present.

2.3   Table Augmentation

Data augmentation is also a common method for acquiring table data. TabAug [11] introduced a novel data augmentation technique that primarily relies on two fundamental operations: Replication and Deletion, applied to rows and columns. Umer et al. [22] developed augmentation techniques that encompass clustering, fusion, and patching of table images. Ichikawa [10] introduced novel label-invariant table augmentation techniques focused on the edge-based region, demonstrating their substantial impact, particularly when training with limited datasets. Liu et al. [13] enhanced existing datasets by employing two types of image distortion algorithms, aiming to simulate distractors introduced by the capture device.

This paper is mainly concerned with the acquisition of large-scale table annotation data encompassing diverse styles. The data augmentation methods mentioned do not significantly alter table styles, hence they differ somewhat from the core issue addressed in this paper.

3   Our Approach for Synthesizing Annotated Tables

If we aim to transform tables from documents into structured data for storage in databases or knowledge graphs within a specific domain, enabling further in-depth analysis and applications, it becomes essential to parse the structure of these tables to grasp their semantics. In many application domains, although the content of some tables across different documents may be very similar, the styles of these tables with similar content often exhibit significant differences. Figure 2 shows six real examples of tables from the Financial Statements, specifically the "Accounts Receivable - Disclosed by Provision Method" category. These tables, belonging to the same category, have similar content but vastly different visual styles. Inspired by this observation, we propose synthesizing new target tables by leveraging the structure and content of existing complex tables, while applying completely different yet realistically plausible styles to these target tables. This approach ensures that the synthesized complex tables more closely resemble real-world scenarios than those produced by methods relying on randomly generated structures and content.

Figure 3 illustrates the overall structure of the table annotation data synthesis method proposed in this paper, which involves extensive handling of text and lines within tables. For tables generated via PDF coding, PDF parsing tools, such as pdfplumber, offer a direct and efficient means to extract texts and lines. In contrast, for scanned tables, OCR can be used to detect and recognize text, and conventional methods such as the Hough Transform [4,15] or LSD [5], as well as deep learning-based line detection techniques [6–8,23,25,29], can be employed to identify lines within the tables.

3.1   Analysis of the Real Table Distribution in the Target Domain

To synthesize table annotation data that closely match the target application domain, a detailed investigation and analysis of the real table data's distribu-

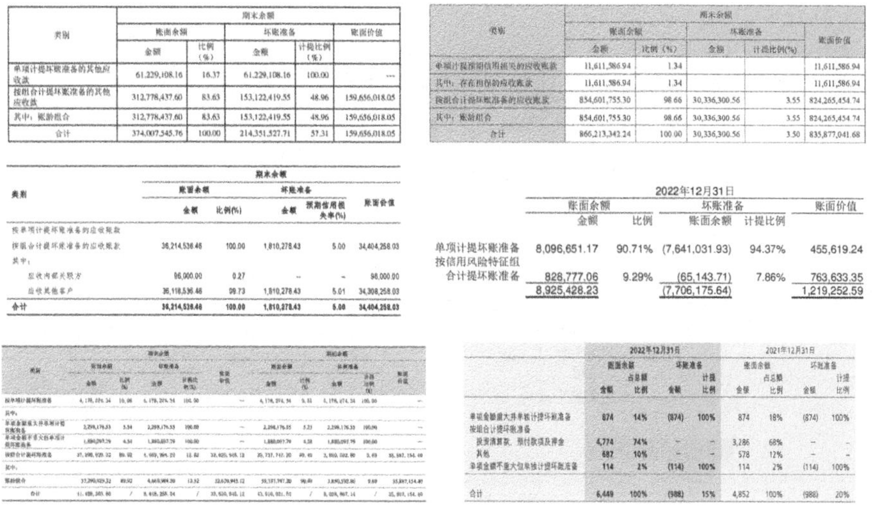

Fig. 2. Real examples of tables that have similar content but display vastly different visual styles.

tion within that domain are essential. In the document repository collected from the target domain, layout analysis and table detection are utilized to detect all included tables. OCR or PDF parsing tools are then used to detect and recognize the text content within these tables. Following this, tables are clustered based on their text content, grouping together those with similar content. Subsequently, manual inspection and adjustment are performed. This approach significantly facilitates the convenient and efficient identification of various critical table types within the application domain. Through clustering analysis and adjustment within the Chinese financial announcement document repository, we have identified a comprehensive array of tables. These include financial statements (such as balance sheets, income statements, cash flow statements, statements of shareholders' equity, and supplementary cash flow statements), notes to financial statements (covering accounts receivable, monetary funds, fair value disclosures, and payroll payable), as well as tables detailing directors, supervisors, senior management members, employee statistics, company profiles (encompassing basic information, contact details, principal accounting data, and financial ratios), shareholder information, and glossaries, among others.

We can extract style-related attributes from each table in the repository. These attributes include font information for each text line, line spacing, and alignment within text blocks contained in each cell. Additionally, we consider the alignment of text blocks within cells across rows or columns, the padding between aligned text blocks and their cell bounding boxes, the cells' background color, and the display mode, style, and color of the borders, both external and internal. For detailed illustrations, refer to the Appendix A.1 and A.2. These

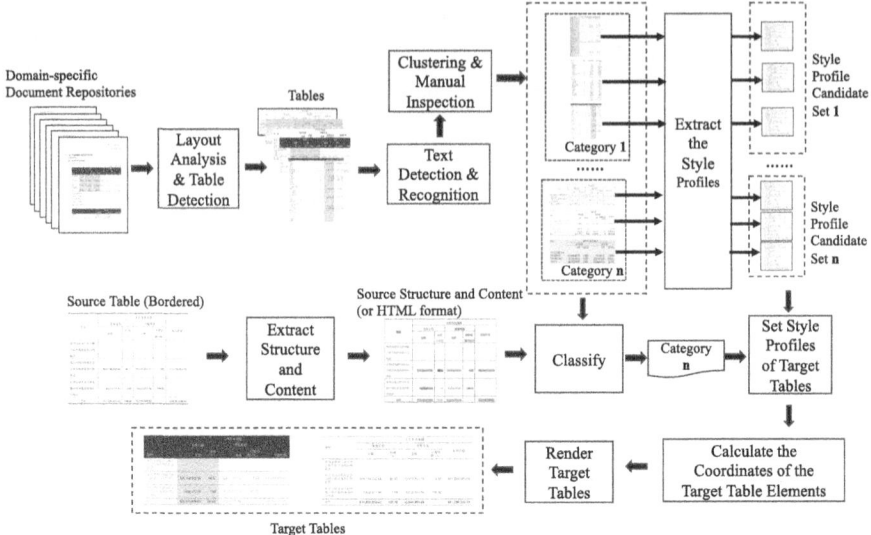

Fig. 3. System Structure of the table annotation data synthesis method

style-related attributes will be stored in the table's style profile, with specific examples provided in the Appendix A.3.

3.2 Transformation of Table Style

For certain datasets, such as FinTabNet, which already contain table structures and content data in HTML format, we can directly use this HTML data as source material to synthesize new tables through style transformation. In the case of other datasets, like those found in Chinese financial announcement document repositories, although the tables lack directly accessible structure and content information, many are bordered. This characteristic makes it relatively straight-forward to deduce the tables' structure by identifying the lines within them and recognizing the text content in each cell. Consequently, these tables can likewise serve as source data for synthesizing tables with new styles.

Based on the content of the current source table, it is possible to match it to a corresponding category. This matching process can be achieved either by training a conventional text classifier or through the application of specific rules, depending on the context. Subsequently, one or more profiles are selected from the Style Profile Candidate Set of the matched category. Minor random adjust-ments are then made to certain attribute values within the selected profiles, which are ultimately defined as the style profiles for the target tables.

3.3 Synthesis and Rendering of Target Tables

Based on the style profile of the target table, coordinates are calculated for regular non-merged cells using a bottom-up approach (text line → text block →

aligned text block bounding box → cell). For illustrations of table elements at various granularities and their corresponding attributes, see Appendices A.1 and A.2. Based on the text line attributes defined in the style profile, calculate the height and width of all text lines. If a cell contains multiple text lines, the total height of the text block can be calculated based on the defined line spacing, along with the vertical relative coordinates of the top-left corner of each text line within the text block. Then, by considering the horizontal alignment of the text lines, calculate the total width of the text block and the horizontal relative coordinates of the top-left corner of each text line within the entire text block. Scan each row of the table to identify the tallest text block among all regular non-merged cells and cells merged only in the horizontal direction in the current row. Use this height as the height for the aligned text block bounding box of that row. Calculate the vertical relative coordinates of the text block's top-left corner within the aligned text block bounding box, based on the text block's vertical alignment settings. Then, based on the cell's padding-top and padding-bottom attributes, calculate both the cell's height and the vertical relative coordinates of the aligned text block bounding box within the cell. Similarly, scan each column of the table to identify the widest text block among all regular non-merged cells and cells merged only in the vertical direction in the current column. Use this width as the width for the aligned text block bounding box of that column. Calculate the horizontal relative coordinates of the top-left corner of the text block within the aligned text block bounding box, according to the text block's horizontal alignment settings. Then, using the cell's padding-left and padding-right attributes, determine the cell's width and the horizontal relative positioning of the aligned text block bounding box within the cell.

However, for merged cells or spanning cells, the calculation process is reversed, proceeding from top to bottom (cell → aligned text block bounding box → text block → text line) to compute coordinates. Once the heights of all regular non-merged cells are determined, the height of vertically merged cells can also be obtained. Subsequently, based on the padding-top and padding-bottom attributes, the height of the aligned text block bounding box within this verti-cally merged cell can be determined. This allows for the calculation of the vertical relative coordinates of the aligned text block bounding box contained in the cell, with the vertical relative coordinates of the text block's top-left corner being calculated based on the text block's vertical alignment settings. Similarly, once the widths of all regular non-merged cells are established, the width of horizon-tally merged cells can be obtained. Based on the padding-left and padding-right attributes, the width of the aligned text block bounding box within this hor-izontally merged cell is determined, enabling the calculation of the horizontal relative coordinates of the aligned text block bounding box contained in the cell. The horizontal relative coordinates of the text block's top-left corner within the aligned text block bounding box are calculated according to the text block's horizontal alignment settings.

Based on the cell sizes calculated in previous steps, the overall size of the table formed by stacking all cells can be determined. By connecting the border lines of cells in the same row and column, the absolute coordinates of the row and column dividers, i.e., the collection of horizontal and vertical lines of the table, are obtained. Calculating the absolute coordinates of the stacked cells, and then based on the relative coordinates of the aligned text block bounding box within the corresponding cell, the relative coordinates of the text block within the aligned text block bounding box, and the relative coordinates of the text lines within the corresponding text block, the absolute coordinates of the text lines within the table can ultimately be calculated.

Based on the table size calculated previously, create a blank canvas of the same dimensions. Then, using each cell's coordinates and background color, draw the table's background color. Next, render the corresponding text lines on the image canvas based on the coordinates, color, and font of each text line. Finally, draw the border lines using drawing tools based on the coordinates, mode, line type, and color of the borders. At the same time, output the annotation data in formats required by various table recognition models.

4   Experiments

To verify the practicality and effectiveness of the method proposed in this paper, we conducted experiments with tables from the financial announcement domain. This choice was made because financial announcements contain a large number of complex tables with varying styles. Recognizing these tables not only presents a significant technical challenge but also holds substantial real-world significance for various financial analysis applications.

Financial announcements in the United States have corresponding HTML format documents for their PDF files, and there already exists a large-scale English table annotation dataset like FinTabNet generated through automatic matching. However, Chinese financial announcements in PDF format do not have corresponding structured documents like HTML, and currently, there is no large-scale table annotation dataset available for them. In response to this situation, we have utilized the method proposed in this paper, leveraging the actual structure and content of tables within Chinese financial announcements, to generate the first large-scale table annotation dataset in the domain.

We collected a total of 5049 annual reports from Chinese listed companies in 2022, from which nearly 1.5 million tables were detected and extracted. The majority of these tables are bordered, with a minority being borderless. For comparison purposes, we sampled approximately the same magnitude of tables from these nearly 1.5 million tables as the English financial announcement dataset FinTabNet, totaling 105,600 bordered tables, to serve as the data source for table synthesis. To better recognize more complex tables with a greater number of merged cells, we increased the proportion of challenging complex tables with multiple merged or spanning cells during the sampling process, as shown in Table 1.

After conducting clustering analysis and manual inspection of the content of these 105,600 bordered tables, they were categorized into 14 categories. Our

Table 1. The data distribution of the 105,600 bordered tables sampled as the data source.

Spanning cell statistics	Number	Percentage
no spanning cell	52208	49.44%
1 spanning cell	4272	4.04%
2 spanning cells	13973	13.23%
3 spanning cells	15582	14.76%
4 spanning cells or more	19565	18.53%

examination revealed that these source tables already encompass a very broad range of styles found in bordered tables within financial announcements. Therefore, when using these 105,600 bordered tables as the source for synthesizing new tables, 50% of the tables were directly retained as part of the final synthesized table collection as bordered tables, without converting them into different styles of bordered tables. Annotating borderless tables is more challenging than annotating bordered tables. In the synthesized dataset, there is a greater need to enhance support for recognizing borderless tables. Therefore, we transformed the other 50% of the sampled 105,600 bordered tables into borderless tables. The bordered tables serving as source tables can be relatively easily processed by identifying their border lines to extract the table's structure and content. Based on this content, the corresponding category is identified, and then a borderless style profile is selected from the Style Profile Candidate Set of that category. Subsequently, random adjustments of up to 10% are made to certain style attribute values to generate the Style Profile for the synthesized target table. Finally, the target table image is synthesized and rendered, and annotation data is generated.

With the synthesized table annotation dataset available, we can train models for recognizing tables in financial announcements based on it. Here, we reproduced two recent deep learning-based end-to-end models, TableMaster [24] and TableFormer [16], both of which are based on an encoder-decoder architecture, particularly utilizing transformer-based decoders, hence both exhibit strong table recognition performance. Compared to the implementations described in the original papers of TableMaster and TableFormer, we increased the maximum dimension of the input images to 640 pixels to accommodate the recognition of more complex tables. TableMaster and TableFormer have very similar underlying architectures. In TableFormer, the Transformer has fewer layers and heads, and it also employs adaptive pooling to reduce the size of the feature map output by the CNN Backbone. Consequently, TableMaster has a higher number of parameters and, correspondingly, a higher computational complexity. Our experiments also show that TableMaster's overall performance is better than that of TableFormer. Given that the experimental results and conclusions of both models are consistent, and in the interest of conserving space and presenting the information more concisely, we only report the experimental results of TableMaster in the subsequent sections.

We employ the Tree-Edit-Distance-based Similarity (TEDS) [27], a metric commonly used in table structure recognition literature, to evaluate the performance of table structure recognition. TEDS assesses the similarity between the tree structures of tables. To utilize the TEDS metric, tables must be represented as tree structures in HTML format. Considering that accounting for errors in the text content of tables could result in unfair comparisons due to the varied text extraction methods or OCR models employed by different table recognition methods, we utilize a modified version of TEDS, named TEDS-Struct. This version focuses on the accuracy of table structure recognition, explicitly disregarding the outcomes from text extraction or OCR processes. We also investigate the performance of text block detection (AP50, MS COCO AP at IoU=.50) [9], which is crucial for the precise matching of each cell to its corresponding text content.

We sampled 2,290 real tables from Chinese financial announcements, ensuring no overlap with the 105,600 tables previously sampled as sources for synthesizing table structures and content. Among these, 1,000 are bordered tables, and 1,290 are borderless tables, aiming to increase focus on borderless tables. Additionally, the selection of tables also prioritized more challenging tables that include multiple merged or spanning cells; for specifics, please refer to Table 2. Using table recognition models trained on synthesized data, we conducted recognition and automatic annotation on these 2,290 tables, followed by manual verification to create the first benchmark for complex tables in the Chinese financial announcement domain. This benchmark can be used to evaluate the table recognition performance of models trained on synthesized data, thereby demonstrating the quality of the synthesized data.

Table 2. The data distribution of real-world table benchmark in the Chinese financial announcement domain.

Spanning cell statistics	All		Bordered		Borderless	
	Number	Percentage	Number	Percentage	Number	Percentage
No spanning cell	1045	45.63%	442	44.20%	603	46.74%
1 spanning cell	82	3.58%	59	5.90%	23	1.78%
2 spanning cells	268	11.70%	145	14.50%	123	9.53%
3 spanning cells	373	16.29%	100	10.00%	273	21.16%
4 spanning cells or more	522	22.79%	254	25.40%	268	20.78%

The FinTabNet table dataset, extracted from English financial announcements, was annotated through automatic matching between PDF and HTML, resulting in a considerable number of errors. The FinTabNet training set includes a large number of tables, and due to time constraints, corrections were not conducted. However, for the FinTabNet test set, which contains 10,635 tables, we manually reviewed and used automated scripts to correct 3,733 tables with inconsistent annotations of leader dots. Additionally, we removed 954 tables that had errors in table positioning, structural annotations, or in text box annotations or shifts. The presence of these errors, resulting from automatic annotation, does not imply that the tables are structurally more complex; thus, removing these tables did not decrease the overall difficulty of the FinTabNet table recognition evaluation task. After our review, the corrected FinTabNet test set now comprises 9,681 tables.

Table 3 presents the evaluation results of TableMaster, trained on the synthesized dataset, on our Chinese financial announcement table benchmark dataset. In comparison to the results listed in Table 4-where TableMaster were trained on the FinTabNet training set and evaluated on the corrected FinTabNet test set-the performance is relatively lower. This discrepancy is attributed to the tables extracted from Chinese financial announcements generally being more complex than those from the FinTabNet dataset, which extracts tables from English financial announcements. For instance, tables from Chinese announcements often contain more cells with multi-line text blocks, cells with very long texts, or a higher density of cells. These tables frequently feature structurally complex spanning cells, significant horizontal alignment deviations of text blocks within the same column, among other factors, all of which significantly increase the difficulty of table recognition in Chinese financial announcements. Furthermore, there is still room for improvement in predicting the positions of text block bounding boxes, which could further enhance the accuracy in matching the recognized table structure with the text content [9].

Table 3. The evaluation results for TableMaster, after being trained on the synthesized dataset, on the Chinese financial announcement table benchmark dataset.

Spanning Cell Statistics	TEDS	TEDS-Struct	AP-50
all tables	0.9091	0.9579	0.482
no spanning cell	0.9216	0.9632	0.501
1 spanning cell	0.9131	0.9503	0.568
2 spanning cells	0.9375	0.9696	0.592
3 spanning cells	0.8785	0.9449	0.505
4 spanning cells or more	0.8908	0.9515	0.458
merged on rows & columns	0.8906	0.9500	0.497

To further validate the practicality of our method, we applied the data synthesis method proposed in this paper to augment the training data for FinTabNet. The original FinTabNet training dataset contains relatively few tables with multiple spanning cells. In augmenting the dataset while keeping the total number of tables roughly the same, a portion of the augmented dataset was obtained directly by sampling from the original FinTabNet training dataset, and another portion was synthesized using the method described in this paper. The augmented dataset increases the proportion of tables containing multiple spanning cells, with specific distribution information of the tables available in Table 5. Table 6 presents the experimental results on the corrected FinTabNet test dataset of models trained using the augmented FinTabNet training data.

Table 4. The evaluation results for TableMaster, after being trained on the FinTabNet training set, on the corrected FinTabNet test set.

Spanning cell statistics	TEDS	TEDS-Struct	AP-50
all tables	0.9758	0.9856	0.619
no spanning cell	0.9727	0.9829	0.631
1 spanning cell	0.9830	0.9922	0.654
2 spanning cells	0.9818	0.9896	0.630
3 spanning cells	0.9657	0.9783	0.557
4 spanning cells or more	0.9342	0.9552	0.511
merged on rows & columns	0.9503	0.9620	0.574

Compared with Table 4, it shows that data augmentation using the method proposed in this paper can comprehensively improve the performance of FinTabNet table recognition, especially in recognizing more complex tables with multiple spanning cells. This demonstrates the effectiveness of our table synthesis method for practical application scenarios.

Table 5. The data distribution of Augmented FinTabNet training dataset.

Table Type	FinTabNet	Augmented dataset		
		Sampled	Synthesized	Sum
no spanning cell	44k+	10000	10000	20000
1 spanning cell	22k+	7500	7500	15000
2 spanning cells	14k+	14439	561	15000
3 spanning cells	4k+	4824	10176	15000
4 spanning cells or more	5k+	5026	14711	19737
merged on rows & columns	2k+	263	7954	8217

Table 6. The evaluation results for TableMaster, after being trained on the Augmented FinTabNet training set, on the corrected FinTabNet test set.

Spanning cell statistics	TEDS	TEDS-Struct	AP-50
all tables	0.9847	0.9971	0.789
no spanning cell	0.9805	0.9971	0.746
1 spanning cell	0.9895	0.9982	0.813
2 spanning cells	0.9912	0.9979	0.827
3 spanning cells	0.9794	0.9906	0.809
4 spanning cells or more	0.9740	0.9906	0.747
merged on rows & columns	0.9684	0.9794	0.841

5 Conclusion and Future Work

Unlike previous methods that rely on automatic matching between PDFs and structured source codes or on random synthesis of table structures and content, this paper introduces a novel approach for synthesizing high-quality tables and annotated data. This method leverages the structure and content of existing real tables to replicate authentic table styles of the target domain. We applied this approach within the financial sector, producing the first extensive table annotation dataset for the Chinese financial announcement domain and enhancing the English financial table dataset, FinTabNet. Our experiments demonstrate the real-world applicability and effectiveness of this table synthesis method.

The experiments utilized end-to-end table recognition methods such as Table-Master, chosen for their relatively simple data preparation process and proven excellence in performance across previous studies. However, these methods encounter considerable challenges when recognizing the structurally complex tables common in Chinese financial announcements. Future work will explore additional methods, particularly those based on segmentation and merging [21], which are anticipated to yield improved results for tables with intricate spanning cell structures. Employing a diverse range of table recognition methods will enable a more thorough assessment of the synthesized data's quality.

Furthermore, we aim to broaden the diversity of table styles by incorporating tables from audit reports, to publicly release our extensive synthesized dataset for the Chinese financial announcement domain, and to enlarge and publicly unveil the benchmark for real complex tables within the same domain.

A Appendix

A.1 Table Elements at Various Granularities

Cells. In Fig. 4, each area filled with a different color represents a cell. Only tables with complete border lines have definite cell coordinates.

Text Lines. In Fig. 5, the areas filled with color represent text lines. Text lines within the same cell share the same color, and a single cell often contains multiple text lines.

Text Blocks. In Fig. 6, each area filled with a different color represents a text block, and the bounding box of a text block covers all text lines within the same cell.

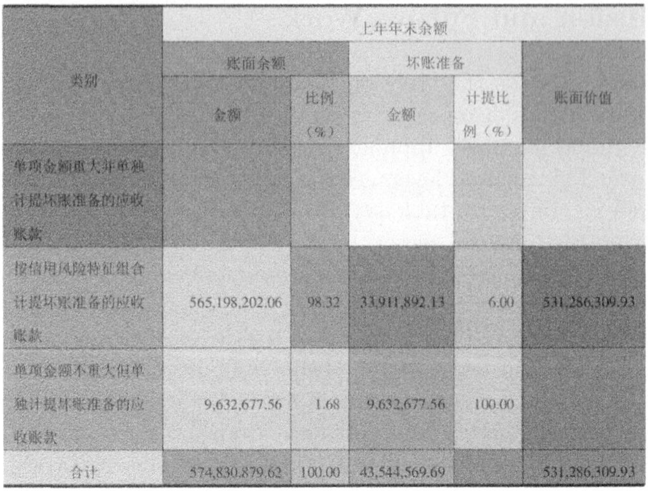

类别	上年年末余额				
	账面余额		坏账准备		账面价值
	金额	比例 (%)	金额	计提比例 (%)	
单项金额重大并单独计提坏账准备的应收账款					
按信用风险特征组合计提坏账准备的应收账款	565,198,202.06	98.32	33,911,892.13	6.00	531,286,309.93
单项金额不重大但单独计提坏账准备的应收账款	9,632,677.56	1.68	9,632,677.56	100.00	
合计	574,830,879.62	100.00	43,544,569.69		531,286,309.93

Fig. 4. Illustration of table cells. (Color figure online)

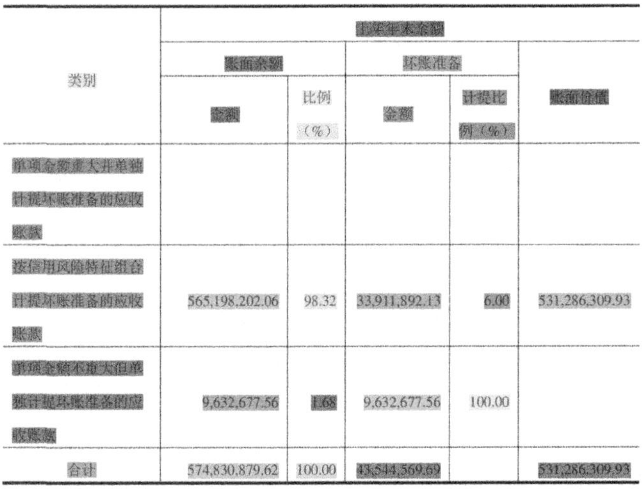

类别	上年年末余额				
	账面余额		坏账准备		账面价值
	金额	比例 (%)	金额	计提比例 (%)	
单项金额重大并单独计提坏账准备的应收账款					
按信用风险特征组合计提坏账准备的应收账款	565,198,202.06	98.32	33,911,892.13	6.00	531,286,309.93
单项金额不重大但单独计提坏账准备的应收账款	9,632,677.56	1.68	9,632,677.56	100.00	
合计	574,830,879.62	100.00	43,544,569.69		531,286,309.93

Fig. 5. Illustration of text lines. (Color figure online)

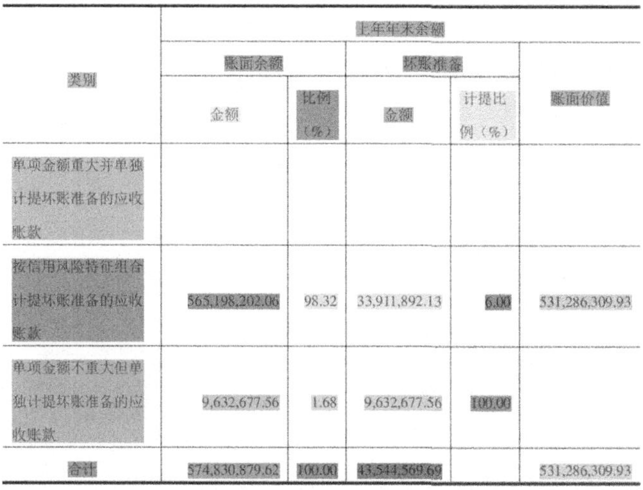

类别	上年年末余额				
	账面余额		坏账准备		账面价值
	金额	比例（%）	金额	计提比例（%）	
单项金额重大并单独计提坏账准备的应收账款					
按信用风险特征组合计提坏账准备的应收账款	565,198,202.06	98.32	33,911,892.13	6.00	531,286,309.93
单项金额不重大但单独计提坏账准备的应收账款	9,632,677.56	1.68	9,632,677.56	100.00	
合计	574,830,879.62	100.00	43,544,569.69		531,286,309.93

Fig. 6. Illustration of text blocks. (Color figure online)

Aligned Text Block Bounding Boxes. In Fig. 7, each area filled with a different color is an aligned text block bounding box. Identify the highest upper bound and the lowest lower bound from all text blocks among all regular non-merged cells and cells merged only in the horizontal direction in the current row to calculate the height, and use this height as the height of the aligned text block bounding box for that row. Identify the leftmost boundary and the rightmost boundary from all text blocks among all regular non-merged cells and cells merged only in the vertical direction in the current column to calculate the width, and use this width as the width of the aligned text block bounding box for that column. The width of the aligned text block bounding box for a horizontally merged cell is determined by the left boundary of the aligned text block bounding box of the leftmost column involved in the merge and the right boundary of the aligned text block bounding box of the rightmost column involved in the merge. The height of the aligned text block bounding box for a vertically merged cell is determined by the upper boundary of the aligned text block bounding box of the topmost row involved in the merge and the lower boundary of the aligned text block bounding box of the bottommost row involved in the merge.

A.2   Attributes of Table Elements

Attributes of Text Lines and Text Blocks. The attributes of text lines include the font, size, and color of the text, all of which can be set at various granularities such as table, row, column, or cell. The attributes of text blocks include the line spacing between the contained text lines and the alignment of text lines within the text block (centered/left-aligned/right-aligned/specified

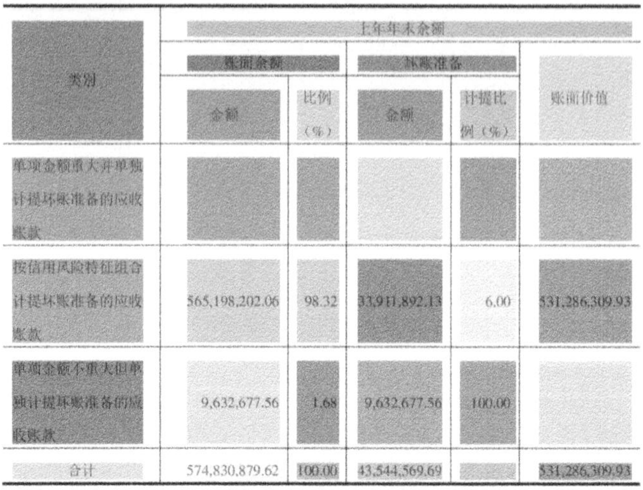

类别	上年年末余额				账面价值
	账面余额		坏账准备		
	金额	比例(%)	金额	计提比例(%)	
单项金额重大并单独计提坏账准备的应收账款					
按信用风险特征组合计提坏账准备的应收账款	565,198,202.06	98.32	33,911,892.13	6.00	531,286,309.93
单项金额不重大但单独计提坏账准备的应收账款	9,632,677.56	1.68	9,632,677.56	100.00	
合计	574,830,879.62	100.00	43,544,569.69		531,286,309.93

Fig. 7. Illustration of Aligned Text Block Bounding Boxes. (Color figure online)

indent distance), all of which can be set at various granularities such as table, row, column, or cell. Figure 8 shows an illustration of the attributes of a text block and the text lines contained within it, extracted from Fig. 6.

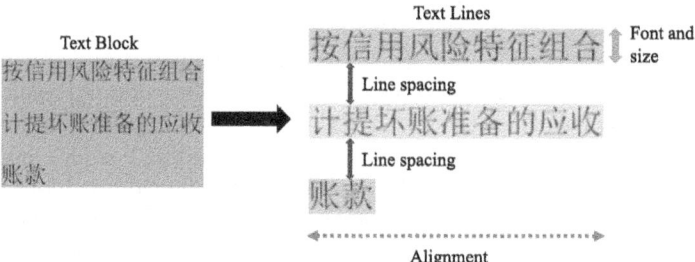

Fig. 8. Illustration of Attributes of text lines and text blocks. (Color figure online)

Attributes of Cells and Corresponding Aligned Text Block Bounding Boxes. The attributes of an aligned text block bounding box mainly involve the alignment of the contained text block, including horizontal alignment (left-aligned, right-aligned, centered, indented by a specified distance to the left or right) and vertical alignment (top-aligned, bottom-aligned, centered, indented by a specified distance from the top or bottom). All of these can be set at various granularities such as table, row, column, or cell. The attributes of a cell include the padding, defined as the distance between the cell bounding box and the aligned text block bounding box, as well as the cell's background color. All of these can be set at various granularities such as table, row, column, or cell. Figure 9 displays a cell and its corresponding aligned text block bounding box, as extracted from Fig. 7.

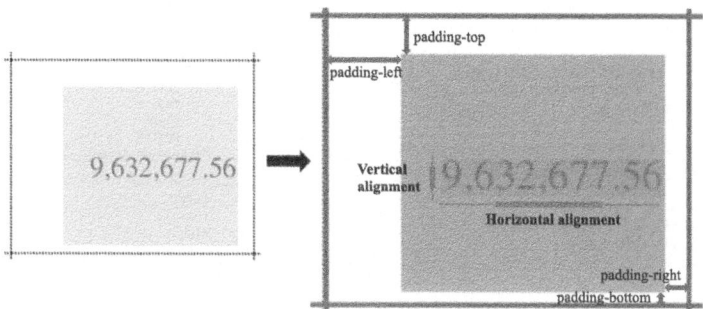

Fig. 9. Illustration of Attributes of Cells and corresponding Aligned Text Block Bounding Boxes. (Color figure online)

Attributes of Borders. In tables, borders are categorized into outer and inner borders. The attributes of outer borders apply to the table as a whole, while inner border attributes can be adjusted at various levels of granularity, such as by table, row, column, or cell. Outer border modes can be fully visible, absent on the sides, absent on the top and bottom, or completely absent. Inner border options encompass fully visible, absent horizontal lines, absent vertical lines, completely absent, partially absent horizontal lines, and partially absent vertical lines. Outer border line types include options like single solid lines and double solid lines, along with specifications for their thickness. Inner border line types feature solid lines, dashed lines, and details regarding line thickness and the spacing in dashed lines. Furthermore, colors can be specified for both outer and inner borders.

A.3   Table Style Profile

Figure 10 shows an example of the format of a Table Style Profile.

```
Style profile format                              <config> format

table:  # Setting at the granularity of the table    text_block:
    - config: <config>                                    text_line_attributes:
                                                              font_type: ......
                                                              font_size: ......
row:  # Setting at the granularity of the row             text_color: ......
    - pattern: <match_condition>                      text_line_relationship:
      config: <config>                                    line_spacing: ......
    - ...                                                 alignment:
                                                              horizontal: ......
column:  # Setting at the granularity of the column   cell:
    - pattern: <match_condition>                          aligned_text_block_attributes:
      config: <config>                                        alignment:
    - ...                                                         horizontal: ......
                                                                  vertical: ......
cell:  # Setting at the granularity of the cell       cell_attributes :
    - pattern: <match_condition>                          padding:
      config: <config>                                        left: ......
    - ...                                                     top: ......
                                                              right: ......
                                                              bottom: ......
                                                          background_color: ......
                                                      border:
<match_condition> format                                  mode:
                                                              left: ......
- index: index_regex                                          top: ......
- html_tag: html_tag_regex                                    right: ......
- text: text_regex                                            bottom: ......
                                                          type:
                                                              left: ......
                                                              top: ......
                                                              right: ......
                                                              bottom: ......
                                                          color:
                                                              left: ......
                                                              top: ......
                                                              right: ......
                                                              bottom: ......
```

Fig. 10. Illustration of an example of the format of a Table Style Profile.

References

1. Chen, L., Huang, C., Zheng, X., Lin, J., Huang, X.J.: TableVLM: multi-modal pre-training for table structure recognition. In: Proceedings of the 61st Annual Meeting of the Association for Computational Linguistics (Volume 1: Long Papers), pp. 2437–2449 (2023)

2. Deng, Y., Rosenberg, D., Mann, G.: Challenges in end-to-end neural scientific table recognition. In: 2019 International Conference on Document Analysis and Recognition (ICDAR), pp. 894–901 (2019)

3. Desai, H., Kayal, P., Singh, M.: TABLEX: a benchmark dataset for structure and content information extraction from scientific tables. In: Lladós, J., Lopresti, D., Uchida, S. (eds.) ICDAR 2021. LNCS, vol. 12822, pp. 554–569. Springer, Cham (2021). https://doi.org/10.1007/978-3-030-86331-9_36

4. Duda, R.O., Hart, P.E.: Use of the hough transformation to detect lines and curves in pictures. Commun. ACM **15**(1), 11–15 (1972)

5. Grompone von Gioi, R., Jakubowicz, J., Morel, J.M., Randall, G.: LSD: a line segment detector. Image Process. Line **2**, 35–55 (2012)

6. Gu, G., Ko, B., Go, S., Lee, S.H., Lee, J., Shin, M.: Towards light-weight and real-time line segment detection. In: Proceedings of the AAAI Conference on Artificial Intelligence, vol. 36, no. 1, pp. 726–734 (2022)

7. Huang, K., Wang, Y., Zhou, Z., Ding, T., Gao, S., Ma, Y.: Learning to parse wireframes in images of man-made environments. In: Proceedings of the IEEE Conference on Computer Vision and Pattern Recognition (CVPR) (2018)

8. Huang, S., Qin, F., Xiong, P., Ding, N., He, Y., Liu, X.: TP-LSD: tri-points based line segment detector. In: Vedaldi, A., Bischof, H., Brox, T., Frahm, J.-M. (eds.) ECCV 2020. LNCS, vol. 12372, pp. 770–785. Springer, Cham (2020). https://doi.org/10.1007/978-3-030-58583-9_46

9. Huang, Y., et al.: Improving table structure recognition with visual-alignment sequential coordinate modeling. In: 2023 IEEE/CVF Conference on Computer Vision and Pattern Recognition (CVPR), pp. 11134–11143. IEEE Computer Society, Los Alamitos, CA, USA (2023)

10. Ichikawa, K.: Image-based relation classification approach for table structure recognition. In: Lladós, J., Lopresti, D., Uchida, S. (eds.) ICDAR 2021. LNCS, vol. 12822, pp. 632–647. Springer, Cham (2021). https://doi.org/10.1007/978-3-030-86331-9_41

11. Khan, U., Zahid, S., Ali, M.A., Ul-Hasan, A., Shafait, F.: TabAug: data driven augmentation for enhanced table structure recognition. In: Lladós, J., Lopresti, D., Uchida, S. (eds.) ICDAR 2021. LNCS, vol. 12822, pp. 585–601. Springer, Cham (2021). https://doi.org/10.1007/978-3-030-86331-9_38

12. Li, M., Cui, L., Huang, S., Wei, F., Zhou, M., Li, Z.: TableBank: table benchmark for image-based table detection and recognition. In: Proceedings of the Twelfth Language Resources and Evaluation Conference, pp. 1918–1925. European Language Resources Association, Marseille, France (2020)

13. Liu, H., Li, X., Liu, B., Jiang, D., Liu, Y., Ren, B.: Neural collaborative graph machines for table structure recognition. In: Proceedings of the IEEE/CVF Conference on Computer Vision and Pattern Recognition, pp. 4533–4542 (2022)

14. Ly, N., Takasu, A., Nguyen, P., Takeda, H.: Rethinking image-based table recognition using weakly supervised methods. In: Proceedings of the 12th International Conference on Pattern Recognition Applications and Methods. SCITEPRESS - Science and Technology Publications (2023)

15. Matas, J., Galambos, C., Kittler, J.: Robust detection of lines using the progressive probabilistic hough transform. Comput. Vis. Image Underst. **78**(1), 119–137 (2000)

16. Nassar, A., Livathinos, N., Lysak, M., Staar, P.: TableFormer: table structure understanding with transformers. In: Proceedings of the IEEE/CVF Conference on Computer Vision and Pattern Recognition (CVPR), pp. 4614–4623 (2022)

17. Qasim, S.R., Mahmood, H., Shafait, F.: Rethinking table recognition using graph neural networks. In: 2019 International Conference on Document Analysis and Recognition (ICDAR), pp. 142–147. IEEE (2019)

18. Qiao, L., et al.: LGPMA: complicated table structure recognition with local and global pyramid mask alignment. In: Lladós, J., Lopresti, D., Uchida, S. (eds.) ICDAR 2021. LNCS, vol. 12821, pp. 99–114. Springer, Cham (2021). https://doi.org/10.1007/978-3-030-86549-8_7

19. Raja, S., Mondal, A., Jawahar, C.V.: Table structure recognition using top-down and bottom-up cues. In: Vedaldi, A., Bischof, H., Brox, T., Frahm, J.-M. (eds.) ECCV 2020. LNCS, vol. 12373, pp. 70–86. Springer, Cham (2020). https://doi.org/10.1007/978-3-030-58604-1_5

20. Smock, B., Pesala, R., Abraham, R.: PubTables-1M: Towards comprehensive table extraction from unstructured documents. In: Proceedings of the IEEE/CVF Conference on Computer Vision and Pattern Recognition (CVPR), pp. 4634–4642 (2022)

21. Tensmeyer, C., Morariu, V.I., Price, B., Cohen, S., Martinez, T.: Deep splitting and merging for table structure decomposition. In: 2019 International Conference on Document Analysis and Recognition (ICDAR), pp. 114–121 (2019)

22. Umer, M., Mohsin, M.A., Ul-Hasan, A., Shafait, F.: PyramidTabNet: transformer-based table recognition in image-based documents. In: Fink, G.A., Jain, R., Kise, K., Zanibbi, R. (eds.) Document Analysis and Recognition - ICDAR 2023, pp. 420–437. Springer Nature Switzerland, Cham (2023). https://doi.org/10.1007/978-3-031-41734-4_26

23. Xu, Y., Xu, W., Cheung, D., Tu, Z.: Line segment detection using transformers without edges. In: 2021 IEEE/CVF Conference on Computer Vision and Pattern Recognition (CVPR), pp. 4255–4264. IEEE Computer Society, Los Alamitos, CA, USA (2021)

24. Ye, J., et al.: PingAn-VCGroup's solution for ICDAR 2021 competition on scientific literature parsing task B: table recognition to HTML (2021)

25. Zhang, Z., et al.: PPGNet: learning point-pair graph for line segment detection. In: Proceedings of the IEEE Conference on Computer Vision and Pattern Recognition (2019)

26. Zheng, X., Burdick, D., Popa, L., Zhong, X., Wang, N.X.R.: Global table extractor (GTE): a framework for joint table identification and cell structure recognition using visual context. In: Proceedings of the IEEE/CVF Winter Conference on Applications of Computer Vision (WACV), pp. 697–706 (January 2021)

27. Zhong, X., ShafieiBavani, E., Jimeno Yepes, A.: Image-based table recognition: data, model, and evaluation. In: Vedaldi, A., Bischof, H., Brox, T., Frahm, J.-M. (eds.) ECCV 2020. LNCS, vol. 12366, pp. 564–580. Springer, Cham (2020). https://doi.org/10.1007/978-3-030-58589-1_34

28. Zhou, W.: Tablegeneration (2022). https://github.com/WenmuZhou/TableGeneration

29. Zhou, Y., Qi, H., Ma, Y.: End-to-end wireframe parsing. In: 2019 IEEE/CVF International Conference on Computer Vision (ICCV), pp. 962–971 (2019)

Multi-cell Decoder and Mutual Learning for Table Structure and Character Recognition

Takaya Kawakatsu$^{(\boxtimes)}$ (iD)

Preferred Networks, Inc., 1–6–1 Otemachi, Chiyoda, Tokyo, Japan
`kat.nii.ac.jp@gmail.com`
`https://researchmap.jp/t.kat`

Abstract. Extracting table contents from documents such as scientific papers and financial reports and converting them into a format that can be processed by large language models is an important task in knowledge information processing. End-to-end approaches, which recognize not only table structure but also cell contents, achieved performance comparable to state-of-the-art models using external character recognition systems, and have potential for further improvements. In addition, these models can now recognize long tables with hundreds of cells by introducing local attention. However, the models recognize table structure in one direction from the header to the footer, and cell content recognition is performed independently for each cell, so there is no opportunity to retrieve useful information from the neighbor cells. In this paper, we propose a multi-cell content decoder and bidirectional mutual learning mechanism to improve the end-to-end approach. The effectiveness is demonstrated on two large datasets, and the experimental results show comparable performance to state-of-the-art models, even for long tables with large numbers of cells.

Keywords: Deep Learning · Table Recognition · Transformer · Mutual Learning

1 Introduction

Information retrieval technology which provides high-quality knowledge to large language models (LLMs) is attracting attention. Many researchers have worked on converting scanned and imaged documents to machine-readable formats such as HTML code [15,37,42] and LaTeX code [3,12]. This initiative has direct and indirect benefits. First, because past literature remains mostly in printed form, it is necessary to convert it into structured electronic documents. This is a direct benefit. Second, the realization of intelligence which recognizes hidden meanings in the layout of published documents that are intended to be read by humans is important from the perspective of human-machine interaction. This is an indirect benefit.

© The Author(s), under exclusive license to Springer Nature Switzerland AG 2024
E. H. Barney Smith et al. (Eds.): ICDAR 2024, LNCS 14804, pp. 389–405, 2024.
https://doi.org/10.1007/978-3-031-70533-5_23

In this paper, we will focus on table recognition, which includes two types of tasks, namely structure recognition and cell content recognition. A simple table has horizontal and vertical borders, and each cell contains characters. Complex tables may contain cells which are merged vertically or horizontally, and/or cells which involve invisible borders. Persons can understand table structure from the cell layout even without explicit boundaries, and this is a challenging problem.

In recent years, following the success of Transformer [31] models in language and visual recognition tasks, many methods [18, 19, 21, 36] based on Transformer have been proposed for the table recognition task. Since we can utilize an external optical character recognition (OCR) system to parse cell contents, we can focus mainly on table structure recognition. This task exploits cross attention between image features and embedded representations of HTML tokens to predict HTML tokens sequentially. In previous studies [18–21, 36], the token prediction was performed in one direction from the header to the footer, from left to right. This impairs the opportunity to focus on the table structure ahead.

Ly and Takasu [19] reported that end-to-end learning of table structure and cell content recognition tasks may improve overall table recognition performance. In addition, tables may contain more than a few hundred cells, and the sequential prediction approaches may suffer from poor performance, which can be improved by introducing local attention [18]. These previous studies performed cell content recognition independently for each detected cell after structure recognition. This impairs the opportunity of obtaining useful information from neighbor cells.

As a solution to the problems, we improve the end-to-end approach [18–20] and propose a method that refers to the recognition results of neighbor cells and a learning mechanism focusing on both previous and following cells. The former is achieved by introducing a cell decoder that infers multiple cells and configuring a hierarchical decoder along with an HTML decoder for structural recognition. The latter is achieved by mutual learning [38] between a forward decoder which reads the table structure from left to right and a backward decoder which reads the table structure from right to left. The effectiveness of the proposed method is demonstrated using two large-scale tabular image datasets.

The main contributions of this paper are: 1) We propose a cell decoder which infers multiple cells and obtains useful information from surrounding cells. 2) We propose a bidirectional mutual learning mechanism to force the proposed model to pay attention to both previous and following cells. 3) Across all experimental results, our proposed method achieved better performance than state-of-the-art models.

2   Related Work

In general, the table recognition task is performed by two subtasks, namely table structure recognition and cell content recognition. Of course, the final output is an HTML [15, 37, 42] or LaTeX [3, 12] document, so there is no need to distinguish between the two subtasks. However, it is better to recognize tags (or commands) and other visible characters using separate models. For cell content recognition task, existing highly accurate OCR systems [17] are available, and thus previous studies [11, 13, 26, 28, 30, 32] have mainly focused on table structure recognition.

Table structure recognition has been studied for a long time, and approaches based on hand-crafted features and heuristic rules [11, 13, 32] were proposed, but their application was limited to simple tables or tables with predefined patterns. With the development of deep learning, methods that automatically learn table structural patterns [26–28, 30, 39] have become mainstream. These studies can be divided into approaches based on object detection and segmentation [27, 30], and approaches based on sequential token prediction [21, 36].

For the detection and segmentation approaches, Schreiber et al. [30] proposed a two-fold system using Faster R-CNN [29] and fully convolutional networks [16] for both table detection and table structure recognition. Raja et al. [28] proposed a two-stage model that estimates the relationships between cells after recognizing their locations. Qiao et al. [27] won his first place in the ICDAR competition [37] by combining text, cell, row and column recognition tasks using Mask R-CNN [5].

For the sequential token prediction approaches, a simple image caption model can be utilized for cell detection because the order of cells is uniquely determined. Ye et al. [36] and Nassar et al. [21] proposed Transformer models with two types of decoders for table structure recognition and cell localization. Peng et al. [25] achieved performance comparable to a model using a deep convolutional encoder while significantly reducing parameters by introducing a convolutional stem.

In the 2020 s, researchers are investigating end-to-end models that learn both table structure and cell content recognition tasks [3]. Zhong et al. [42] proposed a model that uses a ResNet [6] encoder and two LSTM [8] decoders to recognize both table structure and cell contents, but its performance was inferior to models using external OCR.

Ly and Takasu [19] proposed a multi-task model that detects table structure, cell locations, and cell contents. Their model uses a ResNet encoder with global context attention [2] and two Transformer decoders. The first decoder infers the HTML tokens sequentially, and then the second decoder reads the cell contents one by one. This model achieved performance comparable to the models utilizing external OCR. They also proposed weakly supervised learning to reduce the cost of preparing bounding box training data [20] and introduced local attention [1] to effectively recognize tables with a large number of cells [18].

In 2021, the scientific literature parsing competition [37] was held at ICDAR 2021. The competition consisted of document layout recognition task A and table recognition task B. Task B required converting table images to HTML tags with cell contents. The PubTabNet [42] dataset and the final evaluation dataset were provided for this task. The training dataset consists of HTML tokens, cell texts, and cell bounding boxes. There were 30 submissions from 30 teams and most of the top 10 solutions exploited separate OCR models, additional annotation and ensemble techniques.

TabRecSet [35] is a bilingual dataset containing rotated and distorted tables in real photographs for three tasks, namely table detection, structure recognition, and cell content recognition. Detection of such tables is outside the scope of this paper.

3   Background

Similar to previous work [18–20], our proposed model uses a ResNet encoder and an HTML decoder consisting of multiple attention blocks [31] to infer HTML tokens representing table structure. An additional decoder is exploited to infer cell contents. The encoder and two decoders are trained simultaneously using an end-to-end approach. In this section, we introduce some techniques used by the proposed method described in Sect. 4.

3.1   Encoder

Previous studies [18,21,36] used a convolutional neural network (CNN) to extract image features and fed them to the decoder. CNN is useful for recognizing small characters while preserving locality such as character positions, reducing the size of image features, and improving the computational efficiency and performance of the decoder.

The number of convolutional layers contributes to recognition performance, and many derivatives have been explored to increase it. ResNet [6], which consist of a large number of residual blocks of multiple convolutional layers with simple skip connections, has been commonly used. In addition, ResNeXt [34] with group convolution and DenseNet [9] with more complicated skip connections between all convolutional layers were proposed.

One of the weaknesses of CNN is its poor ability to recognize global context by focusing strongly on local features. As a solution, a global context attention (GCA) block [2] was proposed, defined by Eq. (1).

$$\boldsymbol{y}_{ij} = \boldsymbol{x}_{ij} + W_3 \max(0, \text{LayerNorm}(W_2 \sum_i \sum_j \text{SoftMax}(W_1 \boldsymbol{x}_{ij})\, \boldsymbol{x}_{ij})), \quad (1)$$

where i, j are the pixel indices, $\boldsymbol{x}, \boldsymbol{y}$ are the input and output pixels, respectively. W_1, W_2, W_3 are the weight matrices of three linear layers. LayerNorm means layer normalization. The softmax function is defined as follows.

$$\text{SoftMax}(\boldsymbol{z}_{ij}) = \frac{\exp \boldsymbol{z}_{ij}}{\sum_m \sum_n \exp \boldsymbol{z}_{mn}}, \quad (2)$$

where m, n are the pixel indices. The GCA block should be placed between some residual (or dense) blocks.

3.2   Decoder

Transformer [31] achieves superior performance in both language modeling and visual recognition tasks. Compared to recurrent neural networks including long short-term memory [8], a Transformer itself does not involve recursion, allowing parallel processing of sequential input and output data. It should be noted that recurrent, sequential inference is performed unless the prediction length is fixed.

However, Transformer does not require recursion to recognize the context of the sequence, avoiding vanishing gradients and providing better performance.

The key idea of Transformer is called scaled dot product attention. Let X be a sequence of length l_x and d_x channels, and Y be another input sequence. For self attention, X, Y are the same sequence, and the Transformer pays attention to other parts of X in processing X. For cross attention, X and Y are in different domains, and the Transformer pays attention to some parts from Y in processing X. These mechanisms allow Transformer to learn the context of sequential data and the relationship between visual and language domains.

The attention layer first generates a query Q, key K, and value V from X, Y as defined in Eq. (3).

$$
\begin{aligned}
q_i &= W_Q \, x_i, \\
k_j &= W_K \, y_j, \\
v_j &= W_V \, y_j,
\end{aligned}
\tag{3}
$$

where i, j are the sequence indices, q, k, v, x, y are the elements of Q, K, V, X, Y, respectively. W_Q, W_K, W_V are the projection matrices.

The output Z of the attention layer is defined by Eq. (4).

$$
Z_i = W_Z \, \text{SoftMax}(\frac{QK^\top}{\sqrt{d_k}})V,
\tag{4}
$$

where W_Z is the output projection matrix, and d_k is the dimension of k. This is the mechanism of the scaled dot product attention [31].

In practice, the attention layer is divided into several groups, each of which pays attention independently and combines the outputs at the end. This is called multi-head attention [31]. Through the above mechanism, Transformer can focus on specific values of Y and incorporate them into the X series.

3.3   Local Attention

Although Transformer has superior ability to recognize long sequences compared to recurrent neural networks, it is still known to perform poorly upon extremely long sequences. Local attention (LA) [1] is a technique designed to handle such long sequences in Transformer.

Let M_{ij} be a mask to focus on the jth element from ith element of X. The output of the local attention layer is defined by Eq. (5), involving causal masking to prevent leakage from subsequent elements.

$$
Z_i = W_Z \, \text{SoftMax}(\frac{QK^\top}{\sqrt{d_k}} + M_{ij})V.
\tag{5}
$$

The mask matrix M is given by Eq. (6).

$$
M_{ij} = \begin{cases} 0 & 0 \le i - j \le w, \\ -\infty & \text{otherwise}, \end{cases}
\tag{6}
$$

where i, j are the sequence indices, and w is the width of the sliding window.

3.4   Positional Encoding

Transformer [31] itself have poor ability to know the position of each element in the sequences, and the position information must be provided explicitly. Instead of inputting simple position values, two approaches have been proposed, namely positional embedding [4] and positional encoding [31]. In general, the latter works better on small training datasets.

The output $\boldsymbol{p}(n)$ of positional encoding for the index n is defined by Eq. (7).

$$\boldsymbol{p}(n) = \begin{pmatrix} \vdots \\ \sin \dfrac{n}{10000^{\frac{2k}{d}}} \\ \cos \dfrac{n}{10000^{\frac{2k}{d}}} \\ \vdots \end{pmatrix}, \text{ where } k \in \left[0, \frac{d}{2}\right). \tag{7}$$

$\boldsymbol{p}(n)$ must be added directly to the feature vector $\boldsymbol{x}(n)$ with d channels.

If the sequence X has two-dimensional positions (i, j), 2D positional encoding proposed by Zhao et al. [40] may be a better choice. It normalizes the horizontal and vertical coordinates to $[0, 1]$, encodes each with Eq. (7), and then combines them to obtain a single vector. The positional encoding of the i, jth pixel is given by Eq. (8).

$$\boldsymbol{p}_{2\mathrm{D}}(i, j) = \begin{pmatrix} \boldsymbol{p}\left(\frac{i}{H}\right) \\ \boldsymbol{p}\left(\frac{j}{W}\right) \end{pmatrix}, \tag{8}$$

where H, W are the height and width for positional normalization. In this paper, we omitted this normalization.

3.5   Mutual Learning

Ensemble learning is commonly used to improve machine learning generalization performance and fitting accuracy by averaging or complementing the outputs of multiple inference models. However, it is computationally more expensive than single models due to the large number of parameters especially for deep learning approaches.

To achieve similar effects using only a single model, knowledge distillation [7] may be an alternative solution. This is a technique which uses a large, complex neural network, i.e., an ensemble model, as a teacher and a small, simple model as a student to obtain higher performance than simply training a student model using the ground-truth data.

Mutual learning [38] may be another solution. Here, multiple student models are trained simultaneously to teach each other, without training a teacher model in advance. In particular, each student model performs supervised learning using ground-truth data and minimizing Kullback–Leibler (KL) divergence [14] so that the distributions of each other's classification outputs match.

3.6 Metrics

Zhong et al. [42] introduced a tree edit distance based similarity (TEDS) metric for performance evaluation of both table structure and cell content recognition. After converting the recognition results and the ground truth into tree structures of HTML tags, the TEDS score is calculated according to Eq. (9).

$$\mathrm{TEDS}(T_a, T_b) = 1 - \frac{\mathrm{EditDist}(T_a, T_b)}{\max(|T_a|, |T_b|)}, \tag{9}$$

where T_a and T_b are the HTML trees, EditDist is the edit distance function, and $|T|$ is the number of nodes in T.

There are two versions of TEDS, namely structural TEDS and total TEDS. The former is calculated for HTML trees excluding cell contents and represents the recognition performance for table structures only. The latter is computed on complete HTML trees including cell contents and indicates the total recognition performance.

In addition, Zhong [42] classified the tables into two subsets, namely simple tables and complex tables. The former are tables without cells which are merged vertically or horizontally, and the latter are the other tables.

4 Proposal

The proposal consists of a ResNet encoder and two local-attention Transformer decoders. The two decoders infer table structure and cell contents, respectively. An additional output layer estimates the cell bounding boxes.

The two main differences with previous studies [18,19] are 1) introduction of a multi-cell decoder, and 2) introduction of bidirectional mutual learning to the HTML decoder. In addition, 2D positional encoding is employed. We named the proposed method MuTabNet after mutual learning, multi-task learning, and the multi-cell decoder. Figure 1 shows the network architecture.

4.1 Encoder

The encoder consists of a CNN backbone and 2D positional encoding. The CNN extracts image features of 65×65 pixels from an image of 520×520 pixels. For the CNN, we adopted TableResNetExtra [36] with 26 convolutional layers and three GCA blocks. After 2D positional encoding, the image features are flattened into one-dimensional features with 512 channels for cross-attention at the decoders.

4.2 HTML Decoder

The HTML decoder consists of one embedding layer, three local attention blocks, and two output layers. Each attention block accepts a table structure sequence through a self-attention layer in the block. The attention block then incorporates

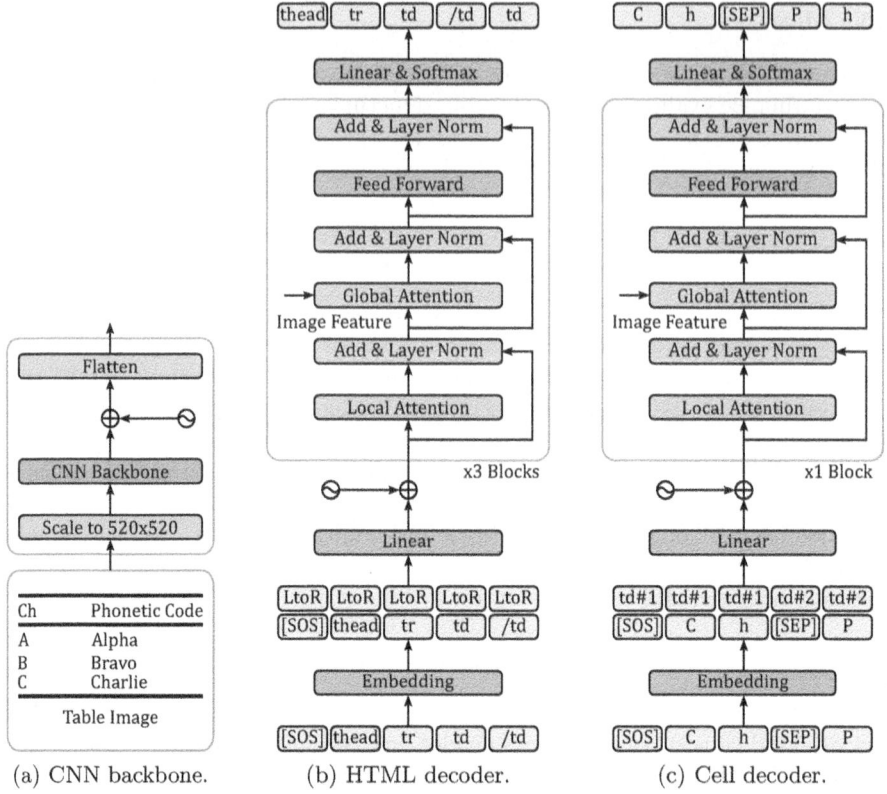

(a) CNN backbone.   (b) HTML decoder.   (c) Cell decoder.

Fig. 1. Proposed network architecture.

image features into the table structure sequence through a cross-attention layer, and outputs the sequence through a feed-forward layer. Several skip connections and layer normalizations are inserted within the block. The output from the last attention block is converted into HTML tokens and cell bounding boxes by the two output layers.

During training, the decoder predicts left or right shifted HTML tokens from the input HTML tokens. The shift direction is specified by an additional one-hot vector. During inference, the decoder predicts the following token and iteratively expands the input sequence to obtain the complete HTML sequence.

In addition to HTML tokens, the decoder accepts some special tokens, namely SOS, EOS, and PAD. SOS is a token that triggers sequential inference and is inserted at the beginning of the tokens. EOS is a token that stops inference and is inserted at the end of the tokens. PAD is inserted after EOS to equalize the lengths of the tokens in the mini batch.

Following previous studies [18,19], the HTML sequence was simply tokenized into HTML tags, except for the `<td>` tag representing the start of a cell. A `<td>` tag is tokenized as '`<td`', '`colspan="2"`', '`rowspan="3"`', '`>`' if it contains colspan or rowspan attributes. Otherwise, the tag is simply tokenized as '<td>'. It should be noted that FinTabNet [41] and PubTabNet [42] described in Sect. 5.1 are publicly available with such tokenization applied. We then merged the <td> and immediately following </td> tokens into one token.

Furthermore, we assigned some special tokens to frequent sequence patterns. If all header cells in the dataset had bold text, we removed the <b> and completed it in post-processing. These methods follow previous studies [18,19].

4.3   Cell Decoder

The cell decoder consists of one embedding layer, one local attention block, and an output layer. Following previous studies [18,19], the embedding layer accepts cell characters one by one. This is because cell contents typically consist of short sentences or unknown words or numbers, making it difficult to utilize pretrained language models.

The basic structure of a cell decoder is similar to that of an HTML decoder, with the following differences. First, a special token SEP is inserted between cell contents to trigger movement to the next cells. Second, the cell decoder accepts a combination of cell contents and their corresponding HTML features extracted from the output of the HTML decoder. These improvements allow the proposal to sequentially read the contents of multiple cells while referring to information from previous cells.

The previous study [18] exploited local attention for the HTML decoder and global attention for the cell decoder. This was because the cell decoder processed each cell independently, and the cell contents were short in general. On the other hand, in our proposed multi-cell decoder, the sequence of cell contents tends to be long. Consequently, we employed local attention.

4.4   Bidirectional Mutual Learning

We propose bidirectional mutual learning inspired by deep mutual learning [38] to train the HTML decoder. Here, two equivalent decoders are trained together to predict table structure in either a left-to-right (LtoR) or right-to-left (RtoL) direction. To reduce model parameters, we implemented the mutual learning in a single decoder by combining an additional one-hot vector that determines the direction with the embedded HTML tokens.

Let $\overrightarrow{x}$ and $\overleftarrow{x}$ be the LtoR and RtoL sequences respectively, and let $p(x)$ be the ground-truth and $q(x)$ the predicted probabilities. The loss $\overrightarrow{\mathcal{L}}$ for the LtoR decoder is defined by Eq. (10).

$$\overrightarrow{\mathcal{L}} = -\frac{1}{N} \sum_{n=1}^{N} p(\overrightarrow{x_n}) \log q(\overrightarrow{x_n}) + \frac{1}{N} \sum_{n=1}^{N} q(\overleftarrow{x_n}) \log \frac{q(\overleftarrow{x_n})}{q(\overrightarrow{x_n})}. \tag{10}$$

Table 1. The statistics of the table image datasets.

Dataset	Training	Validation	Evaluation
FinTabNet	91,596	10,635	10,656
PubTabNet	500,777	9,115	9,064
PubTabNet250	114,111	2,161	–

5   Experiments

To evaluate the effectiveness of the multi-cell decoder and bidirectional mutual learning, we conducted experiments on two public table datasets.

5.1   Datasets

We utilized two large datasets, FinTabNet [41] and PubTabNet [42]. In addition, we used a subset named PubTabNet250 [18] for ablation studies. Table 1 shows the statistics for the datasets.

FinTabNet is a large dataset of table images, including HTML labels and cell bounding boxes, extracted from the annual reports of S&P 500 companies. The dataset contains 112k tables and is divided into training set, validation set, and evaluation set. It should be noted that the original FinTabNet confuses validation and evaluation sets. Following previous studies [21,41], we treated the *validation* set containing 10,656 images as the evaluation set.

PubTabNet is a dataset built by collecting scientific articles from the PubMed central open access subset, containing 568k tables and corresponding structure and cell content annotations and cell bounding boxes. PubTabNet provides the training and validation sets, and the evaluation set was provided for the ICDAR competition [37]. We classified the tables into simple tables and complex tables as described in Sect. 3.6.

PubTabNet250 Ly and Takasu [18] extracted tables with 250 or more HTML tokens from PubTabNet and created a subset named PubTabNet250. They also introduced subsets for tables containing at least 500, 600, and 700 tokens. These subsets were utilized originally [18] to demonstrate the effectiveness of the local attention mechanism. We also utilized these subsets to conduct ablation studies in Sect. 5.4, approximately reducing training time from 179 h to 45 h per model.

Table 2. Table recognition results on FinTabNet evaluation set.

Model		TEDS (%)	
		Structure	Total
EDD	[42]	90.60	–
GTE	[41]	87.14	–
GTE (PT)	[41]	91.02	–
TableFormer	[21]	96.80	–
VAST	[10]	98.63	**98.21**
Ly et al.	[20]	98.72	95.32
Ly and Takasu	[19]	98.79	–
Ly and Takasu	[18]	98.85	95.74
MuTabNet		**98.87**	97.69

5.2 Implementation

The proposed model was implemented in PyTorch using mmcv [22], mmdet [23], and mmocr [24] frameworks and trained on four NVIDIA V100 GPUs with batch size 8 in total. We used Ranger [33] optimizer. The learning rate was initialized to 0.001 for the first 25 epochs, and decreased to 0.0001 and 0.00001 for the next three and last two epochs, respectively.

Each tabular image was normalized and reduced to 520×520 pixels, padding the margins with zeros if necessary. The cell bounding boxes were normalized to have a minimum value of 0 and a maximum value of 1.

HTML tokens and cell contents were converted to 512-dimensional embedded representations. The four attention blocks in the HTML and cell decoders have the same 8-head, 512-channel architecture, and the sliding window size for local attention was set to 300 by default, following previous work [18]. The maximum lengths for table structure sequences and cell content sequences were set to 800 and 8000, respectively, including special tokens. We employed greedy search for sequential prediction.

To ensure a fair comparison with the previous studies, we did not utilize data augmentation or ensemble learning techniques. We also did not take advantage of early stopping.

5.3 Experimental Results

We compared the performance of the proposed model trained on FinTabNet and PubTabNet with the claimed performance of existing models.

FinTabNet We evaluated the experimental results of structure recognition and total recognition using the TEDS metric. Table 2 compares the TEDS scores in the test set between the proposal and previous models. The proposal outperforms

Table 3. Table recognition results on PubTabNet validation set.

Model		TEDS (%)		
		Simple	Complex	Total
EDD	[42]	91.20	85.40	88.30
TabStruct-Net	[28]	–	–	90.10
TableFormer	[21]	95.40	90.10	93.60
SEM	[39]	94.80	92.50	93.70
LGPMA&OCR	[27]	–	–	94.60
VCGroup	[36]	–	–	96.26
VCGroup&ME	[36]	–	–	96.84
VAST	[10]	–	–	96.31
Ly et al.	[20]	97.89	95.02	96.48
Ly and Takasu	[19]	97.92	95.36	96.67
Ly and Takasu	[18]	98.07	95.42	96.77
MuTabNet		**98.16**	**95.53**	**96.87**

the previous models with scores of 98.87% and 97.69%. The inference time using the 4 GPUs was 3.78 h.

The total TEDS score of the proposal was lower than the score of VAST [10], which could be explained by the fact that VAST exploits external OCR for cell content recognition. In contrast, the structural TEDS score of VAST was lower than those of end-to-end approaches [18–20], including the proposal.

PubTabNet We evaluated the experimental results of table recognition on the validation set using the TEDS metric. Table 3 compares the scores between the proposal and previous methods. The proposal outperforms all previous methods with scores of 98.16%, 95.53% and 96.87% on simple tables, complex tables, and all tables, respectively. The inference time using the 4 GPUs was 3.23 h.

We also evaluated our proposal on the evaluation set. Table 4 compares the scores of the proposal with the top 10 solutions of the ICDAR competition [37]. The high scores achieved on both sets indicate high generalization performance of the proposal. The inference time was 3.13 h.

The score of the proposal was higher than the score of VAST [10]. PubTabNet contains a large amount of training data, and the proposed model appears to be well trained for cell content recognition tasks.

It should be noted that VCGroup&ME [36] utilized additional annotation of bounding boxes of text lines within cell contents and ensemble learning of three models. The proposed model outperforms all other non-end-to-end models which utilized additional annotation and ensemble learning even though our model did not utilize such techniques.

Table 4. Table recognition results on PubTabNet evaluation set.

Model		TEDS (%)		
		Simple	Complex	Total
LTIAYN	[37]	97.18	92.40	94.84
anyone	[37]	96.95	93.43	95.23
PaodingAI	[37]	97.35	93.79	95.61
TAL	[37]	97.30	93.93	95.65
DBJ	[37]	97.39	93.87	95.66
YG	[37]	97.38	94.79	96.11
XM	[39]	97.60	94.89	96.27
VCGroup	[36]	97.90	94.68	96.32
Davar-Lab-OCR	[37]	97.88	94.78	96.36
Ly et al.	[20]	97.51	94.37	95.97
Ly and Takasu	[19]	97.60	94.68	96.17
Ly and Takasu	[18]	97.77	94.58	96.21
MuTabNet		**98.01**	**94.98**	**96.53**

5.4 Ablation Studies

We conducted additional experiments for ablation studies using PubTabNet250 dataset for training and PubTabNet subsets for evaluation.

Effectiveness of Multi-cell Decoder and Mutual Learning. We evaluated the effectiveness of the proposed methods, namely multi-cell (MC) decoder and bidirectional mutual learning (BML). We trained two models on the training set and calculated the validation scores as displayed in Table 5. We selected previous experimental results [18] as baselines using exactly the same model architecture and dataset except for MC and BML. LA in the table refers to local attention.

Since the previous study [18] focused on performance for long tables, we also calculated TEDS scores for tables containing at least 500, 600, and 700 structure tokens. The MC decoder outperforms the baselines at all table lengths, and BML further improves table recognition performance.

The effect of BML was unclear in the structural TEDS scores but evident in the total TEDS scores. BML may still have improved the performance of implicit structure recognition and may have made an impact on cell content recognition, which requires precise content locations.

Table 5. Table recognition results with the proposed methods.

Methods			TEDS (%)							
			Structure				Total			
LA	MC	BML	250+	500+	600+	700+	250+	500+	600+	700+
−	−	−	−	−	−	−	93.86	91.16	90.63	88.65
✓	−	−	−	−	−	−	94.28	92.99	91.29	89.61
✓	✓	-	96.60	96.71	96.75	96.67	95.02	94.59	93.73	93.14
✓	✓	✓	97.02	96.70	96.35	96.65	95.81	95.11	94.05	94.02

Table 6. Table recognition results with respect to window sizes.

Size	TEDS (%)								
	Structure				Total				
	250+	500+	600+	700+	0+	250+	500+	600+	700+
100	96.96	96.69	96.98	96.60	75.91	95.70	95.19	94.99	94.35
200	96.79	96.53	96.30	95.83	75.79	95.46	94.66	93.80	92.69
300	97.02	96.70	96.35	96.65	83.15	95.81	95.11	94.05	94.02
400	96.83	96.85	96.48	96.51	82.58	95.40	95.08	94.00	93.50
500	96.97	96.74	97.03	96.54	81.14	95.51	94.46	93.88	92.65

Window Size of Cell Decoder. Ly and Takasu [18] has reported that a window size of 300 was optimal for the HTML decoder, whereas the cell decoder exploited global attention. In this study, we determine the optimal window size for the MC decoder. Table 6 shows the change in TEDS scores for the validation set as the window size varies from 100 to 500, while the window size for the HTML decoder was fixed at 300.

In general, a window size of 300 achieved the highest score, with the exception of tables containing more than 500 tokens, where a window size of 100 achieved the highest score. Tables with many cells tend to have fewer characters per cell, and a shorter window may be sufficient.

It should be noted that we used the PubTabNet250 dataset for training, and the performance for tables with fewer structure tokens was lower than the scores in Table 3. We selected the window size of 300 as the optimal value for the entire PubTabNet dataset containing tables with fewer tokens from the perspective of generalization performance.

6　Conclusion

We improved an end-to-end table recognition model based upon Transformer to achieve performance comparable to state-of-the-art models using external OCR systems. The proposed model consists of a ResNet encoder and two decoders for structure recognition and cell content recognition. After the first decoder infers the structure tokens, the second decoder reads the text within each cell.

We proposed a multi-cell decoder for cell content recognition to exploit useful information from neighbor cells. Furthermore, we proposed bidirectional mutual learning to force the model to pay attention to both previous and following cells. Experimental results using two public datasets demonstrate the effectiveness of the proposed methods.

In future work, we will further consider multitasking models that include the task of recognizing the meaning of tables, which enables deep understanding of printed documents, including table contents, and provides high-quality scientific knowledge for LLMs and question-answering systems.

References

1. Beltagy, I., Peters, M.E., Cohan, A.: Longformer: the long-document transformer (2020)
2. Cao, Y., Xu, J., Lin, S., Wei, F., Hu, H.: GCNet: non-local networks meet squeeze-excitation networks and beyond. In: IEEE/CVF International Conference on Computer Vision Workshop (ICCVW), pp. 1971–1980 (2019)
3. Deng, Y., Rosenberg, D., Mann, G.: Challenges in end-to-end neural scientific table recognition. In: International Conference on Document Analysis and Recognition (ICDAR), pp. 894–901 (2019)
4. Gehring, J., Auli, M., Grangier, D., Yarats, D., Dauphin, Y.N.: Convolutional sequence to sequence learning. In: International Conference on Machine Learning (ICML), pp. 1243–1252 (2017)
5. He, K., Gkioxari, G., Dollár, P., Girshick, R.: Mask R-CNN. In: IEEE International Conference on Computer Vision (ICCV), pp. 2980–2988 (2017)
6. He, K., Zhang, X., Ren, S., Sun, J.: Deep residual learning for image recognition. In: IEEE Conference on Computer Vision and Pattern Recognition (CVPR), pp. 770–778 (2016)
7. Hinton, G., Vinyals, O., Dean, J.: Distilling the knowledge in a neural network (2015)
8. Hochreiter, S., Schmidhuber, J.: Long short-term memory. Neural Comput. **9**, 1735–80 (1997)
9. Huang, G., Liu, Z., Maaten, L.V.D., Weinberger, K.Q.: Densely connected convolutional networks. In: IEEE Conference on Computer Vision and Pattern Recognition (CVPR), pp. 2261–2269 (2017)
10. Huang, Y., Lu, N., Chen, D., Li, Y., Xie, Z., Zhu, S., Gao, L., Peng, W.: Improving table structure recognition with visual-alignment sequential coordinate modeling. In: IEEE/CVF Conference on Computer Vision and Pattern Recognition (CVPR), pp. 11134–11143 (2023)
11. Itonori, K.: Table structure recognition based on textblock arrangement and ruled line position. In: International Conference on Document Analysis and Recognition (ICDAR), pp. 765–768 (1993)
12. Kayal, P., Anand, M., Desai, H., Singh, M.: ICDAR 2021 competition on scientific table image recognition to LaTeX. In: International Conference on Document Analysis and Recognition (ICDAR), pp. 754–766 (2021)
13. Kieninger, T.G.: Table structure recognition based on robust block segmentation. In: Photonics West '98. vol. 3305, pp. 22–32 (1998)

14. Kullback, S., Leibler, R.A.: On information and sufficiency. Ann. Math. Stat. **22**(1) (1951)
15. Li, M., Cui, L., Huang, S., Wei, F., Zhou, M., Li, Z.: TableBank: table benchmark for image-based table detection and recognition. In: Language Resources and Evaluation Conference (LREC), pp. 1918–1925 (2020)
16. Long, J., Shelhamer, E., Darrell, T.: Fully convolutional networks for semantic segmentation. In: IEEE Conference on Computer Vision and Pattern Recognition (CVPR), pp. 3431–3440 (2015)
17. Lu, N., et al.: MASTER: multi-aspect non-local network for scene text recognition. Pattern Recogn. **117**, 107980 (2021)
18. Ly, N.T., Takasu, A.: An end-to-end local attention based model for table recognition. In: International Conference on Document Analysis and Recognition (ICDAR), pp. 20–36 (2023)
19. Ly, N.T., Takasu, A.: An end-to-end multi-task learning model for image-based table recognition. In: International Joint Conference on Computer Vision, Imaging and Computer Graphics Theory and Applications (VISIGRAPP), pp. 626–634 (2023)
20. Ly, N.T., Takasu, A., Nguyen, P., Takeda, H.: Rethinking image-based table recognition using weakly supervised methods. In: International Conference on Pattern Recognition Applications and Methods (ICPRAM), pp. 872–880 (2023)
21. Nassar, A., Livathinos, N., Lysak, M., Staar, P.: TableFormer: table structure understanding with transformers. In: IEEE/CVF Conference on Computer Vision and Pattern Recognition (CVPR), pp. 4604–4613 (2022)
22. OpenMMLab: MMCV. https://github.com/open-mmlab/mmcv
23. OpenMMLab: MMDetection. https://github.com/open-mmlab/mmdetection
24. OpenMMLab: MMOCR. https://github.com/open-mmlab/mmocr
25. Peng, A., Lee, S., Wang, X., Balasubramaniyan, R., Chau, D.H.: High-performance transformers for table structure recognition need early convolutions. In: Table Representation Learning Workshop (2023)
26. Prasad, D., Gadpal, A., Kapadni, K., Visave, M., Sultanpure, K.: CascadeTabNet: an approach for end to end table detection and structure recognition from image-based documents. In: IEEE/CVF Conference on Computer Vision and Pattern Recognition Workshops (CVPRW), pp. 2439–2447 (2020)
27. Qiao, L., et al.: LGPMA: Complicated table structure recognition with local and global pyramid mask alignment. In: International Conference on Document Analysis and Recognition (ICDAR), pp. 99–114 (2021)
28. Raja, S., Mondal, A., Jawahar, C.V.: Table structure recognition using top-down and bottom-up cues. In: Computer Vision – ECCV, pp. 70–86 (2020)
29. Ren, S., He, K., Girshick, R., Sun, J.: Faster R-CNN: towards real-time object detection with region proposal networks. In: Advances in Neural Information Processing Systems. vol. 28 (2015)
30. Schreiber, S., Agne, S., Wolf, I., Dengel, A., Ahmed, S.: DeepDeSRT: deep learning for detection and structure recognition of tables in document images. In: International Conference on Document Analysis and Recognition (ICDAR), pp. 1162–1167 (2017)
31. Vaswani, A., et al.: Attention is all you need. In: Advances in Neural Information Processing Systems. vol. 30, pp. 6000–6010 (2017)
32. Wang, Y., Phillips, I.T., Haralick, R.M.: Table structure understanding and its performance evaluation. Pattern Recogn. **37**(7), 1479–1497 (2004)
33. Wright, L.: Ranger - a synergistic optimizer (2019). https://github.com/lessw2020/Ranger-Deep-Learning-Optimizer

34. Xie, S., Girshick, R.B., Dollár, P., Tu, Z., He, K.: Aggregated residual transformations for deep neural networks. In: IEEE Conference on Computer Vision and Pattern Recognition (CVPR), pp. 5987–5995 (2017)
35. Yang, F., Hu, L., Liu, X., Huang, S., Gu, Z.: A large-scale dataset for end-to-end table recognition in the wild. Sci. Data **10**(1), 110 (2023)
36. Ye, J., et al.: PingAn-VCGroup's solution for ICDAR 2021 competition on scientific literature parsing task B: table recognition to HTML (2021)
37. Yepes, A.J., Zhong, P., Burdick, D.: ICDAR 2021 competition on scientific literature parsing. In: International Conference on Document Analysis and Recognition (ICDAR), pp. 605–617 (2021)
38. Zhang, Y., Xiang, T., Hospedales, T.M., Lu, H.: Deep mutual learning. In: IEEE/CVF Conference on Computer Vision and Pattern Recognition (CVPR), pp. 4320–4328 (2018)
39. Zhang, Z., Zhang, J., Du, J., Wang, F.: Split, embed and merge: an accurate table structure recognizer. Pattern Recogn. **126**, 108565 (2022)
40. Zhao, W., Gao, L., Yan, Z., Peng, S., Du, L., Zhang, Z.: Handwritten mathematical expression recognition with bidirectionally trained transformer. In: International Conference on Document Analysis and Recognition (ICDAR), pp. 570–584 (2021)
41. Zheng, X., Burdick, D., Popa, L., Zhong, X., Wang, N.X.R.: Global table extractor (GTE): a framework for joint table identification and cell structure recognition using visual context. In: IEEE Winter Conference on Applications of Computer Vision (WACV), pp. 697–706 (2021)
42. Zhong, X., ShafieiBavani, E., Yepes, A.J.: Image-based table recognition: Data, model, and evaluation. In: Computer Vision – ECCV, pp. 564–580 (2020)

WikiDT: Visual-Based Table Recognition and Question Answering Dataset

Hui Shi[1(✉)], Yusheng Xie[2], Luis Goncalves[3], Sicun Gao[1], and Jishen Zhao[1]

[1] University of California San Diego, San Diego, USA
{hshi,sicung,jzhao}@ucsd.edu
[2] Amazon AGI, Seattle, USA
yushx@amazon.com
[3] AWS AI Labs, Seattle, USA
luisgonc@amazon.com

Abstract. Companies and organizations grapple with the daily burden of document processing. As manual handling is tedious and error-prune, automating this process is a significant goal. In response to this demand, research on table extraction and information extraction from scanned documents in gaining increasing traction. These extractions are fulfilled by machine learning models that require large-scale and realistic datasets for development. However, despite the clear need, acquiring high-quality and comprehensive dataset can be costly. In this work, we introduce the WikiDT, a TableVQA dataset with hierarchical labels for model diagnosis and potentially benefit the research on sub-tasks, *e.g.* table recognition. This dataset boasts a massive collection of 70,919 images paired with a diverse set of 159,905 tables, providing an extensive corpus for tacking question-answering tasks. The creation of WikiDT is by extending the existing non-synthetic QA datasets, with a fully automated process with verified heuristics and manual quality inspections, and therefore minimizes labeling effort and human errors. A novel focus of WikiDT and its design goal is to answer questions that require locating the target information fragment and in-depth reasoning, given web-style document images. We established the baseline performance on the TableVQA, table extraction, and table retrieval task with recent state-of-the-art models. The results illustrate that WikiDT is yet solved by the existing models that work moderately well on other VQA tasks, and also introduce advanced challenges on table extraction.

1 Introduction

Question answering is widely accepted as an AI-completeness task, while visual question answering (VQA) is an alternative to the visual Turing test [23]. VQA tasks, which require the integration of natural language and image understanding, attract tremendous interest from both computer vision and natural language processing communities. In general, VQA tasks can test a wide range of knowledge and inference skills, provided they can be related to information within an image. While knowledge in the wild world is extensive, VQA tasks typically bound the domain of the images and questions to make the task practical. General VQA tasks restrict the image domain to daily-life images with commonly seen objects, such as GQA [15] and VQA-v2 [12]; Scene-Text

VQAs confine their questions to text information on the images; and document VQAs ask questions on the image-formed documents, for example, OCR-VQA [26], DocVQA [25], and VisualMRC [41].

While most VQA tasks remain primarily in the research domain, document VQA demonstrates significant commercial value for machine-learning-as-a-service providers, generating tangible revenue. Its potential is reflected in the growing demand for intelligent document processing (IDP), with the global market size expected to surge 370% from 1.1 billion USD in 2022 to 5.2 billion in 2027 [1]. IDP solutions offer compelling benefits that reduces manual labor. For instance, insurance companies could use IDP to automatically extract the per-item cost of a claim from user-scanned receipts at a massive scale, significantly reducing the human labor needed. Moreover, tax preparation software could leverage IDP to handle diverse types and formats of income reports and generate tax return documents for a vast customer base.

The research on Document VQA, or IDP, faces several challenges despite its growing demand. Firstly, the differences in the image and questions types between existing document VQA datasets and the real-world applications poses an obstacle. Current dataset often focus on extractive questions with limited candidate answers, while real-world applications usually either have long context or require an answer that can not directly extract from the text. For example, only about 500 out of 30,000 samples in InfographicVQA [24] are non-extractive, which require reasoning or synthesis beyond simply locating information within the image. Even the extractive samples rely on short OCR context compared to the length of actual documents. Secondly, while the diversity of document and element structure is crucial for real-world applications, achieving it in datasets is significantly challenging. Collecting data from a single source could results in highly similar samples, while combining data from difference sources involves additional data cleaning, verifying and unifying procedures. Overcame the challenges in document collection, obtaining annotation from human is costly and can introduce inconsistencies and inaccuracies. Finally, the absence of intermediate labels in the datasets hinders developing models that can generalize to unseen tasks effectively.

Development of well-annotated and diverse dataset is essential for advancing the document VQA, observing that existing dataset's limitations are hard to overcome by current models alone. Recent research in model design highlights the urgent need for fully annotated datasets that explicitly incorporate chains of reasoning from context to answer. The specialization benefit is obvious, despite the prosperity of recent unified models like T5 [33] and GPT-3 [5]. For instance, a unified spatial aware text-to-text model for general document VQA can achieve 39.32 % accuracy [4], while a table-specialist model can achieve 50.97 % accuracy on the same dataset [14]. On the other hand, though end-to-end labeling is cheap to obtain, recent studies show that the intermediate labels or chains of reasoning steps can critically improve the performance and the robustness of the models [35,37,44,45].

To this end, we create a document VQA dataset that specialized in Table QA with abundant intermediate labels, named WikiDT (Wikipedia Document Table). WikiDT is closer to real-world applications that processing table-present documents. The documents in the dataset contain tabular information and natural language questions. The task requires the model or the system to comprehend the tabular structure, retrieve pertaining information, and perform SQL-like operations, e.g. filtering, count, and

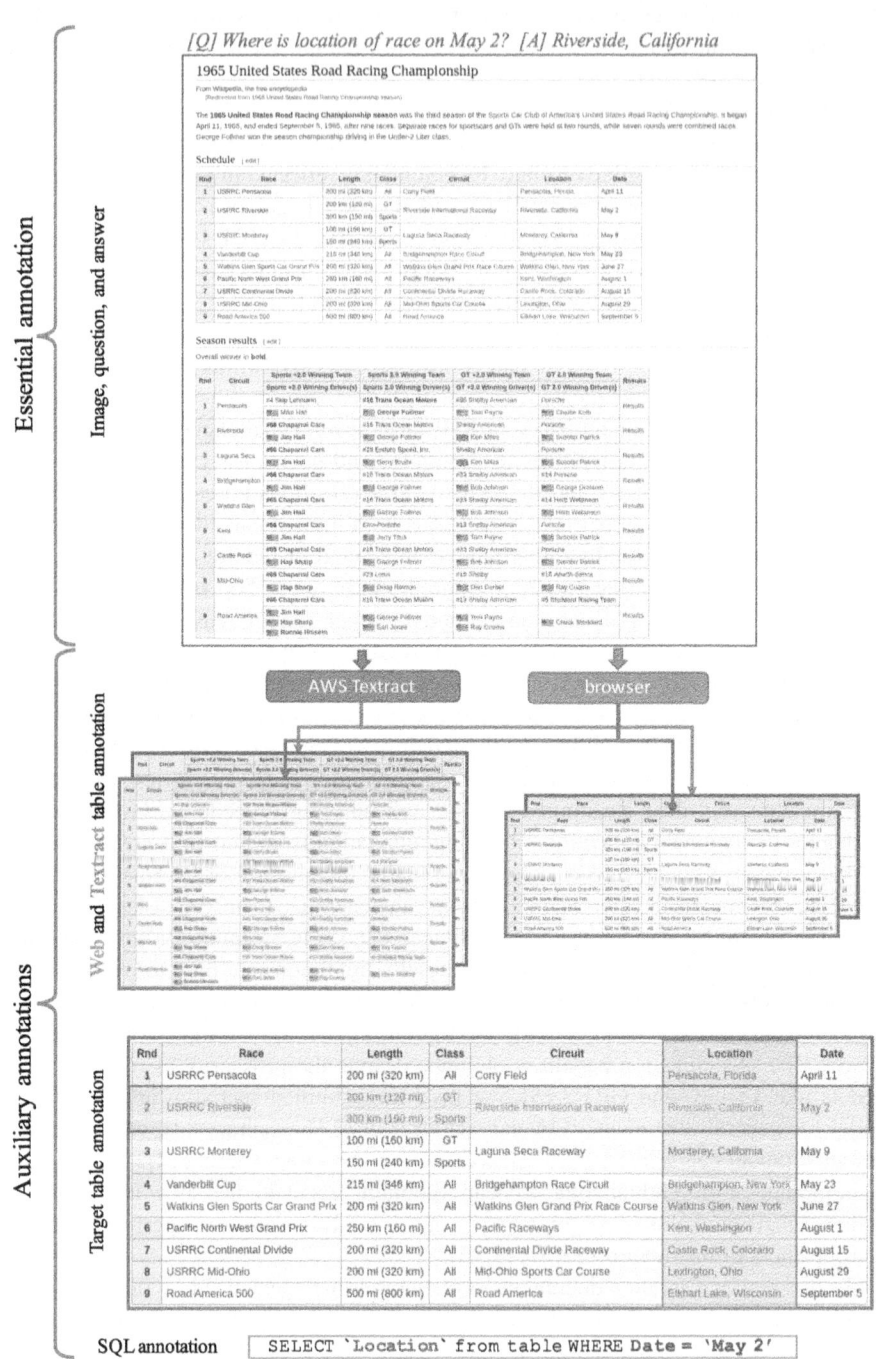

Fig. 1. Example of labels in WikiDT dataset.

summation. Aware of the challenges introduced in our tabular document VQA task, we provide intermediate labels to each of the sub-tasks: table extraction, table retrieval, and question answering. Overall, WikiDT dataset contains 16,887 images, 159,905 tables and their annotations, and 70,652 question-answer pairs with table retrieval labels. Lastly, the experimental results demonstrate that all the sub-tasks in the WikiDT dataset introduce uncovered challenges to the existing models, and the challenges reveal directions for improvement of model designs in the future.

The WikiDT dataset offers the following unique features:

– A specialized table VQA dataset with adequate samples and accurate intermediate labels.
– Unique VQA reasoning challenges, including multi-level span prediction, multi-step reasoning, and a diverse and exponential answering space.
– Multi-level ground-truth labels that facilitate diagnosis of the end-to-end model and enable training for sub-tasks like table recognition and retrieval.
– Challenges in table recognition, such as multiple header rows/columns, and diverse table size and position.

2 Related Work

In this section, we relate WikiDT to other visual question-answering categories and classify WikiDT as a specialized task under the document VQA. Furthermore, WikiDT is compared to the TableQA and table recognition tasks, highlighting its unique diversity and challenges.

VQA Task Categories. VQA task is generally defined as question answering based on any image-type input. Typically the questions are about objects and their attributes, e.g., object type, shape, color, and texture, as well as their spatial relationships. CLEVR [17], GQA [15], VQA-v2 [12] are such general VQA datasets. A new research direction has emerged, focusing on answering questions about textual information within images. For example, given a picture of a grocery store interior, this research aims to answer questions such as "what's the price of the product in the center of the picture?" As images in these datasets are usually natural scenes with text, this type of the VQA task is named Scene-Text VQA. The datasets including e.g., TextVQA [39], ST-VQA [3] and EST-VQA [44]. In Scene-Text VQA, the textual information is unstructured, and one of the key capability requirements is to pair the textual information to the spatial information of objects, e.g., price of good, texts shown in the center. On the contrary, when the texts on the images are structured, e.g., images that have tables and charts, which are commonly seen in documents, books, and websites, the task is usually referred to as Document VQA and emphasizes on understanding the structured information. An example question could be "what is the largest value in a list?" With the diversity of chart types, it's less practical to build a monolithic model that comprehends all kinds of infographics and performs different types of reasoning. Among the charts, tables stand out as a particularly rich source of information. Therefore, we proposed WikiDT, a large-scale document VQA dataset that specialized in TableVQA.

Figure 2 shows the categorization of the VQA tasks with typical datasets for each category, and Table 2 compared the datasets in detail.

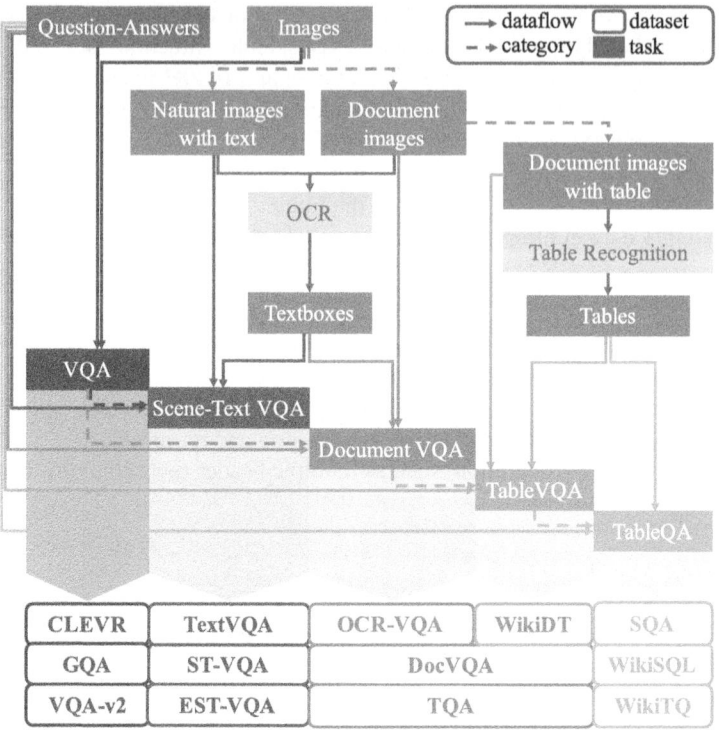

Fig. 2. Task categorizations and relationships.

The table VQA task in DUE [4] is the most similar existing task to WikiDT. However, DUE presents a reduced challenge as its QA context is an image that contains only the target table. In contrast, WikiDT, while similar in question and answer length to DUE and other scene-text VQA and DocVQA datasets, contains significantly more textual information, as shown in Table 6. This extended textual context makes WikiDT more realistic and challenging.

TableQA and TableVQA. Though WikiDT is created from the TableQA dataset, it's not simply a TableQA task that replaces the table with an image. WikiDT is different from its ancestors (*e.g.*, WikiSQL [46] and WikiTQ [31]) in two ways: 1) mainstream of the TableQA tasks focus on semantic parsing, a task translating questions in the natural language into the executable SQL query, where the table schema is definite and strict. TableVQA does not assume the availability of table schema and the rigorousness of the language; 2) WikiDT also requires the model to recognize and identify the table from a noisy document with all other related information, which is more similar to the application of automated document processing.

A similar attempt to WikiDT is DUE [4], which formulates its TableVQA by using the same WikiSQL dataset and replacing the table with an image with only the table (exampled in Fig. 13). In Sect. 4, the results show that giving a noiseless image that contains only the target table makes the TableVQA trivial since the table structure extrac-

Table 1. Comparison of table extraction dataset.

Dataset	Annotations	Size
ICDAR-13 [11]	TD/TSR	0.45k
ICDAR-19 [10]	TD/TSR	3.2k
Marmot[a]	TD	2k
SciTSR [8]	TD/TSR	15k
PubTabNet [47]	TSR	568k
TableBank [20]	TD/TSR	417k
PubTables-1M [40]	TD/TSR	948k
WikiDT-detection	TD	54,032
WikiDT-structure	TSR	159,905

[a] https://www.icst.pku.edu.cn/cpdp/sjzy/.

tion from that images are well-performed. Moreover, contrary to DUE that each image is a single table, images in WikiDT may contain multiple tables.

Comparing to Existing Table Recognition Dataset. Labeling for the table structure recognition dataset is labor intensive and imprecise. Even a 5-row by 5-column small table requires drawing 25 bounding boxes for each of the cells, and the cell boundary is usually just a few pixels from its neighbors. Therefore, recent large-scale table recognition datasets, and also WikiDT, are created from an automated process: given the source of Latex/Word/HTML, we render the source to images meanwhile extract the bounding boxes and texts from the rendering software. When using the dataset for training, the models are only provided with images and bounding box labels, so they could not hack the problem from the source. In addition to diagnostic labels to our TableVQA task, the table recognition labels also would benefit the table recognition research with versatile table shapes and layouts. With the continuous scroll feature of the web pages, compared to PDFs and Words, WikiDT contains tables that do not fit into a single print paper and can be leveraged as a benchmark for table recognition when the table spans multiple pages. An overall comparison to the table recognition dataset is shown in Table 1, and a detailed comparison to the PubTables-1M can be found in Fig. 16 and 17.

3   Dataset Description

The WikiDT dataset contains 16,887 documents with 70,652 QA samples. Each question can be answered from a single document, and the answer should be inferred from one of the tables on the document. As a common practice in Scene-Text and Document VQA, the OCR result on the page is made available to the user. For table VQA model, the input is the document as an image, texts, and bounding boxes of all textual information on the image, and the output should be the answer to the question.

Besides essential input (image and question) and output (answer) for the TableVQA task, WikiDT also provides auxiliary annotation that helps solve the TableVQA task step-by-step. Illustrated in Fig. 1, intuitively, human fulfills the table VQA task by: recognizing the tables, identifying the table that relates to the question, then reasoning on

Table 2. VQA dataset comparisons. † denotes classification tasks. * marks the synthetic datasets that generate the QA and images. ANLS represents Average Normalized Levenshtein Similarity; VisualMRC uses image captioning task metrics including BLEU [30], METEOR [9], ROUGE-L [21], CIDEr [42]; DUE-TableQA and WikiDT use denotation accuracy.

Dataset	Image	Auxiliary input	Metric	Count (QA/Document/Tables)
CLEVR*† [17]	synthetic	logic structure	accuracy	850k/85k/-
GQA† [15]	natural	scene graph	accuracy	22M/110k/-
ST-VQA [3]	natural	OCR	ANLS	32k/23k/-
TQA [19]	textbook	OCR, topic graph	accuracy	26k/1k/-
VisualMRC [41]	webpage	ROI and relevance label	BLEU, etc.	30k/10k/< 1k
DocVQA [25]	documents	OCR	ANLS	50k/13k/11.8k
InforgraphicQA [24]	documents	OCR	ANLS	30k/5.4k/9.5k
DVQA* [18]	synthetic bar charts	chart metadata	accuracy	3.4M/300k/-
DUE-TableVQA [4]	cropped tables	-	accuracy	120k/16k/16k
WikiDT	webpage	various	accuracy	70k/17k/160k

the table to find the answer. To facilitate the diagnosis of end-to-end model and the training of modules in ensemble systems, WikiDT dataset also provides the following intermediate labels as auxiliary annotations:

- Table annotations on both full-length and paged images.
- OCR and AWS Textract table recognition results on paged images to simulate the real-world scenarios.
- Table retrieval labels indicating which table corresponds to each question.
- SQL queries that generate answers from tables.

WikiDT is available at https://huggingface.co/datasets/AmazonScience/WikiDT. Data acquisition and processing are detailed in the supplementary materials.

3.1   Data Acquisition

WikiDT is extended from the WikiSQL [46] and WikiTableQustions (WikiTQ) [31]. The images are rendered from the URL in WikiSQL and WikiTQ's metadata by a browser that also generates the raw table annotations at the same time. The questions in WikiDT are combined from WikiSQL and WikiTQ, in which the questions are all human-annotated. For the questions from the WikiTQ dataset, we directly use the original human-labeled answers, and for the questions from the WikiSQL dataset, we generate answers by executing their human-labeled SQL query on the aforementioned table annotations. Since multiple tables may appear on the same image, we also semi-automatically labeled the target table to the question. While the browser generated *oracle* table annotation is great for decoupling the tasks and diagnosing the end-to-end model, they are assumed unavailable in real-world scenarios. Therefore, following the existing practice, we also provide the OCR and table recognition results on the images from the publicly available service, AWS Textract.

3.2   TableVQA and Sub-Tasks

Table 3. Sub-task formulations.

Task	Input	Output
Table Detection	image	table bbox(es)
Table Structure Recognition	image	row, column, cell bboxes
Table Retrieval	tables, question	target table
TableVQA	image, question	answers

The multi-level intermediate labels allow many possibilities for diagnosing the TableVQA models and training modules for sub-tasks. In this paper, we evaluate TableVQA and two sub-tasks, table recognition and retrieval. Table 3 summarizes the input and output for each (sub-)task.

Table recognition task encompasses two sub-tasks: **table detection (TD)** and **table structure recognition (TSR)**. Table detection involves predicting the table regions within images, typically annotated as rectangular bounding boxes. Following the practice established by PubTables-1M, table structure recognition data is provided as pairs of cropped images and structure annotation. Each image crop contains only a single table with consistent padding, while the structure annotation details the bounding boxes of rows, columns, cells (including multi-row and multi-columns cells, hereinafter, merged cells), and table headers.

Table retrieval aims to identify the specific table within an image that can answer a given question. The retrieval model should not peek at the answer.

TableVQA should answer the question given the full-page image. This end-to-end solution remains a significant challenge for existing models. To simplify this task, we explore leveraging the auxiliary annotation, such as restricting the context OCR to only the table regions.

3.3   Data Analysis

The statistics of the WikiDT are described as follows.

Images and Tables. The *full page* images refer to the original auto-scrolled web screenshots, which usually are long and hard for existing visual models to take as input. A pagination heuristic is applied to segment the full-page screenshot into small segments without cutting through tables and texts. The segmented images are referred to as *subpages*. Table 4 lists the heights of the full pages and subpages images, which re-emphasizes the motivation of the pagination process.

Questions and Answers. Figure 3 shows the type and number of items in the answer, as well as the cumulative probability of answer strings. Most questions in WikiDT expect a single answer, and roughly only 5% answers have more than one entry. A single answer can be a phrase, a word, or a number, while a multi-entry answer is like {*United States*, *Canada*}. For multi-entry answer, denotation accuracy requires the model to answer

Table 4. Image size (in pixel) statistics.

Page	Min	Mean	Max	Variance
Full page	1200	6592	78423	6061
Subpage	15	1761	57622	1028

Table 5. Basic statistics on WikiDT. Standard deviations are shown in the brackets.

Data	Count
Images (full page)	16,887
Sub-page images	54,032
Table annotations	159,905
Question-answer pairs, retrieval labels	70,652
SQL annotations	49k
Single answer questions	68,573
Multiple answer questions	3,524
Mean number of tables per full page	14.7 (21.9)
Mean number of tables per sub-page	3.4 (16.0)
Mean number of words per sub-page	708.9 (517.1)
Mean number of words per table	170.84 (458.7)
Mean number of columns per table	9.08 (48.3)
Mean number of rows per table	12.15 (18.1)

correctly all the entries, regardless of the order. Despite most samples only expect a single entry answer, the diversity of answer suggests the WikiDT TableVQA task should be solved in a generative way. In datasets that can be formulated as classification tasks, such as CLEVR and GQA, the top 1000 answers could cover more than 90% of questions. In the contrary, 1000 most frequent answers could only cover 70% samples.

Social Impact. WikiDT provides diagnostic data that potentially advances multiple machine intelligence areas. The curation of the dataset involves trivial privacy concerns because the data source of WikiDT is Wikipedia pages, where the user, the editor, information is not revealed, and the contents of the pages are publicly available. The potential negative impact of the WikiDT is two-fold: misinformation and outdated information intrinsically from Wikipedia, and the bias towards certain countries (*e.g.*, the *United States* and *Canada* have a much higher frequency than other country names).

4 Experiments

In this section, we evaluate existing neural networks on TableVQA, table recognition, and table retrieval tasks.

4.1 TableVQA

Task Setup. In typical model setup, the entire image and all OCR results are used as input. However, in this work, we restrict the input context by leveraging auxiliary

Table 6. Average lengths in number of words.

Dataset	Question	Answer	Image
ST-VQA	8.8	1.6	7.5
TextVQA	8.1	1.5	12.2
DocVQA	9.5	2.4	182.8
InfographicVQA	11.5	1.6	217.9
WikiDT	11.0	2.0	708.9 (subpage)

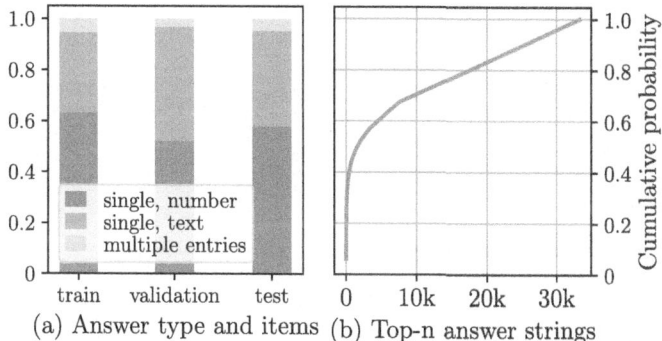

(a) Answer type and items (b) Top-n answer strings

Fig. 3. Analysis of answer items/types and cumulative word probability.

labels. As shown in Table 5, the number of words per subpage image can easily exceed a thousand, and this number further increases after tokenization into subwords. Additionally, a question might associate with more than one subpage, potentially resulting in thousands of tokens, which overload the computation resources. To make the task attackable, we utilize the table detection results and table retrieval labels, such that we can prune irrelevant content from the input context.

Models. The competition between end-to-end and modularized models is yet concluded, so as in the VQA domain. Thus we explore the state-of-the-art performance under the two approaches: LaTr [2] and T5 [33] model as end-to-end representatives, and TAPAS [14] with table recognition module and ground table retrieval labels as the modularized approach. The high-level architectures of the two models are shown in Fig. 4.

– **Modularized model** generally should consist of three components of table recognition, table retrieval, and table question answering. In this experiment, we utilized the auxiliary annotation in WikiDT to substitute the first two modules. The table recognition outcome is from Textract. To the authors' knowledge, there are no table retrieval models for the table question answering is open-sourced, thus we skip this step and use the ground-truth retrieval labels given the Textract tables. The table question answering module is the state-of-the-art TAPAS [14] model. As shown in Fig. 4, the TAPAS model takes the tokenized question, a special token *CLS*, and the flattened table as the input. A flattened table includes the names of columns and the

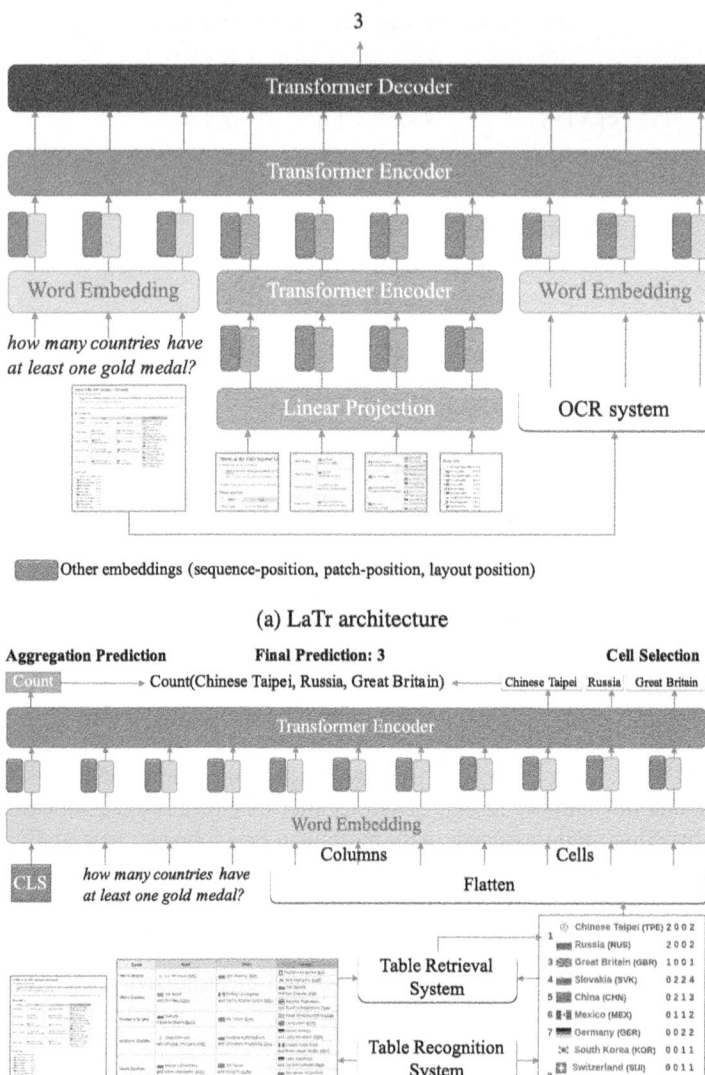

(a) LaTr architecture

(b) TAPAS architecture

Fig. 4. TableVQA model architectures.

content in each table cell. Each token content is embedded by an embedding layer, and the embeddings are then concatenated with extra learnable embeddings for the position in the sequence, column and row ID, rank, and the token position in the table cell. After the transformer encoder, each token has an output vector. The output vector of *CLS* is fed into a classifier to predict the aggregation operation (*e.g.*,

None, Count, Sum, Average), the column name outputs will decide the column to be selected, and the table cell output in the selected column will independently predict if the cell is selected. Lastly, the aggregation is applied to the selected cells and produces the final prediction.

- **Monolithic model:** We compare two similar neural architecture: T5 [33] and LaTr [2]. Both of them receive the question on a subpage that contains the target table, and the OCR results that locate within the region of the target table. The monolithic models are not informed by the table recognition results. LaTr model architecture is shown in Fig. 4 on the top. The input sequence contains the tokenized question, patched and flattened image of the content, and the OCR results on the image. Before feeding into the transformer encoder, the extra encodings of the layout position, token position, and patch position are concatenated to the corresponding embeddings. Unlike the TAPAS model, the LaTr directly generates the answer with a transformer decoder. While the LaTr knows the location of each OCR token, the T5 model is inherently spatially-blind. We use the vanilla T5 model in which the input to the encoders contains only the textual (word embedding) and the sequence order information (positional embedding). Despite T5 knows no spatial information, as table tokens are fed into model by row and column order, T5 can potentially understand the table in this linearized form.

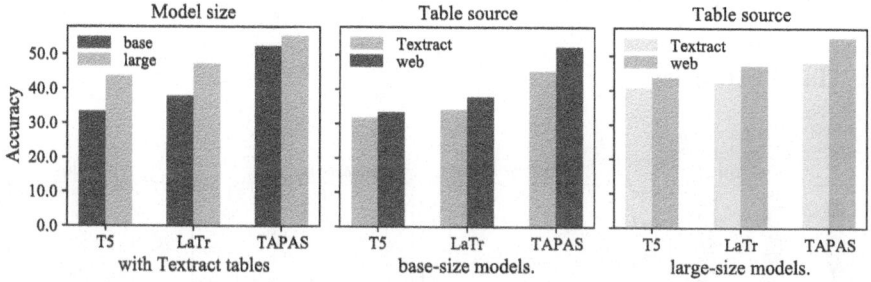

Fig. 5. TableVQA model diagnosis. TAPAS model is more sensitive to table recognition error, and T5/LaTr models are more sensitive to model size. Also, the performance of TAPAS with web tables set the upper bound accuracy of improving table recognition.

Implementation. We finetuned the TAPAS and T5 from their pre-trained models released on Huggingface. For LaTr, we shared the WikiDT with its authors and received the results after fine-tuning the pre-trained LaTr.

Metrics. The performance on Document TableQA is evaluated by denotation accuracy, detailed in [27]. The prediction is correct if 1) it has the same number of entries as in the ground truth answers; 2) every entry in the prediction can non-repeatedly match to an entry in the ground truth answers (ignoring the format variances, *e.g.* 1000 *v.s.* 1,000).

Results. Table 7 summarizes the model performances. TAPAS model achieves the best performance unsurprisingly as it receives the table structure information and is designed to tackle such table-operation tasks. LaTr outperformed the T5 model with only marginal advantage, even the LaTr is visual- and spatial-aware and is pre-trained heavily on the scene-text VQA and document VQA tasks but T5 is only pre-trained on the text-to-text tasks. The results indicate that enhancing document VQA models with table-specific designs is the way to improve their performance in table-based reasoning.

Table 7. Model performance on WikiDT TableVQA task.

Model	Denotation Accuracy (%)		
	Single Answer	Multi-answer	Overall
T5	32.74	1.30	31.67
LaTr	35.29	0.0	34.08
TAPAS	46.24	6.86	45.23

Diagnostic and Ablation Experiment. Owing to the auxiliary annotations, we can analyze the error origins. Figure 5 compares the performance when using the web (ground truth) table annotation against using the Textract annotation, and the impact of model sizes. For monolithic models, the performance improvement with more accurate table annotation is trivial compared with the benefit of increasing the model size, which is contraray to the conclusion on the modularized model of TAPAS.

4.2   Table Extraction

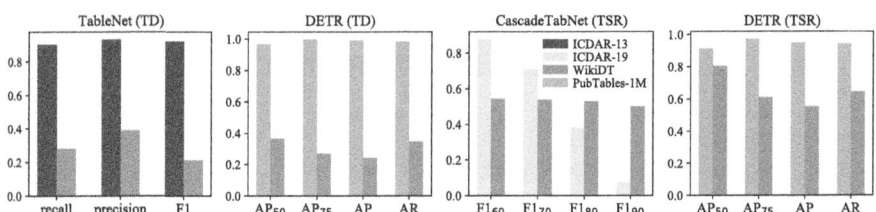

Fig. 6. Performance of pretrained model on WikiDT table recognition. Each pair of bar shows the model's performance difference between its pre-training dataset (ICDAR-13, ICDAR-19, or PubTables-1M) and inference dataset (WikiDT).

The table extraction experiment aims to show that the documents in WikiDT offer different challenges and are complementary to the existing datasets. We demonstrate by 1) evaluating the table extraction performance on models that are trained on other datasets, and showing the model performance is compromised owing to significant domain difference; 2) comparing the performance of the model trained on WikiDT and PubTables-1M dataset and showing that WikiDT contains more challenging table layouts.

Models and Configurations

- TableNet [28] is a CNN using pretrained VGG [38] or DenseNet [16] as encoder and has two up-sampling CNN decoders to predict the table mask and column mask. We use TableNet pretrained on the Marmot dataset with DenseNet-121 as the encoder.
- CascadeTabNet [32] also adopts CNN. Its encoder is a pretrained HRNetV2p-W32 [43], and the decoder uses cascadeCNN-like architecture to predict borderless table mask, cell masks, and bordered table mask. Post-processing uses line detection to translate pixel-level masks to bounding boxes. We evaluate the CascadeTabNet pre-trained on ICDAR 19.
- DETR [6] is a transformer neural network for object detection. In table detection and structure analysis, tables, rows, columns, and table headers are considered as distinct types of objects. We evaluate DETR pretrained on PubTables-1M, and with the same model configuration, we train another DETR from random initialization on WikiDT.

Metrics. Generally, the evaluation metric for TD and TSR is IoU-based precision and recall [22], *e.g.*, Average Precision(AP_{IoU}), Average Recall(AR_{IoU})[1] and F1 scores. We report the metrics used in evaluating the dataset that each model is trained on. The TableNet predicts binary masks instead of bounding boxes, hence the precision, recall, and F1 are reported on the pixel level.

Table 8. DETR performance on PubTables-1M and WikiDT

Task	Table detection				Table structure recognition			
Dataset	AP_{50}	AP_{75}	AP	AR	AP_{50}	AP_{75}	AP	AR
PubTables-1M	96.6	99.5	98.8	98.1	91.2	97.1	94.8	94.2
WikiDT	84.7	72.6	67.5	72.7	88.9	79.6	74.7	82.6

Dataset Difficulty. Table 8 shows the performance of DETR, the current SOTA model for table recognition, on the PubTables-1M and WikiDT datasets. DETR achieved significantly lower performance on the WikiDT dataset, indicating that WikiDT is challenging compared to the existing largest table recognition dataset, in both table detection and table structure recognition. The complexity comes from both the flexibility of table location in the page, table shape, and inner structures.

Domain Transfer. Figure 6 presents the performance gap between models on their pretrained dataset and WikiDT. The results show that WikiDT exhibits substantially different layouts and table styles that models trained on the existing dataset could not generalize to. Especially, the performance decrease more severely in the table detection (TD) than in the table structure recognition, which indicates larger difference in document layout and table size than in table style.

---

[1] The IoU in the subscript is in 10^{-2}, e.g. AP_{75} is the average precision under IoU threshold of 0.75.

Conclusions. The experiments illustrate the WikiDT provides a challenging table recognition dataset that is substantially different from the existing datasets in page layout and table styles. We refer the readers to the supplementary material for a detailed comparison and qualitative examples.

4.3 Table Retrieval

Table 9. Table retrieval performance in Mean Reciprocal Rank (MRR), which is computed on the subset of samples with multiple table candidates.

Method	Web Tables	Textract Tables	Diff.
BM25	0.382	0.389	−0.007
Dense Retrieval	0.587	0.524	0.063

The WikiDT-retrieval task requires retrieving the table relevant to the question from the collection of tables on the same webpage. Viewing from Table 5, in most cases, for each query (question) the candidate tables number is small, especially compared to retrieval in a recommendation system or search engine [29, 36]. However, the challenge is that the tables from the same webpage might have closely related contents that are hard to distinguish.

Dataset Split. The dataset split is the same as other tasks. In the development and test set, the samples having only one table are removed. In the training set, if a training sample contains only a single table, we additionally sample two background tables from all the tables in the dataset. Otherwise, the background tables to a question are other tables on the same page.

Models. We evaluate BM25 [34] and a dense retrieval method on table retrieval task.

- BM25: questions are queries and flattened tables on the entire page are a corpus.
- Dense retrieval: we use the pre-trained TAPAS model with an additional special token R_CLS to predict the likelihood of if the given table is the target table. The dense retrieval method is trained with weighted binary cross-entropy loss. Due to the intrinsic label imbalance, a sample weight of |target_sample| / |background_sample| is applied to all the background samples.

The results are summarized in Table 9. The dense retrieval model significantly outperformed the non-learning method, BM25. Similar to the QA task, the inaccuracy in the table recognition (Textract tables) has a marginal impact on the retrieval results.

5 Conclusions

This paper introduces WikiDT dataset, whose primary goal is to facilitate visual-based table question answering. Featuring multi-modality data and rich intermediate labels in WikiDT, the dataset serves a variety of tasks, encompassing table extraction, table retrieval, and TableVQA. The evaluation of the recent models on each task demonstrates the WikiDT imposes novel challenges that are yet solved. Especially, the current document VQA or scene-text VQA models have great difficulty to solve TableVQA questions that require multi-step reasoning. Proposals for advanced models and approaches to address these challenges are deferred to future work.

Acknowledgement. Research reported in this publication was supported by an Amazon Research Award, Fall 2022 CFP, and Amazon Post-Internship Fellowship. Any opinions, findings, and conclusions or recommendations expressed in this material are those of the author(s) and do not reflect the views of Amazon.

A Data Acquisition and Processing

A.1 Overview

WikiDT is created from Wikipedia, a public online encyclopedia created and updated by its users. We are only interested in those Wikipedia web pages with tables, for purpose of table extraction and table-based question answering. Overall, the data acquisition process is taking the Wikipedia URLs from existing datasets, namely WikiTableQuestions [31], TabFact [7], and WikiSQL [46], rendering the web pages to images while annotating the tables, then re-connecting the rendered images and table annotations to the existing question-answer pairs.

The basic annotations in the WikiDT are {image, question, answer} triplets, which are basic input and output to the end-to-end visual-based table question answering models. Aware of the challenges, WikiDT also provides auxiliary annotations include: ground truth table annotations (web tables); Textract table recognition and OCR annotations; QA target table labels based both on the Textract table and web tables; and executable SQL queries on majority of QA samples.

Although the task may seem straightforward, the creation of the dataset is fraught with numerous challenges. Especially, the web contents has changed dramatically since the original WikiSQL dataset was created. Additionally, the data processing in WikiSQL dataset is unknown and irreversible, which poses great hardship in re-connecting the tables to the question-answer pairs. We detail the approaches to overcome those as below.

A.2 Image Rendering and Pagination

We leverage the Puppeteer[2] to render the screenshots mainly from the URLs. Generally, the rendered images in the continuous scroll mode have large heights that are extremely hard for the popular visual networks to handle. Thus, the original rendered pages (*full-page*) are paginated to several *sub-pages*.

URLs. TabFact provides the URLs to the tables in itself and WikiSQL, and WikiTable-Quesions also provide the URLs to its tables. Despite all the URLs are Wikipedia pages, the URLs in the TabFact is Wikipedia domain followed by article title, which changes drastically over the years, while the WikiTableQuestions provided URLs with archived version IDs, which have the exact content when they created the dataset. So for the first step, we search for the Wikipedia editing history for each TabFact pages, and find the closest version to then the WikiSQL dataset is created. Then both of the recovered TabFact pages and WikiTableQuestions pages are rendered.

Renderer Setting. The width of the page is set to 1600 pixels for a better table layout (as compared in Fig. 9). Less than 0.5% of pages have minimum page width requirements that are larger than 1600 pixels, mainly owing to extremely wide tables, and are rendered with their minimum page width.

Pagination. The extra-long full pages cause huge trouble for the down-stream tasks. For example, the typical width-height ratio of input layer to common visual models is 1:1 for VGG [38] and ResNet [13], and other backbone models trained on ImageNet. Therefore the pagination is critical for usage of the dataset. Fixed-height segmentation may divide a single table to multiple subpages. Thus we use a dynamic-height pagination strategy: first, the blank lines are detect via the slide window; concretely, if the color deviations of a W x H window is 0, we consider this window is a blank line. The W is set to 1200 to ignore the navigation bar on the left of the pages, and H is set to 10. Then a subset of blank lines are selected such that the width-to-height ration of every segment, excepting the last one, is close to 1: 1.

A.3 Table Annotation

The table annotations are derived from HTML table tags, *e.g.*, ⟨table⟩ translates to a table and ⟨tr⟩ translates to a table row. The pseudo-code to process the HTML table annotation with Puppeteer is summarized as below. Puppeteer API returns valid bounding box for visible elements and returns NULL for invisible ones. The annotation process keeps only visible tables and table elements. Additionally, the merged table cells are annotated with non-empty *row_span* and *col_span* values.

---

[2] https://github.com/puppeteer/puppeteer.

function HTML_TO_TABLE(DOM)

Initialize: Annotations = list()
for T in DOM.elements where T.tag=⟨table⟩ **do**
   let CurTable=dictionary()
   for R in T.elements where R.tag=⟨tr⟩ **do**
      let CurRow=dictionary()
      for C in R.elements **do**
         cell ← annotate_cell(C)
         if cell ≠ Null **then**
            CurRow.cells.add(cell)
         end if
      end for
      if c.type=HEADER $\forall c \in$ row.cells **then**
         row.type=HEADER
      end if
      CurTable.rows.add(row)
   end for
   table ← infer_box(CurTable)
   Annotations.add(table)
end for
return Annotation
end function

---

function ANNOTATE_CELL(C)

---

if C is ⟨th⟩ element **then**
   cell.type = HEADER
else if C is ⟨td⟩ element **then**
   cell.type = BODY
else
   return NULL
end if
cell.text ← C.text
cell.row_span ← C.row_span
cell.col_span ← C.col_span
return cell
end function

---

function INFER_BOX(T)

$T.box.x0 \leftarrow \min_{R \in T.rows}(R.box.x0)$
$T.box.y0 \leftarrow \min_{R \in T.rows}(R.box.y0)$
$T.box.x1 \leftarrow \max_{R \in T.rows}(R.box.x1)$
$T.box.y1 \leftarrow \max_{R \in T.rows}(R.box.y1)$
end function

Web Table Filtering. The editors on Wikipedia can use ⟨table⟩ tag to express a broader range of data (*e.g.*, legend as in Fig. 11). We remove those false tables by criteria that a true table should have at least two rows and two columns. Moreover, the nested tables are possible in web page while hardly seen in other document format. We notice that some web pages in our dataset contains nested tables (*e.g.*, Fig. 7), the outer tables of which are usually used to produce certain layout. Therefore, we detect the nested tables by table bounding boxes and keep only the inner-most tables.

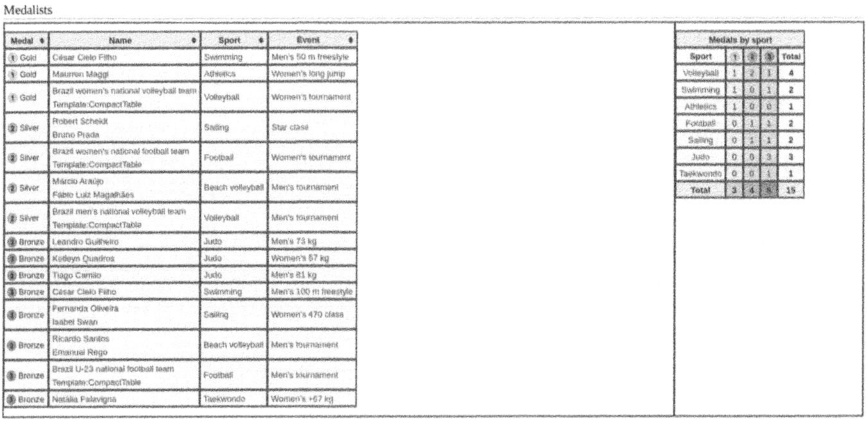

Medalists

Medal	Name	Sport	Event
Gold	César Cielo Filho	Swimming	Men's 50 m freestyle
Gold	Maurren Maggi	Athletics	Women's long jump
Gold	Brazil women's national volleyball team Template:CompactTable	Volleyball	Women's tournament
Silver	Robert Scheidt Bruno Prada	Sailing	Star class
Silver	Brazil women's national football team Template:CompactTable	Football	Women's tournament
Silver	Márcio Araújo Fábio Luiz Magalhães	Beach volleyball	Men's tournament
Silver	Brazil men's national volleyball team Template:CompactTable	Volleyball	Men's tournament
Bronze	Leandro Guilheiro	Judo	Men's 73 kg
Bronze	Ketleyn Quadros	Judo	Women's 57 kg
Bronze	Tiago Camilo	Judo	Men's 81 kg
Bronze	César Cielo Filho	Swimming	Men's 100 m freestyle
Bronze	Fernanda Oliveira Isabel Swan	Sailing	Women's 470 class
Bronze	Ricardo Santos Emanuel Rego	Beach volleyball	Men's tournament
Bronze	Brazil U-23 national football team Template:CompactTable	Football	Men's tournament
Bronze	Natália Falavigna	Taekwondo	Women's +67 kg

Medals by sport

Sport	1	2	3	Total
Volleyball	1	2	1	4
Swimming	1	0	1	2
Athletics	1	0	0	1
Football	0	1	1	2
Sailing	0	1	1	2
Judo	0	0	3	3
Taekwondo	0	0	1	1
Total	3	4	8	15

Fig. 7. Nested tables: the outer tables annotations are removed.

Creating the Table Extraction Task. Figure 8 summarizes the data acquisition and processing of WikiDT table extraction dataset. Following the convention in PubTables-1M, we divide the table extraction to **table detection** which predicts the table bounding boxes from the subpage, and **table structure recognition** which predicts the inner structure in the table from the table crop. The table extraction task data is labeled using Pascal VOC format.

A.4 OCR and Textract Table Recognition

Realistically, in Document VQA task, ground truth OCR annotation is unavailable. Instead, datasets usually provide the OCR results from off-the-shelf systems. Following the same principle, the OCR and table recognition from a publicly available system is

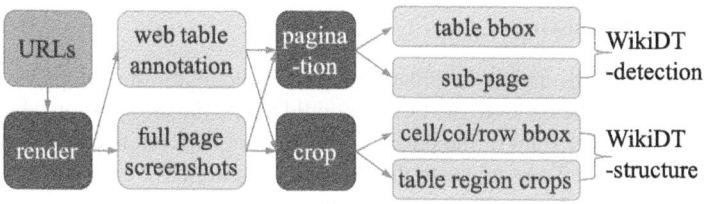

Fig. 8. Data acquisition for WikiDT table extraction task.

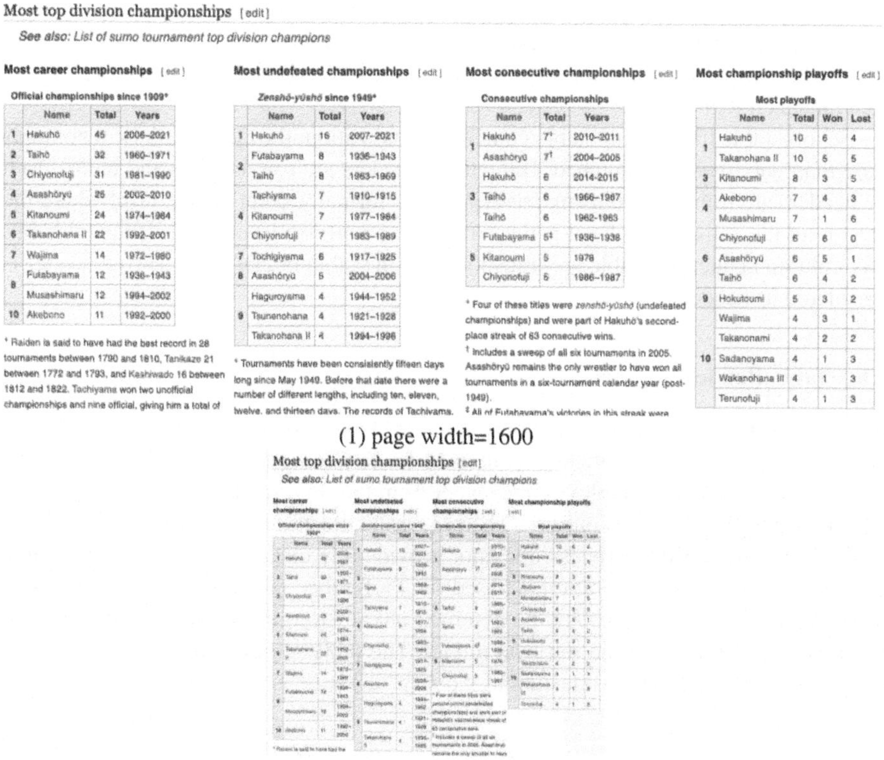

Most top division championships [edit]

See also: List of sumo tournament top division champions

Most career championships [edit]

Official championships since 1909*

	Name	Total	Years
1	Hakuhō	45	2006–2021
2	Taihō	32	1960–1971
3	Chiyonofuji	31	1981–1990
4	Asashōryū	25	2002–2010
5	Kitanoumi	24	1974–1984
6	Takanohana II	22	1992–2001
7	Wajima	14	1972–1980
8	Futabayama	12	1936–1943
	Musashimaru	12	1994–2002
10	Akebono	11	1992–2000

* Raiden is said to have had the best record in 28 tournaments between 1790 and 1810. Tanikaze 21 between 1772 and 1793, and Kashiwado 16 between 1812 and 1822. Tachiyama won two unofficial championships and nine official, giving him a total of

Most undefeated championships [edit]

Zenshō-yūshō since 1949*

	Name	Total	Years
1	Hakuhō	16	2007–2021
2	Futabayama	8	1936–1943
	Taihō	8	1963–1969
	Tachiyama	7	1910–1915
4	Kitanoumi	7	1977–1984
	Chiyonofuji	7	1983–1989
7	Tochigiyama	6	1917–1925
8	Asashōryū	5	2004–2006
	Haguroyama	4	1944–1952
9	Tsunenohana	4	1921–1928
	Takanohana II	4	1994–1996

* Tournaments have been consistently fifteen days long since May 1949. Before that date there were a number of different lengths, including ten, eleven, twelve, and thirteen days. The records of Tachiyama,

Most consecutive championships [edit]

Consecutive championships

	Name	Total	Years
1	Hakuhō	7*	2010–2011
	Asashōryū	7†	2004–2005
3	Hakuhō	6	2014–2015
	Taihō	6	1966–1967
	Taihō	6	1962–1963
	Futabayama	5‡	1936–1938
5	Kitanoumi	5	1978
	Chiyonofuji	5	1986–1987

* Four of these titles were zenshō-yūshō (undefeated championships) and were part of Hakuhō's second-place streak of 63 consecutive wins.

† Includes a sweep of all six tournaments in 2005. Asashōryū remains the only wrestler to have won all tournaments in a six-tournament calendar year (post-1949).

‡ All of Futabayama's victories in this streak were

Most championship playoffs [edit]

Most playoffs

	Name	Total	Won	Lost
1	Hakuhō	10	6	4
	Takanohana II	10	5	5
3	Kitanoumi	8	3	5
4	Akebono	7	4	3
	Musashimaru	7	1	6
	Chiyonofuji	6	6	0
6	Asashōryū	6	5	1
	Taihō	6	4	2
9	Hokutoumi	5	3	2
	Wajima	4	3	1
	Takanonami	4	2	2
10	Sadanoyama	4	1	3
	Wakanohana III	4	1	3
	Terunofuji	4	1	3

(1) page width=1600

Most top division championships [edit]

See also: List of sumo tournament top division champions

(2) page width=500

Fig. 9. Page and table layout in different page width.

provided along with the images for Document VQA tasks. We used the AWS Textract service to acquire the OCR and table extraction results on the subpages. In practice, we also found the Textract results quality on the subpages is substantially better than the quality on full pages.[3],[4]

---

[3] https://aws.amazon.com/textract/.

[4] The annotations are obtained before this update and recognize no merged cell. https://aws.amazon.com/about-aws/whats-new/2022/03/amazon-textract-updates-tables-check-detection/.

A.5   QA Table Processing

Inspected the WikiSQL and WikiTableQuestion samples with the webpage, there are two observations: 1) the tables to the QA pair could not be the information tables, which are shown on the upper right of the page), and could not be reference tables, which are at the end of the page; 2) the tables to the QA pairs can not have only one content row. Therefore, we remove those tables in the QA task.

Furthermore, the content of tables are normalized in the following ways. First, we remove the JavaScript code that are extracted as texts by Puppeteer, which are detected by regular expression matching to keywords (*e.g.*, *mw-parser-output*, *navbar*). Then, if any cell text is longer than 200 characters, it is replaced with a special token $\langle TLDR \rangle$. Lastly, the string normalization method adopted in WikiTableQuestions is applied to each table cell text.

Beyond table filtering and table content normalization, the table headers, if contains more than one rows, are flattened. If the table headers are row-headers and multi-level, the headers are transformed using *level1.level2...* format, except for the level where the text for all the cells in that level is the same, *i.e.*, a uniform level value. If the table has only column header, the table are transposed. If the table is 2D, only the row header is considered as header.

A.6   Table Retrieval Labels

Table retrieval is an essential step if there are many tables on the page. The table retrieval labels are generated automatically, primarily by retrieving the table with the highest recall using the reference tables and answers from the original dataset (*i.e.*, WikiTableQuestions and WikiSQL) as queries. The process is illustrated in Fig. 10. Note that the retrieval label generation process does not provide any cheating method for the table retrieval task: the label generation is based on a reference table, which is not included in the WikiDT dataset and should be assumed unknown to the model, and the answer, which is also unknown to the retrieval model. The table retrieval model of the user of WikiDT should use only the question as a query.

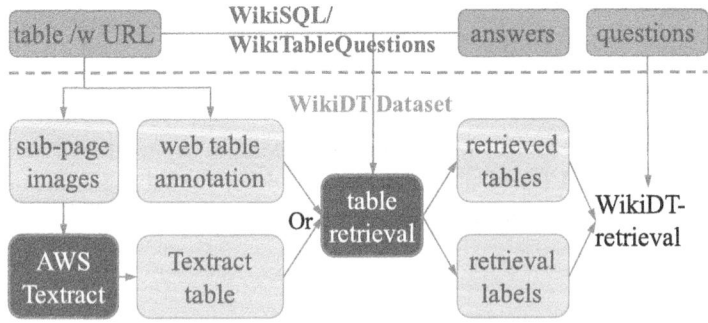

Fig. 10. Data acquisition and processing for WikiDT table retrieval sub-task.

Algorithm 1 shows the table retrieval label generation process. The reference tables and answers from WikiSQL and WikiTableQuestions serve as queries and the candidate tables are regarded as keys. We record the retrieved table by best recall to the reference table and reference keys. In most cases (74%), the two results agree. For the QA experiments, we use the best recall to the reference table results, since the reference answers may not appear in the table owing to the aggregation functions.

Algorithm 1 Retrieval(reference table $\mathcal{R}$, answers set $\mathcal{A}$, candidate tables $[C_j]_1^n$)

$\mathcal{Q} \leftarrow$ set(x.text for $x \in \mathcal{R}$)
for $C_j \in [C_j]_1^n$ **do**
    $\mathcal{K}_j \leftarrow$ set(x.text for $x \in C_i$)
    Table recall $R_j^{table} \leftarrow \mathcal{K}_j \cap \mathcal{Q}$
    Answer recall $R_j^{answer} \leftarrow \mathcal{K}_j \cap \mathcal{A}$
end for
return $arg\max_j R_j^{table}$, $arg\max_j R_j^{answer}$

Team	Pld	W	D	L	GF	GA	GD	Points
Russia	5	4	1	0	148	125	+23	9
South Korea	5	3	1	1	155	127	+28	7
Hungary	5	2	1	2	129	142	-13	5
Sweden	5	2	0	3	123	137	-14	4
Brazil	5	1	1	3	124	137	-13	3
Germany	5	1	0	4	123	134	-11	2

Fig. 11. False table (upper left one) and true table (lower).

A.7 Answer Generation

Inspecting the table retrieval results from web table annotations, we found that the contents of tables in the WikiSQL dataset are crucially different from the contents on the images, presumably due to the web content updates and post-processing during the creation of the WikiSQL dataset. Therefore, we translated the SQL query from the WikiSQL dataset and executed it on the retrieved web table in WikiDT to generate the answers to the questions, shown in Fig. 12. On the other hand, the tables in WikiTable-Questions do not have this issue, thus the answers are still valid given the images and the questions.

B Compare to Existing VQA Datasets

Table 2 compare the input modality and task formulation of the notable VQA datasets, and a concrete examples of the VQA samples are shown in Fig. 13. It can be clearly seen that WikiDT provides a unique type of QA tasks, and are highly challenging in the plenitude of the context information.

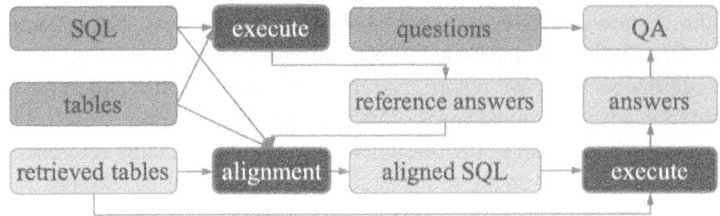

Fig. 12. QA pair generation from WikiSQL dataset from acquired images and tables.

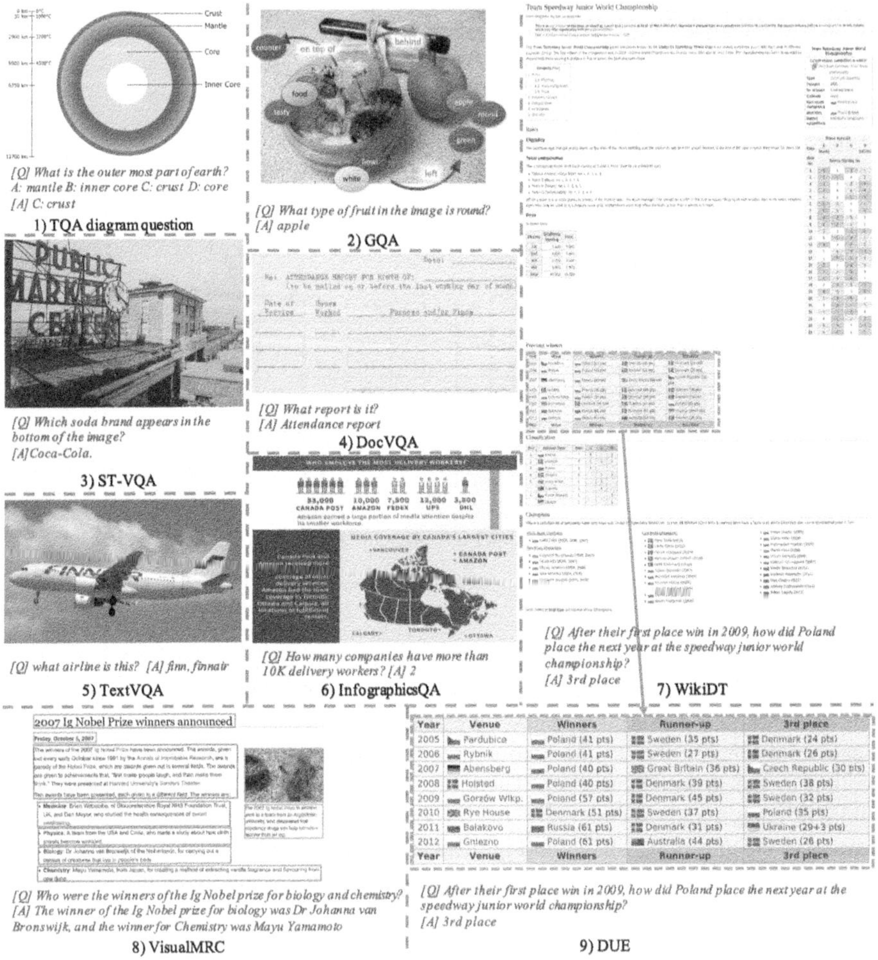

[Q] What is the outer most part of earth?
A: mantle B: inner core C: crust D: core
[A] C: crust
1) TQA diagram question

[Q] What type of fruit in the image is round?
[A] apple
2) GQA

[Q] Which soda brand appears in the
bottom of the image?
[A] Coca-Cola.
3) ST-VQA

[Q] What report is it?
[A] Attendance report
4) DocVQA

[Q] what airline is this? [A] finn, finnair
5) TextVQA

[Q] How many companies have more than
10K delivery workers? [A] 2
6) InfographicsQA

[Q] After their first place win in 2009, how did Poland
place the next year at the speedway junior world
championship?
[A] 3rd place
7) WikiDT

[Q] Who were the winners of the Ig Nobel prize for biology and chemistry?
[A] The winner of the Ig Nobel prize for biology was Dr Johanna van
Bronswijk, and the winner for Chemistry was Mayu Yamamoto
8) VisualMRC

[Q] After their first place win in 2009, how did Poland place the next year at the
speedway junior world championship?
[A] 3rd place
9) DUE

Fig. 13. Illustration of VQA datasets

C   Dataset Description and Analysis

C.1   Tasks and Datasets

WikiDT-Detection Dataset. With the results of pagination, we take the subpage images and the table bounding boxes, ignoring the table structures and content, as a table detection dataset. Images with no tables are discarded.

WikiDT-Structure Dataset. Independent with the pagination, each table region is cropped from the full-page images to form the image inputs to the table structure recognition task. The coordinates of the bounding boxes are translated from full-page coordinates to the cropped image coordinates. The following types of structural objects are annotated directly from raw web table annotation: table body rows, header rows, table cells, and merged cells. Since the HTML tags do not have table column tags while columns are essential labels for many existing table extraction models (*e.g.*, TableNet [28]), we compute the column bounding boxes and include them in the annotation as well.

WikiDT-TableVQA Dataset. The question answering task is to predict the answer to a question given the full page image in an end-to-end manner. Meanwhile, the intermediate labels along the reasoning chains are available for breaking down and diagnosing the models. The intermediate labels include table annotation, table retrieval labels, and SQL queries in most of the samples. The OCR results from Textract are highly accurate and could be leveraged directly to fit into existing document VQA or scene-text VQA frameworks.

C.2   Data Analysis

Images and Subpages. Figure 14 shows the image height distribution before (full page) and after (sub-page) pagination step. It shows the pagination greatly reduced average image heights and extremely long pages. The outliers that still have large image heights may contain long tables that the pagination module is unable to split. Figure 15 shows that the number of tables per image reduces after pagination.

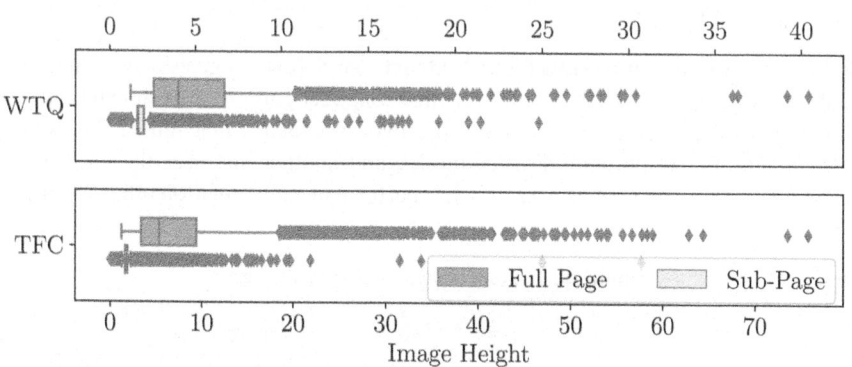

Fig. 14. Image Heights Distributions (x-axis in 10^3).

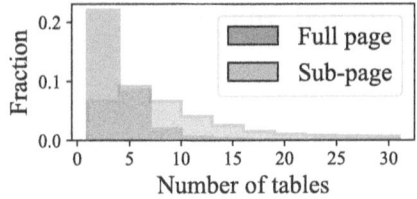

Fig. 15. Number of tables on a single image.

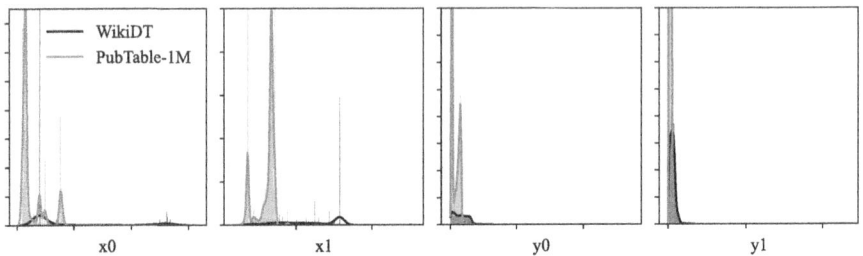

Fig. 16. Table location distribution.

Page Layout. Comparing to existing table recognition dataset created from PDF, the detection task of WikiDT is challenging with diverse table shapes and positions, illustrated in Fig. 17. PubTables-1M images are single or double columned PDFs. The table position and width, especially in the double-columned documents are highly similar. On the contrary, table location and shape are much more arbitrary in WikiDT. Figure 16 plots the distribution of table bounding boxes of both dataset. It's obviously seen that the PubTables-1M has few spikes indicating the few locations that are highly likely to be the table border, while the distribution in WikiDT are almost uniform which indicates a diversity in the table location. Moreover, WikiDT contains list-like text blocks that could be easily mis-detected as tables.

Table Styles. The table structure recognition task is featured with diversified table styles, shown in Fig. 18.

Qualitative Annotation Results from Textract. AWS Textract gives overall high quality table recognition results, but there are still wrong recognition cases shown in Fig. 19. In the experiment section, we compare the two approaches that utilize the Textract Table recognition results with TAPAS model, and another one uses the OCR boxes. Though the TAPAS model has overall better performance, its function relies on the cor-

Table 10. Textract table recognition accuracy.

IOU	0.5			0.75			0.95		
	Precision	Recall	F1	Precision	Recall	F1	Precision	Recall	F1
	0.901	0.558	0.689	0.844	0.523	0.646	0.703	0.435	0.538

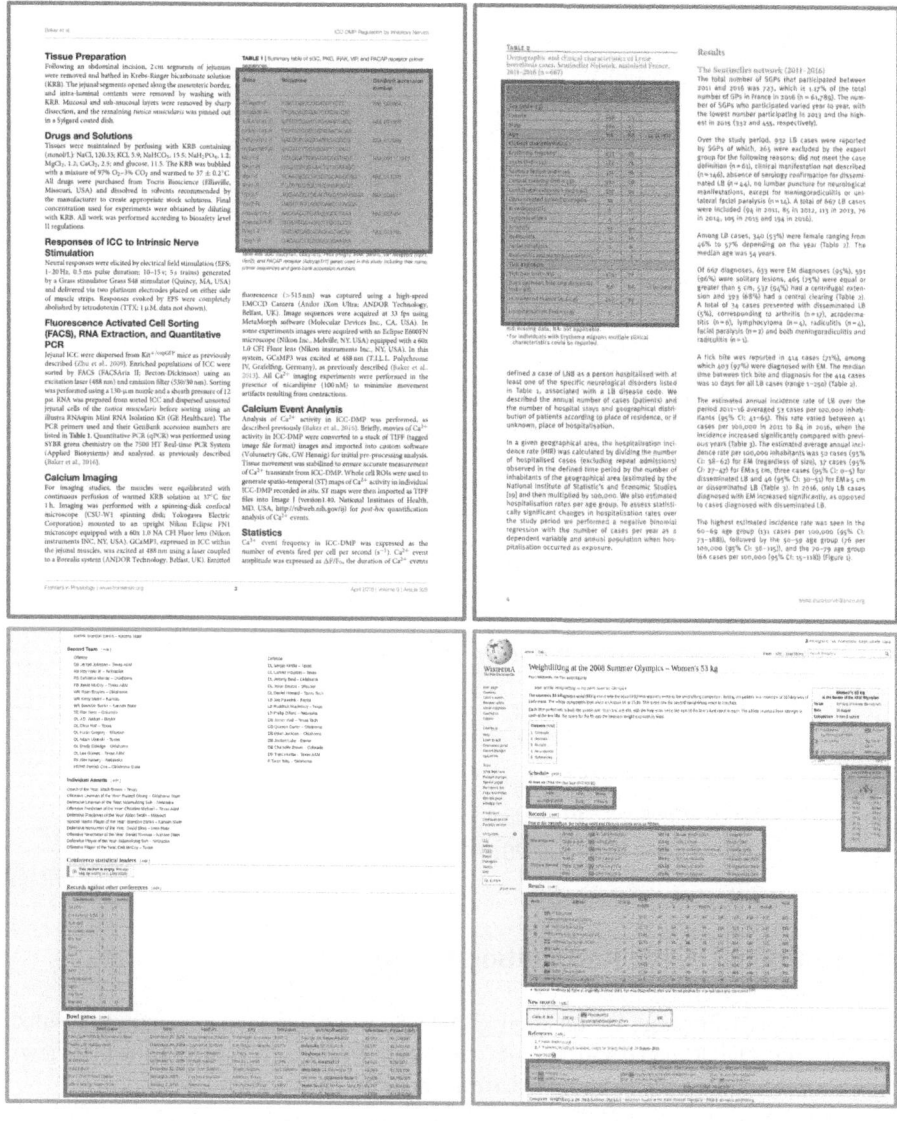

Fig. 17. Table detection samples from PubTables-1M (*upper*) and WikiDT (*lower*).

rect recognition results and could not learn to rectify the table recognition error, like the one in Fig. 19 (4), when the two columns are recognized as one.

Top Answer Items. Table 11 shows the most frequent numeric answers and the text answers. Usually, the numeric answers are produced from some aggregations. For instance, count the number of rows that satisfy certain criteria, or comparing the difference between two values (see Fig. 22 for example). To this end, understanding the implied reasoning from the questions is critical to solve the QA task.

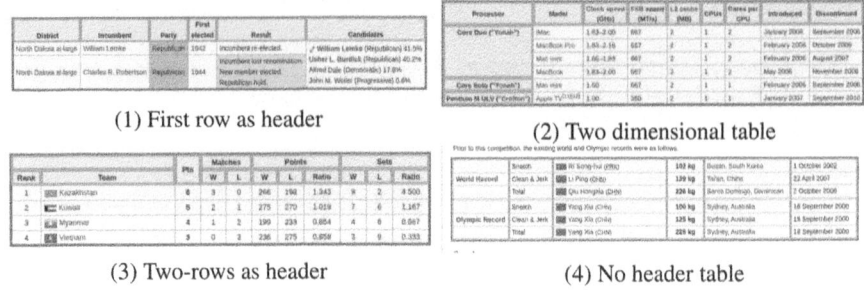

(1) First row as header

(2) Two dimensional table

(3) Two-rows as header

(4) No header table

Fig. 18. Various types of table headers.

Table 11. Top answer items from WikiDT-TableQA/VQA task.

Numbers		Texts	
Item	Count	Item	Count
'1'	4696	yes	654
'0'	1556	no	512
'2'	1532	united states	398
'3'	1289	incumbent re-elected	158
'4'	1048	canada	132
'5'	900	lr	128
'6'	751	1-1	121
'7'	688	race	116
'8'	540	democratic	111
'9'	473	2010	104

D   Additional Table Recognition Results

Figure 19, 20, 21 illustrate the qualitative table detection and table recognition results. Generally, the DETR model and Textract can generate reliable results, despite that viewing from Table 8 the table extraction performance still have potential to improve.

E   TableVQA Additional Results

Figure 22 shows when TAPAS makes correct and wrong predictions. Notice that in the wrong prediction case, the question requires a multi-step aggregation. The QA system needs to count the number of ships wrecked in Lake Huron and in Lake Erie separately, then compare the difference. However, by design, TAPAS could solve questions with no more than one aggregation.

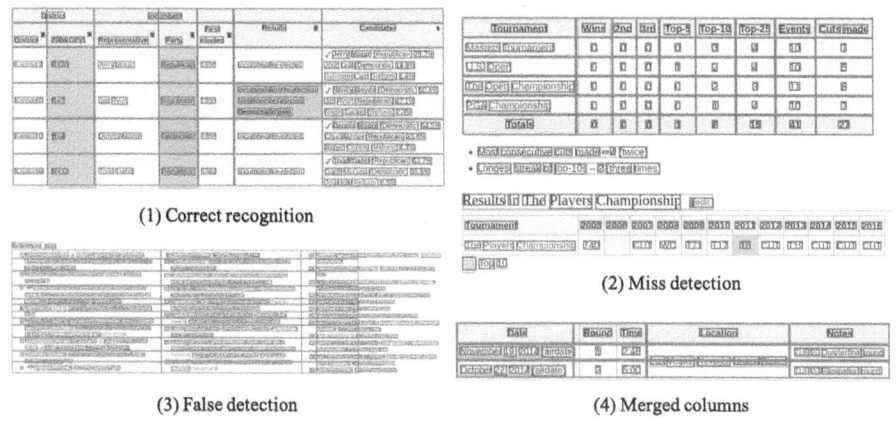

Fig. 19. Textract recognition results. Red boxes show OCR tokens, olive boxes show table structure. (Color figure online)

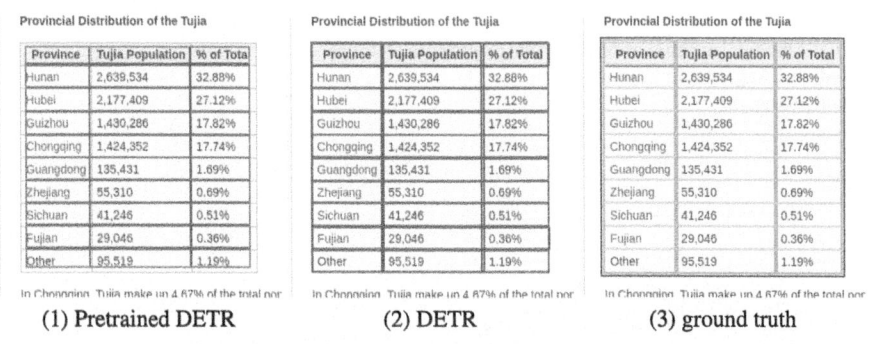

Fig. 20. DETR prediction compared with ground truth.

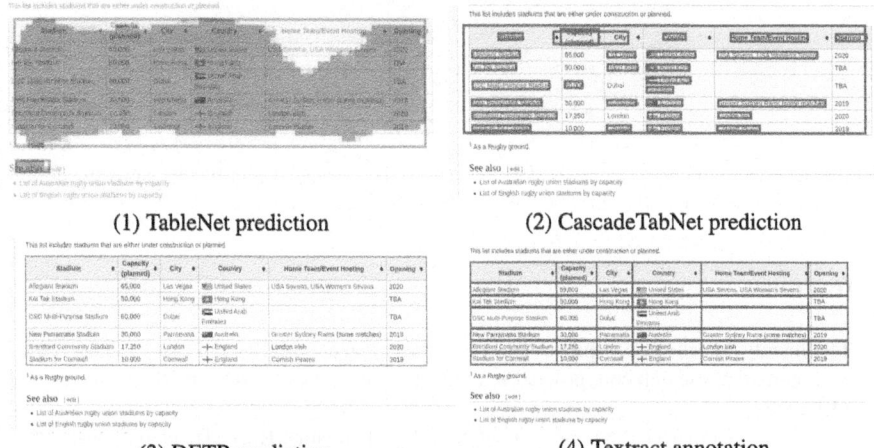

Fig. 21. Table extraction result comparison.

How many more ships were wrecked in Huron than in Erie?
Predicted aggregator: Count
Final prediction: 7

Ship	Type of Vessel	Lake	Location	Lives lost
Argus[23]	Steamer	Lake Huron	25 miles off Kincardine, Ontario	25 lost
James Carruthers[24]	Steamer	Lake Huron	near Kincardine	18 lost
Hydrus[25]	Steamer	Lake Huron	near Lexington, Michigan	28 lost
Leafield[26]	Steamer	Lake Superior		all hands
John A. McGean[27]	Steamer	Lake Huron	near Goderich, Ontario	28 lost
Plymouth[28]	Barge	Lake Michigan		7 lost
Charles S. Price[29]	Steamer	Lake Huron	near Port Huron, Michigan	28 lost
Regina[30]	Steamer	Lake Huron	near Harbor Beach, Michigan	
Issac M. Scott[31]	Steamer	Lake Huron	near Port Elgin, Ontario	28 lost
Henry B. Smith[28]	Steamer	Lake Superior		all hands
Wexford[32]	Steamer	Lake Huron	north of Grand Bend, Ontario	all hands
Lightship No. 82[28]	Lightship	Lake Erie	Point Albino (near Buffalo)	6 lost

What's the total number of festivals that occurred in October?
Predicted Aggregator: Count
Final prediction: 5

Date	Festival	Location	Awards	Link
Feb 2–5, Feb 11	Santa Barbara International Film Festival	Santa Barbara, California USA	Top 11 "Best of the Fest" Selection	sbiff.org
May 21–22, Jun 11	Seattle International Film Festival	Seattle, Washington USA		siff.net
Sep 28	Fantastic Fest	Austin, Texas USA		FantasticFest.com
Oct 9	London Festival of Science Fiction Film	London, England UK	Closing Night Film	Sci-Fi London
Oct 9, Oct 11	Sitges Film Festival	Sitges, Catalonia Spain		Sitges Festival
Oct 1, Oct 15	Gwacheon International SF Festival	Gwacheon, Gyeonggi-do South Korea		gisf.org
Oct 17, Oct 20	Icon TLV	Tel Aviv, Central Israel		icon.org.il
Oct 23	Toronto After Dark	Toronto, Ontario Canada	Best Special Effects, Best Musical Score	torontoafterdark.com
Nov 11	Les Utopiales	Nantes, Pays de la Loire France		utopiales.org

Fig. 22. TAPAS prediction examples (left: correct, right: wrong). Light blocks show the column selection result and the darker blocks show the cell selection.

References

1. Intelligent document processing market size and forecast. https://www.verifiedmarketresearch.com/product/intelligent-document-processing-market/. Accessed 30 Oct 2022
2. Biten, A.F., Litman, R., Xie, Y., Appalaraju, S., Manmatha, R.: LaTr: layout-aware transformer for scene-text VQA. In: CVPR 2022 (2022)
3. Biten, A.F., et al.: Scene text visual question answering. In: Proceedings of the IEEE/CVF International Conference on Computer Vision, pp. 4291–4301 (2019)

4. Borchmann, Ł., et al.: Due: end-to-end document understanding benchmark. In: Thirty-Fifth Conference on Neural Information Processing Systems Datasets and Benchmarks Track (Round 2) (2021)
5. Brown, T., et al.: Language models are few-shot learners. In: Advances in Neural Information Processing Systems, vol. 33, pp. 1877–1901 (2020)
6. Carion, N., Massa, F., Synnaeve, G., Usunier, N., Kirillov, A., Zagoruyko, S.: End-to-end object detection with transformers. In: Vedaldi, A., Bischof, H., Brox, T., Frahm, J.-M. (eds.) ECCV 2020. LNCS, vol. 12346, pp. 213–229. Springer, Cham (2020). https://doi.org/10.1007/978-3-030-58452-8_13
7. Chen, W., et al.: Tabfact: a large-scale dataset for table-based fact verification. arXiv preprint arXiv:1909.02164 (2019)
8. Chi, Z., Huang, H., Xu, H.-D., Yu, H., Yin, W., Mao, X.-L.: Complicated table structure recognition. arXiv preprint arXiv:1908.04729 (2019)
9. Denkowski, M., Lavie, A.: Meteor universal: language specific translation evaluation for any target language. In: Proceedings of the Ninth Workshop on Statistical Machine Translation, pp. 376–380 (2014)
10. Gao, L., et al.: ICDAR 2019 competition on table detection and recognition (CTDAR). In: 2019 International Conference on Document Analysis and Recognition (ICDAR), pp. 1510–1515. IEEE (2019)
11. Göbel, M., Hassan, T., Oro, E., Orsi, G.: ICDAR 2013 table competition. In: 2013 12th International Conference on Document Analysis and Recognition, pp. 1449–1453. IEEE (2013)
12. Goyal, Y., Khot, T., Summers-Stay, D., Batra, D., Parikh, D.: Making the V in VQA matter: elevating the role of image understanding in visual question answering. In: Proceedings of the IEEE Conference on Computer Vision and Pattern Recognition, pp. 6904–6913 (2017)
13. He, K., Zhang, X., Ren, S., Sun, J.: Deep residual learning for image recognition. In: Proceedings of the IEEE Conference on Computer Vision and Pattern Recognition, pp. 770–778 (2016)
14. Herzig, J., Nowak, P.K., Müller, T., Piccinno, F., Eisenschlos, J.M.: Tapas: weakly supervised table parsing via pre-training. arXiv preprint arXiv:2004.02349 (2020)
15. Hudson, D.A., Manning, C.D.: GQA: a new dataset for real-world visual reasoning and compositional question answering. In: Conference on Computer Vision and Pattern Recognition (CVPR) (2019)
16. Iandola, F., Moskewicz, M., Karayev, S., Girshick, R., Darrell, T., Keutzer, K.: Densenet: implementing efficient convnet descriptor pyramids. arXiv preprint arXiv:1404.1869 (2014)
17. Johnson, J., Hariharan, B., Van Der Maaten, L., Fei-Fei, L., Lawrence Zitnick, C., Girshick, R.: CLEVR: a diagnostic dataset for compositional language and elementary visual reasoning. In: CVPR (2017)
18. Kafle, K., Price, B., Cohen, S., Kanan, C.: DVQA: understanding data visualizations via question answering. In: Proceedings of the IEEE Conference on Computer Vision and Pattern Recognition, pp. 5648–5656 (2018)
19. Kembhavi, A., Seo, M., Schwenk, D., Choi, J., Farhadi, A., Hajishirzi, H.: Are you smarter than a sixth grader? Textbook question answering for multimodal machine comprehension. In: Proceedings of the IEEE Conference on Computer Vision and Pattern Recognition, pp. 4999–5007 (2017)
20. Li, M., Cui, L., Huang, S., Wei, F., Zhou, M., Li, Z.: Tablebank: a benchmark dataset for table detection and recognition. arXiv preprint arXiv:1903.01949 (2019)
21. Lin, C.-Y.: Rouge: a package for automatic evaluation of summaries. In: Text Summarization Branches Out, pp. 74–81 (2004)
22. Lin, T.-Y., et al.: Microsoft COCO: common objects in context. In: Fleet, D., Pajdla, T., Schiele, B., Tuytelaars, T. (eds.) ECCV 2014. LNCS, vol. 8693, pp. 740–755. Springer, Cham (2014). https://doi.org/10.1007/978-3-319-10602-1_48

23. Manmadhan, S., Kovoor, B.C.: Visual question answering: a state-of-the-art review. Artif. Intell. Rev. **53**(8), 5705–5745 (2020)
24. Mathew, M., Bagal, V., Tito, R., Karatzas, D., Valveny, E., Jawahar, C.V.: Infographicvqa. In: Proceedings of the IEEE/CVF Winter Conference on Applications of Computer Vision, pp. 1697–1706 (2022)
25. Mathew, M., Tito, R., Karatzas, D., Manmatha, R., Jawahar, C.V.: Document visual question answering challenge 2020 (2020)
26. Mishra, A., Shekhar, S., Singh, A.K., Chakraborty, A.: OCR-VQA: visual question answering by reading text in images. In: 2019 International Conference on Document Analysis and Recognition (ICDAR), pp. 947–952. IEEE (2019)
27. Monroe, W., Wang, Y.: Dependency parsing features for semantic parsing (2014)
28. Paliwal, S.S., Vishwanath, D., Rahul, R., Sharma, M., Vig, L.: Tablenet: deep learning model for end-to-end table detection and tabular data extraction from scanned document images. In: 2019 International Conference on Document Analysis and Recognition (ICDAR), pp. 128–133. IEEE (2019)
29. Pan, F., Canim, M., Glass, M., Gliozzo, A., Fox, P.: CLTR: an end-to-end, transformer-based system for cell level table retrieval and table question answering. arXiv preprint arXiv:2106.04441 (2021)
30. Papineni, K., Roukos, S., Ward, T., Zhu, W.-J.: Bleu: a method for automatic evaluation of machine translation. In: Proceedings of the 40th Annual Meeting of the Association for Computational Linguistics, pp. 311–318 (2002)
31. Pasupat, P., Liang, P.: Compositional semantic parsing on semi-structured tables. arXiv preprint arXiv:1508.00305 (2015)
32. Prasad, D., Gadpal, A., Kapadni, K., Visave, M., Sultanpure, K.: Cascadetabnet: an approach for end to end table detection and structure recognition from image-based documents (2020)
33. Raffel, C., et al.: Exploring the limits of transfer learning with a unified text-to-text transformer. J. Mach. Learn. Res. **21**(140), 1–67 (2020)
34. Robertson, S., Walker, S., Jones, S., Hancock-Beaulieu, M.M., Gatford, M.: Okapi at TREC-3. In: Overview of the Third Text REtrieval Conference (TREC-3), pp. 109–126. NIST, Gaithersburg, MD (1995)
35. Shi, H., Gao, S., Tian, Y., Chen, X., Zhao, J.: Learning bounded context-free-grammar via LSTM and the transformer: difference and the explanations. In: Proceedings of the AAAI Conference on Artificial Intelligence, vol. 36, pp. 8267–8276 (2022)
36. Shi, H., Gu, Y., Zhou, Y., Zhao, B., Gao, S., Zhao, J.: Every preference changes differently: neural multi-interest preference model with temporal dynamics for recommendation. arXiv preprint arXiv:2207.06652 (2022)
37. Shi, H., Zhang, Y.: Deep symbolic superoptimization without human knowledge. In: ICLR 2020 (2020)
38. Simonyan, K., Zisserman, A.: Very deep convolutional networks for large-scale image recognition. arXiv preprint arXiv:1409.1556 (2014)
39. Singh, A., et al.: Towards VQA models that can read. In: Proceedings of the IEEE/CVF Conference on Computer Vision and Pattern Recognition (CVPR) (2019)
40. Smock, B., Pesala, R., Abraham, R.: PubTables-1M: towards comprehensive table extraction from unstructured documents. arXiv preprint arXiv:2110.00061 (2021)
41. Tanaka, R., Nishida, K., Yoshida, S.: Visualmrc: machine reading comprehension on document images. In: Proceedings of the AAAI Conference on Artificial Intelligence, vol. 35, pp. 13878–13888 (2021)
42. Vedantam, R., Lawrence Zitnick, C., Parikh, D.: Cider: consensus-based image description evaluation. In: Proceedings of the IEEE Conference on Computer Vision and Pattern Recognition, pp. 4566–4575 (2015)

43. Wang, J., et al.: Deep high-resolution representation learning for visual recognition. IEEE Trans. Pattern Anal. Mach. Intell. **43**(10), 3349–3364 (2020)
44. Wang, X., et al.: On the general value of evidence, and bilingual scene-text visual question answering. In: Proceedings of the IEEE/CVF Conference on Computer Vision and Pattern Recognition (CVPR) (2020)
45. Wei, J., et al.: Chain of thought prompting elicits reasoning in large language models. arXiv preprint arXiv:2201.11903 (2022)
46. Zhong, V., Xiong, C., Socher, R.: Seq2sql: generating structured queries from natural language using reinforcement learning. arXiv preprint arXiv:1709.00103 (2017)
47. Zhong, X., ShafieiBavani, E., Jimeno Yepes, A.: Image-based table recognition: data, model, and evaluation. arXiv preprint arXiv:1911.10683 (2019)

DTSM: Toward Dense Table Structure Recognition with Text Query Encoder and Adjacent Feature Aggregator

Xinhong Chen, Bangdong Chen, Chenfan Qu, Dezhi Peng, Chongyu Liu, and Lianwen Jin(✉)

South China University of Technology, Guangzhou, China
202221012605@mail.scut.edu.cn, eelwjin@scut.edu.cn

Abstract. In recent years, significant progress has been made in table structure recognition, yet the recognition of structures within dense tables remains a challenge that has been largely overlooked. To address this gap, we introduce DenseTab, a new dataset consisting of 16,575 dense tables with comprehensive annotation information that includes physical position, logical position, and text content of each cell, along with the HTML sequence. To tackle the challenge of dense table structure recognition, we propose a new method called **D**ense **T**able **S**plitting and **M**erging Model (DTSM). DTSM includes a novel text query encoder to leverage layout information associated with the text's location, and an adjacent feature aggregator to enhance the prediction of cell merging information. Experimental results demonstrate that our proposed method achieves state-of-the-art performance in recognizing dense table structures. The dataset and code is available at https://github.com/TenMilesLotus/DTSM.

Keywords: Table Structure Recognition · Dense Table · Dataset

1 Introduction

Tables are commonly used in our daily life to record and summarize important data for quick and better visualization of information [10,17,38]. As information technology advances, the number of tables requiring processing by individuals has substantially increased. Particularly, dense tables that contain numerous cells pose significant challenges including the intricate task of precisely detecting each individual cell, predicting complex inter-cell relationships, and dealing with a substantial number of spanning cells. Therefore, it is highly demanded to automatically extract and parse table structures from dense tables.

Given a table image, Table Structure Recognition (TSR) aims at extracting the structure, locating their cells, and obtaining the row-column information in the image. Previously, this problem is studied as TSR focusing on

---

Bangdong Chen: Equal Contribution.

E. H. Barney Smith et al. (Eds.): ICDAR 2024, LNCS 14804, pp. 438–452, 2024.
https://doi.org/10.1007/978-3-031-70533-5_25

regular tables, which typically consist of a limited number of rows, columns, and cells. Early pioneering works, e.g. [4,11], tackled the TSR problem in a bottom-up manner by heuristically grouping detected cells based on low-level cues (e.g., lines, boundaries, and word regions). Recently, deep learning-based approaches [1,2,10,12,14,17–19,21–25,27,28,31,32,36,38,39] are presented to avoid the heuristic grouping scheme design and resort to developing end-to-end models. However, limited by the training datasets [3,6,7,13,20,40] used for TSR, they still addressed this problem under the assumption of a low number of cells in tables. For a practical requirement of recognizing table structures from dense tables, the existing state-of-the-art (SOTA) approaches [10,17–19,22,27,36,39] are prone to fail as the commonly used assumption of tables no longer holds. Specifically, the tables in the widely used datasets (e.g., ICDAR- 2013 [7], SciTSR [3]) are usually with a limited number of cells. In addition, the abundance of cells within dense tables poses challenges in precisely determining their exact physical and logical positions. Consequently, existing approaches can only handle TSR problem in a relatively simple scenario.

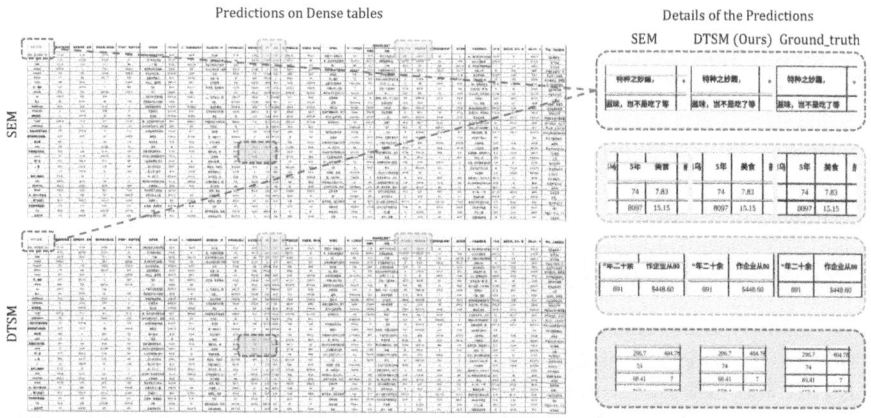

Fig. 1. The visualization comparison between the prediction results of SEM [39] and DTSM. The prediction results for a sample of the DenseTab dataset are presented on the left, while a zoomed-in comparison of the prediction results is shown on the right. Our analysis reveals that the prediction generated by our proposed DTSM method are more precise than those obtained using SEM. In particular, we found that SEM is prone to errors in details when tasked with dense table prediction, whereas DTSM demonstrates superior ability in accurately predicting row separators, column separators and merging information for each grid.

Given the absence of a publicly available benchmark for recognizing dense table structures, our primary objective is to introduce a dataset, termed as DenseTab, to faciliate the research in this area. DenseTab is a large-scale dataset, which comprises a diverse collection of dense tables with varying styles, such as wired tables and wireless tables, many of which contain multiple spanning cells.

Specifically, the DenseTab contains 16,575 images with the annotated information of physical position, logical position, text content of each cell, and the corresponding HTML sequence. DenseTab consists of 11,204 training samples and 5,371 testing samples. As shown in Fig. 1, the tables in the DenseTab dataset are very different from the regular tables, thus posing a new challenging problem to the TSR task. Specifically, existing approaches [39] are demonstrated difficult to recognize dense tables due to the many rows, columns, and cells that are presented in dense tables.

To address the problem of dense table structure recognition (DTSR), we further propose a novel simple-yet-effective approach, termed Dense Table Splitting and Merging Model (DTSM). The DTSM first utilizes a text detector [15] to detect the text in a given table image. Subsequently, it generates a text mask of the table and extracts the feature maps of the table image and text mask individually using visual encoders. To leverage the layout information regarding the text's position within the table, we fuse the two feature maps using a text query encoder. The splitting module employs the fused feature maps to predict the row and column separators, thereby generating the split grid map. To efficiently and effectively manage the vast quantity of cells within dense tables, we employ RoIAlign [8] and self-attention [33] to obtain and enhance the visual attributes of each grid. Moreover, we adopt adjacenct feature aggregator to create a more comprehensive and visually appealing representation. Lastly, to mitigate computational redundancy resulting from the abundance of cells in dense tables, we utilize a fully connected layer to predict the merging information of each grid, as opposed to determining the adjacency between each pair of grids. Our experiments demonstrate that DTSM surpasses existing SOTA TSR methods in recognizing dense table structures.

Our contributions are summarized as follows:

- We identify the challenge of DTSR, which has not received significant attention in the existing literature. Our work sheds light on the importance of DTSR and lays the groundwork for future research in this area.
- We introduce DenseTab, a comprehensive dataset of 16,575 dense tables with detailed annotation information, aiming to provide a valuable benchmark for future research in this area.
- We propose a novel Dense Table Splitting and Merging Model (DTSM), a split-merge based model that leverages a novel text query encoder and adjacent feature aggregator to address the challenge of recognizing dense table structures.
- Experimental results on DenseTab demonstrate that our method significantly outperforms the existing methods on recognizing dense table structures.

2   Related Works

2.1   Table Structure Recognition Datasets

The ICDAR-2013 dataset [7] is an early dataset for Table Structure Recognition (TSR), comprising a mere 156 tables. Later on, the SciTSR dataset [3] was introduced and gained widespread usage as a benchmark. PubTabNet [40] emerged as the first large-scale TSR dataset, featuring 568k tables accompanied by HTML annotations. However, the aforementioned datasets solely encompass flat document tables. To offer a more varied range of table data found in real-world scenarios, the WTW dataset [20] was introduced. Despite the existence of numerous public TSR datasets, they are all restricted to regular tables, with only a sparse representation of dense tables. In order to address this limitation for dense table structure recognition (DTSR), we present DenseTab, a dataset comprising 16,575 dense tables. This dataset aims to facilitate research on DTSR.

2.2   Table Structure Recognition Methods

The TSR methods before the advent of deep learning [30,34] primarily relied on hand-extracted features and heuristic rules. These methods were limited to recognizing simple tables with finite cells. However, with the emergence of deep learning, deep learning-based methods have become the mainstream approach. These methods are more generalizable and do not require hand-extracted features, enabling them to recognize not only simple tables but also more complex tables that contain cross-cells or are captured in the wild with distortion and skew interference. Deep learning-based methods for TSR can be classified into three categories: detection-based methods [14,18,19,26,36], sequence-based methods [10,12,21,24,27,38,40], and split-merge-based methods [17,23,25,31,39]. Detection-based methods involve detecting cells and then determining the adjacency between the cells. In the structure recognition of dense tables, the prediction of relationships between detected cells becomes highly challenging due to the presence of numerous cells in dense tables. Sequence-based methods transform the TSR problem into an image-to-sequence problem by encoding the table image using a convolutional model and feeding it into an autoregressive decoder to output a HTML sequence. However, attention drift is an inevitable issue that cannot be overlooked when dealing with dense tables, given the substantial quantity of cells they contain. Split-merge-based methods involve the initial partitioning of the table into grids, followed by the merging of these grids to ultimately establish the desired structure of the table. This type of method usually results in more accurate cell bounding boxes. For example, TSR-Former [17] and RobusTabNet [23] effectively predicted the relationship between grid pairs during the merge phase, yielding satisfactory performance in regular tables. However, the computational redundancy arises due to the extensive number of cell pairs in dense tables. In contrast, the SPLERGE model [31] employs heuristic rules for mesh merging information, eliminating the need for additional training. Nonetheless, this approach often produces unsatisfactory results when dealing with dense tables. On the other hand, the SEM model [39] incorporates GRU for sequential decoding in the merging stage, aligning more closely with human thinking patterns. Nevertheless, the inference speed experiences a substantial decline when dealing with a high volume of cells in dense tables.

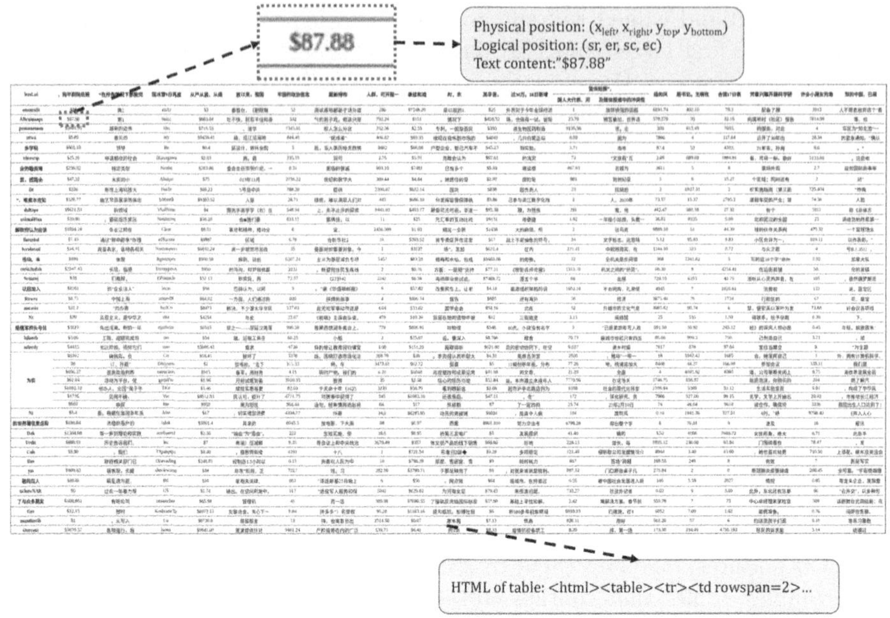

Physical position: $(x_{left}, x_{right}, y_{top}, y_{bottom})$
Logical position: (sr, er, sc, ec)
Text content: "$87.88"

$87.88

HTML of table: `<html><table><tr><td rowspan=2>...`

Fig. 2. Each table image is annotated with the physical position, logical position, and text content of each cell, as well as the HTML sequence corresponding to the table. Specifically, the start-row, end-row, start-column, and end-column are denoted as sr, er, sc, and ec, respectively. Zoom in for better view.

3   Problem Definition

For a given dense table image, our objective is to extract the table structure, encompassing both the physical and logical aspects. Specifically, the physical structure consists of a set of bounding boxes $B = \{\mathbf{b}^1, \mathbf{b}^2, ..., \mathbf{b}^n\}$, where each bounding box $\mathbf{b} = (x_{left}, x_{right}, y_{top}, y_{bottom})$, represents the horizontal and vertical coordinates of the top-left and bottom-right corners. On the other hand, the logical structure comprises a set of cells $L = \{\mathbf{l}^1, \mathbf{l}^2, ..., \mathbf{l}^n\}$, where each cell $\mathbf{l} = (sr, er, sc, ec)$, denotes the start-row, end-row, start-colmun and end-column of the cell.

4   The DenseTab Dataset

In this section, we present DenseTab, a dataset consisting exclusively of dense tables. DenseTab particularly aims at the problem of dense table structure recognition that is quite under-explored in existing approaches and benchmarks. Specifically, DenseTab contains 16,575 samples in total. The training set comprises a total of 11,204 samples. The testing set is partitioned into two subsets: testing set A contains 4,796 samples, while testing set B comprises 575 samples.

Table 1. Comparison with public TSR datasets.

Dataset	Format	Wired Table	Wireless Table	Tables	Dense Tables
PubTabNet [40]	Image	✓	✓	568k	560
SciTSR [3]	Image+PDF	✓	✓	15k	29
WTW [20]	Image	✓	✗	14,581	787
FinTabNet [13]	PDF	✓	✓	112k	81
ICDAR-2013 [7]	PDF	✓	✓	156	2
ICDAR-2019 [6]	Image	✓	✓	3.6k	137
TabRecSet [37]	Image	✓	✓	38.17k	9
TableBank [12]	Image	✓	✓	145k	85
PubTables-1M [29]	Image	✓	✓	948k	1,065
SynthTabNet [24]	Image	✓	✓	600k	0
DenseTab (ours)	Image	✓	✓	16,575	16,575

The samples in training set and testing set A are synthetic. To assess the generalization ability of methods, we collect real dense tables from public datasets [3,7,40] and combine them to create the testing set B.

4.1 Dataset Settings

For training set and testing set A, the HTML sequences of the dense tables are generated based on the following three conditions: a cell count greater than 500, a row count greater than 10, and a column count greater than 10. A table is considered dense if it meets all three conditions. To enhance the resemblance of the synthetic tables to real ones, we incorporated spanning cells into the design. Each cell may contain numerical data, English or Chinese language text, or be left empty. The number of rows, columns, spanning cells, and their corresponding number of cells spanned, as well as cell contents are all randomly determined. Additionally, diverse styles are added to DenseTab through the randomized choice between wired and wireless tables with equal probability. After browser rendering, the images of the dense tables are obtained, with automatic labeling accomplished by the script. The label includes the cell's logical location, physical location, text content, and the HTML sequence for the entire table as illustrated in Fig. 2.

For testing set B, we filter dense tables with the same criteria: a cell count greater than 500, a row count greater than 10, and a column count greater than 10. To ensure the diversity of DenseTab, we choose dense tables from PubTabNet [40], SciTSR [3], and ICDAR-2013 [7].

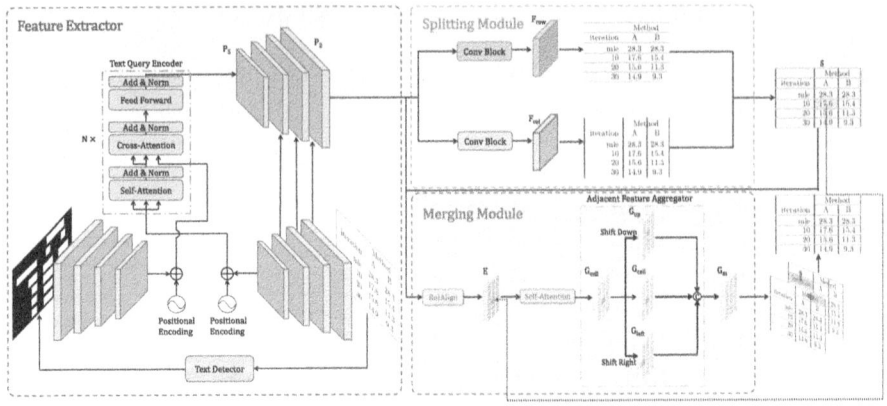

Fig. 3. The architecture of DTSM. Zoom in for better view.

4.2   DenseTab Evaluation

DenseTab stands out from other TSR datasets in that it exclusively contains dense tables. Table 1 provides a comparison of DenseTab with other datasets. Although numerous datasets for TSR exist, they are primarily focused on regular tables and contain only seldom dense tables.

5   Method

Figure 3 illustrates the overall architecture of the proposed model. We propose a novel model called Dense Table Splitting and Merging Model (DTSM) that utilizes a split-merge framework. DTSM comprises of three modules: a feature extractor for extracting visual features, a splitting module for predicting row separators and column separators, and a merging module for predicting grid merging information. Given a table image, the feature extractor extracts feature maps, which are then used by the splitting module to predict row and column separators, resulting in a split grid map. The split grid map and feature map are then passed to the merging module to predict the merging information of grids. To handle a large number of cells in a dense table more efficiently, we use fully connected layers to predict grid merging information instead of sequential decoding [39] or predicting grid pair relationships [17]. This significantly speeds up the model's inference process.

5.1   Feature Extractor

The task of recognizing table structures often relies on layout information embedded within the text. Therefore, we employ DBNet [15] to detect the position of the text and produce the corresponding text mask. Next, we adopt ResNet50-FPN [9,16] as the backbone network for extracting multi-scale feature maps

$\{I_2, I_3, I_4, I_5\}$ from the table images and $\{M_2, M_3, M_4, M_5\}$ from the text masks, and add positional encoding [33] to these features. Drawing inspiration from [35], we propose a text query encoder to fuse the features of I_5 and M_5 to get P_5. Specifically, we add a multi-head cross-attention layer after the self-attention layer. We utilize the image feature I_5 as the key and value, and the text feature M_5 as the query. The text-query cross-attention mechanism can be formulated as:

$$CrossAttention(M_5, I_5, I_5) = softmax(\frac{M_5 I_5^T}{\sqrt{d_k}})I_5, \tag{1}$$

where d_k is the scaling factor. Since text location contains rich layout information, we use the text mask as a query to make the text seek the image features of interest. Furthermore, the model can accurately pay attention to the cells positions of table in the image. The text query encoder is composed of a stack of N blocks, each of which consists of a self-attention layer, a cross-attention layer, and a feed-forward layer. The query of each cross-attention layer is M_5.

To thoroughly leverage the layout information acquired from the text positions, we employ P_5, which is rich in semantic information, to produce feature maps that can be utilized in subsequent modules. Following the approach in FPN [16], we perform up-sampling on P_5 and integrate P_5 with I_2, I_3, and I_4 to generate a feature map P_2 of the same size as the input image.

5.2 Splitting Module

The splitting module receives the feature map P_2 as input and processes it through both the row and column splitting modules. As a result, the row and column splitting modules consist of two 3×3 convolutional layers with strides of 1 and ReLU activations. The resulting row splitting map $F_{row} \in \mathbb{R}^{H \times W \times 1}$ and column splitting map $F_{col} \in \mathbb{R}^{H \times W \times 1}$ both have the same size as the input feature map. We calculate the row separator and column separator by averaging the outputs of the row splitting module and the column splitting module. Subsequently, we combine these separators to obtain the split grid map S.

The loss function for the splitting module is defined as follows:

$$L_s^{row} = \frac{1}{|S_{row}|} \sum_{(x,y) \in S_{row}} BCE(F_{row}(x,y), F_{row}^*(x,y)), \tag{2}$$

$$L_s^{col} = \frac{1}{|S_{col}|} \sum_{(x,y) \in S_{col}} BCE(F_{col}(x,y), F_{col}^*(x,y)), \tag{3}$$

where S_{row} denotes the set of pixels from F_{row}, $F_{row}(x,y)$ and $F_{row}^*(x,y)$ represent the predicted and ground-truth labels for the pixel (x,y) in S_{row} respectively. S_{col}, F_{col}, and F_{col}^* are defined in a similar manner.

5.3 Merging Module

Unlike the methodology employed in [17,39], our objective is to develop a merging module that is uncomplicated, efficient, and accurately emulated human perceptual mechanisms. Our merging module takes as input the feature map P_2 obtained from the CNN backbone and the split grid map S. Firstly, a RoIAlign operation [8] is applied to each grid in the feature map P_2 based on the split grid map S, resulting in a feature representation $E \in \mathbb{R}^{M \times N \times D}$ for all grids. Here, M, N, and D represent the number of rows, columns, and channels for each grid feature, respectively. Next, the self-attention mechanism [33] is utilized to enhance the grid features and outputs $G_{cell} \in \mathbb{R}^{M \times N \times D}$. The merging module anticipates the optimal direction for merging each grid, and we contend that the incorporation of neighboring grid features could enhance the accuracy of merging prediction. Therefore, we adopt the approach of adjacent feature aggregator. This involves the integration of G_{cell} with $G_{up} \in \mathbb{R}^{M \times N \times D}$ and $G_{left} \in \mathbb{R}^{M \times N \times D}$, resulting in the generation of the ultimate grid feature G_m. Specifically, G_{up}, G_{left} and G_m are defined as follows, respectively:

$$G_{up}(i,j) = \begin{cases} 0, & i = 0 \\ G_{cell}(i-1,j), & otherwise \end{cases}, \tag{4}$$

$$G_{left}(i,j) = \begin{cases} 0, & j = 0 \\ G_{cell}(i,j-1), & otherwise \end{cases}, \tag{5}$$

$$G_m = Concatenate(G, G_{up}, G_{left}), \tag{6}$$

where $i \in (0, M-1)$ denotes the row index of the grid feature, while $j \in (0, N-1)$ denotes the column index of the grid feature.

To predict the merging information of each grid, we employ two fully connected layers that take in the grid features G_m. The first layer has a ReLU activation layer, while the second layer uses a Softmax activation function to determine a grid's merging information. Specifically, merged upwards, merged leftwards , merged both upwards and leftwards or not merged. Notably, our merging module does not rely on an autoregressive approach or need to caculate a large number of adjacent relationships between every pair of grids. As a result, our merging module exhibits excellent performance and inference speed.

The loss function for the merging module is defined as follows:

$$L_{merge} = \frac{1}{K} \sum_{i=1}^{K} CrossEntropy(T_i, T_i^*), \tag{7}$$

where T_i denotes the predicted label of the i-th grid, T_i^* denotes the ground-truth label of the i-th grid, and K denotes the total number of grids.

6 Experiments

6.1 Evaluation Metrics

To ensure a thorough and unbiased evaluation of the performance of each TSR method on dense tables, we employ Precision, Recall, and F1-score [7] , as well as TEDS-S [40] as evaluation metrics. Precision, Recall, and F1-score are evaluation

Table 2. Results on DenseTab. P, R, F1 represent Precision, Recall, F1-Score, respectively. TEDS-S only considers the table structures. **Bold** denotes SOTA, while <u>Underline</u> indicates the second best.

Method	Venue	Testing set A				Testing set B				
		P	R	F1	TEDS-S	P	R	F1	TEDS-S	FPS
SPLERGE [31]	ICDAR 2019	**97.00**	45.27	57.28	46.98	<u>89.68</u>	1.58	2.96	2.96	0.67
TGRNet [36]	ICCV 2021	87.40	65.41	74.44	70.75	77.08	42.88	54.03	28.21	0.85
TableMaster [38]	arXiv 2021	75.49	80.71	78.00	87.19	33.31	**99.54**	49.42	52.12	0.02
Multitype-TD-TSR [5]	KI 2021	93.21	87.27	89.69	84.81	75.96	51.87	55.10	<u>55.73</u>	**1.43**
SEM [39]	PR 2022	95.38	<u>93.94</u>	<u>94.65</u>	<u>94.92</u>	**91.15**	67.90	<u>77.82</u>	55.70	0.22
DTSM (ours)	ICDAR 2024	<u>96.59</u>	**94.41**	**95.49**	**95.74**	81.63	<u>77.13</u>	**79.32**	**64.47**	<u>1.12</u>

metrics utilized in ICDAR-2013, which are grounded on both cell text content and logical position, and can effectively reflect the TSR method's capability to predict the table structures. The Tree Edit Distance Similarity (TEDS) is a metric that assesses the similarity of the HTML sequence that represents a table. On the other hand, TEDS-S is a variation of TEDS that disregards the textual content of the cells and only evaluates the accuracy of the table structure prediction.

6.2   Implementation Details

In our DTSM, the feature dimension of all attention layers is set to 256, with 8 heads for each layers. We set the channel number of each grid feature in merging module to 256. During the training of the DTSM model on the DenseTab dataset, we employ the Adam optimizer with an initial learning rate of 0.0001, a cosine annealing strategy to adjust the learning rate for 10 epochs with a batch size of 4. Prior to inputting the image into the model, we scale each table image to the original image scale, with the long side fixed at 500.

6.3   Comparison with SOTAs

Table 2 presents the performance and inference speed of DTSM compared to other methods on DenseTab. For a fair comparison, those methods are retrained on the DenseTab dataset with the author-provided hyperparameter settings. DTSM achieve a F1-score of 95.49%, a TEDS-S of 95.74%, a FPS of 1.12 on the testing set A, and a F1-score of 79.32%, a TEDS-S of 64.47% on testing B. To assess the inference time, an experimental evaluation is carried out by utilizing tables from the testing set A on a Nvidia Titan Xp GPU with a batch size of 1, and subsequently calculating the mean FPS. With the integration of our proposed text query encoder and adjacent feature aggregator, which help model to perceive richer visual information, DTSM outperforms the previous SOTA method [39] and achieves a stunning 5x faster inference speed, significantly enhancing its real-world application value. Split-merge based methods [31,39]

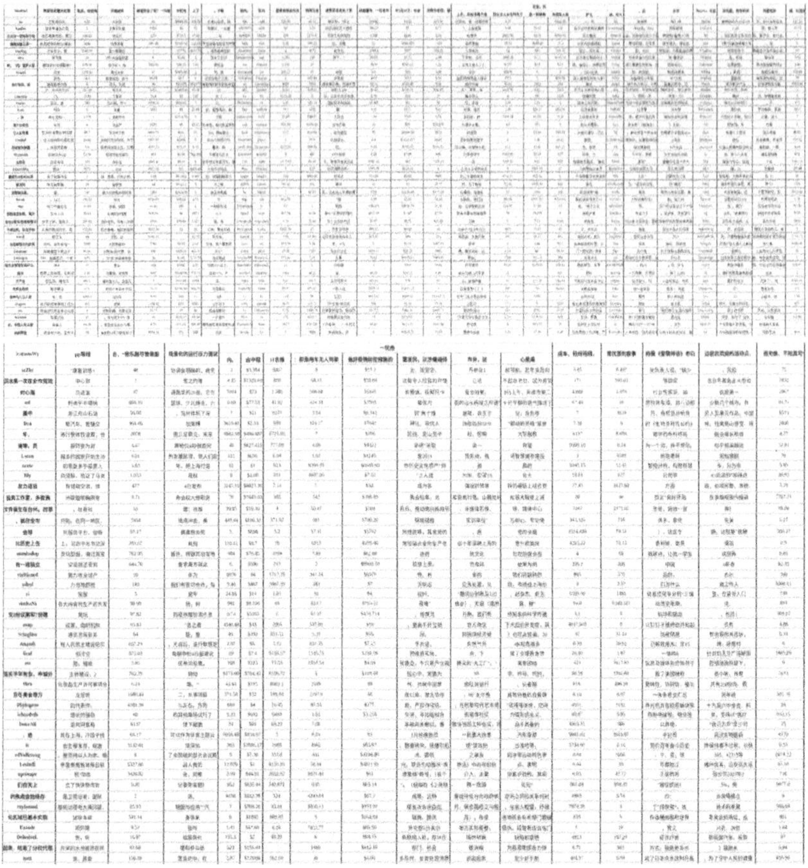

(a) Images from testing set A

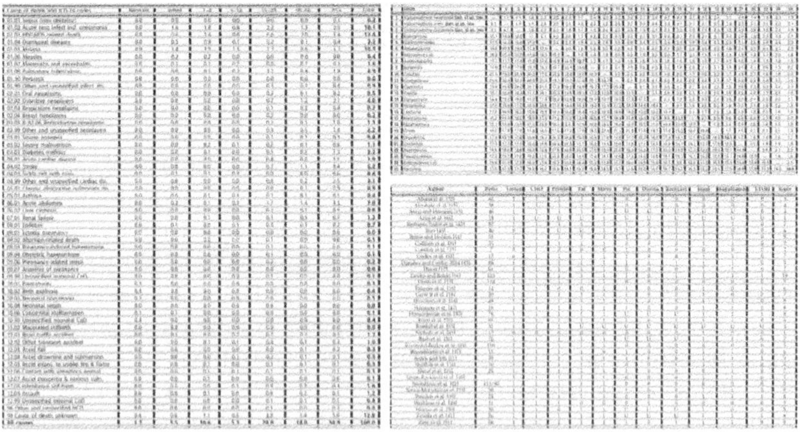

(b) Images from testing set B

Fig. 4. Illustration of dense table structure recognition results produced by DTSM. Zoom in for better view.

Table 3. This section presents the findings of ablation experiments conducted on the design of the text query encoder and the adjacent feature aggregator. The text query encoder and adjacent feature aggregator are denoted as TQE and AFA, respectively.

TQE	AFA	P	R	F1
✓		94.37	91.89	93.12
	✓	95.73	94.39	95.05
✓	✓	**96.59**	**94.41**	**95.49**

demonstrate strengths in addressing structure recognition in dense tables, with SPLERGE [31] achieving a high precision and SEM [39] having a better F1-score than other previous methods. However, these methods failed to adequately address the intricacies surrounding inter-cell adjacency relationships presented in dense tables containing a significant number of cells, making them less effective and efficient. The proposed DTSM effectively reduces the risk of over-fitting by text query encoding, aggregating adjacent table grid features and greatly improves inference speed by parallel decoding.

Figure 4 demonstrates qualitative results of structure recognition on DenseTab. The figure illustrates that DTSM effectively handles dense tables by accurately predicting the row and column separators of table as well as the merging information of each grid.

6.4   Ablation Study

The Text Query Encoder (TQE) is developed to capture the layout information contained in the location of text. The proposed Adjacent Feature Aggregator (AFA) involves comparing features in adjacent grids to enhance the quality of visual information. To evaluate the effectiveness of TQE and AFA, ablation experiments are conducted on the testing set A of DenseTab, with the results presented in Table 3. We evaluate the effectiveness of TQE by conducting experiments where we replace each cross-attention layer in TQE with a self-attention layer. In other words, we substitute the query in each cross-attention layer, originally sourced from the feature map of the text mask, with the feature map of the table image. It can be seen that TQE alone leads to an improvement of approximately 0.44% in F1-score, providing evidence that it aids in the enhanced detection of table row and column separators. To assess the effectiveness of the AFA, we conducted experiments wherein the AFA was not introduced. Instead, we directly employed the grid features acquired from self-attention to predict grid merging information. The results demonstrate a 1.93% increase in F1-score due to the AFA module, highlighting its crucial role in accurately predicting the merging information of grids.

7   Conclusion

This paper introduces a novel problem concerning the structure recognition of dense tables. To mitigate the lack of comprehensive dense table dataset, we introduce a large-scale dataset called DenseTab, which consists of 16,575 dense tables. Furthermore, we propose a novel approach named DTSM for the recognition of dense table structures. DTSM consists of three main modules: a feature extractor for extracting features from images, a splitting module responsible for predicting row/column separators, and a merging module designed to predict grid merging information. To capture the layout information embedded within text locations, we introduce a text query encoder. This module enhances the model's ability to acquire more detailed visual information, leading to improved accuracy of the subsequent splitting and merging modules. Additionally, we incorporate an adjacent feature aggregator into the merging module to simulate human cognitive processes, thus improving the visual perception of grids and enhancing accuracy in grid merging. Experimental results demonstrate that our proposed method surpasses existing approaches, achieving SOTA performance with significantly faster inference speed. Our work contributes to advancing the research on dense table structure recognition, and we hope that this will inspire new research topics and facilitate the development of more effective methods for handling dense tables.

Acknowledgments. This research is supported in part by National Natural Science Foundation of China (Grant No.: 62441604, 61936003).

References

1. Chen, B., Peng, D., Zhang, J., Ren, Y., Jin, L.: Complex table structure recognition in the wild using transformer and identity matrix-based augmentation. In: International Conference on Frontiers in Handwriting Recognition, pp. 545–561. Springer (2022). https://doi.org/10.1007/978-3-031-21648-0_37
2. Chen, L., Huang, C., Zheng, X., Lin, J., Huang, X.J.: TableVLM: multi-modal pre-training for table structure recognition. In: Annual Meeting of the Association for Computational Linguistics, pp. 2437–2449 (2023)
3. Chi, Z., Huang, H., Xu, H.D., Yu, H., Yin, W., Mao, X.L.: Complicated table structure recognition. arXiv preprint arXiv:1908.04729 (2019)
4. Doermann, D., Tombre, K. (eds.): Handbook of Document Image Processing and Recognition. Springer, London (2014). https://doi.org/10.1007/978-0-85729-859-1
5. Fischer, P., Smajic, A., Abrami, G., Mehler, A.: Multi-type-td-TSR–extracting tables from document images using a multi-stage pipeline for table detection and table structure recognition: from OCR to structured table representations. In: Advances in Artificial Intelligence, pp. 95–108 (2021)
6. Gao, L., et al.: ICDAR 2019 competition on table detection and recognition (cTDaR). In: International Conference on Document Analysis and Recognition, pp. 1510–1515 (2019)

7. Göbel, M., Hassan, T., Oro, E., Orsi, G.: ICDAR 2013 table competition. In: International Conference on Document Analysis and Recognition, pp. 1449–1453 (2013)

8. He, K., Gkioxari, G., Dollár, P., Girshick, R.: Mask R-CNN. In: IEEE International Conference on Computer Vision, pp. 2961–2969 (2017)

9. He, K., Zhang, X., Ren, S., Sun, J.: Deep residual learning for image recognition. In: IEEE Conference on Computer Vision and Pattern Recognition, pp. 770–778 (2016)

10. Huang, Y., et al.: Improving table structure recognition with visual-alignment sequential coordinate modeling. In: IEEE Conference on Computer Vision and Pattern Recognition, pp. 11134–11143 (2023)

11. Itonori, K.: Table structure recognition based on textblock arrangement and ruled line position. In: International Conference on Document Analysis and Recognition, pp. 765–768 (1993)

12. Li, M., Cui, L., Huang, S., Wei, F., Zhou, M., Li, Z.: TableBank: table benchmark for image-based table detection and recognition. In: Twelfth Language Resources and Evaluation Conference, pp. 1918–1925 (2020)

13. Li, Y., Huang, Z., Yan, J., Zhou, Y., Ye, F., Liu, X.: GFTE: graph-based financial table extraction. In: International Conference on Pattern Recognition, pp. 644–658 (2021)

14. Li, Z., Peng, F., Xue, Y., Hao, N., Jin, L.: Scene table structure recognition with segmentation and key point collaboration. In: International Conference on Document Analysis and Recognition, pp. 295–310. Springer (2023). https://doi.org/10.1007/978-3-031-41679-8_17

15. Liao, M., Wan, Z., Yao, C., Chen, K., Bai, X.: Real-time scene text detection with differentiable binarization. In: AAAI Conference on Artificial Intelligence. vol. 34, pp. 11474–11481 (2020)

16. Lin, T.Y., Dollár, P., Girshick, R., He, K., Hariharan, B., Belongie, S.: Feature pyramid networks for object detection. In: IEEE Conference on Computer Vision and Pattern Recognition, pp. 2117–2125 (2017)

17. Lin, W., Sun, Z., Ma, C., Li, M., Wang, J., Sun, L., Huo, Q.: TSRFormer: table structure recognition with transformers. In: ACM International Conference on Multimedia, pp. 6473–6482 (2022)

18. Liu, H., Li, X., Liu, B., Jiang, D., Liu, Y., Ren, B.: Neural collaborative graph machines for table structure recognition. In: IEEE Conference on Computer Vision and Pattern Recognition, pp. 4533–4542 (2022)

19. Liu, H., et al.: Show, read and reason: table structure recognition with flexible context aggregator. In: ACM International Conference on Multimedia, pp. 1084–1092 (2021)

20. Long, R., et al.: Parsing table structures in the wild. In: IEEE International Conference on Computer Vision, pp. 944–952 (2021)

21. Ly, N.T., Takasu, A.: An end-to-end local attention based model for table recognition. In: International Conference on Document Analysis and Recognition, pp. 20–36. Springer (2023). https://doi.org/10.1007/978-3-031-41679-8_2

22. Lyu, P., et al.: GridFormer: towards accurate table structure recognition via grid prediction. In: ACM International Conference on Multimedia, pp. 7747–7757 (2023)

23. Ma, C., Lin, W., Sun, L., Huo, Q.: Robust table detection and structure recognition from heterogeneous document images. Pattern Recogn. **133**, 109006 (2023)

24. Nassar, A., Livathinos, N., Lysak, M., Staar, P.: TableFormer: table structure understanding with transformers. In: IEEE Conference on Computer Vision and Pattern Recognition, pp. 4614–4623 (2022)
25. Nguyen, N.Q., Le, A.D., Lu, A.K., Mai, X.T., Tran, T.A.: Formerge: recover spanning cells in complex table structure using transformer network. In: International Conference on Document Analysis and Recognition, pp. 522–534. Springer (2023). https://doi.org/10.1007/978-3-031-41734-4_32
26. Qiao, L., et al.: LGPMA: complicated table structure recognition with local and global pyramid mask alignment. In: International Conference on Document Analysis and Recognition, pp. 99–114 (2021)
27. Shen, H., et al.: Divide rows and conquer cells: towards structure recognition for large tables. In: International Joint Conference on Artificial Intelligence, pp. 1369–1377 (2023)
28. Siddiqui, S.A., Fateh, I.A., Rizvi, S.T.R., Dengel, A., Ahmed, S.: Deeptabstr: deep learning based table structure recognition. In: International Conference on Document Analysis and Recognition, pp. 1403–1409 (2019)
29. Smock, B., Pesala, R., Abraham, R.: Pubtables-1m: towards comprehensive table extraction from unstructured documents. In: IEEE Conference on Computer Vision and Pattern Recognition, pp. 4634–4642 (2022)
30. Tengli, A., Yang, Y., Ma, N.L.: Learning table extraction from examples. In: International Conference on Computational Linguistics, pp. 987–993 (2004)
31. Tensmeyer, C., Morariu, V.I., Price, B., Cohen, S., Martinez, T.: Deep splitting and merging for table structure decomposition. In: International Conference on Document Analysis and Recognition, pp. 114–121 (2019)
32. Umer, M., Mohsin, M.A., Ul-Hasan, A., Shafait, F.: PyramidTabNet: transformer-based table recognition in image-based documents. In: International Conference on Document Analysis and Recognition, pp. 420–437. Springer (2023). https://doi.org/10.1007/978-3-031-41734-4_26
33. Vaswani, A., et al.: Attention is all you need. In: International Conference on Neural Information Processing Systems, pp. 6000–6010 (2017)
34. Wang, Y., Phillips, I.T., Haralick, R.M.: Table structure understanding and its performance evaluation. Pattern Recogn. **37**(7), 1479–1497 (2004)
35. Xie, X., Fu, L., Zhang, Z., Wang, Z., Bai, X.: Toward understanding wordart: corner-guided transformer for scene text recognition. In: European Conference on Computer Vision, pp. 303–321. Springer (2022)
36. Xue, W., Yu, B., Wang, W., Tao, D., Li, Q.: TGRNet: a table graph reconstruction network for table structure recognition. In: IEEE International Conference on Computer Vision, pp. 1295–1304 (2021)
37. Yang, F., Hu, L., Liu, X., Huang, S., Gu, Z.: A large-scale dataset for end-to-end table recognition in the wild. Sci. Data **10**(1), 110 (2023)
38. Ye, J., .: PingAn-VCGroup's solution for ICDAR 2021 competition on scientific literature parsing task B: table recognition to HTML. arXiv preprint arXiv:2105.01848 (2021)
39. Zhang, Z., Zhang, J., Du, J., Wang, F.: Split, embed and merge: an accurate table structure recognizer. Pattern Recogn. **126**, 108565 (2022)
40. Zhong, X., ShafieiBavani, E., Jimeno Yepes, A.: Image-based table recognition: data, model, and evaluation. In: European Conference on Computer Vision, pp. 564–580 (2020)

ChartReformer: Natural Language-Driven Chart Image Editing

Pengyu Yan$^{(\boxtimes)}$, Mahesh Bhosale , Jay Lal , Bikhyat Adhikari,
and David Doermann

University at Buffalo, Buffalo, NY 14228, USA
{pyan4,mbhosale,jayashok,bikhyata,doermann}@buffalo.edu

Abstract. Chart visualizations are essential for data interpretation and communication; however, most charts are only accessible in image format and lack the corresponding data tables and supplementary information, making it difficult to alter their appearance for different scenarios of application. To eliminate the need for original underlying data and information to perform chart editing, we propose ChartReformer, a natural language-driven chart image editing solution that directly edits the charts from the input images with the given instruction prompts. Instead of predicting the plotting code, the key in this method is that we allow the model to comprehend the chart and reason over the prompt to generate the corresponding underlying data table and visual attributes for new charts, enabling a precise and stable editing result. To generalize ChartReformer, we define and standardize the chart editing category and generate the ChartCraft dataset, covering style, layout, format, and data-centric edits. The experiments show promising results for the natural language-driven chart image editing. Our datasets and model are available at: https://github.com/pengyu965/ChartReformer.

Keywords: Chart Editing · Chart Appearance Editing · Chart Data Extraction · Chart Understanding · Visual Language Model

1 Introduction

Charts are designed with specific aesthetics and formats to effectively visualize tabular data. However, a given visualization may only be ideal for a specific scenario or purpose. Modifying chart images would allow them to be adapted for diverse applications by enabling the highlighting of specific data segments, amplifying the distinctions between data points, converting charts into different formats, or editing the appearance of the graph style. These are significant and can enhance accessibility for readers.

In the evolving landscape of data analysis, charts and graphs play an indispensable role in deciphering complex datasets and facilitating informed decision-making. The ability to effectively visualize tabular data through charts is crucial, yet the specificity of a visualization's design to its initial context limits adaptability for broader applications. This necessitates the development of methods

E. H. Barney Smith et al. (Eds.): ICDAR 2024, LNCS 14804, pp. 453–469, 2024.
https://doi.org/10.1007/978-3-031-70533-5_26

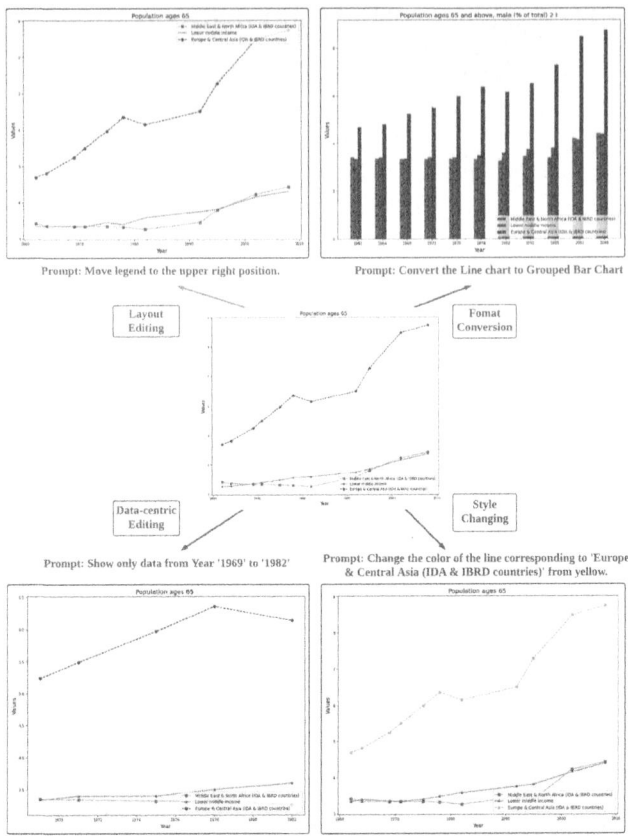

Fig. 1. Examples of chart editing results from our methods. In total, our methods define and cover four types of chart editing: style, layout, format and data-centric edit.

to modify chart images, enabling them to highlight particular data segments, enhance distinctions between data points, or improve accessibility for diverse readerships. However, traditional chart-editing methods are fraught with challenges. These processes often require significant manual intervention, a deep understanding of the plotting software's parameters, and access to the original data tables. These limitations become particularly acute in scenarios where source data are lost or unavailable, highlighting the need for more flexible and accessible editing techniques (Fig. 1).

Recent advances in computer vision and natural language processing (NLP) have opened new avenues for understanding charts. As multi-modal tasks, chart understanding related research topics such as data extraction, question answering, and chart summarization are tackled by utilizing visual language model in [2,5,10,11,15]. However, for chart editing tasks, [12] still rely on input visualization code and resource data table, while ChartLlama [5] fails to cover the full spectrum of possible edits, such as data manipulation for input charts. To close this gap, we introduce ChartReformer, which edits chart images from natural language prompts without any prior knowledge of the underlying data and original plot settings. Training on our dataset allows it to cover a wide range of edits from style, format, layout to data-centric edits. In our methods, we decompose the input charts and reasoning over the prompts for a new corresponding data table and visual attributes, allowing for detailed, comprehensive, and accurate chart editing. This approach predicting and adjusting the embedded visual attributes and data under original chart images enables the creation of customized chart images through a re-plotter without explicitly predicting the plotting code, producing and delivering robust and stable chart editing results.

Overall the main contributions of our work can be summarized as follows:

- The first work thoroughly discusses the chart editing tasks. Define and standardize the editing category. Provide a detailed taxonomy of the edits. Suitable evaluation metrics are designed for such tasks.
- Provide ChartCraft datasets that span major edit categories, including style, format, layout, and data-centric edits. The datasets contain 100K pairs of original and edited chart images with corresponding underlying data tables, visual attributes, and instruction prompts.
- Present ChartReformer pipeline with a visual language model trained on our dataset from deplot's checkpoint, and empirically demonstrate the effectiveness of our system in experiments.

2 Related Work

2.1 Natural-Language-Driven Visualization

The intersection of Natural Language Processing (NLP) and data visualization within the field of human-computer interaction (HCI) has become increasingly prominent [20]. This surge in interest, especially within the deep learning community, is driven by advances in natural language understanding [17]. Tools such as VegaLite [18] and ChartDialogs [19] demonstrate the ability to generate and adjust visual charts in natural language, the former utilizing JSON for chart specifications and the latter applying Seq2Seq models for editing in natural language dialogues. Similarly, VizGpt(https://vizgpt.ai/) employs GPT models to react with human language instructions for visualization and styling. These approaches are practical, yet different from ours; they rely on available underlying tabular data and do not directly alter visualized images.

2.2 Chart Comprehension

Datasets. Editing charts is a nuanced task that necessitates a grasp of both the chart's visual features and the data it represents. Contemporary multimodal models such as GPT-4V [23] and LLaVA-1.5 face challenges in analyzing and extracting the underlying data of the charts [4], while chart manipulation is even more difficult. To facilitate the model in understanding the charts, several datasets have been introduced. Some assess understanding via straightforward question-and-answer formats with human-annotated QA pairs - for example, ChartQA [14], or utilize templates from crowd-sourced platforms - like PlotQA [16], or employ synthetic examples created by Large Language Models (LLMs) - as seen in ChartLlama [5]. Other dataset categories, such as chart-to-text [7], measure comprehension through summarization. However, to our knowledge, there is no publicly available datasets for chart editing.

Models. Many existing methods [1,8,13,22] analyze the component of the charts and extract the underlying data by relating the results of component detection. This routine relies heavily on the intermediate results and is potentially vulnerable. Utilizing the large visual language model, direct understanding and reasoning can be performed on input charts without the need for intermediate steps. Pix2Struct [9] is a such model on similiar task that extracts visually-situated language from web page screenshot into structured text. Then Matcha [11] adapts Pix2Struct for chart reasoning by pre-training the model on plot deconstruction and numerical reasoning. Deplot [10] fine-tune the matcha for chart-to-table conversion, while this conversion will loss the crucial apperance references, vital for chart editing. Both Matcha [11] and ChartLlama [5] can de-render chart images into tables and plot code. However, none of these models is explicitly capable of predicting and adjusting the charts' visual attributes and corresponding underlying data, which is vital to editing charts accurately. Meanwhile, predicting code is a challenging way to deliver the final edit results, since they have a higher chance of failure in code compilation.

3   Problem Statement

3.1   Chart Edit Taxonomy

Surveying the typical edits performed on chart images, we can categorize them into four distinct classes: style, layout, format, and data-centric, as detailed in Table 1. Furthermore, most advance chart edits can be obtained by chaining these individual edits. The following subsections elaborate on each of the different edit categories.

Style Edits: Style in chart images encompasses essential low-level features such as plot colors, styles (including line style, marker style, and bar pattern), and font characteristics (type, size, etc.). Modifications to a chart's style strictly alter these foundational visual attributes of individual elements without tampering with the data or visualization format. Such appearance-based tweaks are crucial for various applications. Considering accessibility: A color-differentiated chart may not be discernible to color-blind individuals, and transitioning to a different color palette can significantly enhance the chart comprehension for them.

Table 1. Chart Edit Taxonomy

Edit Category	Aspect	Attributes
Style	Colors	Plot Color
	Line	Line Style, Marker
	Pattern	Bar Pattern
	Text	Font, size
Layout	Axes	Grid Lines
	Legend	Position, Internal Layout
Format	Plot	*Line ↔ Bar*
Data Centric	Data Filtering	Range-based
		Series-based
	Data Addition	Add/Update Data-point, Add/Update Data-series

Layout Re-composition: Our layout considerations encompass two primary aspects: axes grids and legends. Tweaks layouts in chart editing can ensure that each element is represented systematically in the modified chart. Furthermore, specific layout changes can improve data readability, such as the inclusion or exclusion of grid lines.

Format Conversion: Different chart types highlight specific aspects of the data. Line charts are particularly effective at showcasing trends over time, allowing viewers to quickly discern patterns and changes. Bar charts, on the other hand, excel in comparing quantities among different groups or categories. More specifically, grouped bar charts show mostly the absolute value comparison, while stacked bar charts reflect the ratio. Transitioning between these formats can offer more comprehensive data views from different aspects.

Data-Centric Modifications: Data-centric chart modifications enable precise manipulation of the chart's underlying data, facilitating tailored adjustments to the visualization's displayed information. These edits, which range from subtle alterations to significant additions, can dramatically shift the narrative and insights gleaned from a chart. Critical operations include range filtering to focus on specific data segments, series filtering to highlight particular data series, and adding new data series or data points to enrich the visualization. Such direct interactions with the chart data empower users to create custom views and explore data in novel and informative ways.

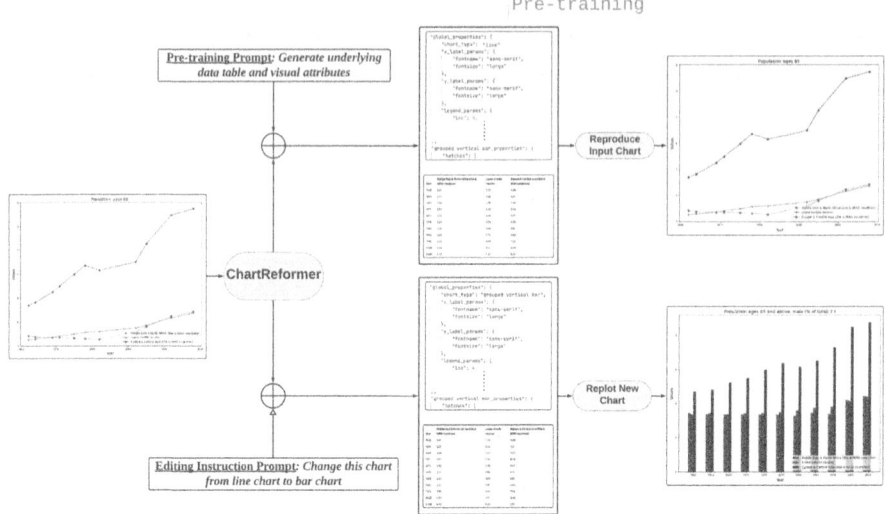

Fig. 2. A Chart Image and edit-prompt are taken as input by the ChartReformer model, which predicts visual attributes and data for the corresponding edited chart. A Replotter software takes in these predicted parameters and generates the edited chart-image

4   ChartReformer

Chart images are inherently complex. Hence, accurate chart comprehension is a prerequisite to successful chart editing. The structural decomposition of the chart image has been seen as a potential approach to the problem. Matcha [11] and ChartLlama [5] try to predict the Python visualization code corresponding to the chart. However, predicting accurate plotting code for the wide variety of charts proves to be a challenging task and easy to fail in success code compilation. We propose ChartReformer, a method that de-renders charts into underlying data and visual attributes to address these limitations for chart editing. Figure 2 shows how ChartReformer predicts a decomposed chart image that a replotter software can effectively utilize to construct the edited charts. To allow the model to reason over the prompt accurately generating the visual attributes and underlying data, a dataset aligning our pipeline is required. The following sections introduce our dataset and pipeline in detail.

4.1   Dataset

This section provides insight into the data creation process for Chart Edits. We synthesize paired chart images using data tables from existing chart datasets to obtain a sizeable chart editing dataset. Our primary emphasis is on line and bar charts, including both grouped and stacked vertical/horizontal bar variations, given their prevalent use in real-world charting scenarios.

Data Source: We utilized data tables from AdobeSynth-19 data previously released as part of the ICPR ChartInfographics competition [3]. This dataset originally consists of synthetically generated images using Matplotlib [6], however, the underlying data is derived from real-world sources such as World Development Indicators and Gender Statistics (World Bank), among others.

Chart Image Generation: We developed a custom software using Matplotlib to synthesize chart images with varying visual attributes given as input. The tool supports all the edits specified in Table 1 while allowing sampling from a comprehensive pool of visual attributes. A thorough parameter pool is provided in the Appendix Table 5. Parameters are randomly selected from this pool, a strategy that introduces significant diversity in the visual appearance of the generated charts. Since plotting parameters are explicitly defined, storing them in a modifiable JSON format becomes straightforward, facilitating further adjustments and reuse. Appendix Fig. 6 shows an example for such a JSON specifying all parameters.

Edit Pair Synthesis: The chart editing data generation software produces, for every sample, edited chart pairs and corresponding edit text prompts. Each chart in the pair consists of visual attributes, a data table, and the synthesized image. The details of generating pairs for each editing category are described as follows:

- **Style and Layout Edits:** For each style or layout modification outlined in the edit specifications, we modify the relevant plotting parameter in the parameter JSON file.
- **Format Edits:** This edit category facilitates the conversion between line and bar charts, and vice versa. We employ the identical data table, generating a new chart type and substituting the original plotting parameters with those corresponding to the new chart, the color of each data series before and after conversion remains the same, while we allow the style like hatches or markers for new plot to be different, e.g., from line chart to bar chart, the line color and bar color are corresponding to each other while bar pattern can be randomly chosen.
- **Data Edits:** Here, we alter the data table according to the given prompt and tailor the original plotting parameters accordingly. This ensures that while the visual attributes of the chart remain consistent, the data in the charts are updated.

For each pair of edits, 5–7 varied prompts are generated from one base prompt to help the model capture the diversity of natural language expressions.

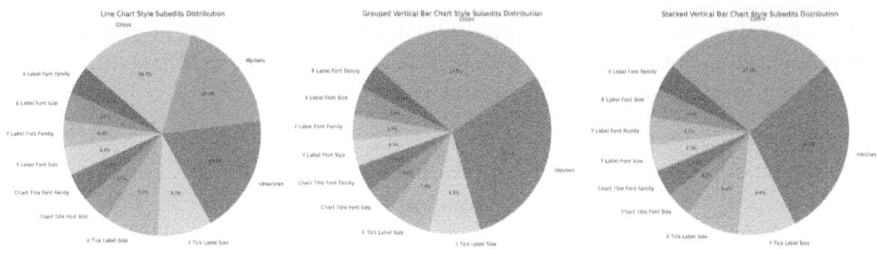

Fig. 3. Distribution of Samples for Style Edits across Chart Categories

Dataset Statistics: Table 2 outlines the statistics of our datasets, categorizing them into four principal types of edits. Overall, our dataset encompasses approximately 100k paired samples. Data-centric edit has significantly more number of samples due to the task challenges.

Table 2. Dataset Statistics: Number of paired samples by Edit Types across Chart Categories

Chart Type	Style	Layout	Format	Data-centric	Total
Line	4775	1910	3820	11460	21965
Grouped Vertical Bar	4775	1910	3820	11460	21965
Grouped Horizontal Bar	4775	1910	3820	11460	21965
Stacked Vertical Bar	4775	1910	3820	11460	21965
Stacked Horizontal Bar	4775	1910	3820	11460	21965
Total	23875	9550	19100	57300	109825

The distribution of detailed edits for style, layout, format and data-centric editing is shown in Fig,3. The generation of chart-editing dataset is based on the manipulation of visual attribute parameters. There are overall two types of visual attribute parameters, chart-type-relevant parameters (line styles/markers, bar patterns, etc.), and chart-type-irrelevant parameters (font size, axis labels orientation, etc.). The edits based on the first type of parameters are oversampled than the second type, since it helps model to learn more challenging part – identifying and manipulating the plotting graph.

4.2 Our Pipeline

To address the challenge of chart editing, we adapt the visual-language encoder-decoder-based transformer [9] to our chart editing tasks. We break down the training into two stages: pre-training for accurate chart de-rendering and fine-tuning for chart editing.

Chart De-Rendering. Accurately de-rendering the chart to visual attributes and underlying data is a prerequisite. We pre-train ChartReformer on our dataset with unpaired images to enable accurate visual attributes and underlying data extraction. Simultaneously predicting the underlying data and visual attributes allows the model to learn the mapping between them. We initialize the model with the checkpoint from [10] and train it on our dataset with 100K samples (sampled from each side of the paired charts). To avoid text distortion and blur during the input resizing of chart images, we opt for a larger input image size of $(800, 800)$ with padding to maintain the original aspect ratio. The maximum output sequence length is 1024, allowing sufficient prediction of all parameters and data tables.

Chart Editing. We fine-tune the pretrained model on paired images and edit prompts, totally 88k samples, enabling it to interpret prompts and adjust data and visual properties accordingly. Edited charts are then replotted with predicted data and plotting parameters. To enhance real-world plotting success, we suggest using JSON repair and default plotting parameters for incomplete predictions, though our evaluations eschew repairs for unbiased performance assessment.

5 Experiments

We use ChartLlama as a baseline for comparison. To the best of our knowledge, no existing dataset related to chart editing is publicly available[1]. Hence, we perform an evaluation exclusively on our dataset.

5.1 Metrics

Image-Based Evaluation. To facilitate a model-agnostic comparison, we utilize Structural Similarity Index Measure (SSIM) [21] to assess the quality of the generated image relative to the edited image from the ground truth. SSIM offers a nuanced perspective on the degree to which the edited chart mirrors the expected outcome, capturing subtle and critical edits speaking to structural similarity aspective. We also calculate a success rate, which reflects the proportion of edits where the edited image was successfully generated, accounting for instances of plotting failures due to inaccurate or incomplete structure prediction. For comparison method ChartLlama, the success rate measures the ratio of samples that the predicted code can be successfully compiled.

Evaluating Edit Correctness. To offer a more precise evaluation of the performance of our methods on the chart editing task, we use the predicted visual attributes and underlying data table for evaluation. Our replotting process is

---

[1] Based on current information, the dataset developed by ChartLlama [5] has not yet been formally published.

Table 3. ChartReformer performance across different types of edits. The first two rows represent Visual Attributes and Data-Table scores for edits, whereas the last two represent image-level comparison metrics

Metrics	Style			Layout			Format			Data-Centric			Total		
	P	R	F1	P	R	F1	P	R	F1	P	R	F1	P	R	F1
VAE	93.08	93.01	93.04	97.20	97.15	97.17	83.61	83.58	83.59	89.75	89.69	89.72	85.14	85.09	85.11
RMS	81.80	81.50	81.58	79.90	79.44	79.60	85.76	85.34	85.51	79.34	78.50	78.76	82.22	81.77	81.92
SSIM	87.39			86.06			74.19			83.07			83.05		
Success Rate	99.81			100			96.9			99.77			98.99		

heurstic, therefore, comparing those with ground truth reflect the performance well. We use the Relative Mapping Similarity (RMS) from [10], and Visual Attribute Edit score (VAES) to evaluate the accuracy of underlying data table and visual attributes prediction, respectively.

$$S_{changed} = \frac{1}{|X_c|} \sum_{e \in X_c} S(e, g)$$
$$S_{unchanged} = \frac{1}{|X_u|} \sum_{e \in X_u} S(e, g)$$
(1)

$$S_f = \frac{2 \cdot S_{changed} \cdot S_{unchanged}}{S_{changed} + S_{unchanged}}$$
(2)

VAES is calculated by grouping attributes into two groups: attributes should be edited $S_{changed}$, and attributes should remain unchanged $S_{unchanged}$, as shown in Eq. 1. e and g represent the edited attribute value and the corresponding ground truth value. First, we calculate a similarity matrix to match the key/value between the ground truth and the prediction. Then the score for each key is calculated based on the value type: the categorical value are based on an exact match; and numeric value are scores from 0 to 1 based on threshold of 0.4. Finally, the visual attribute edit score (VAES) S_f is calculated using the harmonic mean among them as Eq. 2. This prevents biased evaluation, as only a minor part of the visual attributes will change. The precision, recall and F1 score are obtained based on the matching result.

Table 4. Comparison with ChartLlama across different edits on a subset of the test-set (550 samples, 10% sub-test samples)

	Style		Layout		Format		Data-Centric		Total	
	SSIM	Success Rate	SSIM	Success Rate	SSIM	Success Rate	SSIM	Success Rate	SSIM	Success Rate
ChartLlama	73.09	10.63	64.77	26.47	64.32	8.04	67.68	19.33	67.46	16.95
ChartReformer	83.3	100	82.55	100	84.22	100	81.43	100	82.39	100

5.2   Results

Table 3 shows the results of ChartReformer on our test set consisting of 5.5K samples across different edit types. VAES and RMS measure edit-correctness corresponding to visual attributes and data, respectively, whereas SSIM measures image-level similarity with the ground-truth edited image. The results show that data-centric edits appear to be the hardest as they require a precise understanding of the existing data and manipulation based on the prompt while the format editing is easier since it requires least change in the visual attributes json based on the method setup. Overall, the model performs reasonably well on chart edits while maintaining high fidelity with the original chart image, as also seen in Fig. 5.

Compared to recent work ChartLlama [5], since ChartLlama follows a different methodology than ours, we could only compare with ChartLlama with SSIM and success rate. From Table 4, ChartReformer performs better across all edit categories. This performance gap could be attributed to the lack of training on a comprehensive edit dataset. Further, the success rate of our method is better as we do not predict the visualization python code, which is harder to get correct. We concede that in the current setup, we did not perform prompt engineering for ChartLlama, which will likely drag the performance of ChartLlama (Fig. 4).

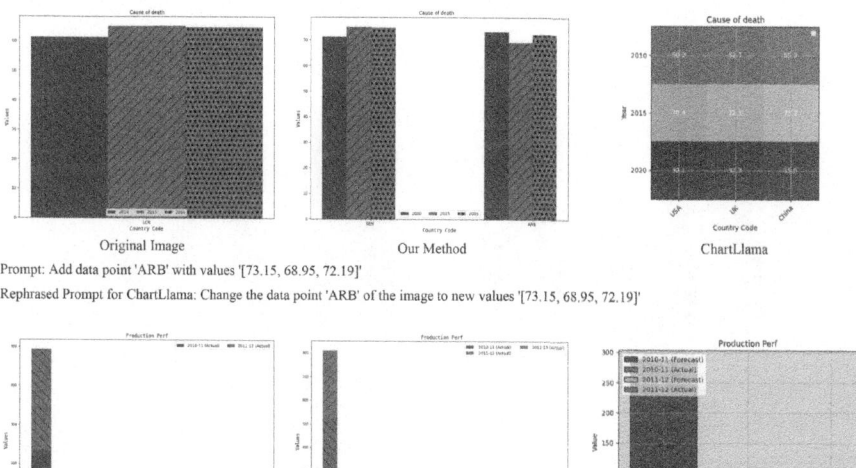

Original Image   Our Method   ChartLlama

Prompt: Add data point 'ARB' with values '[73.15, 68.95, 72.19]'

Rephrased Prompt for ChartLlama: Change the data point 'ARB' of the image to new values '[73.15, 68.95, 72.19]'

Original Image   Our Method   ChartLlama

Prompt: Add data series '2012-13 (Actual)' to the chart with values '[289.03, 34.09, 9.5, 0.93]'

Rephrased Prompt for ChartLlama: Change the image by adding data series '2012-13 (Actual)' with values '[289.03, 34.09, 9.5, 0.93]'

Fig. 4. Qualitative Results for ChartLlama

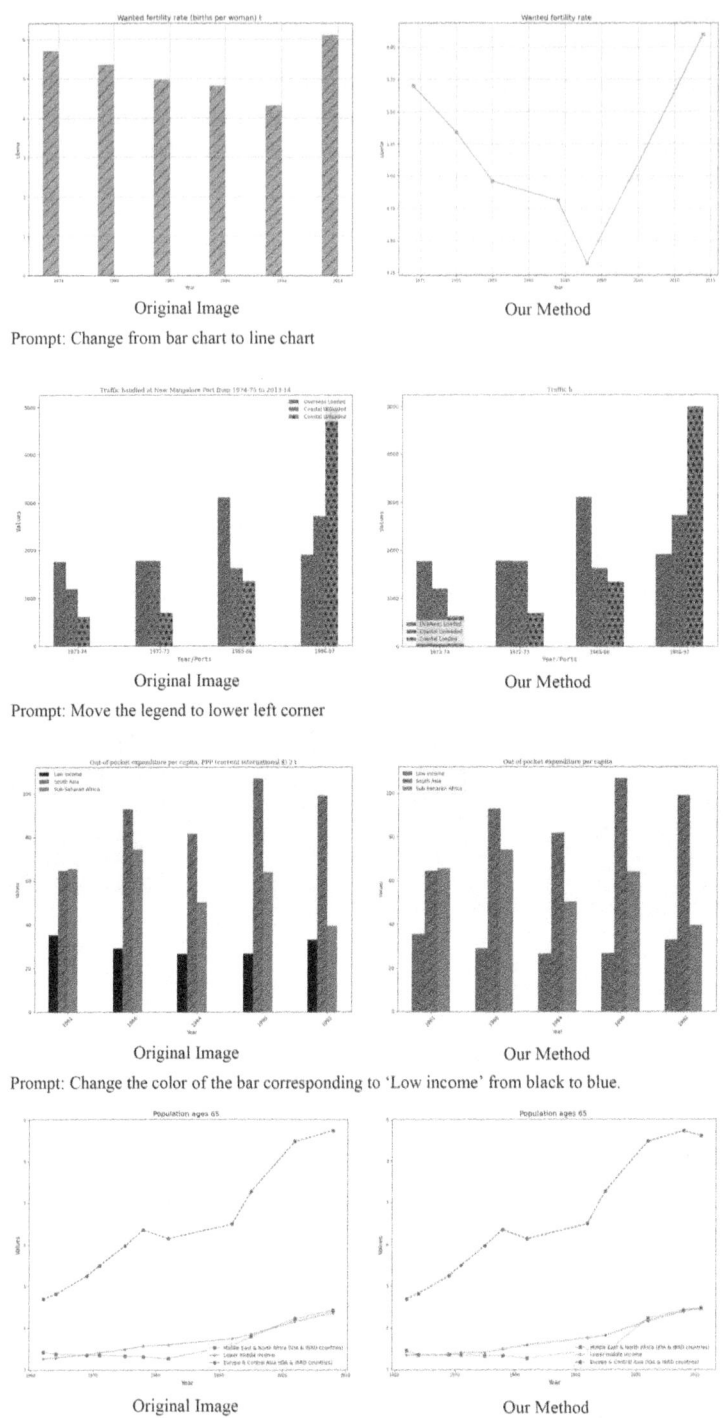

Fig. 5. Qualitative Results for ChartReformer

6 Discussion and Limitations

As shown in Table 3, ChartReformer can successfully extract and alternate visual attributes accordingly for all types of chart edits. In the experiments, we noticed that the overall edit performance is heavily dependent on the data extraction accuracy. Therefore, a more accurate data extraction approach would result in more precise data editing and would be a promising avenue for further research. The proposed chart-edit dataset covers a wide range of edits, yet real-world edit instructions could be abstract and arbitrarily complex, e.g., 'Modify the chart color palette to make it accessible to colorblind people'. One way to handle such queries is to use a preprocessing module (for instance, a decoder-only language model), which could be used to better interpret and simplify them into a series of simply chained edit prompts that ChartReformer can handle.

7 Conclusion

In this work, we present and standardize the chart editing task and generate a large dataset, namely ChartCraft. ChartReformer presents a novel approach to chart editing, allowing for modifications directly from chart images without the need for underlying data tables or supplementary information. By generating edited charts in a decomposed form that includes both the data table and visual attributes, ChartReformer enables precise, natural language-driven edits across style, layout, format, and data-centric modifications. Our experiments demonstrate promising results, highlighting ChartReformer's potential to enhance chart accessibility and adaptability for diverse applications.

8 Appendix

8.1 Dataset

Matplotlib Visual Attributes Pool: Table 5 describes the Matplotlib parameters that we randomly sample from to generate our dataset have more diverse charts instead of relying on Matplotlib's default selection of these properties.

Table 5. Matplotlib Property Pool

Scope	Property	Pool	Editable
Line	Color	["b", "g", "r", "c", "m", "y", "k"]	Yes
Line	Marker	["o", "⌃", "s", "*", "None"]	Yes
Line	Line-style	["solid", "dashed", "dotted", "dense dotted", "loose dotted", "dense dashed", "loose dashed"]	Yes
Bar	Hatch	["xx", ".", "*", "/", "\", "None"]	Yes
Bar	Color	["b", "g", "r", "c", "m", "y", "k"]	Yes
Global	X-axis Label Font Name	["monospace", "Serif", "sans-serif", "Arial Black"]	Yes
Global	X-axis Label Font Size	["medium", "large", "x-large"]	Yes
Global	Y-axis Label Font Name	["monospace", "Serif", "sans-serif", "Arial Black"]	Yes
Global	Y-axis Label Font Size	["medium", "large", "x-large"]	Yes
Global	Legend Location	[0, 1, 2, 3, 4, 8, 9]	Yes
Global	Legend Columns	[1, 2, 3]	No
Global	Title Font Name	["monospace", "Serif", "sans-serif", "Arial Black"]	Yes
Global	Title Font Size	["medium", "large", "x-large"]	Yes
Global	X-tick Label Size	["x-small", "small", "medium", "large"]	Yes
Global	X-tick Rotation	[0, 45]	No
Global	Y-tick Label Size	["x-small", "small", "medium", "large"]	Yes
Global	Grid Visibility	[True, False]	Yes
Global	Grid Axis	['both', 'x', 'y']	No
Global	Grid Line-style	['solid', 'dashed']	No

Example of JSON Properties. Figure 6 shows one example of our visual attribute and underlying data JSON format of a line chart. The underlying data consists of three values: 'data_table', 'chart_title', 'x_axis_title', 'y_axis_title'. We put the data table in front of the visual attributes since the model's text generation is more stable at front than back, while the data is more sensitive and valuable than visual attributes speaking to real application.

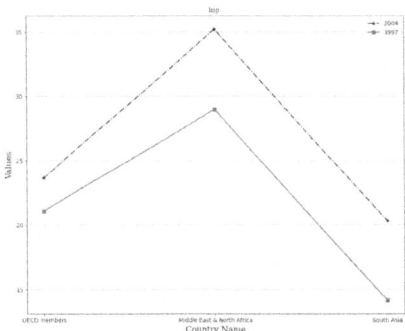

Fig. 6. An example showcasing the JSON configuration for a line chart alongside the generated chart itself. It is important to note that the JSON includes the plotting parameters and the underlying data table. This inclusion ensures a clear and discriminative association between the different data series and their respective visual attributes.

```
{
    ''data_table'': ''Country Name | 2004 | 1997 <0x0A>
    OECD members | 23.66 | 21.08 <0x0A>
    Middle East & North Africa | 35.21 | 29.0
    <0x0A> South Asia | 20.33 | 14.15 <0x0A> '',

    ''chart_title'': ''Imp'',
    ''x_axis_title'': ''Country Name'',
    ''y_axis_title'': ''Values'',
    ''global_properties'': {
        ''chart_type'': ''line'',
        ''x_label_params'': {
            ''fontname'': ''Serif'',
            ''fontsize'': ''x-large''
        },
        ''y_label_params'': {
            ''fontname'': ''Serif'',
            ''fontsize'': ''x-large''
        },
        ''legend_params'': {
            ''loc'': 1,
            ''ncol'': 1
        },
        ''chart_title_params'': {
            ''fontname'': ''Serif'',
            ''fontsize'': ''medium'',
            ''rotation'': 0
        },
        ''x_tick_params'': {
            ''axis'': ''x'',
            ''which'': ''major'',
            ''rotation'': 0,
            ''labelsize'': ''medium'',
            ''labelfontfamily'': ''sans-serif''
        },
        ''y_tick_params'': {
            ''axis'': ''y'',
            ''which'': ''major'',
            ''rotation'': 0,
            ''labelsize'': ''medium'',
            ''labelfontfamily'': ''sans-serif''
        },
        ''grid_params'': {
            ''visible'': true,
            ''axis'': ''y'',
            ''linestyle'': ''dashed''
        }
    },
    ''line_properties'': {
        ''linestyles'': [
            ''dashdot'',
            ''solid''
        ],
        ''markers'': [
            ''*'',
            ''s''
        ],
        ''colors'': [
            ''k'',
            ''r''
        ]
    }
}
```

References

1. Ahmed, S., Yan, P., Doermann, D., Setlur, S., Govindaraju, V.: SpaDen: sparse and dense keypoint estimation for real-world chart understanding. In: Fink, G.A., Jain, R., Kise, K., Zanibbi, R. (eds.) International Conference on Document Analysis and Recognition, vol. 14188, pp. 77–93. Springer, Cham (2023). https://doi.org/10.1007/978-3-031-41679-8_5
2. Cheng, Z.Q., Dai, Q., Li, S., Sun, J., Mitamura, T., Hauptmann, A.G.: ChartReader: a unified framework for chart derendering and comprehension without heuristic rules. arXiv preprint arXiv:2304.02173 (2023)
3. Davila, K., Tensmeyer, C., Shekhar, S., Singh, H., Setlur, S., Govindaraju, V.: ICPR 2020 - competition on harvesting raw tables from infographics. In: Del Bimbo, A., et al. (eds.) ICPR 2021. LNCS, vol. 12668, pp. 361–380. Springer, Cham (2021). https://doi.org/10.1007/978-3-030-68793-9_27
4. Guan, T., et al.: HallusionBench: an advanced diagnostic suite for entangled language hallucination & visual illusion in large vision-language models (2023)
5. Han, Y., et al.: ChartLlama: a multimodal LLM for chart understanding and generation. arXiv preprint arXiv:2311.16483 (2023)
6. Hunter, J.D.: Matplotlib: A 2D graphics environment. Comput. Sci. Eng. 9(3), 90–95 (2007). https://doi.org/10.1109/MCSE.2007.55
7. Kantharaj, S., et al.: Chart-to-text: a large-scale benchmark for chart summarization. In: Muresan, S., Nakov, P., Villavicencio, A. (eds.) Proceedings of the 60th Annual Meeting of the Association for Computational Linguistics (Volume 1: Long Papers), pp. 4005–4023. Association for Computational Linguistics, Dublin, Ireland (2022). https://doi.org/10.18653/v1/2022.acl-long.277, https://aclanthology.org/2022.acl-long.277
8. Lal, J., Mitkari, A., Bhosale, M., Doermann, D.: LineFormer: line chart data extraction using instance segmentation. In: Fink, G.A., Jain, R., Kise, K., Zanibbi, R. (eds.) International Conference on Document Analysis and Recognition, vol. 14191, pp. 387–400. Springer, Cham (2023). https://doi.org/10.1007/978-3-031-41734-4_24
9. Lee, K., et al.: Pix2Struct: screenshot parsing as pretraining for visual language understanding. In: Proceedings of the 40th International Conference on Machine Learning. ICML 2023, JMLR.org (2023)
10. Liu, F., et al.: DePlot: one-shot visual language reasoning by plot-to-table translation. arXiv preprint arXiv:2212.10505 (2022)
11. Liu, F., et al.: MatCha: enhancing visual language pretraining with math reasoning and chart derendering. arXiv preprint arXiv:2212.09662 (2022)
12. Liu, H., Li, C., Li, Y., Lee, Y.J.: Improved baselines with visual instruction tuning (2023)
13. Luo, J., Li, Z., Wang, J., Lin, C.Y.: ChartOCR: data extraction from charts images via a deep hybrid framework. In: Proceedings of the IEEE/CVF Winter Conference on Applications of Computer Vision, pp. 1917–1925 (2021)
14. Masry, A., Do, X.L., Tan, J.Q., Joty, S., Hoque, E.: ChartQA: a benchmark for question answering about charts with visual and logical reasoning. In: Muresan, S., Nakov, P., Villavicencio, A. (eds.) Findings of the Association for Computational Linguistics: ACL 2022, pp. 2263–2279. Association for Computational Linguistics, Dublin, Ireland (2022). https://doi.org/10.18653/v1/2022.findings-acl.177, https://aclanthology.org/2022.findings-acl.177

15. Masry, A., Kavehzadeh, P., Do, X.L., Hoque, E., Joty, S.: UniChart: a universal vision-language pretrained model for chart comprehension and reasoning. arXiv preprint arXiv:2305.14761 (2023)
16. Methani, N., Ganguly, P., Khapra, M.M., Kumar, P.: PlotQA: reasoning over scientific plots. In: 2020 IEEE Winter Conference on Applications of Computer Vision (WACV), pp. 1516–1525 (2020). https://doi.org/10.1109/WACV45572. 2020.9093523
17. Narechania, A., Srinivasan, A., Stasko, J.: NL4DV: A toolkit for generating analytic specifications for data visualization from natural language queries. IEEE Trans. Vis. Comput. Graphics (TVCG) (2020). https://doi.org/10.1109/TVCG. 2020.3030378
18. Satyanarayan, A., Moritz, D., Wongsuphasawat, K., Heer, J.: Vega-lite: a grammar of interactive graphics. IEEE Trans. Vis. Comput. Graphics **23**(1), 341–350 (2017). https://doi.org/10.1109/TVCG.2016.2599030, https://doi.org/10. 1109/TVCG.2016.2599030
19. Shao, Y., Nakashole, N.: ChartDialogs: plotting from natural language instructions. In: Proceedings of the 58th Annual Meeting of the Association for Computational Linguistics, pp. 3559–3574. Association for Computational Linguistics, Online (2020). https://doi.org/10.18653/v1/2020.acl-main.328, https://aclanthology.org/ 2020.acl-main.328
20. Srinivasan, A., Stasko, J.: Orko: facilitating multimodal interaction for visual exploration and analysis of networks. IEEE Trans. Visual Comput. Graphics **24**(1), 511–521 (2018). https://doi.org/10.1109/TVCG.2017.2745219
21. Wang, Z., Bovik, A., Sheikh, H., Simoncelli, E.: Image quality assessment: from error visibility to structural similarity. IEEE Trans. Image Process. **13**(4), 600–612 (2004). https://doi.org/10.1109/TIP.2003.819861
22. Yan, P., Ahmed, S., Doermann, D.: Context-aware chart element detection. In: Fink, G.A., Jain, R., Kise, K., Zanibbi, R. (eds.) Document Analysis and Recognition - ICDAR 2023, pp. 218–233. Springer Nature Switzerland, Cham (2023). https://doi.org/10.1007/978-3-031-41676-7_13
23. Yang, Z., et al.: The dawn of LMMs: preliminary explorations with GPT-4V(ision) (2023)

DocTabQA: Answering Questions from Long Documents Using Tables

Haochen Wang[1], Kai Hu[2], Haoyu Dong[3], and Liangcai Gao[1(✉)]

[1] Peking University, Beijing, China
wanghaochen326@stu.pku.edu.cn, gaoliangcai@pku.edu.cn
[2] University of Science and Technology of China, Hefei, China
hk970213@mail.ustc.edu.cn
[3] Microsoft Corporation, Chennai, India
hadong@microsoft.com

Abstract. We study a new problem setting of question answering (QA), referred to as DocTabQA. Within this setting, given a long document, the goal is to respond to questions by organizing the answers into structured tables derived directly from the document's content. Unlike traditional QA approaches which predominantly rely on unstructured text to formulate responses, DocTabQA aims to leverage structured tables as answers to convey information clearly and systematically, thereby enhancing user comprehension and highlighting relationships between data points. To the best of our knowledge, this problem has not been previously explored. In this paper, we introduce the QTabA dataset, encompassing 300 financial documents, accompanied by manually annotated 1.5k question-table pairs. Initially, we leverage Large Language Models (LLMs) such as GPT-4 to establish a baseline. However, it is widely acknowledged that LLMs encounter difficulties when tasked with generating intricate, structured outputs from long input sequences. To overcome these challenges, we present a two-stage framework, called DocTabTalk, which initially retrieves relevant sentences from extensive documents and subsequently generates hierarchical tables based on these identified sentences. DocTabTalk incorporates two key technological innovations: AlignLLaMA and TabTalk, which are specifically tailored to assist GPT-4 in tackling DocTabQA, enabling it to generate well-structured, hierarchical tables with improved organization and clarity. Comprehensive experimental evaluations conducted on both QTabA and RotoWire datasets demonstrate that our DocTabTalk significantly enhances the performances of the GPT-4 in our proposed DocTabQA task and the table generation task. The code and dataset are available at https://github.com/SmileWHC/DocTabQA for further research.

Keywords: Question Answering · Table Generation · Large Language Model · Retrieval Augmented Generation · Dataset

E. H. Barney Smith et al. (Eds.): ICDAR 2024, LNCS 14804, pp. 470–487, 2024.
https://doi.org/10.1007/978-3-031-70533-5_27

1   Introduction

The task of question answering (QA) has long been a cornerstone in the field of information retrieval [15] and natural language processing (NLP), serving as a fundamental way for machines to demonstrate their understanding of human language. At its core, QA systems aim to automatically answer questions posed by humans, typically by locating and presenting the relevant information extracted from a given text. Over the years, QA has evolved from simple factoid-based questions [11] to more complex, context-dependent questions [4] that require deep understanding and reasoning over extensive documents. Despite the significant advancements in language models and information retrieval techniques, effectively extracting and presenting answers from lengthy and dense documents remains a challenging endeavor. As we enter an era where data is increasingly vast and complex, there is a pressing need for QA systems to not only understand the content of documents but also to structure the retrieved information in a way that is both accessible and informative for end-users. In this context, we introduce a new problem setting of QA, referred to as DocTabQA, which revolutionizes the output format of QA tasks by transforming textual responses into structured tables, thereby enhancing the clarity and usability of the extracted information for decision-making processes.

While QA systems have greatly diversified in terms of content input, ranging from short-text snippets [6,17,33] to long documents [14], and from purely textual data to mixed media such as images (VQA [2], ChartQA [1], TextVQA [36]), document images (DocVQA [24]) and videos (VideoQA [51]), the format of their outputs has remained predominantly unchanged. Traditional QA models have consistently produced answers in the form of plain text, irrespective of the complexity or the nature of the content being queried. This approach, however, often neglects the inherent structure and the relationships between pieces of information, which can be essential for users to thoroughly comprehend the context and make well-informed decisions. Structured tables, on the other hand, offer an attractive option by providing a clear and organized visual representation of data, making complex information more digestible and actionable. The advantages of table-based outputs are manifold; they can efficiently summarize key information, highlight relationships between data points, and facilitate comparisons across different dimensions. Figure 1 illustrates the transformation from a traditional text-based response to a structured table, showcasing how this format can encapsulate and convey intricate details more effectively.

Building upon the foundational concept of DocTabQA, we introduce a new dataset, QTabA, to facilitate the exploration of this novel QA paradigm. This dataset is comprised of 300 financial documents, accompanied by manually annotated 1.5k question-table pairs. Unlike the datasets used in the text-to-table [47] task, which typically focuses on extracting key-value pairs from short input texts to construct flat tables, QTabA presents a more complex challenge. Our dataset aims to generate contextually relevant tables based on varying questions, where both the content and structure of the generated tables are dynamically controllable. Moreover, the input in our dataset is long documents that are rich

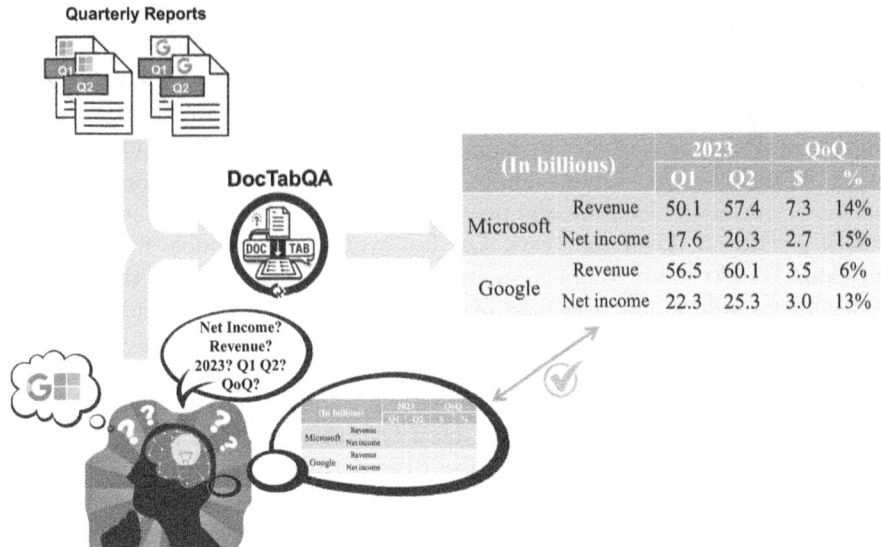

Fig. 1. An illustrative example of DocTabQA. Input documents: 2023 Q1 and Q2 quarterly reports from Microsoft and Google. Input question: "Could you compare the revenue and net income of Microsoft and Google for Q1 and Q2 of 2023, and provide the Quarter-over-Quarter growth in both dollar amounts and percentages?".

in content and may describe numerous tables. The task demands not just the extraction of pertinent details but also the assembly of these details into hierarchical tables.

In this paper, we commence our exploration by utilizing Large Language Models (LLMs) such as GPT-4 [27] to establish a baseline for our task. Our initial approach involves designing a task-specific prompt, to which we feed a (document, question, table) triple, serving as an example for in-context learning. Following this, we input the target document and question, prompting GPT-4 to generate the corresponding table as an answer. However, we observe that LLMs, including GPT-4, encounter significant challenges when tasked with generating intricate, structured outputs from lengthy input sequences. They often struggle to consistently present data in a structured format that adheres to the requirements when dealing with lengthy financial documents.

Inspired by the retrieval augmented generation (RAG) paradigm [7], we present a two-stage framework, called DocTabTalk, which initially retrieves relevant sentences from extensive documents and subsequently generates hierarchical tables based on these identified sentences. DocTabTalk incorporates two key technological innovations: AlignLLaMA and TabTalk, which are specifically tailored to assist GPT-4 in tackling DocTabQA, enabling it to generate well-structured, hierarchical tables with improved organization and clarity. Drawing on the query rewriting techniques [23] employed in RAG, our AlignLLaMA fine-tunes a LLaMA model to rewrite input questions and document sentences to achieve semantic alignment between them. Subsequently, these rewritten questions and document sentences are leveraged to retrieve relevant text sequences. Following this, our TabTalk provides a chain-of-table generation prompt guiding the LLM through the creation of row headers, column headers, and table body cells in a sequential manner. Contrary to previous prompts that required the LLM to generate tables in one go, our TabTalk first creates the hierarchical structure of the table, including hierarchical row headers and column headers, and then fills in the content of each table body cell sequentially. Extensive experiments conducted on both QTabA and RotoWire datasets demonstrate that our DocTabTalk significantly enhances the performance of the GPT-4 in our proposed DocTabQA task and the table generation task.

The main contributions of this paper can be summarized as follows:

- We are the first to present the DocTabQA task, designed to answer questions derived from long documents using tables.
- To support research in this emerging question-answering paradigm, we introduce a new dataset, called QTabA, which consists of 300 financial documents, accompanied by manually annotated 1.5k question-table pairs.
- To address this challenge, we introduce a novel two-stage framework, named DocTabTalk. DocTabTalk marries two critical technological advancements: AlignLLaMA and TabTalk, which improve the capabilities of the GPT-4 in executing our proposed DocTabQA task and the table generation task.

2 Related Work

2.1 Question Answering

In the realm of Natural Language Processing (NLP), question answering (QA) stands as one of the most pivotal research areas and has been extensively studied for many years. This domain focuses on developing systems capable of providing precise answers to questions posed by users, drawing from a variety of underlying content sources. Traditionally, QA tasks have been categorized based on the nature of their content inputs, encompassing several distinct forms such as short-document QA [6,17,33], long-document QA [14], Knowledge Graph QA (KGQA) [10], Table QA [54], Visual QA (VQA) [2], Document Visual QA (DocVQA) [24], and Video QA [51]. These diverse QA tasks, each addressing a unique set of challenges and complexities, have given rise to a rich body of literature [12,28,38,48,55]. However, despite the extensive exploration of these areas, previous methodologies have consistently yielded answers in plain text format, regardless of the complexity or nature of the content in question. This conventional approach often falls short in structuring the retrieved information in an accessible and informative manner for end-users. To bridge this gap, we propose DocTabQA, a novel paradigm that transcends the limitations of text-based responses by utilizing structured tables to answer questions derived from long documents.

2.2 Table Generation

Generating structured tables from textual data has garnered considerable attention in the field of information extraction [20,29,47]. This research direction involves extracting information from unstructured text and presenting it in tabular form, thereby facilitating more accessible data interpretation and utilization. Wu et al. [47] approached the text-to-table conversion as a seq2seq problem, conceptualizing it as the reverse of the table-to-text generation task [22]. Subsequent research has built upon this foundation, with studies [8,29,35,45] employing pretrained language models, including BART [18] and T5 [32], for the generation of tables from the text. Recent studies [26,39] have explored the effectiveness of LLMs in generating structured outputs. Ni et al. [26] demonstrated the use of LLMs for information extraction, specifically in the generation of key-value pairs within a structured context. Tang et al. [39] provided a comparative analysis of various LLMs regarding their ability to generate complex structured data. Despite the significant progress made by previous text-to-table methodologies, these approaches are predominantly data-driven, extracting information to generate tables in an uncontrollable manner, which can impede their applicability in downstream tasks. Furthermore, prior methods have typically focused on inputs consisting of just a few sentences rather than entire documents, and the output tables are often simple, lacking the complexity of hierarchical structures that can represent multi-layered relationships within the data. To address these limitations, we introduce the DocTabQA task, which requires the generation of complex tables to answer user queries based on the input of long documents. The novelty of DocTabQA lies in its ability to produce tables that are not only more controllable but also inherently more usable for end users.

2.3 Retrieval Augmented Generation

RAG (Retrieval-Augmented Generation), introduced by Lewis et al. [19], has emerged as a novel paradigm within the domain of Large Language Models (LLMs), significantly enhancing generative tasks. RAG specifically includes an initial retrieval step where LLMs query an external dataset to obtain pertinent information before commencing question-answering or text generation. The retrieval phase involves various research directions, such as enhancing semantic representations of chunks [5,21]; aligning queries and documents [23,41] and aligning retriever and LLMs [49]. Inspired by the query rewriting techniques [23], which aim to align the semantics between a query and its corresponding document, we introduce AlignLLaMA, which fine-tunes a LLaMA model to rewrite input questions and document sentences to achieve semantic alignment, thereby enhancing the precision of the retrieval process.

2.4 Prompt Engineering

A prompt is a textual instruction that delineates the task an AI is expected to execute [31]. The landscape of prompting strategies is diverse, encompassing methods such as chain-of-thought [44], least-to-most [53], and decomposed prompting [13]. The chain-of-thought technique enables LLMs to navigate through a sequence of intermediate steps, thereby constructing a pathway to the final answer. Diverging from this, the least-to-most approach incrementally addresses problems, starting from the simplest to the most complex, culminating in the resolution of the entire question. Decomposed prompting, distinct from the aforementioned strategies, does not confine the decomposition of tasks to a linear difficulty gradient; instead, it iteratively generates subsequent steps that various systems can implement. While much of the existing research concentrates on reasoning to address QA challenges, our focus deviates to the domain of table generation. Our TabTalk prompting strategy utilizes a chain-of-table generation prompt that first constructs the structure of the table, and then extracts relevant information to populate into the table body cells, significantly enhancing the accuracy of the generated tabular representations.

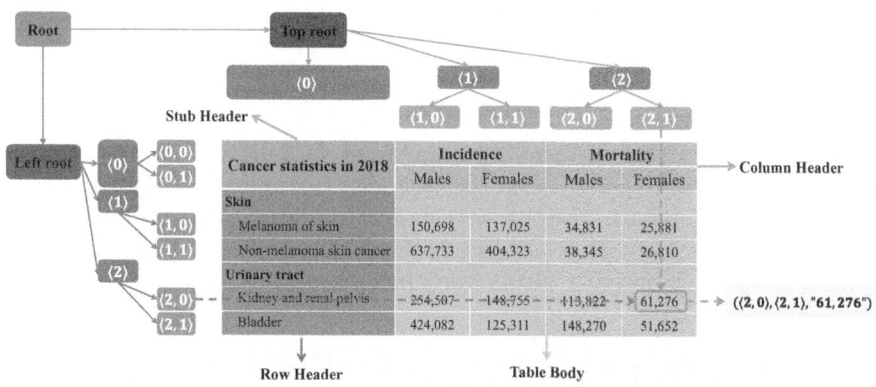

Fig. 2. An example of a hierarchical table using bi-dimensional tree coordinates.

3   DocTabQA

3.1   Problem Definition

As shown in Fig. 1, given a long document as the input content for QA, the goal of DocTabQA is to answer users' specific questions by employing structured tables, thereby enhancing the clarity and usability of the extracted information.

Formally, the input comprises a question, represented as Q, and a long document, denoted as $D = \{s_1, s_2, \ldots, s_{|D|}\}$, where each s_i represents a sentence within the document D. The desired output is a single table, represented as T. Unlike previous text-to-table tasks [47], which focused solely on generating *flat* tables, our DocTabQA necessitates the creation of *hierarchical* tables that are tailored to represent the intricate hierarchical relationships among the data. Following [43], we use the *bi-dimensional coordinate tree* to structure and systematically define cell location in generally structured tables considering both spatial and hierarchical information. As shown in Fig. 2, each hierarchical table consists of a stub header, a hierarchical row header, a hierarchical column header, and a table body. The hierarchical row header forms the *left coordinate tree* of the bi-dimensional coordinate system, while the hierarchical column header constitutes the *top coordinate tree*. Each cell within the table body can precisely pinpoint its location using a unique set of bi-dimensional tree coordinates. For example, the bi-dimensional tree coordinates for the cell "61,276" is $(\langle 2, 0\rangle, \langle 2, 1\rangle)$, where $\langle 2, 0\rangle$ is the left tree coordinate and $\langle 2, 1\rangle$ is the top tree coordinate. Therefore, each cell within Table T is defined by a triplet: $(\langle left\_tree\_coord\rangle, \langle top\_tree\_coord\rangle, text)$. The goal of DocTabQA is to generate such a hierarchical table T as an answer, given an input document D and a question Q.

3.2   QTabA Dataset

We introduce a new dataset, QTabA, to facilitate the research of our proposed DocTabQA. This section delineates the procedures for document collection and the annotation pipeline employed in the construction of the dataset.

Document Collection. To construct the QTabA dataset, we download approximately 300 financial reports issued over the past two years from the Securities and Exchange Commission (SEC)[1] in PDF format. Subsequently, we utilize the Adobe PDF Services API[2] to extract both text and table contents from the reports for further processing within our annotation pipeline.

Annotation Pipeline. As shown in Fig. 3, given the downloaded text contents and the corresponding tables, we manually annotate the (document, question, table) triples utilizing the following pipeline:

- **Relevant Text Annotation:** The preliminary step in our annotation pipeline for each table involves the identification of document sentences that are pertinent to the table's description. This task is initiated by deploying regular expressions to accurately locate sentences that contain text matching the content in the cells of the table. However, it is possible for multiple sentences to match the content within a specific cell. In these cases, a meticulous manual review of these sentences is conducted to ascertain their relevance to the table in question.
- **Table Filtering:** Nevertheless, certain cells within a table may not have corresponding descriptive sentences in the document. In scenarios where 30% or more of a table's cells are devoid of associated descriptive statements, we exclude the table from our dataset. This measure is taken to ensure the high quality and integrity of the data.

---

[1] https://www.sec.gov/.
[2] https://developer.adobe.com/document-services/apis/pdf-services/.

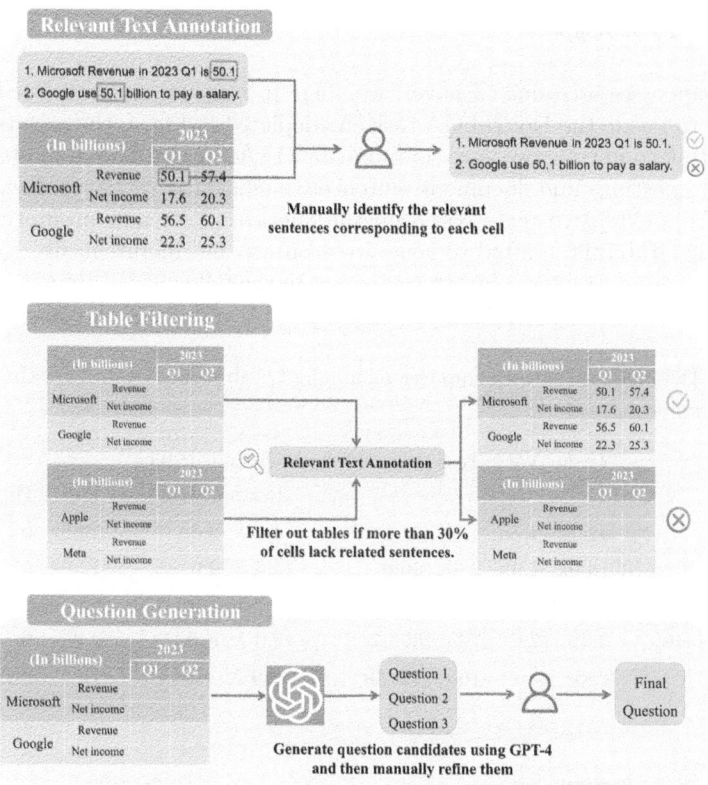

Fig. 3. QTabA dataset annotation pipeline.

- **Question Generation**: Finally, we employ GPT-4 to generate an initial set of questions based on the input table. Detailed prompting instructions are provided in Appendix A.1. Subsequently, these questions undergo a manual refinement process to ensure their direct alignment with the table's content.

Dataset Statistics. Detailed statistical comparisons between our QTabA dataset and existing text-to-table datasets are provided in Table. 1. As indicated in Table 1, the input text sequences in our QTabA consist of long documents, which are substantially lengthier than those found in previous datasets. In addition, the tables generated by our QTabA are significantly more complex than those in prior text-to-table datasets. Previous datasets typically generated tables with only two columns, primarily consisting of key-value pair tables where each row corresponds to an individual key-value pair. Although the RotoWire dataset is much larger, it contains only flat tables. In contrast, tables from QTabA feature numerous hierarchical structures.

4   Methodology

In this paper, we introduce a novel two-stage framework, dubbed DocTabTalk, designed to tackle the DocTabQA task. As depicted in Fig. 4, the architecture of DocTabTalk comprises two core components: (1) An AlignLLaMA module aligns the input questions and document sentences to efficiently retrieve relevant information from extensive texts based on the queries; (2) A new prompting strategy, termed TabTalk, crafted to generate accurate and hierarchically structured tables from the extracted relevant sentences. In the following sections, we provide an in-depth exploration of both the AlignLLaMA and TabTalk components.

Table 1. Detailed statistical comparisons among QTabA dataset and existing text-to-table datasets.

Dataset	Train	Val	Test	Input Tokens	Output Tables		
					Rows & Columns	Flat	Hierarchical
E2E [25]	42.1k	4.7k	4.7k	24.9	4.6 & 2.0	49.8k	0
WikiTableText [3]	10.0k	1.3k	2.0k	20.0	4.3 & 2.0	12.0k	0
WikiBio [16]	582.7k	72.8k	72.7k	122.3	4.2 & 2.0	655.4k	0
RotoWire [46]	3.4k	727	728	373.7	7.3 & **8.8**	4.8k	0
QTabA	1.4k	-	160	**19.4k**	**8.4** & 4.1	1k	**532**

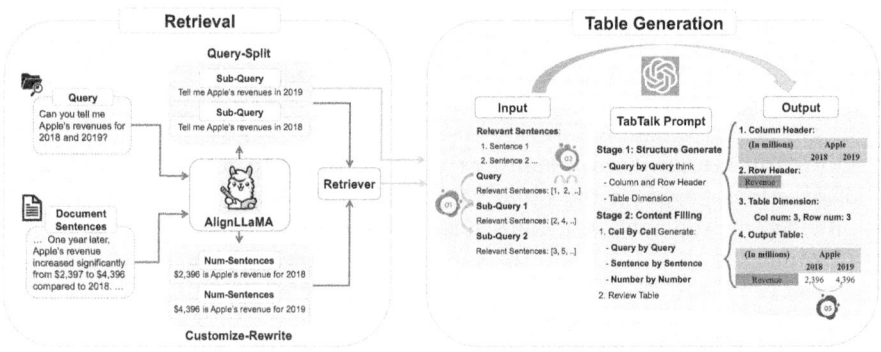

Fig. 4. Overview of DocTabTalk.

4.1   AlignLLaMA

Inspired by the query rewriting techniques [23] employed in RAG, we fine-tune a LLaMA model, which we named AlignLLaMA, to rewrite input questions and document sentences for semantic alignment. Specifically, AlignLLaMA initially decomposes each input question into a set of sub-questions. These sub-questions are then leveraged to perform multiple parallel retrievers to retrieve relevant document sentences. This strategy proves advantageous when addressing complex questions comprising multiple sub-problems. Additionally, we utilize Align-LLaMA to rewrite document sentences so that all sentences are structured with specific data as the subject, which makes them more akin to answers to the question, thereby enhancing the accuracy and recall of the retrieval process. Finally, we utilize a Sentence-BERT model [34] as our retrieval mechanism, calculating the cosine similarity between each rewritten sub-question and the rewritten document sentences. We then extract the top-K sentences with the highest similarity measures, identifying them as the most relevant sentences and feeding them into our table generation stage.

4.2   TabTalk

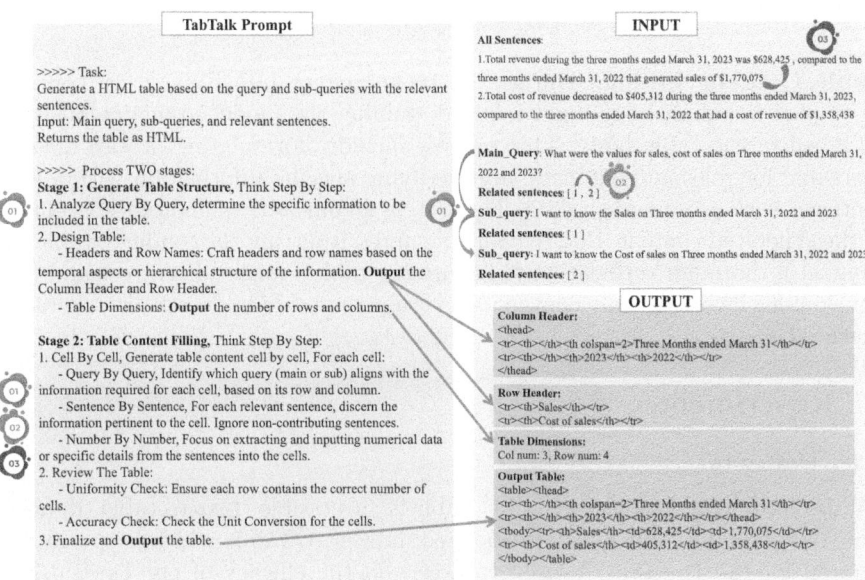

Fig. 5. A schematic view of the proposed TabTalk prompt.

During the table generation phase, we introduce a novel prompting mechanism dubbed TabTalk. This approach is designed to accurately synthesize hierarchically structured tables by interpreting input questions and extracted relevant sentences. As illustrated in Fig. 5, TabTalk systematically deconstructs the intricate process of table generation into a two-stage operation. In the initial stage, TabTalk focuses on the hierarchical architecture of the table, establishing both the hierarchical row and column headers, thereby setting up a *bi-dimensional tree coordinates* for the table. In the subsequent stage, TabTalk fills the table body with data, aligning the cell contents with the information presented in the sentences to finalize the construction of the table. This strategy not only facilitates the model's thought process and the accumulation of information in an orderly progression from simple to complex but also encourages self-reflection at each step of output. This ensures that the resulting table structure is complete and the content is accurate.

Table Structure Generation. During the table structure generation stage, we employ a chain-of-thought [44] approach to progressively delve into the process. This strategy, which progresses from sub-queries to the main query, determines the information that the table should include in a part-to-whole sequence, thereby accurately establishing the content for row headers and column headers. Subsequently, we instruct the model to sequentially produce the row headers, column headers, and dimensions for the constructed table structure. This process not only facilitates self-reflection within the model but also markedly enhances the structural precision of the table.

Table Content Filling. During the table content filling stage, we continue to employ the chain-of-thought method, guiding the model to meticulously fill each cell in the table body. This process includes formulating precise queries, searching for relevant sentences, and verifying specific information within each sentence, such as numerical data. These steps significantly enhance the accuracy of the generated content. Before finalizing and outputting the complete table, we conduct a thorough verification to ensure the table's format is correct and that any content involving unit conversions within the table body cells is accurately executed.

5   Experiments

5.1   Datasets

We perform evaluations using a publicly accessible text-to-table dataset, RotoWire [46], and our newly introduced dataset, QTabA, which is specifically curated for the DocTabQA task, to ascertain the efficacy of our DocTabTalk framework.

RotoWire is a dataset designed for the text-to-table conversion, comprising textual descriptions of basketball matches along with comprehensive statistical tables. For each description, two tables can be generated: one representing the team scores and the other representing individual player scores. We focus on generating the player table due to its higher complexity. This dataset has been randomly divided into training, validation, and test subsets, containing 3398, 727, and 728 text-table pairs, respectively.

QTabA is our proposed dataset, tailored for the DocTabQA problem. It encompasses 300 financial reports from the past two years and is randomly partitioned into training and test subsets, with 240 and 60 documents each. For the training and test subsets, we manually annotate 1.4K and 160 question-table pairs, respectively.

5.2   Evaluation Metrics

Our DocTabTalk framework operates in a two-stage manner, utilizing specific metrics for each stage to assess the performance of our proposed technologies.

During the retrieval stage, we utilize the top-K recall metric to evaluate the capability of various models to retrieve and rank relevant sentences effectively.

In the table generation stage, in line with [39], we decompose the similarity assessment between the two tables into structural and content components. Given that a table's structure can be represented as a bi-dimensional tree, as illustrated in Fig. 2, we propose to use the tree edit distance similarity (TEDS) as the metric for evaluating structural similarity. For assessing table content similarity, our approach diverges from previous text-to-table methodologies, such as those outlined in [47], which typically assess similarity at the individual cell level within a table. Instead, we utilize the *bi-dimensional tree coordinates* to transform each cell in the table body into a distinct key-value pair. Take the cell containing *"61, 276"* as an example. This cell is converted into a key-value pair represented as ($\langle$*"Urinary tract"*, *"Kidney and renal pelvis"*$\rangle$, $\langle$*"Mortality"*, *"Females"*$\rangle$, *"61, 276"*). We then measure the content similarity by comparing these key-value pairs, which are derived from the generated tables, against the key-value pairs from the ground-truth tables. To evaluate table content similarity, we employ a suite of classical metrics that are well-established for assessing textual similarity, including character n-gram F-score (Chrf) [30] and BERTScore [50]. Following [39], we additionally engage GPT-4 to evaluate the generated tables, tasking it with scoring the similarity of both content and structure. These evaluations are referred to as the GPT-Score. Detailed information can be found in [39].

5.3   Implementation Details

Our AlignLLaMA utilizes the conventional instruction tuning method to fine-tune LLaMA2-7B [40] with LoRA [9]. This instruction tuning empowers LLaMA2 to rewrite input questions and document sentences, achieving better alignment between them. For the fine-tuning process of the LLaMA-7B model, we employ a dataset comprising 1.4k tables and 10k sentences, all of which feature significant numeric information extracted from the documents of the training set. The ground truth of rewritten questions and documents in this fine-tuning dataset is generated by GPT-4. Detailed prompting instructions are provided in Appendix A.2. The fine-tuning is carried out on a workstation equipped with 2 NVIDIA Tesla A800 GPUs (80 GB of memory). In the retrieval phase, we prioritize the top-30 relevant document sentences, which are then fed into the subsequent table generation stage. During the table generation phase, our approach diverges from previous text-to-table methods that produce tables in Markdown format, which lack the capability to represent hierarchical structures. Instead, we opt to generate tables in HTML format, enabling the depiction of complex, hierarchical table structures.

5.4   Comparisons with Prior Arts

In our DocTabTalk framework, there are two key technological innovations: AlignLLaMA and TabTalk. We conduct comparisons of these two modules against state-of-the-art methods respectively.

Table 2. Performance comparison of retrieval stage on QTabA. (R@K denotes the recall rate for the top-K retrieved results, in %)

Model	R@10	R@20	R@30	R@40	R@50	R@60
MiniLM [42]	62.47	73.87	79.51	81.55	84.19	85.82
MPNet [37]	58.89	72.22	76.72	81.85	85.3	88.01
GPT-4 [27]	60.28	72.80	76.57	77.84	78.61	78.84
AlignLLaMA	**69.64**	**80.37**	**85.22**	**88.41**	**92.01**	**94.05**

Retrieval Stage. During the retrieval phase, we implement several baseline methodologies, including MiniLM [42], MPNet [37], and GPT-4 [27], to fetch document sentences pertinent to the posed questions. MiniLM and MPNet accomplish this by computing the semantic similarity between the queries and document sentences, thus retrieving sentences of relevance. In contrast, GPT-4 adopts in-context learning, utilizing examples to infer the relevance between the input document sentences and the query, subsequently generating a similarity score for these entities. Detailed prompting instructions are provided in Appendix A.3. Our AlignLLaMA initially rewrites both questions and document sentences before employing Sentence-BERT to assess the similarity among these modified sentences. This approach markedly enhances the top-K recall. As evidenced in Table 2, AlignLLaMA achieves an impressive recall rate of 85.22% at top-30, significantly surpassing other methods.

Table 3. Performance comparison of table generation stage on QTabA. Metrics include BERT-Score (BERT, %), GPT-Score (GPT, on a scale of 0–10), and Tree Edit Distance Similarity (TEDS, %). (R&C Header represents row header and column header)

#	Model	Content Similarity				Structure Similarity	
		Table Body		R&C Header		TEDS	GPT
		BERT	GPT	BERT	GPT		
1	GPT-4	62.59	4.16	63.20	4.43	92.15	4.29
2	+ Two Stage	67.14	4.26	64.31	4.66	92.97	4.36
3	+ AlignLLaMA	70.08	4.32	64.70	4.92	94.15	4.47
4	+ TabTalk	**74.76**	**5.90**	**66.36**	**5.44**	**96.33**	**6.59**

Table Generation Stage. After retrieving relevant sentences, the goal of our table generation stage is to generate a hierarchically structured table from these sentences as an answer to the input question. We validate the effectiveness of our TabTalk on two datasets, QTabA and RotoWire.

QTabA. In our initial experiment, we establish a straightforward baseline by using a simple prompting strategy to assess the one-shot in-context learning capabilities of GPT-4. This involves employing a single illustrative example and a prompt that clearly outlines the task. The prompt includes a (document, question, table) triplet to guide GPT-4 toward generating tabular results after processing the entire document text and the corresponding query. Detailed prompting instructions are provided in Appendix A.4.

As shown in row #1 of Table 3, the baseline approach does not perform optimally, primarily due to GPT-4's challenges with processing lengthy documents and maintaining context across extended texts. We then apply a two-stage framework as indicated in row #2 of Table 3, which involves an initial retrieval phase using GPT-4 to extract relevant sentences, followed by a generation phase. The results demonstrate a significant improvement over the baseline method, underscoring the effectiveness of the two-stage strategy.

Further enhancements are observed when we introduce our AlignLLaMA module as the retrieval mechanism, as depicted in row #3 of Table 3. The performance is notably better, especially when our TabTalk prompting strategy is implemented. This strategy leads to marked improvements in the generated table structures, with a 2.18% increase in the TEDS metric, indicating that our method is more effective at generating complex hierarchical tables. Overall, the comparative analysis shows that our DocTabTalk framework, with the integration of AlignLLaMA and TabTalk prompts, significantly enhances GPT-4's ability to structure and present information in tabular form for complex QA tasks involving long documents.

Table 4. Performance comparison of text-to-table on RotoWise. Metrics include character n-gram F-score (Chrf, %), BERT-Score (BERT, %), and Tree Edit Distance Similarity (TEDS, %). (R Header and C Header represent row header and column header, respectively. ICL represents in-context learning.)

#	Model	Content Similarity						Structure
		Table Body		R Header		C Header		TEDS
		Chrf	BERT	Chrf	BERT	Chrf	BERT	
Supervised	Sent-level RE [52]	83.42	85.35	93.00	90.98	89.38	93.07	-
	Seq2Seq-c [47]	84.74	88.97	94.0	93.71	91.26	94.41	-
	Seq2Seq&set [20]	**85.75**	**90.93**	94.48	96.43	**91.60**	**95.08**	-
ICL	GPT-4	82.48	82.40	96.06	97.46	67.62	77.20	100.0
	+ TabTalk	**85.75**	84.92	**97.53**	**98.68**	71.30	80.97	100.0

RotoWire. We apply our TabTalk to the text-to-table task, and as shown in Table 4, our method significantly improves GPT-4's ability to generate structured tables. Our method achieves comparable results with supervised methods in the table body and even better on row headers with only a few examples for in-context learning. On column headers, our approach improves GPT-4 but still falls short compared to supervised methods, which is due to the limited examples provided by the in-context learning. Additionally, since the tables in this dataset are all simple, flat tables, the similarity in table structure for the generated tables is always 100%.

6   Limitations

Despite our QTabA bringing new challenges (DocTabQA) to the Document-based Question Answering (DocQA) domain and our DocTabTalk showing promising results, there are still several limitations in both our dataset and approach that need to be addressed. For the dataset, its exclusive focus on English sources and single-document input restricts the breadth of our study and its potential cross-linguistic application. Expanding the dataset to accommodate multilingual content and multi-document inputs would greatly enhance the versatility and depth of the DocQA framework. Regarding our method, the initial retrieval stage is relatively simplistic, primarily assessing the relevance of document content to the question without deeply engaging the model's inferential capabilities. This superficial process may overlook complex relationships that require advanced reasoning. Therefore, improving our method's capacity for inference is crucial, ensuring that it can more accurately discern and utilize relevant information for question answering in future research.

7   Conclusion and Future Work

In this paper, we make a significant contribution to the field of Document-based Question Answering by introducing DocTabQA, a novel problem setting that transforms answers from textual responses into structured tables. Through the development and evaluation of the QTabA dataset, we propose a two-stage framework, called DocTabTalk, to improve the performance of GPT-4. DocTabTalk incorporates two key technological innovations: AlignLLaMA and TabTalk, which are specifically tailored to assist GPT-4 in tackling DocTabQA, enabling it to generate well-structured, hierarchical tables with improved organization and clarity. The experimental results on the QTabA and the RotoWire dataset convincingly demonstrate the effectiveness of our method. Our approach marks a substantial step forward in presenting information succinctly and systematically.

In the future, we will aim to address the limitations highlighted in our study. We plan to expand the dataset to include multilingual documents and multi-document inputs, which will challenge and potentially improve the robustness of DocQA systems. Moreover, we will focus on enhancing the inferential capabilities of our method, enabling it to grasp the subtleties and complexities of document content beyond the superficial level. We will also work on improving the model's reasoning abilities, which are essential for accurately determining the relevance of information in response to user queries. By tackling these challenges, we expect to advance the generation of structured summaries and the overall effectiveness of question answering systems for complex, long-form documents.

Acknowledgement. This work is supported by the projects of National Science and Technology Major Project (2021ZD0113301) and National Natural Science Foundation of China (No. 62376012), which is also a research achievement of Key Laboratory of Science, Technology and Standard in Press Industry (Key Laboratory of Intelligent Press Media Technology).

References

1. Ahmed, S., Jawade, B., Pandey, S., Setlur, S., Govindaraju, V.: Realcqa: scientific chart question answering as a test-bed for first-order logic. In: ICDAR, pp. 66–83. Springer (2023). https://doi.org/10.1007/978-3-031-41682-8_5
2. Antol, S., et al.: VQA: visual question answering. In: ICCV, pp. 2425–2433 (2015)
3. Bao, J., et al.: Table-to-text: describing table region with natural language. In: AAAI (2018)
4. Choi, E., et al.: QUAC: question answering in context. In: EMNLP, pp. 2174–2184 (2018)
5. Dai, Z., et al.: Promptagator: few-shot dense retrieval from 8 examples. In: ICLR (2023)
6. Fan, A., Jernite, Y., Perez, E., Grangier, D., Weston, J., Auli, M.: ELI5: long form question answering. In: ACL, pp. 3558–3567 (2019)
7. Gao, Y., et al.: Retrieval-augmented generation for large language models: a survey. arXiv preprint arXiv:2312.10997 (2023)
8. He, Y., Hu, J., Tang, B.: Revisiting event argument extraction: can EAE models learn better when being aware of event co-occurrences? In: ACL, pp. 12542–12556 (2023)
9. Hu, E.J., et al.: LORA: low-rank adaptation of large language models. In: ICLR (2022)
10. Huang, X., Zhang, J., Li, D., Li, P.: Knowledge graph embedding based question answering. In: WSDM, pp. 105–113 (2019)
11. Iyyer, M., Boyd-Graber, J., Claudino, L., Socher, R., Daumé III, H.: A neural network for factoid question answering over paragraphs. In: EMNLP, pp. 633–644 (2014)
12. Jin, N., Siebert, J., Li, D., Chen, Q.: A survey on table question answering: recent advances. In: CCKS, pp. 174–186. Springer (2022). https://doi.org/10.1007/978-981-19-7596-7_14
13. Khot, T., et al.: Decomposed prompting: a modular approach for solving complex tasks. arXiv preprint arXiv:2210.02406 (2022)
14. Kočiský, T., et al.: The NarrativeQA reading comprehension challenge. Trans. Assoc. Comput. Linguist. **6**, 317–328 (2018)
15. Kwok, C.C., Etzioni, O., Weld, D.S.: Scaling question answering to the web. In: WWW, pp. 150–161 (2001)

16. Lebret, R., Grangier, D., Auli, M.: Neural text generation from structured data with application to the biography domain. In: EMNLP, pp. 1203–1213 (2016)
17. Lelkes, A.D., Tran, V.Q., Yu, C.: Quiz-style question generation for news stories. In: WWW, pp. 2501–2511 (2021)
18. Lewis, M., et al.: BART: denoising sequence-to-sequence pre-training for natural language generation, translation, and comprehension. In: ACL, pp. 7871–7880 (2020)
19. Lewis, P., et al.: Retrieval-augmented generation for knowledge-intensive NLP tasks. In: NeurIPS, pp. 9459–9474 (2020)
20. Li, T., Wang, Z., Shao, L., Zheng, X., Wang, X., Su, J.: A sequence-to-sequence&set model for text-to-table generation. In: ACL-Findings, pp. 5358–5370 (2023)
21. Li, X., Li, J.: Angle-optimized text embeddings. arXiv preprint arXiv:2309.12871 (2023)
22. Liu, T., Wang, K., Sha, L., Chang, B., Sui, Z.: Table-to-text generation by structure-aware seq2seq learning. In: AAAI (2018)
23. Ma, X., Gong, Y., He, P., Zhao, H., Duan, N.: Query rewriting for retrieval-augmented large language models. In: EMNLP, pp. 5303–5315 (2023)
24. Mathew, M., Karatzas, D., Jawahar, C.: DOCVQA: a dataset for VQA on document images. In: WACV, pp. 2200–2209 (2021)
25. Nayak, T., Ng, H.T.: Effective modeling of encoder-decoder architecture for joint entity and relation extraction. In: AAAI, pp. 8528–8535 (2020)
26. Ni, X., Li, P.: Unified text structuralization with instruction-tuned language models. arXiv preprint arXiv:2303.14956 (2023)
27. OpenAI: Gpt-4 technical report. arXiv preprint arXiv:2303.08774 (2023)
28. Pandya, H.A., Bhatt, B.S.: Question answering survey: Directions, challenges, datasets, evaluation matrices. arXiv preprint arXiv:2112.03572 (2021)
29. Pietruszka, M., et al.: Stable: Table generation framework for encoder-decoder models. arXiv preprint arXiv:2206.04045 (2022)
30. Popović, M.: CHRF: character n-gram F-score for automatic MT evaluation. In: WMT, pp. 392 395 (2015)
31. Radford, A., Wu, J., Child, R., Luan, D., Amodei, D., Sutskever, I., et al.: Language models are unsupervised multitask learners. OpenAI blog **1**(8), 9 (2019)
32. Raffel, C., et al.: Exploring the limits of transfer learning with a unified text-to-text transformer. J. Mach. Learn. Res. **21**(1) (2020)
33. Rajpurkar, P., Jia, R., Liang, P.: Know what you don't know: unanswerable questions for SQuAD. In: ACL, pp. 784–789 (2018)
34. Reimers, N., Gurevych, I.: Sentence-BERT: sentence embeddings using Siamese BERT-networks. In: EMNLP-IJCNLP, pp. 3982–3992 (2019)
35. Rossiello, G., Chowdhury, M.F.M., Mihindukulasooriya, N., Cornec, O., Gliozzo, A.M.: Knowgl: knowledge generation and linking from text. In: AAAI, pp. 16476–16478 (2023)
36. Singh, A., et al.: Towards VQA models that can read. In: CVPR, pp. 8317–8326 (2019)
37. Song, K., Tan, X., Qin, T., Lu, J., Liu, T.Y.: MPNET: masked and permuted pre-training for language understanding. In: NeurIPS, pp. 16857–16867 (2020)
38. Sun, G., Liang, L., Li, T., Yu, B., Wu, M., Zhang, B.: Video question answering: a survey of models and datasets. Mob. Netw. Appl. 1–34 (2021)
39. Tang, X., Zong, Y., Zhao, Y., Cohan, A., Gerstein, M.: Struc-bench: Are large language models really good at generating complex structured data? arXiv preprint arXiv:2309.08963 (2023)

40. Touvron, H., et al.: Llama 2: Open foundation and fine-tuned chat models. arXiv preprint arXiv:2307.09288 (2023)
41. Wang, L., Yang, N., Wei, F.: Query2doc: query expansion with large language models. In: EMNLP, pp. 9414–9423 (2023)
42. Wang, W., Wei, F., Dong, L., Bao, H., Yang, N., Zhou, M.: Minilm: deep self-attention distillation for task-agnostic compression of pre-trained transformers. In: NeurIPS, pp. 5776–5788 (2020)
43. Wang, Z., et al.: TUTA: tree-based transformers for generally structured table pre-training. In: KDD, pp. 1780–1790 (2021)
44. Wei, J., et al.: Chain-of-thought prompting elicits reasoning in large language models. In: NeurIPS, pp. 24824–24837 (2022)
45. Whitehouse, C., Vania, C., Aji, A.F., Christodoulopoulos, C., Pierleoni, A.: WebIE: faithful and robust information extraction on the web. In: ACL, pp. 7734–7755 (2023)
46. Wiseman, S., Shieber, S., Rush, A.: Challenges in data-to-document generation. In: EMNLP, pp. 2253–2263 (2017)
47. Wu, X., Zhang, J., Li, H.: Text-to-table: a new way of information extraction. In: ACL, pp. 2518–2533 (2022)
48. Yani, M., Krisnadhi, A.A.: Challenges, techniques, and trends of simple knowledge graph question answering: a survey. Information 12(7), 271 (2021)
49. Yu, Z., Xiong, C., Yu, S., Liu, Z.: Augmentation-adapted retriever improves generalization of language models as generic plug-in. In: ACL, pp. 2421–2436 (2023)
50. Zhang, T., Kishore, V., Wu, F., Weinberger, K.Q., Artzi, Y.: Bertscore: evaluating text generation with BERT. In: ICLR (2020)
51. Zhong, Y., Ji, W., Xiao, J., Li, Y., Deng, W., Chua, T.S.: Video question answering: datasets, algorithms and challenges. In: EMNLP, pp. 6439–6455 (2022)
52. Zhong, Z., Chen, D.: A frustratingly easy approach for entity and relation extraction. In: NAACL, pp. 50–61 (2021)
53. Zhou, D., et al.: Least-to-most prompting enables complex reasoning in large language models. arXiv preprint arXiv:2205.10625 (2022)
54. Zhu, F., Lei, W., Huang, Y., Wang, C., Zhang, S., Lv, J., Feng, F., Chua, T.S.: TAT-QA: a question answering benchmark on a hybrid of tabular and textual content in finance. In: ACL-IJCNLP, pp. 3277–3287 (2021)
55. Zou, Y., Xie, Q.: A survey on VQA: datasets and approaches. In: CIT, pp. 289–297. IEEE (2020)

Author Index

GPSR Compliance

The European Union's (EU) General Product Safety Regulation (GPSR) is a set of rules that requires consumer products to be safe and our obligations to ensure this.

If you have any concerns about our products, you can contact us on ProductSafety@springernature.com

In case Publisher is established outside the EU, the EU authorized representative is:

Springer Nature Customer Service Center GmbH
Europaplatz 3
69115 Heidelberg, Germany

The manufacturer's authorised representative in the EU is Springer
Nature Customer Service Centre GmbH, Europaplatz 3, 69115 Heidelberg,
Germany. If you have any concerns regarding our products, please
contact ProductSafety@springernature.com

Printed and bound by CPI Group (UK) Ltd, Croydon, CR0 4YY
05/05/2026
02103537-0007